Applied Mathematical Sciences
Volume 165

T0205429

Applied Mathematical Sciences

(continued after index)

Brian Straughan

Stability and Wave Motion
in Porous Media

 Springer

Brian Straughan
Durham University
Department of Mathematical Sciences
Durham
UK
brian.straughan@durham.ac.uk

Editors

S.S. Antman
Department of Mathematics
and
Institute for Physical Science
and Technology
University of Maryland
College Park, MD 20742-4015
USA
ssa@math.umd.edu

J.E. Marsden
Control and Dynamical
Systems 107-81
California Institute of
Technology
Pasadena, CA 91125
USA
marsden@cds.caltech.edu

L. Sirovich
Laboratory of Applied
Mathematics
Department of
Biomathematical Sciences
Mount Sinai School
of Medicine
New York, NY 10029-6574
USA
chico@camelot.mssm.edu

ISBN 978-1-4419-2626-5   e-ISBN 978-0-387-76543-3
DOI: 10.1007/978-0-387-76543-3

Mathematics Subject Classification (2000): 76S05, 76E06, 76E05, 76E30, 74F10, 74F05, 74J30, 35B30, 35B35, 35B40, 35L60, 35Q35

Printed on acid-free paper

9 8 7 6 5 4 3 2 1

springer.com

To
Carole

Preface

This book presents an account of theories of flow in porous media which have proved tractable to analysis and computation. In particular, the theories of Darcy, Brinkman, and Forchheimer are presented and analysed in detail. In addition, we study the theory of voids in an elastic material due to J. Nunziato and S. Cowin. The range of validity of each theory is outlined and the mathematical properties are considered. The questions of structural stability, where the stability of the model itself is under consideration, and spatial stability are investigated. We believe this is the first such account of these topics in book form. Throughout, we include several new results not published elsewhere.

Temporal stability studies of a variety of problems are included, indicating practical applications of each. Both linear instability analysis and global nonlinear stability thresholds are presented where possible. The mundane, important problem of stability of flow in a situation where a porous medium adjoins a clear fluid is also investigated in some detail. In particular, the chapter dealing with this problem contains some new material only published here. Since stability properties inevitably end up requiring to solve a multi-parameter eigenvalue problem by computational means, a separate chapter is devoted to this topic. Contemporary methods for solving such eigenvalue problems are presented in some detail.

Nonlinear acceleration waves in classes of porous media are also studied. The connection with this and sound propagation in porous media is analysed. The nonlinear wave analysis is performed for a class of simplified mixture-like theories and for the Nunziato-Cowin theory of elastic materials with voids.

It is a pleasure to thank Achi Dosanjh of Springer for her advice with editorial matters, and also to thank Frank Ganz of Springer for his help in sorting out latex problems for me.

This research was in part supported by a Research Project Grant of the Leverhulme Trust, reference number F/00 128/AK. This support is very gratefully acknowledged.

Brian Straughan   Durham

Contents

1

Introduction

1.1 Porous media

1.1.1 Applications, examples

Porous media is a subject well known to everyone. Such materials occur everywhere and influence all of our lives. There are numerous types of porous media and almost limitless applications of and uses for porous media. The theory of porous media is driven by the need to understand the nature of the many such materials available and to be able to use them in an optimum way.

A key terminology in the theory of porous media is the concept of *porosity*. The porosity is the ratio of the void fraction in the porous material to the total volume occupied by the porous medium. The void fraction is usually composed of air or some other liquid and since both liquids may be described as fluids we define the *porosity* at position $\mathbf{x}$ and time t, $\phi(\mathbf{x}, t)$ by

$$\phi = \frac{\text{fluid volume}}{\text{total volume of porous medium}}. \tag{1.1}$$

Clearly, $0 \leq \phi \leq 1$. However, in mundane situations ϕ may be as small as 0.02 in coal or concrete, see e.g. (Nield and Bejan, 2006), whereas ϕ is close to 1 in some animal coverings such as fur or feathers, (Du et al., 2007), or in man-made high porosity metallic foams, (Zhao et al., 2004).

We include photographs of some well known porous materials. Figure 1.1 shows animal fur which is a good example of a porous medium with high

B. Straughan, *Stability and Wave Motion in Porous Media*,
DOI: 10.1007/978-0-387-76543-3_1, © Springer Science+Business Media, LLC 2008

Figure 1.1. Animal fur is a good example of a high porosity material, as seen in this kitten

porosity, i.e. close to 1. Figure 1.2 displays a highly grained piece of wood (elm) and this shows that a porous medium may be highly anisotropic. Figure 1.3 shows lava from Mount Etna in Sicily, while figure 1.4 dislays sandstone which is another type of porous rock, but one with a very different structure from lava.

In addition to these we can cite other examples of porous media, such as biological tissues, e.g. bone, skin; building materials such as sand, cement, plasterboard, brick; man-made high porosity metallic foams such as those based on copper oxide or aluminium, and other materials in everyday use such as ceramics. The types of porous materials we can think of is virtually limitless.

Applications of porous media in real life are likewise very many. We could list a great many, but simply quote some to give an idea of the vastness of porous media theory. Use of copper based foams and other porous materials in heat transfer devices such as heat pipes used to transfer heat from such as computer chips is a field influencing everyone, see e.g. (Amili and Yortsos, 2004), (Calmidi and Mahajan, 2000), (Doering and Constantin, 1998), (Nield and Bejan, 2006), (Nield and Kuznetsov, 2001), (Pestov, 1998), (Salas and Waas, 2007), (Vadasz et al., 2005a), (Zhao et al., 2004), also in combustion heat transfer devices where the porous medium is employed with a liquid fuel in a porous combustion

Figure 1.2. Wood is a very good example of a porous medium which exhibits a strong anisotropy. The grain effect is clearly visible here. This is a material which is approximately transversely isotropic

heater, see (Jugjai and Phothiya, 2007). The use of acoustic techniques in non-destructive testing of materials such as foams or ceramics is highly efficient since the specimen may be examined intact, see e.g. (Ayrault et al., 1999), (Diebold, 2005), (Johnson et al., 1994), (Ouellette, 2004), (Raiser et al., 1994), (Saggio-Woyansky et al., 1992). Another interesting use of acoustic microscopy is to ultra sound testing in medical applications, see e.g. (Ouellette, 2004). Yet a further use of acoustic waves is in the drying of foodstuffs, such as apples, see (Simal et al., 1998).

The applications in geophysical situations are also numerous. For example, salt movement underground which may have a direct effect on water supplies is studied by e.g. (Bear and Gilman, 1995), (Gilman and Bear, 1996), (Wooding et al., 1997a; Wooding et al., 1997b). Contaminant transport and underground water flow is of relevance to us all, see e.g. (Boano et al., 2007), (Das et al., 2002), (Das and Lewis, 2007), (Discacciati et al., 2002), (El-Habel et al., 2002), (Ewing et al., 1994), (Ewing and Weekes, 1998), (Miglio et al., 2003), (Riviere, 2005). Global warming is very topical and porous media are involved there in connection with topics such as ice melting, or carbon dioxide storage, see e.g. (Bogorodskii and Nagurnyi, 2000), (Carr, 2003a; Carr, 2003b), (Ennis-King et al., 2005), (Xu et al., 2006; Xu et al., 2007). Sand boils, earthquakes, and landslides also have

Figure 1.3. Lava from Mount Etna, Sicily. Photograph taken at Capomulini, June 2007

a major effect on human life, see e.g. (Kolymbas, 1998), (Rajapakse and Senjuntichai, 1995), (Wang et al., 1991).

Sound propagation in buildings, building materials, or porous gels is of immediate environmental concern, see e.g. (Brusov et al., 2003), (Buishvili et al., 2002), (Ciarletta and Straughan, 2006), (Garai and Pompoli, 2005), (Jordan, 2005a; Jordan, 2006), (Meyer, 2006), (Mouraille et al., 2006), (Moussatov et al., 2001), (Singer et al., 1984), (Wilson, 1997). Noise reduction is also important in exhaust systems such as those of cars, or motor bikes, and metallic foams are employed here, see e.g. (Maysenhölder et al., 2004), who test silencers made from aluminium foam and compare the results with a mineral wool fibre - absorber. Another interesting use of sound waves is to the detection of mines nearly submerged in the seabed, (Feuillade, 2007).

Many foodstuffs are porous materials. Modern technology is involved in such as microwave heating, (Dincov et al., 2004), or drying of foods or other natural materials, see e.g. (Gigler et al., 2000a; Gigler et al., 2000b), (Mitra et al., 1995), (Sanjuán et al., 1999), (Vedavarz et al., 1992), (Zorrilla and Rubiolo, 2005a; Zorrilla and Rubiolo, 2005b). There are many other diverse application areas of porous materials, such as heat retention in birds or animals, (Du et al., 2007), bone modelling, (Eringen, 2004b), or the manufacture of composite materials, increasingly in use in aircraft or

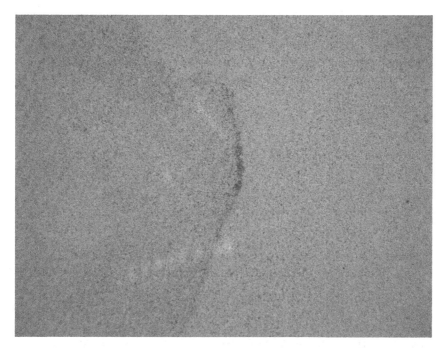

Figure 1.4. Sandstone. Note the much smoother texture of this compared to the lava. Photograph taken at Rickaby House, Low Pittington, August 2007

motor car production, see e.g. (Blest et al., 1999). A novel use of a theory for a porous material is by (Tadj et al., 2007) who model a crop in a greenhouse as a porous medium when assessing whether heating pipes should be placed above the crop, at mid-crop height, or at ground level.

We have mentioned some of the many important areas of application of porous materials. A theoretical understanding of such materials via mathematical models is clearly desirable. In this book we present some relatively simple mathematical models to describe porous media behaviour. We also present analyses of these models since mathematical and numerical analyses of them coupled with experimental observations will allow us to assess how useful a theory is.

We do not attempt to describe the specialist and highly important area of plasticity in porous materials, see e.g. (Borja, 2004), and the references therein. Also, we do not deal with the exciting but specialist areas of honeycomb and auxetic materials. Information on honeycomb materials may be found at e.g. (Scarpa and Tomlinson, 2000), (Schwingshackl et al., 2006), (Jung, 2008), (Vasilevich and Alexandrovich, 2008), while information on auxetic materials in general is contained on the excellent website of Professor R.S. Lakes, (Lakes, 2008), where many pertinent references may be found.

1.1.2 Notation, definitions

Standard indicial notation is used throughout this book together with the
Einstein summation convention for repeated indices. Standard vector or
tensor notation is also employed where appropriate. For example, we write

$$u_x \equiv \frac{\partial u}{\partial x} \equiv u_{,x} \qquad u_{i,t} \equiv \frac{\partial u_i}{\partial t} \qquad u_{i,i} \equiv \frac{\partial u_i}{\partial x_i} \equiv \sum_{i=1}^{3} \frac{\partial u_i}{\partial x_i}$$

$$u_j u_{i,j} \equiv u_j \frac{\partial u_i}{\partial x_j} \equiv \sum_{j=1}^{3} u_j \frac{\partial u_i}{\partial x_j}, \qquad i = 1, 2 \text{ or } 3.$$

In the case where a repeated index sums over a range different from 1 to 3
this will be pointed out in the text. Note that

$$u_j u_{i,j} \equiv (\mathbf{u} \cdot \nabla)\mathbf{u} \qquad \text{and} \qquad u_{i,i} \equiv \operatorname{div} \mathbf{u}.$$

As indicated above, a subscript t denotes partial differentiation with respect
to time. When a superposed dot is used it means the material derivative,
i.e.

$$\dot{u}_i \equiv \frac{\partial u_i}{\partial t} + u_j \frac{\partial u_i}{\partial x_j},$$

where u_i in the equation above is the velocity field. (A special definition
of a material derivative following a constituent particle is introduced in
sections 1.9.1 and 1.9.2. The rest of the book employs the standard material
derivative.)

The letter Ω will denote a fixed, bounded region of 3-space with boundary, Γ, sufficiently smooth to allow applications of the divergence theorem.
When we are dealing with convection problems we handle motion in a plane
layer, say $\{(x, y) \in \mathbb{R}^2\} \times \{z \in (0, d)\}$. In this case, we usually refer to functions that have an (x, y) behaviour which is repetitive in the (x, y) direction,
such as regular hexagons. The periodic cell defined by such a shape and its
Cartesian product with $(0, d)$ will be denoted by V. The boundary of the
period cell V will be denoted by ∂V.

The symbols $\| \cdot \|$ and $(\cdot, \cdot)$ will denote, respectively, the L^2 norm on Ω or
V, and the inner product on $L^2(\Omega)$ or $L^2(V)$, where the context will define
whether Ω or V is to be used, e.g.,

$$\int_V f^2 dV = \|f\|^2 \quad \text{and} \quad (f, g) = \int_V f g \, dV,$$

with equivalent definitions for Ω. We sometimes have recourse to use the
norm on $L^p(\Omega)$, $1 < p < \infty$, and then we write

$$\|f\|_p = \left(\int_\Omega |f|^p dx \right)^{1/p}.$$

We introduce the ideas of stability and instability in the context of the
porous medium equation, see section 1.2.1, (which would be defined with

suitable boundary conditions):

$$\frac{\partial u}{\partial t} - \Delta \Phi(u) = 0, \tag{1.2}$$

where Φ is a known nonlinear function, where $\mathbf{x} \in \Omega \subset \mathbb{R}^3$, and where $\Delta = \partial^2/\partial x^2 + \partial^2/\partial y^2 + \partial^2/\partial z^2$ is the Laplace operator.

We introduce notation in the context of a steady solution to (1.2), namely a solution $\bar{u}$ satisfying

$$\Delta \Phi(\bar{u}) = 0. \tag{1.3}$$

(We could equally deal with the stability of a time-dependent solution, but many of the problems encountered here are for stationary solutions and at this juncture it is as well to keep the ideas as simple as possible.) Let w be a perturbation to (1.3), i.e. put $u = \bar{u} + w(\mathbf{x}, t)$. Then, it is seen from (1.2) and (1.3) that w satisfies the system

$$\frac{\partial w}{\partial t} - \left\{ \Delta \big[\Phi(\bar{u} + w)\big] - \Delta \big[\Phi(\bar{u})\big] \right\} = 0. \tag{1.4}$$

To discuss linearized instability we linearize (1.4) which means we keep only the terms which are linear in w. From a Taylor series expansion of Φ we have

$$\Phi(\bar{u} + w) = \Phi(\bar{u}) + w\Phi'(\bar{u}) + O(w^2). \tag{1.5}$$

Then, using (1.3), (1.5) in (1.4) we derive the linearized equation satisfied by w, namely

$$\frac{\partial w}{\partial t} - \Delta \big[w\Phi'(\bar{u})\big] = 0. \tag{1.6}$$

Since (1.6) is a linear equation we may introduce an exponential time dependence in w so that $w = e^{\sigma t}s(\mathbf{x})$. Then (1.6) yields

$$\sigma s - \Delta \big[s\Phi'(\bar{u})\big] = 0. \tag{1.7}$$

We say that the steady solution $\bar{u}$ to (1.3) is *linearly unstable* if

$$Re(\sigma) > 0,$$

where $Re(\sigma)$ denotes the real part of σ. Equation (1.7) (together with appropriate boundary conditions) is an eigenvalue problem for σ. For many of the problems discussed in this book the eigenvalues may be ordered so that

$$Re(\sigma_1) > Re(\sigma_2) > \dots$$

For linear instability we then need only ensure $Re(\sigma_1) > 0$.

Let $w_0(\mathbf{x}) = w(\mathbf{x}, 0)$ be the initial data function associated to the solution w of equation (1.4). The steady solution $\bar{u}$ to (1.3) is *nonlinearly stable* if and only if for each $\epsilon > 0$ there is a $\delta = \delta(\epsilon)$ such that

$$\|w_0\| < \delta \Rightarrow \|w(t)\| < \epsilon \tag{1.8}$$

and there exists γ with $0 < \gamma \leq \infty$ such that

$$\|w_0\| < \gamma \Rightarrow \lim_{t \to \infty} \|w(t)\| = 0. \qquad (1.9)$$

If $\gamma = \infty$, we say the solution is *unconditionally* nonlinearly stable (or simply refer to it as being asymptotically stable), otherwise for $\gamma < \infty$ the solution is *conditionally* (nonlinearly) stable. For nonlinear stability problems it is an important goal to derive parameter regions for unconditional nonlinear stability, or at least conditional stability with a finite initial data threshold (i.e. finite, non-vanishing, radius of attraction). It is important to realise that the linearization as in (1.6) and (1.7) can only yield linear *instability*. It tells us nothing whatsoever about stability. There are many equations for which nonlinear solutions will become unstable well before the linear instability analysis predicts this, cf. (Lu and Shao, 2003). Also, when an analysis is performed with $\gamma < \infty$ in (1.9) this yields conditional nonlinear stability, i.e. nonlinear stability for only a restricted class of initial data.

We have only defined stability with respect to the $L^2(\Omega)$ norm in (1.8) and (1.9). However, sometimes it is convenient to use an analogous definition with respect to some other norm or positive-definite solution measure. It will be clear in the text when this is the case. When we refer to continuous dependence on the initial data we mean a phenomenon like (1.8). Thus, a solution w to equation (1.4) depends continuously on the initial data if a chain of inequalities like (1.8) holds.

Throughout the book we make frequent use of inequalities. In particular, we often use the Cauchy-Schwarz inequality for two functions f and g, i.e.

$$\int_\Omega f g \, dx \leq \left(\int_\Omega f^2 dx \right)^{1/2} \left(\int_\Omega g^2 dx \right)^{1/2}, \qquad (1.10)$$

or what is the same in L^2 norm and inner product notation,

$$(f, g) \leq \|f\| \, \|g\|. \qquad (1.11)$$

The arithmetic-geometric mean inequality (with a constant weight $\alpha > 0$) is, for $a, b \in \mathbb{R}$,

$$ab \leq \frac{1}{2\alpha} a^2 + \frac{\alpha}{2} b^2, \qquad (1.12)$$

and this is easily seen to hold since

$$\left(\frac{a}{\sqrt{\alpha}} - \sqrt{\alpha} b \right)^2 \geq 0.$$

Another inequality we frequently have recourse to is Young's inequality, which for $a, b \in \mathbb{R}$ we may write as

$$ab \leq \frac{|a|^p}{p} + \frac{|b|^q}{q}, \qquad \frac{1}{p} + \frac{1}{q} = 1, \quad p, q \geq 1. \qquad (1.13)$$

1.1.3   Overview

The layout of the book is now briefly described. In sections 1.2 – 1.5 we discuss the classical models of Darcy, Forchheimer and Brinkman, on which a major part of this book is based. Sections 1.6 and 1.7 discuss how effects like temperature and a salt field are incorporated into porous media, and the relevant boundary conditions to be used. In section 1.8 we refer to another theoretical description of porous media where the motion of the elastic matrix is considered by including a void distribution in the theory of nonlinear elasticity. Finally, the introduction concludes by examining two theories for a mixture of an elastic solid with one or more fluids.

Chapter 2 is, we believe, new in book form. This investigates the important problem of whether the porous model itself is stable. This concept of stability, known as structural stability, is very important. Several new results are included here. Chapter 3 is another chapter we have never seen before in book form. This considers the aspect of spatial decay of a solution in porous media. Again, several results not published elsewhere are included here.

Chapters 4 and 5 consider a variety of stability problems in porous media flow. There is very little overlap with the material of my earlier book (Straughan, 2004a) or the book by (Nield and Bejan, 2006). Chapter 4 concentrates on thermal convection problems in porous media, whereas chapter 5 reviews recent work by a variety of writers.

Chapter 6 considers another relatively new topic, that of stability of several flows in a fluid which overlies a saturated porous medium. The sections where the fluid overlies a transition layer composed of a Forchheimer porous material, or a Brinkman-Forchheimer porous material, which in turn overlies a Darcy porous medium, are new. The last section of this chapter studies wave motion when the wave is incident from a fluid but is reflected / refracted by a saturated porous medium below. This leads naturally into the next two chapters, chapters 7 and 8 where nonlinear wave motion is considered in a porous body.

Chapter 7 investigates nonlinear waves propagating in a nonlinear elastic body which has a distribution of voids throughout. Temperature effects and temperature waves are also considered in detail. Chapter 8 studies what are known as equivalent fluid theories where a compressible fluid saturates a porous medium, but the solid matrix is regarded as fixed. Again, consideration is also given to temperature wave propagation.

The monograph is completed with chapter 9 which deals with three methods for the numerical solution of differential equation eigenvalue problems. The techniques discussed are the compound matrix method, the Chebyshev tau method, and a Legendre - Galerkin technique. While the first two are discussed in my book (Straughan, 2004a), the third is not. The third technique has some useful advantages which are outlined here. However, the examples given for the other two methods are different from those in (Straughan, 2004a).

We briefly touch on the important topic of stability of flow in unsaturated porous media in section 5.7. This is a relatively new topic which is increasingly occupying attention. Another area which is highly topical, is stability of flow of a viscoelastic fluid in a porous medium, and this is considered in section 5.2. In fact, even the development of an appropriate theory to model flow of a viscoelastic fluid in a porous medium is nontrivial, as may be witnessed from the papers, for example, of (Lopez de Haro et al., 1996) and (Wei and Muraleetharan, 2007). Nevertheless, due to problems involving flow of oils in soils / rocks, this is an important area.

1.2 The Darcy model

The celebrated Darcy equation is believed to originate from work of (Darcy, 1856), and this equation is discussed in detail in (Nield and Bejan, 2006), section 1.4. Darcy's law basically states that the flow rate of a fluid in a porous material is proportional to the pressure gradient. In current terminology, if the flow is in the $x-$direction and the speed in that direction is u then this may be represented as

$$\mu u = -k\,\frac{dp}{dx},\qquad(1.14)$$

where μ, k are viscosity and permeability with p being the pressure of the fluid in the porous medium. Despite its apparent simplicity, equation (1.14) has been very successful in providing a theoretical description of flow in porous media.

Equation (1.14) is generalized to three spatial dimensions and we allow for external forces (such as gravity). Thus, if we denote the velocity field in the porous medium by $\mathbf{v}$, where $\mathbf{v} = (u, v, w)$, and denote the body force by $\mathbf{f}$, we have a three-dimensional version of equation (1.14) which takes form

$$0 = -\frac{\partial p}{\partial x_i} - \frac{\mu}{k}\,v_i + \rho f_i,\qquad(1.15)$$

where ρ is the density of the fluid, cf. (Joseph, 1976a). Equation (1.15) describes flow of a fluid in a saturated porous medium, provided the flow rate is sufficiently low. Throughout this book we make extensive use of equation (1.15), often coupled to other equations, for the temperature, or salt concentration, for example.

If the fluid in the porous medium is incompressible then we must couple equation (1.15) with the incompressibility condition

$$\frac{\partial v_i}{\partial x_i} = 0.\qquad(1.16)$$

Derivations of Darcy's law based on various assumptions may be found in many places in the literature, cf. (Nield and Bejan, 2006), section 1.4. Very interesting accounts based on homogenization and on asymptotic expansions are provided by (Firdaouss et al., 1997) and by (Giorgi, 1997), respectively, see also (Whitaker, 1986). In this book we are concerned with uses of Darcy's law and related models. The history of porous media model development may be found in (de Boer, 1999).

1.2.1   The Porous Medium Equation

There is an equation which frequently appears in the mathematical analysis literature known as the porous medium equation, cf. (Alikakos and Rostamian, 1981), (Aronson and Peletier, 1981), (Di Benedetto, 1983), (Aronson and Caffarelli, 1983), (Flavin and Rionero, 1995). If the fluid is compressible, i.e. equation (1.16) does not hold, then the density instead satisfies the equation

$$\frac{\partial \rho}{\partial t} + \frac{\partial}{\partial x_i}(\rho v_i) = 0. \tag{1.17}$$

If we couple this equation to equation (1.15) with $f_i = 0$, then since p is a function of ρ,

$$v_i = -\frac{k}{\mu}\frac{\partial p}{\partial x_i}. \tag{1.18}$$

Substitute in equation (1.17) for v_i and we obtain the equation

$$\frac{\partial \rho}{\partial t} - \frac{k}{\mu}\frac{\partial}{\partial x_i}\left(\rho\frac{\partial p}{\partial x_i}\right) = 0$$

or

$$\frac{\partial \rho}{\partial t} - \frac{k}{\mu}\frac{\partial}{\partial x_i}\left(\rho p'(\rho)\frac{\partial \rho}{\partial x_i}\right) = 0.$$

Let $\Phi(\rho)$ be a potential (integral) for the function $(k\rho p'/\mu)(\partial \rho/\partial x_i)$, i.e.

$$\frac{\partial \Phi}{\partial x_i} = \Phi'(\rho)\frac{\partial \rho}{\partial x_i} = \frac{k\rho p'(\rho)}{\mu}\frac{\partial \rho}{\partial x_i},$$

and then the last equation may be rewritten as

$$\frac{\partial \rho}{\partial t} - \Delta\Phi(\rho) = 0. \tag{1.19}$$

Equation (1.19) is that equation often referred to as the porous medium equation.

Some of the very interesting early articles on the porous medium equation are those of (Alikakos and Rostamian, 1981), (Aronson and Peletier, 1981), (Di Benedetto, 1983), (Aronson and Caffarelli, 1983), and a recent very interesting article dealing with a novel application of the porous medium equation to image contour enhancement in image processing is by (Barenblatt and Vazquez, 2004). Further interesting pointwise stability results and finite time blow-up results for a solution to equations like (1.19) may be found in (Flavin and Rionero, 1998; Flavin and Rionero, 2003), (Flavin, 2006), for various choices of nonlinear function Φ. The work of (Rionero and Torcicollo, 2000) studies continuous dependence in a weighted L^2 norm for a similar problem and (Rionero, 2001) contains some interesting asymptotic results.

1.3   The Forchheimer model

If the flow rate exceeds a certain value then it is believed that the linear relationship of (1.14) or (1.15) will be inadequate to describe the velocity field accurately. (Forchheimer, 1901) (see also (Dupuit, 1863)) proposed modifying the linear velocity / pressure gradient law and replacing it with a nonlinear one. According to (Firdaouss et al., 1997), (Forchheimer, 1901) proposes replacing the left hand side of equation (1.14) by one of three formulae,

$$\alpha = au + bu^2, \qquad \alpha = mu^n, \qquad \alpha = au + bu^2 + cu^3,$$

where α denotes the left hand side of equation (1.14). In this book we pay particular attention to the first of these. The generalization of equation (1.15) which is consistent with this results in the Forchheimer model

$$0 = -\frac{\partial p}{\partial x_i} - \frac{\mu}{k} v_i - b|\mathbf{v}|v_i + \rho f_i. \qquad (1.20)$$

For incompressible flow we couple this with equation (1.16).

Rigorous justifications of equation (1.20) and generalizations may be found in the papers of (Whitaker, 1996), (Firdaouss et al., 1997), (Giorgi, 1997), (Bennethum and Giorgi, 1997).

1.4   The Brinkman model

If the porosity, ϕ, of the porous medium is close to 1, i.e. the solid skeleton occupies little of the total volume, or if the porous medium is adjacent to a solid wall, then there is a belief that neither of the models of sections 1.2 nor 1.3 will prove sufficient. Indeed, when $\phi \approx 1$, one might expect the higher derivatives of the Laplacian in the Navier-Stokes equations to play a role. In fact, there is evidence that equation (1.15) should be replaced by the following

$$0 = -\frac{\partial p}{\partial x_i} - \frac{\mu}{k} v_i + \lambda \Delta v_i + \rho f_i. \qquad (1.21)$$

Equation (1.21) is usually associated with (Brinkman, 1947) and this equation is discussed in some detail by (Nield and Bejan, 2006), section 1.5.3. The coefficient λ is usually referred to as an equivalent viscosity.

Throughout this book we make an investigation into properties of the solution to the Darcy model, (1.15) together with (1.16), the Forchheimer model, (1.20) together with (1.16), or the Brinkman model, (1.21) together with (1.16).

1.5 Anisotropic Darcy model

The models discussed in sections 1.2, 1.3, 1.4 are frequently very useful when dealing with the flow in a porous medium when the situation is isotropic, i.e. the response is the same in all directions. However, many porous media exhibit strongly anisotropic characteristics. For example, wood behaves very differently along the grain to the way it does across the grain. Rock strata is another example of highly anisotropic porous media. When one is interested in modelling flow in anisotropic porous media, then we should modify each of the models in sections 1.2, 1.3, 1.4, accordingly.

In this section we indicate how we may modify the model in section 1.2. Anisotropic modifications for the other models follow in a similar manner.

Typically the permeability, i.e. the ease with which the fluid flows, will vary if the solid fraction of the porous medium displays a strong anisotropy. To account for this we replace the permeability k in (1.15) by a tensor K_{ij}. Thus, we may replace equation (1.15) by

$$K_{ij}\frac{\partial p}{\partial x_j} = -\mu v_i + \rho K_{ij}f_j. \tag{1.22}$$

If we introduce a generalized inverse tensor to K_{ij}, say M_{ij}, then we may recast equation (1.22) in a form not dissimilar to (1.15). To do this we suppose

$$\mathbf{MK} = c\mathbf{I}, \qquad \text{i.e.} \qquad \mathbf{M} = c\mathbf{K}^{-1},$$

for $c > 0$ a constant, and then equation (1.22) is equivalent to

$$0 = -\frac{\mu}{c}M_{ij}v_j - \frac{\partial p}{\partial x_i} + \rho f_i. \tag{1.23}$$

Equation (1.23) is to be coupled with equation (1.16). The precise form for $\mathbf{K}$ depends on what the structure is of the underlying solid matrix in the porous medium.

As a specific example we consider the case where the permeability in the vertical direction is different from that in the horizontal directions. Then, for $k \neq k_3$, $k, k_3 > 0$,

$$\mathbf{K} = \begin{pmatrix} k & 0 & 0 \\ 0 & k & 0 \\ 0 & 0 & k_3 \end{pmatrix}$$

If we select $c = k_3$, we need

$$\mathbf{M} = \begin{pmatrix} k_3/k & 0 & 0 \\ 0 & k_3/k & 0 \\ 0 & 0 & 1 \end{pmatrix}$$

A porous convection problem with such a permeability is analysed by (Carr and de Putter, 2003). A further example of application to thermal convection in a porous medium with transversely isotropic permeability at an angle oblique to the vertical is given in (Straughan and Walker, 1996a), see also (Straughan, 2004a), p. 338, and sections 4.1.3 and 4.2.6 of this book. The permeability for this situation is not simply a diagonal matrix and this severely complicates the analysis and numerical calculations.

1.6   Equations for other fields

1.6.1   Temperature

It is typical in flow through a porous medium that we may also wish to determine the temperature at a point, or the concentration of a chemical, say salt. A convenient derivation of the equation for these fields is given by (Joseph, 1976a). We follow his derivation.

In a given (small) volume, $\tilde{\Omega}$, containing the point $\mathbf{x}$, we denote the solid (porous matrix) part by s while f denotes the fluid. The individual fluid and solid volumes within $\tilde{\Omega}$ are denoted by Ω_f and Ω_s, respectively. The "small" volume is such that a typical length scale is sufficiently larger than the pore scale of the porous material, but the same length scale is much smaller than the overall flow domain. ((Nield and Bejan, 2006), pp. 2,3, refer to such a volume as a representative elementary volume.) The thermal diffusivities are κ_s, κ_f, the densities are $\rho_{0\alpha}$, $\alpha = s$ or f, c_s is the specific heat of the solid, and c_{pf} denotes the specific heat at constant pressure of the fluid. Then, let v_i be the average velocity of the fluid at point $\mathbf{x}$ (the seepage velocity) which appears in (1.15) (v_i is the average of the real fluid velocity over the whole of $\tilde{\Omega}$). Let also V_i, defined by $v_i = \phi V_i$, be the pore average velocity (i.e. the fluid velocity averaged over Ω_f). For the separate solid and fluid components we have for the temperature field, T,

$$(\rho_0 c)_s \frac{\partial T}{\partial t} = \kappa_s \Delta T, \tag{1.24}$$

$$(\rho_0 c_p)_f \left(\frac{\partial T}{\partial t} + V_i \frac{\partial T}{\partial x_i} \right) = \kappa_f \Delta T. \tag{1.25}$$

We add $(1 - \phi)(1.24)$ and $\phi(1.25)$ to see that

$$
\left[\phi(\rho_0 c_p)_f + (\rho_0 c)_s(1 - \phi)\right]\frac{\partial T}{\partial t}
$$
$$
+ (\rho_0 c_p)_f \phi V_i \frac{\partial T}{\partial x_i} = \left[\kappa_s(1 - \phi) + \kappa_f \phi\right]\Delta T. \tag{1.26}
$$

Denote by $M = (\rho_0 c_p)_f/(\rho_0 c)_m$ where $(\rho_0 c)_m = \phi(\rho_0 c_p)_f + (\rho_0 c)_s(1 - \phi)$, and by $\kappa = k_m/(\rho_0 c_p)_f$, where $k_m = \kappa_s(1 - \phi) + \kappa_f \phi$. Then, equation (1.26) may be rewritten

$$
\frac{1}{M}\frac{\partial T}{\partial t} + v_i \frac{\partial T}{\partial x_i} = \kappa \Delta T. \tag{1.27}
$$

Equation (1.27) is the equation we employ to govern the temperature field in a porous medium. If we couple it with equations (1.15), (1.16), then we must specify how temperature enters equation (1.15). This is usually done via some equation of state for $\rho(T)$, cf. section 4.1.

1.6.2 Salt field

If we have a fluid with a salt dissolved in it in a porous medium we may use a similar procedure to derive an equation for the salt concentration, C. Suppose the salt is not absorbed by the solid matrix. Then, in the fluid part, the salt concentration, C, satisfies the differential equation

$$
\frac{\partial C}{\partial t} + V_i \frac{\partial C}{\partial x_i} = k_c \Delta C. \tag{1.28}
$$

Since $V_i = v_i/\phi$, the equation governing C is

$$
\phi \frac{\partial C}{\partial t} + v_i \frac{\partial C}{\partial x_i} = \phi k_c \Delta C. \tag{1.29}
$$

If we were to non-dimensionalize with time, $\mathcal{T} = d^2/M\kappa$, velocity, $U = \kappa/d$, d being a length scale, then one may show the appropriate non-dimensional form of (1.29) is

$$
M\phi \frac{\partial C}{\partial t} + v_i \frac{\partial C}{\partial x_i} = \frac{1}{Le}\Delta C. \tag{1.30}
$$

Here, $Le = \phi k_c/\kappa$ is a Lewis number.

We see that some care must be exercised to derive the correct coefficients in a non-dimensional form of the equations governing non-isothermal flow of a salt laden fluid in a porous medium.

1.7 Boundary conditions

For the Darcy model of section 1.2 or the Forchheimer model of section 1.3 we do not have derivatives present in the v_i term. Thus, we need to prescribe the normal component of v_i on the boundary, Γ, of a volume Ω, i.e. we prescribe $v_i n_i$, n_i being the unit outward normal to Γ. However, when dealing with the Brinkman model of section 1.4 the presence of the Δv_i term means we need to prescribe v_i on the whole of Γ (assuming we are dealing with no-slip boundary conditions; slip boundary conditions could also be handled, cf. (Webber, 2007; Webber, 2006)). If the porous medium borders a fluid then the correct form of boundary condition is a contentious matter. This topic is treated in detail in chapter 6.

For the temperature field, T, we may prescribe T on Γ if the temperature is measurable there. If, on the other hand, we can measure the heat flux on the boundary Γ, then since the heat flux $\mathbf{q}$ is usually given by $q_i = -\kappa T_{,i}$, $\kappa > 0$, we can prescribe $q_i n_i$ on Γ. In other words, we may prescribe $\partial T/\partial n$ on Γ. Perhaps a combination boundary condition may also be appropriate, i.e. one of form

$$\frac{\partial T}{\partial n} + \gamma T = a,$$

where a is a prescribed function. If radiation heating is the dominant effect, e.g. a surface directly in sunlight, then γ is likely to be small. If, however, there is little radiant heating, it may be more appropriate to assign T on Γ. The combination boundary condition is between both extremes and may hold for several real problems. Surface radiation may have a significant effect on thermal convection and stability, see e.g. (Jaballah et al., 2007).

For a salt field, if there is zero flux of the salt out of the boundary, we may assume $\partial C/\partial n = 0$ there. However, there are some instances where it is possible to control the concentration field on Γ, cf. (Krishnamurti, 1997) and in this instance we would have a boundary condition of form

$$C = C_G, \qquad \mathbf{x} \in \Gamma,$$

where C_G is a known function.

1.8   Elastic materials with voids

1.8.1   Nunziato-Cowin theory

Another class of theories which may be thought of as describing certain properties of porous media were derived by (Nunziato and Cowin, 1979). The key idea is to suppose there is an elastic body which has a distribution of voids throughout. The voids are gaps full of air, water, or some other fluid. This theory provides equations for the displacement of the elastic matrix of the porous medium and the void fraction occupied by the fluid. This is thus very different from the equations considered thus far in the introduction where we are interested in determining the velocity field for the fluid in the porous medium. Nevertheless, we believe the voids theory has a large potential, especially in wave propagation problems. We do not describe explicitly a voids theory at this point, although the equations are derived in sections 7.2.2 and 7.2.3 of chapter 7, where they are analysed in connection with acceleration wave propagation.

The theory of an elastic body containing voids essentially generalizes the classical theory of nonlinear elasticity by adding a function $\nu(\mathbf{X}, t)$ to describe the void fraction within the body. Here $\mathbf{X}$ denotes a point in the reference configuration of the body. Thus, in addition to the momentum equation for the motion $x_i = x_i(\mathbf{X}, t)$ as time evolves, one needs to prescribe an evolution equation for the void fraction ν. For a non-isothermal situation one also needs an energy balance law which effectively serves to determine the temperature field $T(\mathbf{X}, t)$. The original theory is due to (Nunziato and Cowin, 1979) and the temperature field development was largely due to D. Iesan, see details in chapter 1 of (Iesan, 2004). This theory has much in common with the continuum theory for granular materials, cf. (Massoudi, 2005; Massoudi, 2006a; Massoudi, 2006b).

In this book we also consider temperature field development in a voids theory firstly by adding the time derivative of temperature as a constitutive variable, cf. (Ciarletta and Straughan, 2007b) and section 7.3 of this book. We then add temperature effects via the (Green and Naghdi, 1991) thermal displacement variable $\alpha(\mathbf{X}, t) = \int_0^t T(\mathbf{X}, s)ds$. This theory is due to (De Cicco and Diaco, 2002) and acceleration waves are considered by (Ciarletta et al., 2007), see also section 7.4 of this book. We also include the temperature field via a (Green and Naghdi, 1991; Green and Naghdi, 1992) type III theory development in section 7.5. We have not seen this development before.

1.8.2   Microstretch theory

(Eringen, 1990; Eringen, 2004b) develops a voids theory which has a richer structure than the (Nunziato and Cowin, 1979) model. This is achieved by incorporating an equation for the spin at each point of the body. Again, this theory is likely to have rich application in wave propagation problems. We describe this theory in connection with nonlinear wave motion in section 7.6.

In addition to the fields ν, x_i, T, the (Eringen, 1990; Eringen, 2004b) microstretch theory adds a variable ϕ_i which is a microrotation vector. (Iesan and Scalia, 2006) study nonlinear singular surfaces in this theory, although we also address some new questions regarding singular surfaces in section 7.6 of this book.

A detailed account of many properties of elastic bodies containing voids may also be found in the book by (Iesan, 2004), chapters 1 to 3.

1.9   Mixture theories

The theories of sections 1.2 – 1.5 account for the velocity field and pressure of the fluid in a saturated porous material. They do not allow one to account for dynamic modelling of the deformation of the solid porous matrix. The theory of voids mentioned in section 1.8 allows us to calculate the dynamic behaviour of the elastic matrix and of the void fraction (fluid fraction). However, it does not allow us to specify the constitutive properties of the fluid filling the voids, nor does it allow us to calculate the fluid velocity field. In order to have a complete theory in which we may determine both the dynamic behaviour of the elastic matrix and of the fluid filling the matrix we need a more complete (and inevitably more complicated) theory. One such avenue open to us is to employ a continuum mixture theory of an elastic solid and an appropriate fluid, for example, a Newtonian fluid. By assuming each point of the body is simultaneously occupied by a fluid and a solid, one can produce a suitable continuum theory. The literature is full of such theories, which one may refer to as poroelasticity, and we do not attempt to review them all. The origins of the theory of poroelasticity are normally attributed to Terzaghi in 1923 and the classical linear equations are usually associated with Biot in 1941 (see e.g. (Lewis and Schrefler, 1998) or (Senjuntichai and Rajapakse, 1995)). Several mixture theory approaches, particularly those useful for numerical and engineering applications are discussed at length in the book of (Lewis and Schrefler, 1998), and in the articles of (Biot, 1956a; Biot, 1956b), (Hassanizadeh and Gray, 1990; Hassanizadeh and Gray, 1993), (de Boer et al., 1993), (Jiang and Rajapakse, 1994), (Zhou et al., 1998), (dell'Isola and Hutter, 1998; dell'Isola and Hutter, 1999), (Albers, 2003), (Albers and Wilmansky, 2005), (Weinstein and Bennethum, 2006), where many other relevant references are quoted.

Of the many available mixture theories which lead to a theoretical description of the behaviour of a porous medium, we choose to give a brief exposition of two. One is that developed by (Eringen, 1994; Eringen, 2004a). This leads to a concise nonlinear theory. The other is due to (Bowen, 1982). The latter is chosen because it allows for a variable porosity treatment through an internal variable approach. In fact, both theories are used in this book. Eringen's theory is employed in section 8.1.2 to derive the Jordan - Darcy equations which allow for a description of acoustic waves in a porous medium. The theory of Bowen is used in section 6.8 to study the effect of variable porosity on the transmission and reflection of an acoustic wave at the sea bed.

1.9.1   Eringen's theory

We do describe a theory of mixtures derived by (Eringen, 1994; Eringen, 2004a). To do this we need some general notation appropriate to the continuum theory of mixtures. (Eringen, 1994) develops his theory for a three component mixture of a fluid, a gas, and an elastic solid. These constituents are denoted respectively by f, g, s. We suppose that each constituent occupies a continuous body in three - dimensions. Each body is assigned to a fixed reference configuration. The motion of each point in the constituent body satisfies a mapping

$$x_i^\alpha = x_i^\alpha(X_A^\alpha, t), \qquad \alpha = f, g, s,$$

where X_A and x_i refer to the positions in a reference configuration and a current configuration, respectively. At the point x_i^α at time t, each constituent is present. The velocity and acceleration of each constituent are represented by

$$v_i^\alpha = {'x_i^\alpha} = \frac{\partial x_i^\alpha}{\partial t}\Big|_{X_A^\alpha}, \qquad a_i^\alpha = {''x_i^\alpha} = \frac{\partial^2 x_i^\alpha}{\partial t^2}\Big|_{X_A^\alpha}.$$

A material derivative of a function $\psi(x_i^\alpha, t)$ may be defined following the motion of the αth constituent, by

$${'\psi^\alpha} = \frac{\partial \psi}{\partial t}\Big|_{X_A^\alpha} = \frac{\partial \psi}{\partial t}(x_i^\alpha, t) + {'x_j^\alpha} \frac{\partial \psi}{\partial x_j^\alpha}(x_i^\alpha, t).$$

In what follows the $'$ or $''$ notion refers to the first or second material derivative following the motion of a particular constituent. Each constituent has a mass density ρ^α ($\alpha = f, g, s$) and then the mixture density, $\rho(x_i, t)$, and mixture velocity, $v_i(x_j, t)$, are given by

$$\rho(x_i, t) = \rho^f + \rho^g + \rho^s,$$

$$v_i(x_j, t) = \dot{x}_i = \frac{1}{\rho}(\rho^f v_i^f + \rho^g v_i^g + \rho^s v_i^s).$$

A superposed dot denotes the material time derivative, i.e.

$$\dot{\psi} = \frac{\partial \psi}{\partial t}(x_j, t) + v_i \frac{\partial \psi}{\partial x_i}(x_j, t).$$

The gradient of deformation of X_A^α may be defined for each constituent as

$$F_{iA}^\alpha = \frac{\partial x_i^\alpha}{\partial X_A^\alpha}, \qquad \alpha = f, g, s.$$

Eringen's theory is based on the equations of conservation of mass for each constituent, i.e.

$$\frac{\partial \rho^f}{\partial t} + \frac{\partial}{\partial x_i}(\rho^f v_i^f) = 0, \tag{1.31}$$

$$\frac{\partial \rho^g}{\partial t} + \frac{\partial}{\partial x_i}(\rho^g v_i^g) = 0, \tag{1.32}$$

$$\frac{\partial \rho^s}{\partial t} + \frac{\partial}{\partial x_i}(\rho^s v_i^s) = 0, \tag{1.33}$$

although equation (1.33) may be written in the perhaps more convenient form

$$\rho^s |\det(F_{iA}^s)| = \rho_0^s ,\tag{1.34}$$

where ρ_0^s denotes the solid density in the reference configuration. (In fact, equations (1.31) and (1.32) may also be written in a form similar to (1.34). However, the fluid components frequently are best solved from equations (1.31) and (1.32).)

The balance of momentum equations have form

$$\rho^f \, ''x_i^f = \frac{\partial t_{ij}^f}{\partial x_j} + \rho^f b_i^f - p_i^f ,\tag{1.35}$$

$$\rho^g \, ''x_i^g = \frac{\partial t_{ij}^g}{\partial x_j} + \rho^g b_i^g - p_i^g ,\tag{1.36}$$

$$\rho^s \, ''x_i^s = \frac{\partial t_{ij}^s}{\partial x_j} + \rho^s b_i^s - p_i^s ,\tag{1.37}$$

where $t_{ij}^f, t_{ij}^g, t_{ij}^s$ are the stress tensors, b_i^f, b_i^g, b_i^s are the body force terms, and p_i^f, p_i^g, p_i^s are the momentum supplies (the interaction forces). In addition there are balance of moment of momentum equations, (Eringen, 1994), equation (2.16). (We do not specifically use these here.)

(Eringen, 1994; Eringen, 2004a) writes separate energy balance laws for each constituent. However, since he subsequently assumes a common temperature, θ, for each constituent, it is sufficient to work with a single energy balance law, and this is

$$\rho \dot{\epsilon} - \frac{\partial q_i}{\partial x_i} - \sum_\alpha (t_{ij}^\alpha v_{i,j}^\alpha + p_i^\alpha v_i^\alpha) - \rho h = 0,\tag{1.38}$$

where the $\sum_\alpha$ is over $\alpha = f, g$ and s, and where ϵ is the internal energy, q_i is the heat flux vector, and h is the external supply of energy.

Eringen's theory is developed in terms of a Helmholtz free energy function $\psi = \epsilon - \eta\theta$, η being the entropy. The constitutive theory of (Eringen, 1994) is such that

$$\psi, \eta, q_i, t_{ij}^\alpha, m_{ij}^\alpha, p_i^\alpha, \qquad \alpha = f, g, s,$$

depend on the independent variables

$$\theta, \rho^g, \rho^f, F_{iA}^s, \dot{\theta}, \theta_{,i}, d_{ij}^f, w_{ij}^f, v_i^f, v_i^g, v_i^s,\tag{1.39}$$

where t_{ij}^α and m_{ij}^α now denote the symmetric and skew-symmetric parts of the stress tensors, $\alpha = f, g, s$, and d_{ij}^f, w_{ij}^f are the symmetric and skew-symmetric parts of the fluid velocity gradient $v_{i,j}^f$, i.e. $d_{ij}^f = (v_{i,j}^f + v_{j,i}^f)/2$, $w_{ij}^f = (v_{i,j}^f - v_{j,i}^f)/2$. This constitutive theory is thus appropriate for a mixture of a viscous fluid, an inviscid fluid (gas), and an elastic solid.

By employing a suitable entropy inequality, (Eringen, 1994; Eringen, 2004a) reduces his constitutive theory. He splits the independent variables into the first four in the list (1.39), with the remainder representing a dissipative part. In fact, he shows that

$$\psi = \psi(\theta, \rho^g, \rho^f, C_{KL}), \qquad (1.40)$$

where

$$C_{KL} = \frac{\partial x_i}{\partial X_K^s} \frac{\partial x_i}{\partial X_L^s},$$

and introduces a dissipation function Φ with

$$\Phi = \Phi(I_1, \ldots, I_5^a, \dot{\theta}; \theta, \rho^g, \rho^f, C_{KL}), \qquad a = f, g,$$

where $I_1, \ldots, I_5^a$ are the following invariants

$$I_1 = (d_{ii}^f)^2, \quad I_2 = d_{ij}^f d_{ji}^f, \quad I_3 = \theta_{,i} \theta_{,i}/\theta^2 \,,$$
$$I_4^{ff} = (v_i^f - v_i^s)(v_i^f - v_i^s), \quad I_4^{fg} = I_4^{gf} = (v_i^f - v_i^s)(v_i^g - v_i^s),$$
$$I_4^{gg} = (v_i^g - v_i^s)(v_i^g - v_i^s),$$
$$I_5^f = (v_i^f - v_i^s)\theta_{,i}/\theta, \quad I_5^g = (v_i^g - v_i^s)\theta_{,i}/\theta \,.$$

The governing equations are then (1.31) – (1.33), (1.35) – (1.37) and (1.38), with the constitutive equations derived by Eringen having form

$$\eta = -\frac{\partial \psi}{\partial \theta} - \frac{1}{\rho}\frac{\partial \Phi}{\partial \dot{\theta}},$$
$$q_i = \frac{2}{\theta}\frac{\partial \Phi}{\partial I_3}\theta_{,i} + \frac{\partial \Phi}{\partial I_5^f}(v_i^f - v_i^s) + \frac{\partial \Phi}{\partial I_5^g}(v_i^g - v_i^s),$$
$$t_{ij}^g = -\delta_{ij}\pi^g, \qquad \pi^g = \rho\rho^g\frac{\partial \psi}{\partial \rho^g},$$
$$t_{ij}^f = \left(-\pi^f + 2\frac{\partial \Phi}{\partial I_1}d_{kk}^f\right)\delta_{ij} + 4\frac{\partial \Phi}{\partial I_2}d_{ij}^f, \quad \pi^f = \rho\rho^f\frac{\partial \psi}{\partial \rho^f},$$
$$t_{ij}^s = \rho\left(\frac{\partial \psi}{\partial C_{AB}} + \frac{\partial \psi}{\partial C_{BA}}\right)\frac{\partial x_i}{\partial X_A^s}\frac{\partial x_j}{\partial X_B^s}, \qquad (1.41)$$
$$p_i^f = \frac{1}{\theta}\frac{\partial \Phi}{\partial I_5^f}\theta_{,i} + 2\frac{\partial \Phi}{\partial I_4^{ff}}(v_i^f - v_i^s) + 2\frac{\partial \Phi}{\partial I_4^{fg}}(v_i^g - v_i^s)$$
$$p_i^g = \frac{1}{\theta}\frac{\partial \Phi}{\partial I_5^g}\theta_{,i} + 2\frac{\partial \Phi}{\partial I_4^{gg}}(v_i^g - v_i^s) + 2\frac{\partial \Phi}{\partial I_4^{gf}}(v_i^f - v_i^s)$$
$$p_i^s = -p_i^f - p_i^g \,.$$

Once the functions ψ and Φ are prescribed, we then have a consistent theory which can, in principle, be solved. I am unaware of any specific problems which have been solved using the fully nonlinear theory, equations (1.31) – (1.33), (1.35) – (1.37), (1.38), and (1.41), but this would be a very useful avenue to pursue, especially if a good choice of ψ and Φ were made, based on sound physical principles.

To make progress, (Eringen, 1994) considers a theory in which ψ is quadratic in appropriate variables, and then develops a linearized set of governing equations for the mixture.

We record the linearized equations for an isothermal, isotropic mixture, linearized about reference densities $\rho_0^f, \rho_0^g, \rho_0^s$. In these equations u_i^α $\alpha = f, g, s$, denote the displacements from a point x_i^α. The three sets of equations become

$$\rho_0^g \ddot{u}_i^g = -\sigma^{gg} u_{j,ji}^g - \sigma^{gf} u_{j,ji}^f - \sigma^g u_{j,ji}^s$$
$$- \xi^{gg}(\dot{u}_i^g - \dot{u}_i^s) - \xi^{gf}(\dot{u}_i^f - \dot{u}_i^s) + \rho_0^g b_i^g,$$

$$\rho_0^f \ddot{u}_i^f = -\sigma^{fg} u_{j,ji}^g - \sigma^{ff} u_{j,ji}^f - \sigma^f u_{j,ji}^s - \xi^{gf}(\dot{u}_i^g - \dot{u}_i^s)$$
$$- \xi^{ff}(\dot{u}_i^f - \dot{u}_i^s) + (\lambda_v + \mu_v)\dot{u}_{j,ji}^f + \mu_v \dot{u}_{i,jj}^f + \rho_0^f b_i^f,$$

$$\rho_0^s \ddot{u}_i^s = -\sigma^g u_{j,ji}^g - \sigma^f u_{j,ji}^f + (\lambda + \mu)u_{j,ji}^s + \mu u_{i,jj}^s$$
$$+ (\xi^{gf} + \xi^{gg})(\dot{u}_i^g - \dot{u}_i^s) + (\xi^{gf} + \xi^{ff})(\dot{u}_i^f - \dot{u}_i^s) + \rho_0^s b_i^s,$$

where the coefficients are constants satisfying inequalities given by (Eringen, 1994). The linear equations governing the behaviour of a fluid and a solid, or a gas and a solid, are also given by (Eringen, 1994), and these are easily deduced from the above system of equations. Various decay and related properties of solutions to these equations are studied by (Quintanilla, 2002d; Quintanilla, 2002a; Quintanilla, 2002c; Quintanilla, 2003; Quintanilla, 2004) and by (Gales, 2003).

(Eringen, 1994) also considers the passage from his theory to classical diffusion theories. From the equations for p_i^f and the fluid momentum equation (1.35) he shows that if one neglects the inertia coefficient in (1.35), then one deduces

$$v_i^f - v_i^s = \frac{1}{3}\left(-\frac{\partial \pi^f}{\partial x_i} + \rho^f b_i^f - \gamma\theta_{,i}\right). \tag{1.42}$$

If the elastic part of the mixture (the solid matrix) is fixed and the temperature is constant, then (1.42) reduces to

$$v_i^f = -\frac{1}{3}\frac{\partial \pi^f}{\partial x_i} - \frac{\rho^f}{3}b_i^f. \tag{1.43}$$

Equation (1.43) is just Darcy's law, which has then been deduced by (Eringen, 1994) from his mixture theory.

1.9.2 Bowen's theory

An interesting development of mixture theory to porous media is due to (Bowen, 1982). He is interested in a mixture in which the constituents are not necessarily miscible and he employs what is effectively an internal variable theory to achieve this.

The governing equations of (Bowen, 1982) are (1.31) – (1.33), (1.35) – (1.37), and (1.38), (although his notation is different). However, his theory is different from that of (Eringen, 1994), one of the reasons being that he essentially develops a variable porosity mixture theory by employing an internal variable theory. While (Bowen, 1982) considers a mixture composed of a solid plus $N-1$ fluids, we restrict attention to an elastic solid, constituent denoted by s, with two fluids denoted by sub or superscripts 2 and 3.

Let $\bar{\rho}_s, \bar{\rho}_2, \bar{\rho}_3$ denote the actual densities of the solid and fluid constituents before mixing. The volume fractions ϕ_s, ϕ_2, ϕ_3 are introduced as

$$\phi_s = \frac{\rho_s}{\bar{\rho}_s}, \quad \phi_2 = \frac{\rho_2}{\bar{\rho}_2}, \quad \phi_3 = \frac{\rho_3}{\bar{\rho}_3}. \tag{1.44}$$

Due to volume additivity,

$$\phi_s + \phi_2 + \phi_3 = 1. \tag{1.45}$$

(Bowen, 1982) argues that by including dependence on ϕ_s, ϕ_2, ϕ_3 in the constitutive theory one may account for a difference between mixture volumes and constituent volumes. Thus, by incorporating $\phi_\alpha, \alpha = s, 2, 3$, into the constitutive theory he argues that the volume fractions will have an affect on the mixture response.

(Bowen, 1982) introduces another class of free energies by the definitions

$$\Psi_\alpha = \rho^\alpha \psi_\alpha, \quad \alpha = s, 2, 3.$$

His constitutive theory involves supposing Ψ_α, η, depend on the independent variables

$$\theta, C^s_{AB}, \rho^\alpha, \phi_\alpha, \quad \alpha = 2, 3,$$

while the interaction forces p^α_i, the heat flux, and the stress tensors depend on the variables

$$\mathcal{L} \equiv \theta, \theta_{,A}, C^s_{AB}, C^s_{AB,K}, \rho^\alpha, \phi_\alpha, \rho^\alpha_{,A}, \phi_{\alpha,A}, v^\alpha_i,$$

where $\alpha = 2, 3$, and $\mathcal{L}$ stands for the list of independent variables.

Since ϕ_α are also included as independent variables, these quantities require a further set of equations for their determination. Due to equation (1.45) one only needs two equations. (Bowen, 1982) supposes ϕ_2, ϕ_3 satisfy rate-like equations of form

$$'\phi_2 = \omega_2(\mathcal{L}), \quad '\phi_3 = \omega_3(\mathcal{L}), \tag{1.46}$$

where ω_2, ω_3 are functional forms which would need to be given to make the theory determinate. Equations (1.46) are typical of those which govern internal variables in continuum mechanics. Hence, the notion that Bowen's theory is an internal variable one.

Making use of thermodynamic arguments, (Bowen, 1982) reduces his constitutive equations. He defines the free energy function Ψ_I as

$$\Psi_I = \Psi_s + \Psi_2 + \Psi_3.$$

In addition to equations (1.31) – (1.33), and equations (1.46), (Bowen, 1982) shows the momentum balance equations reduce to

$$\rho_2{''x_i^2} = -\rho_2\mu_{,i}^2 - \sigma_2\phi_{,i}^2 + \frac{\partial \Psi_2}{\partial \theta}\theta_{,i} + f_i^2 + \rho^2 b_i^2\,, \tag{1.47}$$

$$\rho_3{''x_i^3} = -\rho_3\mu_{,i}^3 - \sigma_3\phi_{,i}^3 + \frac{\partial \Psi_3}{\partial \theta}\theta_{,i} + f_i^3 + \rho^3 b_i^3\,, \tag{1.48}$$

and

$$\rho_s{''x_i^s} = -\rho_2{''x_i^2} - \rho_3{''x_i^3} + T_{ji,j}^I + \rho b_i\,. \tag{1.49}$$

Here T_{ji}^I is defined by

$$T_{ji}^I = \Psi_I\delta_{ij} + 2F_{iA}^s\frac{\partial \Psi_I}{\partial C_{AB}^s}F_{Bj}^s - \rho^2\frac{\partial \Psi_2}{\partial \rho^2}\delta_{ij} - \rho^3\frac{\partial \Psi_3}{\partial \rho^3}\delta_{ij}\,,$$

and f_i^α is the interaction force p_i^α with added contributions involving the derivatives of the functions Ψ_α, specifically

$$f_i^\alpha = p_i^\alpha - \mu_\alpha\rho_{,i}^\alpha + \frac{\partial \Psi_\alpha}{\partial \rho^2}\rho_{,i}^2 + \frac{\partial \Psi_\alpha}{\partial \rho^3}\rho_{,i}^3$$

$$+ (F_{iA}^{-1})^T\frac{\partial \Psi_\alpha}{\partial C_{AB}^s}C_{AB,K}^s + \sigma_\alpha\phi_{,i}^\alpha + \frac{\partial \Psi_\alpha}{\partial \phi^2}\phi_{,i}^2 + \frac{\partial \Psi_\alpha}{\partial \phi^3}\phi_{,i}^3\,.$$

The functions μ_α and σ^σ are given by

$$\mu_\alpha = \frac{\partial \Psi_I}{\partial \rho^\alpha}\,, \qquad \sigma_\alpha = -\frac{\partial \Psi_I}{\partial \phi^\alpha}\,.$$

In addition, (Bowen, 1982) writes his energy equation in terms of Ψ_α, the heat flux, the functions f_i^α, and the functions ω_2, ω_3, so that

$$\theta\left\{\rho\dot{\eta} - \frac{\partial}{\partial x_i}\left[\sum_{a=1}^{3}\frac{\partial \Psi_a}{\partial \theta}('x_i^a - \dot{x}_i)\right]\right\} = -\frac{\partial m_i}{\partial x_i} + \sigma_2\omega_2 + \sigma_3\omega_3$$

$$+ \rho r - v_i^2 f_i^2 - v_i^3 f_i^3\,, \tag{1.50}$$

where

$$\eta = -\frac{1}{\rho}\left|\det\left[(F_{iA}^s)^{-1}\right]\right|\frac{\partial W}{\partial \theta}\,,$$

with

$$W = |\det(F_{iA}^s)|\Psi_I,$$

and

$$m_i = \sum_{a=1}^{3}\left[q_i^a + \rho_a\epsilon_a('x_i^a - \dot{x}_i) - \Psi_a('x_i^a - \dot{x}_i) + \theta\frac{\partial \Psi_a}{\partial \theta}('x_i^a - \dot{x}_i)\right].$$

Thus, once the functions Ψ_s, Ψ_2, Ψ_3 are prescribed, together with the functional forms for ω_2, ω_3, one may fully determine the nonlinear governing partial differential equations, given the interaction forces and heat flux. The fully nonlinear system of equations is then comprised of equations (1.31) – (1.33), (1.46) – (1.49), and (1.50). This is a very interesting mixture theory. I am unaware of any solutions or analysis of the fully nonlinear equations. (Bowen, 1982) further reduces his theory by assuming linear dependence in the dissipation terms and by assuming the body has a centre of symmetry. He then derives equations when the solid remains fixed and indicates how other porous media models may be derived from his equations. In addition he then develops from the complete theory a set of equations for linear poroelasticity and applies these to wave propagation problems. Further use of the (Bowen, 1982) theory to wave reflection-refraction at the sea bed is described in section 6.8 of this book.

While the majority of the work in this book centres on the more tractable theories of porous media outlined in sections 1.2 – 1.8 we do employ the Eringen mixture theory and that of Bowen in sections 8.1.2 and 6.8. These sections investigate acoustic waves in a porous medium and at the sea bed, respectively. Future work involving such theories as those of sections 1.9.1, 1.9.2, will undoubtedly prove rewarding, especially if solutions to problems can be found employing the fully nonlinear equations.

2
Structural Stability

2.1 Structural stability, Darcy model

Structural stability is the study of stability of the model itself. The classical definition of stability involves continuous dependence of the solution on changes in the initial data, cf. section 1.1.2. However, it is increasingly being realised that continuous dependence on changes in the coefficients, in the model, in boundary data, or even in the partial differential equations themselves, is very important. This aspect of continuous dependence, or stability, is what we refer to as structural stability. (Hirsch and Smale, 1974) were prominent in introducing the ideas of structural stability. In chapter 16 of their book (Hirsch and Smale, 1974) ask, ... "What effect does changing the differential equation itself have on the solution? ... This is the problem of structural stability." The book of (Hirsch and Smale, 1974) gives an authoritative account of structural stability in an ordinary differential equation context. Structural stability is also emphasized in the books by (Bellomo and Preziosi, 1995), (Doering and Gibbon, 1995), (Drazin and Reid, 1981), and (Flavin and Rionero, 1995), although the topic of porous media is not specifically addressed in the context of structural stability in these works. In this chapter we focus on examples of structural stability in the context of the equations of porous media. It is extremely important, because if a small change in the equations, or a coefficient in an equation, causes a major change in the solution it may well say something about how accurate the model is as a vehicle to describe flow in porous media.

B. Straughan, *Stability and Wave Motion in Porous Media*,
DOI: 10.1007/978-0-387-76543-3_2, © Springer Science+Business Media, LLC 2008

Early articles dealing with structural stability questions in porous flows are those of (Ames and Payne, 1994), (Franchi and Straughan, 1993a; Franchi and Straughan, 1996), and (Payne and Straughan, 1996) investigates in some detail the continuous dependence of the solution on changes in the initial-time geometry. We do not describe the work of (Payne and Straughan, 1999a), but this paper establishes continuous dependence on the coefficients of Forchheimer and of Brinkman, and also investigates how the solution to the Brinkman equations converges to that of the Darcy equations as the Brinkman coefficient tends to zero. We focus on examples which illustrate various different effects, and the sections on continuous dependence on the Dufour, Krishnamurti, and Vadasz coefficients are new.

We commence with a result of (Payne and Straughan, 1998b) which establishes continuous dependence on the cooling coefficient for Newton's law of cooling in a Darcy porous material. (Franchi and Straughan, 1996) proved a similar result for a Brinkman porous material, but their method is inadequate to deal with the less dissipative Darcy system. (Payne and Straughan, 1998b) were able to prove a priori continuous dependence in three space dimensional problems without having to restrict the size of the time interval or the size of the initial data. In contrast, when one considers the Navier-Stokes equations, such a restriction is evidently necessary, (Ames and Payne, 1997)

We do not consider in this chapter structural stability questions for the porous medium equation model based on a distribution of voids in an elastic body, see section 7.2. However, this topic is investigated in (Chirita et al., 2006). (Chirita and Ciarletta, 2008) develop the structural stability analysis further by including temperature effects in the model.

A class of nonlinear models which possess properties not dissimilar to those of the model in section 2.1.1 are those studied by (Payne and Straughan, 1999c). These writers investigated continuous dependence on the spatial geometry for a Stokes' flow system when the nonlinearity in the temperature equation was regarded as important. This class of Stokes' flow is called a nonlinear Stokes' problem by (Duka et al., 2007). The paper by (Duka et al., 2007) derives interesting bounds for a solution to a nonlinear Stokes' system for thermal convection in a horizontal annulus.

2.1.1   Newton's law of cooling

The Darcy equations for non-isothermal flow in a porous medium are as in chapter 1, sections 1.2, 1.6.1, namely,

$$v_i = -\frac{\partial p}{\partial x_i} + g_i T, \qquad (2.1)$$

$$\frac{\partial v_i}{\partial x_i} = 0, \qquad (2.2)$$

$$\frac{\partial T}{\partial t} + v_i \frac{\partial T}{\partial x_i} = \Delta T, \tag{2.3}$$

where v_i, T, p and g_i are the velocity, temperature, pressure and the gravity vector. The density ρ in equation (1.15) has been assumed linear in T with the body force $f_i = g_i$, the constant part of the body force being absorbed in the pressure term. In this section equations (2.1) – (2.3) hold on a bounded spatial domain Ω with boundary Γ, for positive time. On the boundary Γ we suppose v_i and T satisfy the conditions

$$v_i n_i = 0, \qquad \text{and} \qquad \frac{\partial T}{\partial n} = -\kappa \big(T - T_a(\mathbf{x}, t)\big), \tag{2.4}$$

where $\kappa(> 0)$ is the cooling coefficient, $T_a(\mathbf{x}, t)$ is the temperature outside of the porous body at the boundary, n_i is the outward unit normal to Γ, and $\partial/\partial n$ denotes the outward normal derivative. The initial condition is

$$T(\mathbf{x}, t) = T_0(\mathbf{x}), \tag{2.5}$$

for T_0 given.

To investigate continuous dependence on κ we let (v_i, T, p) be a solution to (2.1) – (2.5) with a cooling coefficient κ_2, and we let (u_i, S, q) be another soultion to (2.1) – (2.5) for the same T_a and initial data, but for a different cooling coefficient κ_1. We wish to derive an a priori estimate for a measure of $T - S$ and $v_i - u_i$ in terms of the difference $\kappa_2 - \kappa_1$. To this end let w_i, θ, π and κ be the difference variables

$$w_i = v_i - u_i, \quad \theta = T - S, \quad \pi = p - q, \quad \kappa = \kappa_2 - \kappa_1, \tag{2.6}$$

and then from (2.1) – (2.5) we see that (w_i, θ, π) satisfies the partial differential equations

$$w_i = -\frac{\partial \pi}{\partial x_i} + g_i \theta, \tag{2.7}$$

$$\frac{\partial w_i}{\partial x_i} = 0, \tag{2.8}$$

$$\frac{\partial \theta}{\partial t} + u_i \frac{\partial \theta}{\partial x_i} + w_i \frac{\partial T}{\partial x_i} = \Delta \theta. \tag{2.9}$$

The boundary and initial conditions are

$$n_i w_i = 0, \quad \frac{\partial \theta}{\partial n} = -\kappa_1 \theta - \kappa(T - T_a), \qquad \text{on } \Gamma \times [0, \mathcal{T}], \tag{2.10}$$

$$\theta(\mathbf{x}, 0) = 0, \quad x \in \Omega, \tag{2.11}$$

where $\mathcal{T} < \infty$ is an arbitrary (but preassigned) time.

We assume, without loss of generality, that

$$|\mathbf{g}| \le 1. \tag{2.12}$$

To establish continuous dependence we multiply (2.7) by w_i and integrate over Ω, and using the Cauchy - Schwarz inequality, one finds

$$\|\mathbf{w}\|^2 = g_i(\theta, w_i) \leq \|\theta\| \|\mathbf{w}\|,$$

and so

$$\|\mathbf{w}\| \leq \|\theta\|. \tag{2.13}$$

Next, multiply (2.9) by θ and integrate over Ω to derive

$$\frac{1}{2}\frac{d}{dt}\|\theta\|^2 = \int_\Omega w_i T\theta_{,i} dx - \|\nabla\theta\|^2 - \kappa_1 \oint_\Gamma \theta^2 dA - \kappa \oint_\Gamma \theta(T - T_a) dA. \tag{2.14}$$

Next, employ the arithmetic-geometric mean inequality to see that

$$-\kappa \oint_\Gamma \theta(T_1 - T_a) dA \leq \frac{\kappa}{2\alpha} \oint_\Gamma \theta^2 dA + \frac{\kappa\alpha}{2} \oint_\Gamma (T_1 - T_a)^2 dA, \tag{2.15}$$

for $\alpha > 0$ arbitrary. We select $\alpha = \kappa/2\kappa_1$, and then use (2.15) in (2.14). In this manner we derive

$$\frac{1}{2}\frac{d}{dt}\|\theta\|^2 \leq \int_\Omega w_i T\theta_{,i} dx - \|\nabla\theta\|^2 + \frac{\kappa^2}{4\kappa_1} \oint_\Gamma (T - T_a)^2 dA. \tag{2.16}$$

2.1.2 A priori bound for T

To proceed we require an *a priori* bound for $|T|$. We establish such a bound for a function T satisfying (2.2) and (2.3), following (Payne and Straughan, 1998b). We simply use T, v_i and κ, rather than T, v_i and κ_2. Multiply (2.3) by T^{p-1} for $p > 1$ (we assume the temperature is scaled to be non-negative). Thus,

$$\frac{d}{dt}\int_\Omega T^p dx = -p(p-1)\int_\Omega T^{p-2}|\nabla T|^2 dx - \kappa p \oint_\Gamma T^{p-1}(T - T_a) dA.$$

With the aid of Young's inequality we have

$$\kappa p T^{p-1} T_a \leq \kappa p T^p + \kappa T_a^p \left(\frac{p-1}{p}\right)^{p-1}.$$

Employing this in the previous inequality allows us to show that

$$\frac{d}{dt}\int_\Omega T^p dx \leq -p(p-1)\int_\Omega T^{p-2}|\nabla T|^2 dx + \kappa \left(\frac{p-1}{p}\right)^{p-1} \oint_\Gamma T_a^p dA.$$

This inequality is integrated after discarding the first term on the right, to deduce

$$\left[\int_\Omega T^p dx\right]^{1/p} \leq \left[\int_\Omega T_0^p dx + \kappa \left(\frac{p-1}{p}\right)^{p-1} \int_0^t ds \oint_\Gamma T_a^p dA\right]^{1/p}. \tag{2.17}$$

Now, let $p \to \infty$ in (2.17) to see that

$$\sup_\Omega |T| \leq T_m, \tag{2.18}$$

where the constant T_m is given by

$$T_m = \max\left\{\sup_{\Omega} |T_0|, \sup_{\Gamma \times [0,T]} |T_a|\right\}.$$

Equipped with the estimate (2.18) for T (maximum principle), we bound the first term on the right of (2.16),

$$\int_{\Omega} w_i T\theta_{,i} dx \leq T_m \|\mathbf{w}\| \|\nabla\theta\|,$$

$$\leq T_m \|\theta\| \|\nabla\theta\|,$$

where (2.13) has been used, and then after further use of the arithmetic-geometric mean inequality,

$$\int_{\Omega} w_i T\theta_{,i} dx \leq \frac{T_m^2}{4} \|\theta\|^2 + \|\nabla\theta\|^2. \tag{2.19}$$

Upon utilizing (2.19) in (2.16) we find

$$\frac{d}{dt}\|\theta\|^2 \leq \frac{T_m^2}{2} \|\theta\|^2 + A\kappa^2, \tag{2.20}$$

where the function A is defined by

$$A(t) = \frac{1}{2\kappa_1} \oint_{\Gamma} (T_m - T_a)^2 dA.$$

In deriving (2.20), bound (2.18) has been extended to the boundary by continuity. Inequality (2.20) may be integrated by an integrating factor method to see that

$$\|\theta(t)\|^2 \leq R(t)\kappa^2, \tag{2.21}$$

where R is defined as

$$R(t) = \int_0^t A(s) \exp\left[\frac{1}{2}T_m^2(t-s)\right] ds.$$

The bound (2.21) is our continuous dependence estimate for θ. Now, from (2.13) we also find

$$\|\mathbf{w}(t)\|^2 \leq R(t)\kappa^2, \tag{2.22}$$

which establishes continuous dependence of v_i on the cooling coefficient. Continuous dependence on the cooling coefficient κ is established, since $R(t)$ is a priori because it only depends on data and the geometry of Ω.

2.2 Structural stability, Forchheimer model

In this section we describe work of (Franchi and Straughan, 2003) who consider the isothermal Forchheimer equations with quadratic and cubic terms, namely

$$\frac{\partial u_i}{\partial t} = -au_i - b|\mathbf{u}|u_i - c|\mathbf{u}|^2 u_i - p_{,i}, \qquad \frac{\partial u_i}{\partial x_i} = 0, \qquad (2.23)$$

where u_i is the average fluid velocity in the porous medium, a is the Darcy coefficient (viscosity divided by permeability), b and c are the Forchheimer coefficients, and p is the pressure.

2.2.1 Continuous dependence on b

We commence with a study of continuous dependence on the coefficient b. Therefore let u_i and v_i solve the following boundary initial value problems for different Forchheimer coefficients b_1 and b_2, but for the same second Forchheimer coefficient c,

$$\begin{aligned}
&\frac{\partial u_i}{\partial t} = -au_i - b_1|\mathbf{u}|u_i - c|\mathbf{u}|^2 u_i - p_{,i}, \quad \frac{\partial u_i}{\partial x_i} = 0, \ \text{in } \Omega \times \{t>0\}, \\
&n_i u_i = 0, \qquad \text{on } \Gamma \times \{t>0\}, \\
&u_i(x,0) = f_i(x), \qquad x \in \Omega,
\end{aligned} \qquad (2.24)$$

$$\begin{aligned}
&\frac{\partial v_i}{\partial t} = -av_i - b_2|\mathbf{v}|v_i - c|\mathbf{v}|^2 v_i - q_{,i}, \quad \frac{\partial v_i}{\partial x_i} = 0, \quad \text{in } \Omega \times \{t>0\}, \\
&n_i v_i = 0, \qquad \text{on } \Gamma \times \{t>0\}, \\
&v_i(x,0) = f_i(x), \qquad x \in \Omega.
\end{aligned} \qquad (2.25)$$

In these problems Ω is a bounded domain in $\mathbb{R}^3$ with boundary Γ, n_i is the unit outward normal to Γ, and f_i is the given initial data.

The difference variables w_i, π, b are defined by

$$w_i = u_i - v_i, \qquad \pi = p - q, \qquad b = b_1 - b_2. \qquad (2.26)$$

By subtraction we see that w_i satisfies the boundary initial value problem

$$\begin{aligned}
&\frac{\partial w_i}{\partial t} = -aw_i - \big(b_1|\mathbf{u}|u_i - b_2|\mathbf{v}|v_i\big) - c\big(|\mathbf{u}|^2 u_i - |\mathbf{v}|^2 v_i\big) - \pi_{,i}, \\
&\frac{\partial w_i}{\partial x_i} = 0, \quad \text{in } \Omega \times \{t>0\}, \\
&n_i w_i = 0, \qquad \text{on } \Gamma \times \{t>0\}, \\
&w_i(x,0) = 0, \qquad x \in \Omega.
\end{aligned} \qquad (2.27)$$

The first step involves rearranging the b_1 and b_2 terms as

$$b_1|\mathbf{u}|u_i - b_2|\mathbf{v}|v_i = \frac{b}{2}(|\mathbf{u}|u_i + |\mathbf{v}|v_i) + \tilde{b}(|\mathbf{u}|u_i - |\mathbf{v}|v_i), \qquad (2.28)$$

where $\tilde{b} = (b_1 + b_2)/2$, and observing (Payne and Straughan, 1999a) show that

$$(|\mathbf{u}|u_i - |\mathbf{v}|v_i)w_i = \frac{1}{2}(|\mathbf{u}| + |\mathbf{v}|)w_i w_i + \frac{1}{2}(|\mathbf{u}| - |\mathbf{v}|)^2(|\mathbf{u}| + |\mathbf{v}|). \qquad (2.29)$$

Next, multiply $(2.27)_1$ by w_i and integrate over Ω, to find with the aid of (2.28) and (2.29),

$$\frac{d}{dt}\frac{1}{2}\|\mathbf{w}\|^2 = -a\|\mathbf{w}\|^2 - \frac{b}{2}\int_\Omega (|\mathbf{u}|u_iw_i + |\mathbf{v}|v_iw_i)dx$$
$$-\frac{\tilde{b}}{2}\int_\Omega (|\mathbf{u}| + |\mathbf{v}|)w_iw_i dx$$
$$-\frac{\tilde{b}}{2}\int_\Omega (|\mathbf{u}| - |\mathbf{v}|)^2(|\mathbf{u}| + |\mathbf{v}|)dx$$
$$-c\int_\Omega (|\mathbf{u}|^2u_i - |\mathbf{v}|^2v_i)w_i dx. \tag{2.30}$$

(Franchi and Straughan, 2003) show that

$$(|\mathbf{u}|^2u_i - |\mathbf{v}|^2v_i)w_i = \frac{1}{2}|\mathbf{u}|^2(u_i - v_i + v_i)w_i - \frac{1}{2}|\mathbf{v}|^2v_iw_i$$
$$+ \frac{1}{2}|\mathbf{u}|^2u_iw_i + \frac{1}{2}|\mathbf{v}|^2w_i(u_i - v_i - u_i)$$
$$= \frac{1}{2}(|\mathbf{u}|^2 + |\mathbf{v}|^2)w_iw_i + \frac{1}{2}(u_i + v_i)w_i(|\mathbf{u}|^2 - |\mathbf{v}|^2)$$
$$= \frac{1}{2}(|\mathbf{u}|^2 + |\mathbf{v}|^2)w_iw_i + \frac{1}{2}(|\mathbf{u}|^2 - |\mathbf{v}|^2)^2. \tag{2.31}$$

This expression is employed in (2.30) to obtain

$$\frac{d}{dt}\frac{1}{2}\|\mathbf{w}\|^2 \leq -a\|\mathbf{w}\|^2 - \frac{b}{2}\int_\Omega (|\mathbf{u}|u_iw_i + |\mathbf{v}|v_iw_i)dx$$
$$-\frac{\tilde{b}}{2}\int_\Omega (|\mathbf{u}| + |\mathbf{v}|)w_iw_i dx - \frac{c}{2}\int_\Omega (|\mathbf{u}|^2 + |\mathbf{v}|^2)w_iw_i dx. \tag{2.32}$$

We suppose $c > 0$. The case where $c = 0$ is covered in (Franchi and Straughan, 2003). We use the Cauchy-Schwarz and arithmetic-geometric mean inequalities to see that

$$-\frac{b}{2}\left|\int_\Omega (|\mathbf{u}|u_iw_i + |\mathbf{v}|v_iw_i)\right|dx \leq \frac{b^2}{8c}\int_\Omega (u_iu_i + v_iv_i)dx$$
$$+ \frac{c}{2}\int_\Omega (|\mathbf{u}|^2 + |\mathbf{v}|^2)w_iw_i dx. \tag{2.33}$$

Now use this inequality in (2.32) and discard the $\tilde{b}$ term to derive

$$\frac{d}{dt}\frac{1}{2}\|\mathbf{w}\|^2 \leq -a\|\mathbf{w}\|^2 + \frac{b^2}{8c}\int_\Omega (u_iu_i + v_iv_i)dx. \tag{2.34}$$

From equations (2.24) and (2.25) one shows

$$\|\mathbf{u}\|^2 \leq \|\mathbf{f}\|^2 \exp(-2at) \quad \text{and} \quad \|\mathbf{v}\|^2 \leq \|\mathbf{f}\|^2 \exp(-2at). \tag{2.35}$$

These bounds are now used in (2.34) to arrive at

$$\frac{d}{dt}\frac{1}{2}\|\mathbf{w}\|^2 + a\|\mathbf{w}\|^2 \le \frac{b^2}{4c}\exp(-2at)\|\mathbf{f}\|^2.$$

With the aid of an integrating factor and integration one sees that

$$\|\mathbf{w}(t)\|^2 \le b^2 \frac{\|\mathbf{f}\|^2}{2c} t\exp(-2at). \tag{2.36}$$

Inequality (2.36) establishes continuous dependence on b when $c > 0$.

2.2.2   Continuous dependence on c

In this subsection we establish continuous dependence on the coefficient c. Let now (u_i, p) and (v_i, q) solve the boundary initial value problems (2.24) and (2.25) for the same b but with c_1 and c_2 different.

Define in this case

$$w_i = u_i - v_i, \quad \pi = p - q, \quad c = c_1 - c_2.$$

Then (w_i, π) satisfies the boundary initial value problem

$$\frac{\partial w_i}{\partial t} = -aw_i - b(|\mathbf{u}|u_i - |\mathbf{v}|v_i) - c_1|\mathbf{u}|^2 u_i + c_2|\mathbf{v}|^2 v_i - \pi_{,i},$$

$$\frac{\partial w_i}{\partial x_i} = 0, \quad \text{in } \Omega \times \{t > 0\}, \tag{2.37}$$

$$n_i w_i = 0, \quad \text{on } \Gamma \times \{t > 0\},$$

$$w_i(x, 0) = 0, \quad x \in \Omega.$$

(Franchi and Straughan, 2003) use the rearrangement

$$c_1|\mathbf{u}|^2 u_i - c_2|\mathbf{v}|^2 v_i = \frac{c}{2}(|\mathbf{u}|^2 u_i + |\mathbf{v}|^2 v_i) + \tilde{c}(|\mathbf{u}|^2 u_i - |\mathbf{v}|^2 v_i), \tag{2.38}$$

where $\tilde{c} = (c_1 + c_2)/2$.

Now multiply $(2.37)_1$ by w_i and integrate over Ω. We employ the rearrangements (2.29), (2.38) and (2.31) and then show

$$\frac{1}{2}\frac{d}{dt}\|\mathbf{w}\|^2 = -a\|\mathbf{w}\|^2 - \frac{b}{2}\int_\Omega (|\mathbf{u}| + |\mathbf{v}|)w_i w_i dx$$

$$- \frac{b}{2}\int_\Omega (|\mathbf{u}| - |\mathbf{v}|)^2(|\mathbf{u}| + |\mathbf{v}|)dx - \frac{c}{2}\int_\Omega (|\mathbf{u}|^2 u_i w_i + |\mathbf{v}|^2 v_i w_i)dx$$

$$- \frac{\tilde{c}}{2}\int_\Omega (|\mathbf{u}|^2 + |\mathbf{v}|^2)w_i w_i dx - \frac{\tilde{c}}{2}\int_\Omega (|\mathbf{u}|^2 - |\mathbf{v}|^2)^2 dx.$$

The two b terms and the $\tilde{c}$ term involving $(|\mathbf{u}|^2 - |\mathbf{v}|^2)^2$ are discarded to derive

$$\frac{1}{2}\frac{d}{dt}\|\mathbf{w}\|^2 + a\|\mathbf{w}\|^2 \leq -\frac{c}{2}\int_\Omega (|\mathbf{u}|^2 u_i w_i + |\mathbf{v}|^2 v_i w_i)dx$$
$$-\frac{\tilde{c}}{2}\int_\Omega (|\mathbf{u}|^2 + |\mathbf{v}|^2)w_i w_i dx. \tag{2.39}$$

Next, the Cauchy-Schwarz and arithmetic-geometric mean inequalities are employed to see that

$$\frac{c}{2}\int_\Omega (|\mathbf{u}|^2 u_i w_i + |\mathbf{v}|^2 v_i w_i)dx \leq \frac{c^2}{8\tilde{c}}\int_\Omega (|\mathbf{u}|^4 + |\mathbf{v}|^4)dx$$
$$+\frac{\tilde{c}}{2}\int_\Omega (|\mathbf{u}|^2 + |\mathbf{v}|^2)w_i w_i dx. \tag{2.40}$$

Upon use of (2.40) in (2.39) we see after integration,

$$\|\mathbf{w}\|^2 + 2a\int_0^t \|\mathbf{w}\|^2 ds \leq \frac{c^2}{4\tilde{c}}\int_0^t\int_\Omega (|\mathbf{u}|^4 + |\mathbf{v}|^4)dx\,ds. \tag{2.41}$$

The right hand side of (2.41) is estimated by multiplying (2.24) by u_i, (2.25) by v_i, and integrating over $\Omega \times (0,t)$ to show that

$$\int_0^t\int_\Omega (|\mathbf{u}|^4 + |\mathbf{v}|^4)dx\,ds \leq \left(\frac{c_1 + c_2}{2c_1 c_2}\right)\|\mathbf{f}\|^2.$$

Upon using this inequality in (2.41) one finds

$$\|\mathbf{w}\|^2 + 2a\int_0^t \|\mathbf{w}\|^2 ds \leq \frac{\|\mathbf{f}\|^2}{4c_1 c_2}c^2. \tag{2.42}$$

Inequality (2.42) establishes continuous dependence on c. A further bound for w_i may be obtained from (2.42) with the use of an integrating factor, this is

$$\int_0^t \|\mathbf{w}\|^2 ds \leq \frac{\|\mathbf{f}\|^2}{8ac_1 c_2}\left(1 - e^{-2at}\right)c^2.$$

2.2.3 Energy bounds

Interesting upper and lower bounds for $\|\mathbf{u}\|$ are obtained by (Franchi and Straughan, 2003) who follow the method of (Payne and Straughan, 1999a). To derive these estimates we suppose u_i is a solution to (2.24) with b_1 replaced by b, so u_i satisfies the boundary initial value problem

$$\frac{\partial u_i}{\partial t} = -au_i - b|\mathbf{u}|u_i - c|\mathbf{u}|^2 u_i - \pi_{,i}, \quad \frac{\partial u_i}{\partial x_i} = 0, \quad \text{in } \Omega \times \{t > 0\},$$
$$n_i u_i = 0, \quad \text{on } \Gamma \times \{t > 0\},$$
$$u_i(x,0) = f(x), \quad x \in \Omega. \tag{2.43}$$

Multiply (2.43) by u_i and integrate over Ω to find

$$\frac{1}{2}\frac{d}{dt}\|\mathbf{u}\|^2 = -a\|\mathbf{u}\|^2 - b\int_\Omega |\mathbf{u}|^3 dx - c\int_\Omega |\mathbf{u}|^4 dx. \qquad (2.44)$$

We first derive a lower bound for $\|\mathbf{u}\|$, and set $\Phi(t) = \|\mathbf{u}(t)\|^2$. From (2.44)

$$\frac{d\Phi}{dt} = -2a\|\mathbf{u}\|^2 - 2b\int_\Omega |\mathbf{u}|^3 dx - 2c\int_\Omega |\mathbf{u}|^4 dx. \qquad (2.45)$$

Define the function χ by

$$\chi(t) = -2a\|\mathbf{u}\|^2 - \frac{4}{3}b\int_\Omega |\mathbf{u}|^3 dx - c\int_\Omega |\mathbf{u}|^4 dx, \qquad (2.46)$$

and observe that $\chi \le 0$. From (2.46) and (2.45) $d\Phi/dt \le \chi$, and then

$$\Phi\frac{d\chi}{dt} = 4\|\mathbf{u}\|^2(u_{i,t}, u_{i,t}) \ge \left(\frac{d\Phi}{dt}\right)^2 \ge \left(-\frac{d\Phi}{dt}\right)(-\chi). \qquad (2.47)$$

Hence, $(d\chi/dt)/\chi \le (d\Phi/dt)/\Phi$, which after integration and rearrangement yields

$$-\chi(t) \le \Phi(t)\frac{\{-\chi(0)\}}{\|\mathbf{f}\|^2}. \qquad (2.48)$$

We may now show $2\chi \le d\Phi/dt \le \chi$, and so with the aid of (2.48) we deduce

$$\frac{1}{2}\frac{d\Phi}{dt} \ge \chi(t) \ge \Phi(t)\frac{\chi(0)}{\|\mathbf{f}\|^2}.$$

After integration we obtain

$$\|\mathbf{u}(t)\|^2 \ge \|\mathbf{f}\|^2 \exp\left[\frac{-2\{-\chi(0)\}t}{\|\mathbf{f}\|^2}\right]. \qquad (2.49)$$

From inequality (2.49) one sees that u_i cannot vanish identically in a finite time.

We may use the Cauchy-Schwarz inequality to show

$$-\int_\Omega |\mathbf{u}|^4 dx \le -\frac{\|\mathbf{u}\|^4}{m},$$

where $m = m(\Omega)$ is the measure of Ω. If this inequality is utilized in (2.45) one may show

$$\frac{d}{dt}\|\mathbf{u}\|^2 + 2a\|\mathbf{u}\|^2 + \frac{2c}{m}\|\mathbf{u}\|^4 \le 0.$$

Now since u_i cannot vanish in a finite time we divide by $\|\mathbf{u}\|^4$ and solve the resulting inequality for $\|\mathbf{u}\|^{-2}$. This leads to the upper bound

$$\|\mathbf{u}(t)\|^2 \le \frac{\|\mathbf{f}\|^2}{e^{2at} + c\|\mathbf{f}\|^2(e^{2at} - 1)/am}. \qquad (2.50)$$

If we combine (2.50) and (2.49) we find the estimates for $\|\mathbf{u}(t)\|$,

$$\frac{\|\mathbf{f}\|^2}{\exp\left[\left(4a + 8b\int_\Omega |\mathbf{f}|^3 dx/3\|\mathbf{f}\|^2 + 2c\int_\Omega |\mathbf{f}|^4 dx/\|\mathbf{f}\|^2\right)t\right]}$$
$$\leq \|\mathbf{u}(t)\|^2$$
$$\leq \frac{\|\mathbf{f}\|^2}{e^{2at} + c\|\mathbf{f}\|^2(e^{2at} - 1)/am}. \tag{2.51}$$

2.2.4 Brinkman-Forchheimer model

(Celebi et al., 2006) study structural stability for a version of the Brinkman-Forchheimer equations, namely, they study the boundary - initial value problem,

$$\frac{\partial u_i}{\partial t} = \gamma\Delta u_i - au_i - b|\mathbf{u}|^\alpha u_i - \pi_{,i},$$
$$\frac{\partial u_i}{\partial x_i} = 0, \quad \text{in } \Omega \times \{t > 0\}, \tag{2.52}$$
$$u_i = 0, \qquad \text{on } \Gamma \times \{t > 0\},$$
$$u_i(x, 0) = f(x), \qquad x \in \Omega,$$

where γ is a Brinkman coefficient and $\alpha \in [1, 2]$ is a constant.

(Celebi et al., 2006) establish existence and uniqueness of a solution to (2.52), and show that there is a constant D, depending on f and the coefficients in (2.52), such that

$$\sup_{0 \leq t \leq T} \|\nabla\mathbf{u}(t)\| \leq D, \qquad \int_0^T \left\|\frac{\partial\mathbf{u}}{\partial t}(t)\right\|^2 dt \leq D,$$

for any $T > 0$. They also show that the solution u_i depends continuously on the Forchheimer coefficient b, and on the Brinkman coefficient γ. This is an interesting paper and the proofs employ the Sobolev inequality in a non-trivial manner.

2.3 Forchheimer model, non-zero boundary conditions

(Payne et al., 1999) studied continuous dependence on changes in the viscosity for a Forchheimer and a Brinkman model. The motivation of (Payne et al., 1999) was to analyse mathematically a model for the process of salinization, whereby salts are transported upwards in soils in dry regions. A model for this was developed by (Gilman and Bear, 1996) and this model has a strong viscosity - concentration dependence. The work of (Gilman and

Bear, 1996) involves a nonlinear set of equations, and similar models are studied in (Wooding et al., 1997a; Wooding et al., 1997b) and in (van Duijn et al., 2002). (Payne et al., 1999) analyses the manner in which the velocity and concentration depend on changes in the viscosity. The reason for the need to study continuous dependence on the viscosity is that (Gilman and Bear, 1996) point out that the viscosity dependence on concentration is 1.5 to 3 times greater than that of pure water. By comparison the variation in density is only of order 0.15 to 0.30 times greater. Certainly such a strong variation indicates that convective motion of salt in a porous medium ought to take into account viscosity dependence on salt concentration.

The model based on Darcy's law studied by (Payne et al., 1999) is now presented. If we let u_i, c and p denote the fields of velocity, concentration and pressure, the Forchheimer equations for flow in a porous medium studied by (Payne et al., 1999) are

$$bu_i|\mathbf{u}| + (1 + \gamma_1 c)u_i = -p_{,i} + g_i c,$$
$$\frac{\partial u_i}{\partial x_i} = 0, \tag{2.53}$$
$$\frac{\partial c}{\partial t} + u_i \frac{\partial c}{\partial x_i} = \Delta c,$$

where γ_1 and b are positive constants, $g_i(\mathbf{x})$ is a gravity field which we again assume satisfies

$$|\mathbf{g}| \leq 1. \tag{2.54}$$

Equations (2.53) hold on the region $\Omega \times (0, \mathcal{T})$ for Ω a bounded domain in $\mathbb{R}^3$ and for some time $\mathcal{T}, 0 < \mathcal{T} < \infty$. The viscosity variation is represented by the term $1+\gamma_1 c$, i.e. we allow a linear variation in c so that the viscosity μ has form $\mu = \mu_1(1+\gamma_1 c)$. The $g_i c$ term represents a linear variation in c for the density, i.e. a Boussinesq like approximation. Since c is a concentration it is reasonable to assume that it is non-negative, although if we knew a priori that u_i is bounded then $c \geq 0$ would follow from the maximum principle.

On the boundary Γ (of Ω) the conditions imposed are

$$u_i n_i = f(\mathbf{x}, t), \quad c = h(\mathbf{x}, t), \quad \mathbf{x} \in \Gamma, \tag{2.55}$$

for known functions f and h. The initial condition is that concentration is prescribed at $t = 0$, i.e.

$$c(\mathbf{x}, 0) = c_0(\mathbf{x}), \quad \mathbf{x} \in \Omega, \tag{2.56}$$

c_0 given.

We note in passing that existence and uniqueness questions of solutions to systems like that studied here may be answered by the methods of (Ly and Titi, 1999) or those of (Rodrigues, 1986; Rodrigues, 1992).

The work of (Payne et al., 1999) relies on establishing an upper bound for c. We now give very brief details of how this is achieved.

2.3.1 A maximum principle for c

To derive a maximum principle for c (Payne et al., 1999) use the method of (Payne and Straughan, 1998a).

They introduce a function H by

$$\Delta H(\mathbf{x},t) = 0 \quad \text{in} \quad \Omega \times (0,\mathcal{T}),$$
$$H(\mathbf{x},t) = h^{2p-1}(\mathbf{x},t) \quad \text{on} \quad \Gamma \times (0,\mathcal{T}).$$

The analysis commences with the identity

$$\int_0^t ds \int_\Omega (H - c^{2p-1})\{c_{,t} + u_i c_{,i} - \Delta c\}\, dx = 0.$$

An integration by parts and rearrangement leads to

$$\int_\Omega c^{2p} dx + \frac{2(2p-1)}{p} \int_0^t ds \int_\Omega c^p_{,i} c^p_{,i} dx = \int_\Omega c_0^{2p} dx$$
$$+ 2p(H,c) - 2p(H_0,c_0) - 2p \int_0^t ds \int_\Omega H_{,s} c\, dx$$
$$+ 2p \int_0^t ds \int_\Omega H u_i c_{,i} dx + 2p \int_0^t ds \oint_\Gamma \frac{\partial H}{\partial n} h\, dA$$
$$- \int_0^t ds \oint_\Gamma f c^{2p} dA. \tag{2.57}$$

The remainder of the proof of the maximum principle for c is from this point very technical. The purpose of this section is to describe continuous dependence on γ_1 and so we refer to (Payne et al., 1999) or (Payne and Straughan, 1998a) for full details. After many steps the proof arrives at an inequality of form

$$\|c\|_{2p} \le \left[\|c_0\|_{2p}^{2p} + \left(\sum_{i=1}^5 r_i \right) h_m^{2p} \right]^{1/2p}, \tag{2.58}$$

where $\| \cdot \|_{2p}$ is the norm on $L^{2p}(\Omega)$, r_i involve h or c_0, and $h_m = \max_{\Gamma \times [0,\mathcal{T}]} |h|$. Taking the limit $2p \to \infty$ leads to the a priori bound

$$\sup_{\Omega \times [0,\mathcal{T}]} |c| \le \max \left\{ |c_0|_m, \sup_{[0,\mathcal{T}]} h_m \right\} = c_m \tag{2.59}$$

where $|c_0|_m = \max_\Omega |c_0|$, and c_m is defined as indicated.

2.3.2 Continuous dependence on the viscosity

To investigate continuous dependence on the viscosity coefficient γ_1 in (2.53) suppose (u_i, c_1, p) and (v_i, c_2, q) are solutions to (2.53) – (2.56) for the same data functions f, h and c_0, but for different viscosity coefficients,

γ_1 and γ_2, respectively. The difference solution (w_i, ϕ, π) is introduced as

$$w_i = u_i - v_i, \quad \phi = c_1 - c_2, \quad \pi = p - q, \quad \gamma = \gamma_1 - \gamma_2. \quad (2.60)$$

By calculation (w_i, ϕ, π) is seen to satisfy the boundary-initial value problem

$$b[u_i|\mathbf{u}| - v_i|\mathbf{v}|] + w_i + \gamma c_1 u_i + \gamma_2 \phi u_i + \gamma_2 c_2 w_i = -\pi_{,i} + g_i \phi,$$
$$w_{i,i} = 0, \quad (2.61)$$
$$\phi_{,t} + w_i c_{1,i} + v_i \phi_{,i} = \Delta\phi,$$

in $\Omega \times (0, \mathcal{T})$, with the boundary and initial conditions

$$w_i = \phi = 0 \quad \text{on } \Gamma, \qquad \phi(\mathbf{x}, 0) = 0, \quad \mathbf{x} \in \Omega. \quad (2.62)$$

It is convenient to also rearrange $(2.61)_1$ in the form

$$b[u_i|\mathbf{u}| - v_i|\mathbf{v}|] + w_i + \gamma_1 c_1 w_i + \gamma c_1 v_i + \gamma_2 \phi v_i = -\pi_{,i} + g_i \phi. \quad (2.63)$$

The proof starts by multiplying $(2.61)_1$ by w_i and integrating to find

$$b \int_\Omega (u_i|\mathbf{u}| - v_i|\mathbf{v}|) w_i dx + \int_\Omega (1 + \gamma_2 c_2) w_i w_i dx$$
$$= g_i(\phi, w_i) - \gamma \int_\Omega c_1 u_i w_i dx - \gamma_2 \int_\Omega \phi u_i w_i dx. \quad (2.64)$$

The right hand side is estimated using the maximum principle and Hölder's inequality. Identity (2.29) is used on the first term on the left and we drop a term to derive

$$\frac{b}{2} \int_\Omega (|\mathbf{u}| + |\mathbf{v}|) w_i w_i dx + \int_\Omega (1 + \gamma_2 c_2) w_i w_i dx$$
$$\leq \|\phi\| \|\mathbf{w}\| + \gamma c_m \|\mathbf{u}\| \|\mathbf{w}\| + \gamma_2 \left(\int_\Omega |\mathbf{u}| w_i w_i dx \right)^{1/2} \left(\int_\Omega |\mathbf{u}| \phi^2 dx \right)^{1/2}$$
$$\leq \frac{1}{2\alpha} \|\phi\|^2 + \left(\frac{\alpha}{2} + \frac{\beta}{2} \right) \|\mathbf{w}\|^2 + \frac{\gamma^2 c_m^2}{2\beta} \|\mathbf{u}\|^2$$
$$+ b \int_\Omega |\mathbf{u}| w_i w_i dx + \frac{\gamma_2^2}{4b} \left(\int_\Omega |\mathbf{u}|^3 dx \right)^{1/3} \left(\int_\Omega |\phi|^3 dx \right)^{2/3}, \quad (2.65)$$

where $\alpha, \beta > 0$ are constants to be chosen. We next use the Sobolev inequality

$$\int_\Omega \phi^4 dx \leq k^2 \left(\int_\Omega \phi^2 dx \right)^{1/2} \left(\int_\Omega |\nabla\phi|^2 dx \right)^{3/2},$$

for $k > 0$ constant, together with the Cauchy-Schwarz inequality in (2.65) to obtain

$$\frac{b}{2}\int_\Omega (|\mathbf{u}| + |\mathbf{v}|)w_i w_i dx + \int_\Omega (1 + \gamma_2 c_2)w_i w_i dx$$
$$\leq \frac{1}{2\alpha}\|\phi\|^2 + \frac{1}{2}(\alpha + \beta)\|\mathbf{w}\|^2 + \gamma^2 \frac{c_m^2}{2\beta}\|\mathbf{u}\|^2$$
$$+ b\int_\Omega |\mathbf{u}|w_i w_i dx + \frac{\gamma_2^2 k^{2/3}}{4b}\|\mathbf{u}\|_3 \|\phi\| \|\nabla\phi\|. \tag{2.66}$$

An analogous procedure starting from (2.64) leads to

$$\frac{b}{2}\int_\Omega (|\mathbf{u}| + |\mathbf{v}|)w_i w_i dx + \int_\Omega (1 + \gamma_1 c_1)w_i w_i dx$$
$$\leq \frac{1}{2\alpha}\|\phi\|^2 + \frac{1}{2}(\alpha + \beta)\|\mathbf{w}\|^2 + \gamma^2 \frac{c_m^2}{2\beta}\|\mathbf{v}\|^2$$
$$+ b\int_\Omega |\mathbf{v}|w_i w_i dx + \frac{\gamma_2^2 k^{2/3}}{4b}\|\mathbf{v}\|_3 \|\phi\| \|\nabla\phi\|. \tag{2.67}$$

Upon addition of (2.66) and (2.67) we see that

$$\int_\Omega (2 + \gamma_1 c_1 + \gamma_2 c_2)w_i w_i dx \leq \frac{1}{\alpha}\|\phi\|^2 + (\alpha + \beta)\|\mathbf{w}\|^2$$
$$+ \gamma^2 \frac{c_m^2}{2\beta}(\|\mathbf{u}\|^2 + \|\mathbf{v}\|^2) + \frac{\gamma_2^2 k^{2/3}}{4b}(\|\mathbf{u}\|_3 + \|\mathbf{v}\|_3)\|\phi\| \|\nabla\phi\|. \tag{2.68}$$

A further use of the arithmetic-geometric mean inequality shows that, for a constant $\epsilon > 0$ to be chosen

$$[2 - (\alpha + \beta)]\|\mathbf{w}\|^2 \leq \left[\frac{1}{\alpha} + \frac{\gamma_2^2 k^{2/3}}{64b^2 \epsilon}(\|\mathbf{u}\|_3 + \|\mathbf{v}\|_3)^2\right]\|\phi\|^2$$
$$+ \gamma^2 \frac{c_m^2}{2\beta}(\|\mathbf{u}\|^2 + \|\mathbf{v}\|^2) + \epsilon\|\nabla\phi\|^2. \tag{2.69}$$

Directly from (2.53) we may deduce for a constant d involving data

$$\|\mathbf{u}\|^2 \leq 4\|c_1\|^2 + d, \qquad \|\mathbf{u}\|_3 \leq \frac{1}{b^{1/3}}(4\|c_1\|^2 + d)^{1/3},$$
$$\|\mathbf{v}\|^2 \leq 4\|c_2\|^2 + d, \qquad \|\mathbf{v}\|_3 \leq \frac{1}{b^{1/3}}(4\|c_2\|^2 + d)^{1/3}. \tag{2.70}$$

Employing (2.70) in (2.69) yields for computable constants $\beta_1, \ldots, \beta_3$, dependent only on data, choosing $\alpha = \beta = 1/2$, an inequality of form

$$\|\mathbf{w}\|^2 \leq \beta_1\|\phi\|^2 + \beta_2 + \beta_3\|\nabla\phi\|^2. \tag{2.71}$$

To estimate the $\|\phi\|$ and $\|\nabla\phi\|$ terms we multiply $(2.61)_3$ by ϕ and integrate to find

$$\frac{1}{2}\|\phi\|^2 + \int_0^t \|\nabla\phi\|^2 ds = \int_0^t ds \int_\Omega w_i c_{1} \phi_{,i} dx.$$

Bounding c_1 and using the Cauchy-Schwarz inequality yields

$$\|\phi\|^2 + \int_0^t \|\nabla\phi\|^2 ds \le c_m^2 \int_0^t \|\mathbf{w}\|^2 ds. \tag{2.72}$$

Use of (2.72) in (2.71) shows that after integration

$$\int_0^t \|\mathbf{w}\|^2 ds \le k_1 \int_0^t (t-s)\|\mathbf{w}\|^2 ds + k_2(t)\gamma^2, \tag{2.73}$$

where k_1, k_2 depend only on data. From this inequality we may establish the estimates

$$\int_0^t (t-s)\|\mathbf{w}\|^2 ds \le k_3(t)\gamma^2, \quad \text{and} \quad \int_0^t \|\mathbf{w}\|^2 ds \le k_4\gamma^2, \tag{2.74}$$

for k_3 and k_4 computable data bounds. These are continuous dependence estimates for w_i. An analogous estimate for ϕ follows from (2.72), of the form

$$\|\phi(t)\|^2 + \int_0^t \|\nabla\phi\|^2 ds \le k_4 c_m^2 \gamma^2. \tag{2.75}$$

The inequalities (2.74) and (2.75) demonstrate continuous dependence on the viscosity coefficient γ_1. They are truly a *priori* since the coefficients of γ^2 depend only on boundary and initial data, and on the geometry of Ω.

2.4   Brinkman model, non-zero boundary conditions

In this section we review work of (Payne et al., 1999) which establishes continuous dependence on the viscosity coefficient γ_1 for the following Brinkman system,

$$-\Delta u_i + (1 + \gamma_1 c)u_i = -p_{,i} + g_i c,$$

$$\frac{\partial u_i}{\partial x_i} = 0, \tag{2.76}$$

$$\frac{\partial c}{\partial t} + u_i \frac{\partial c}{\partial x_i} = \Delta c,$$

on $\Omega \times (0, \mathcal{T})$. The boundary and initial conditions in this case are

$$u_i = f_i(\mathbf{x}, t), \quad c = h(\mathbf{x}, t), \qquad \mathbf{x} \in \Gamma \times \{t > 0\}, \tag{2.77}$$

$$c(\mathbf{x}, 0) = c_0(\mathbf{x}), \qquad \mathbf{x} \in \Omega. \tag{2.78}$$

(Payne et al., 1999) first compare the solution u_i to (2.76) with a solution a_i which solves the Stokes' flow problem in Ω, namely

$$\Delta a_i = \rho_{,i}, \qquad \frac{\partial a_i}{\partial x_i} = 0 \qquad \text{in } \Omega,$$

$$a_i = f_i \qquad \text{on } \Gamma \tag{2.79}$$

where ρ is a pressure term. For a data term d_0 they go via a_i to show that

$$\|\mathbf{u}\|^2 \le 5\|c\|^2 + d_0. \tag{2.80}$$

Continuous dependence on γ_1 proceeds via letting (u_i, c_1, p) and (v_i, c_2, q) solve (2.76) – (2.78) for the same data functions f_i, h and c_0, but for different viscosity coefficients γ_1 and γ_2, respectively. The difference variables (w_i, ϕ, π) and γ are defined as in equations (2.60). The boundary-initial value problem is

$$- \Delta w_i + (1 + \gamma_2 c_2)w_i + \gamma c_1 u_i + \gamma_2 \phi u_i = -\pi_{,i} + g_i \phi,$$

$$\frac{\partial w_i}{\partial x_i} = 0,$$

$$\frac{\partial \phi}{\partial t} + w_i \frac{\partial c_1}{\partial x_i} + v_i \frac{\partial \phi}{\partial x_i} = \Delta \phi,$$

$$w_i = \phi = 0 \text{ on } \Gamma, \quad \phi(\mathbf{x}, 0) = 0, \ \mathbf{x} \in \Omega. \tag{2.81}$$

By using inequality estimates (Payne et al., 1999) show that one may compute data constants α_1 and α_2 such that

$$\|\mathbf{w}(t)\|^2 + \|\nabla \mathbf{w}(t)\|^2 \le \alpha_1 \gamma^2, \qquad \|\phi\|^2 \le \alpha_2 \gamma^2. \tag{2.82}$$

Inequalities (2.82) are a priori bounds which demonstrate continuous dependence of the solution on the viscosity coefficient γ_1. Note that the stronger dissipation in the Brinkman model allows continuous dependence to be proven in the $\|\mathbf{w}\|$ and $\|\nabla \mathbf{w}\|$ measures.

Further novel structural stability results for the Brinkman equations may be found in (Lin and Payne, 2007a; Lin and Payne, 2007b). Also, interesting structural stability results for the Brinkman-Forchheimer equations are established by (Celebi et al., 2006).

2.5 Convergence, non-zero boundary conditions

(Payne et al., 1999) also consider the question of convergence of the solution to an equivalent Darcy system to (2.53) to the case where $\gamma_1 = 0$. That is, (Payne et al., 1999) also consider the viscosity variation in (2.53), but they neglect the b (Forchheimer) term. Their goal is to investigate the behaviour as $\gamma_1 \to 0$. To state this result let (u_i, c_1, p) satisfy the following boundary-initial value problem, where γ_1 has been replaced by γ,

$$(1 + \gamma c_1)u_i = -p_{,i} + g_i c_1, \qquad \frac{\partial u_i}{\partial x_i} = 0,$$

$$\frac{\partial c_1}{\partial t} + u_i \frac{\partial c_1}{\partial x_i} = \Delta c_1, \tag{2.83}$$

in $\Omega \times (0, \mathcal{T})$, with

$$u_i n_i = f, \qquad c_1 = h \qquad \text{on } \Gamma \times (0, \mathcal{T}),$$

$$c_1(\mathbf{x}, 0) = c_0(\mathbf{x}), \ \mathbf{x} \in \Omega, \tag{2.84}$$

i.e. the equivalent Darcy system to (2.53). We let (v_i, c_2, q) satisfy the analogous Darcy system when $\gamma = 0$, i.e.

$$v_i = -q_{,i} + g_i c_2, \qquad \frac{\partial v_i}{\partial x_i} = 0,$$

$$\frac{\partial c_2}{\partial t} + v_i \frac{\partial c_2}{\partial x_i} = \Delta c_2, \tag{2.85}$$

in $\Omega \times (0, \mathcal{T})$, with

$$v_i n_i = f, \qquad c_2 = h \qquad \text{on } \Gamma \times (0, \mathcal{T}),$$

$$c_2(\mathbf{x}, 0) = c_0(\mathbf{x}), \ \mathbf{x} \in \Omega. \tag{2.86}$$

By defining $w_i = u_i - v_i$ (Payne et al., 1999) show that

$$\int_0^t \|\mathbf{w}\|^2 ds \le \alpha_3 \gamma^2, \tag{2.87}$$

for a data term α_3.

Inequality (2.87) demonstrates convergence of u_i to v_i as $\gamma \to 0$ in the measure indicated. (Payne et al., 1999) also obtain convergence of w_i in $L^2(\Omega)$ norm and convergence of $\phi = c_1 - c_2$ in $L^2(\Omega)$ and $H^1(\Omega)$ norms.

2.6   Continuous dependence, Vadasz coefficient

(Vadasz, 1995; Vadasz, 1996; Vadasz, 1997; Vadasz, 1998a; Vadasz, 1998b) has made an extensive investigation of convection in a porous medium when the layer of saturated porous medium is rotating about a fixed axis. (Vadasz, 1998a) is a very interesting contribution. In this paper he employs linear instability and weakly nonlinear analysis to investigate the instability mechanisms governing convection in a rotating porous layer. Of particular interest is the fact that he discovers that if the inertia term is left in the momentum equation, then convection may commence by oscillatory convection. This is a striking result which implies that the inertia term plays a predominant role in determining the character of convection. In view of this we now examine how the solution to the equations for convection in a saturated porous material depends on the coefficient of the inertia term. The coefficient of the inertia term is denoted by $1/Va$, where Va is the Vadasz number. The usual Darcy law is recovered by letting $Va \to \infty$.

If we let u_i, T and p be the velocity, temperature and pressure, then the equations for non-isothermal flow in a saturated porous medium, taking inertia into account may be taken to be, cf. (Vadasz, 1998a), (Straughan, 2001b),

$$\frac{1}{Va}\frac{\partial u_i}{\partial t} = -\frac{\partial p}{\partial x_i} - u_i + g_i T, \tag{2.88}$$

$$\frac{\partial u_i}{\partial x_i} = 0, \tag{2.89}$$

$$\frac{\partial T}{\partial t} + u_i\frac{\partial T}{\partial x_i} = \Delta T. \tag{2.90}$$

These equations hold on $\Omega \times (0, T)$, $\Omega \subset \mathbb{R}^3$ bounded, and g_i, $|\mathbf{g}| \leq 1$, is the gravity vector. The boundary conditions we consider are

$$u_i n_i = 0 \qquad \text{and} \qquad T = h(\mathbf{x}, t), \tag{2.91}$$

where $\mathbf{n}$ is the unit outward normal to Γ, the boundary of Ω. The initial conditions are that

$$u_i(\mathbf{x}, 0) = u_i^0(\mathbf{x}), \qquad T(\mathbf{x}, 0) = T_0(\mathbf{x}). \tag{2.92}$$

It is convenient to employ $\alpha = 1/Va$ in (2.88), so this equation is rewritten as

$$\alpha\frac{\partial u_i}{\partial t} = -p_{,i} - u_i + g_i T. \tag{2.93}$$

In this section we study the continuous dependence of the solution on the coefficient α. To achieve this we need a maximum principle for T.

2.6.1   A maximum principle for T

A weak maximum principle for T is established by (Payne et al., 2001) (see also (Temam, 1988)) and we outline their proof. For a test function ϕ which vanishes on Γ, T satisfies the equation

$$\int_\Omega (T_{,t}\phi - u_i T\phi_{,i} + T_{,i}\phi_{,i})dx = 0. \tag{2.94}$$

Note that equation (2.94) may be obtained from (2.90) by multiplying that equation by ϕ and integrating over Ω. Define the number T_m by

$$T_m = \max\left\{\sup_\Omega |T_0|, \sup_{\Omega \times [0,T]} |h|\right\}. \tag{2.95}$$

The function ϕ is chosen as

$$\phi = [T - T_m]^+ = \sup(T - T_m, 0).$$

Since $\phi_{,i} = T_{,i}$ when $T > T_m$, $\phi_{,i} = 0$ for $T \le T_m$, (2.94) reduces to, after integration

$$\frac{1}{2} \int_0^t ds \int_\Omega |[T - T_m]^+|_{,s}^2 dx + \int_0^t ds \int_\Omega |\nabla[T - T_m]^+|^2 dx = 0.$$

(Note that $\int_\Omega u_i T \phi_{,i} dx = 0$.) From the last inequality we deduce that $[T - T_m]^+ = 0$, or $T \le T_m$.

Next, select $\phi = [-T - T_m]^+$ in (2.94). A similar calculation to the above shows $T \ge -T_m$. Thus,

$$|T| \le T_m, \quad (\mathbf{x}, t) \in \Omega \times [0, \mathcal{T}]. \tag{2.96}$$

2.6.2 Continuous dependence on α.

Let (u_i, T, p) be a solution to (2.89) – (2.93) with coefficient α_1 and let (v_i, S, q) be a solution to (2.89) – (2.93) for the same boundary and initial functions h, u_i^0, T_0 in (2.91), (2.92), but for a different Vadasz coefficient α_2. Define the difference variables w_i, θ and π, and the difference of the Vadasz coefficients α by

$$w_i = u_i - v_i, \quad \theta = T - S, \quad \pi = p - q, \quad \alpha = \alpha_1 - \alpha_2. \tag{2.97}$$

From equations (2.89) – (2.93) we find (w_i, θ, π) satisfy the boundary-initial value problem

$$\alpha_1 \frac{\partial w_i}{\partial t} + \alpha \frac{\partial v_i}{\partial t} = -\frac{\partial \pi}{\partial x_i} + g_i \theta - w_i,$$

$$\frac{\partial w_i}{\partial x_i} = 0, \tag{2.98}$$

$$\frac{\partial \theta}{\partial t} + w_i \frac{\partial T}{\partial x_i} + v_i \frac{\partial \theta}{\partial x_i} = \Delta \theta,$$

these equations holding on $\Omega \times (0, \mathcal{T})$, with

$$w_i n_i = 0, \quad \theta = 0, \quad \text{on } \Gamma \times [0, \mathcal{T}], \tag{2.99}$$

$$w_i(\mathbf{x}, 0) = 0, \quad \theta(\mathbf{x}, 0) = 0, \quad \mathbf{x} \in \Omega. \tag{2.100}$$

The analysis begins by multiplying (2.98)$_1$ by w_i and integrating over Ω to find, with the aid of (2.98)$_2$ and (2.99),

$$\|\mathbf{w}\|^2 + \frac{\alpha_1}{2} \frac{d}{dt} \|\mathbf{w}\|^2 = (w_i, g_i \theta) - \alpha(v_{i,t}, w_i),$$

$$\le \frac{1}{2\zeta} \|\mathbf{w}\|^2 + \frac{\zeta}{2} \|\theta\|^2 + \frac{\alpha^2}{2\beta} (v_{i,t}, v_{i,t}) + \frac{\beta}{2} \|\mathbf{w}\|^2, \tag{2.101}$$

where the arithmetic-geometric mean inequality has been employed and $\beta, \zeta > 0$ are to be chosen. Next, multiply (2.98)$_3$ by θ and integrate over Ω

to obtain with the aid of $(2.98)_2$ and (2.99),

$$\frac{d}{dt}\frac{1}{2}\|\theta\|^2 = (w_i T, \theta_{,i}) - \|\nabla\theta\|^2,$$

$$\leq T_m \|\mathbf{w}\| \|\nabla\theta\| - \|\nabla\theta\|^2,$$

$$\leq \frac{T_m^2}{4}\|\mathbf{w}\|^2, \tag{2.102}$$

where T has been bounded using (2.96), and the Cauchy-Schwarz and arithmetic-geometric mean inequalities have been employed. Integrate (2.102) over $(0, t)$ and use (2.100) to find

$$\|\theta(t)\|^2 \leq \frac{T_m^2}{2}\int_0^t \|\mathbf{w}\|^2 ds. \tag{2.103}$$

We next integrate (2.101) over $(0, t)$ and pick $\beta/2 + 1/2\zeta = 1$, e.g. $\beta = \zeta = 1$. This yields

$$\alpha_1 \|\mathbf{w}\|^2 \leq \int_0^t \|\theta\|^2 ds + \alpha^2 \int_0^t \|v_{i,s}\|^2 ds. \tag{2.104}$$

To bound the first term on the right we integrate (2.103) to obtain

$$\int_0^t \|\theta\|^2 ds \leq \frac{T_m^2 \mathcal{T}}{2}\int_0^t \|\mathbf{w}\|^2 ds.$$

Thus, from (2.104) we may derive,

$$\alpha_1 \|\mathbf{w}(t)\|^2 \leq \frac{T_m^2 \mathcal{T}}{2}\int_0^t \|\mathbf{w}\|^2 ds + \alpha^2 \int_0^t \|v_{i,s}\|^2 ds. \tag{2.105}$$

To estimate the $v_{i,t}$ term we multiply the equivalent v_i equation from (2.93) by $v_{i,t}$ and integrate over Ω then $(0, t)$ to find

$$\alpha_2 \|v_{i,t}\|^2 + \frac{1}{2}\frac{d}{dt}\|\mathbf{v}\|^2 = (g_i S, v_{i,t}),$$

$$\alpha_2 \int_0^t \|v_{i,s}\|^2 ds + \frac{1}{2}\|\mathbf{v}\|^2$$

$$\leq \frac{1}{2}\|\mathbf{v_0}\|^2 + \frac{\alpha_2}{2}\int_0^t \|v_{i,s}\|^2 ds + \frac{1}{2\alpha_2}\int_0^t \|S\|^2 ds, \tag{2.106}$$

where the arithmetic-geometric mean inequality has been employed. From (2.106) we see that

$$\alpha_2 \int_0^t \|v_{i,s}\|^2 ds \leq \|\mathbf{v_0}\|^2 + \frac{1}{\alpha_2}\int_0^t \|S\|^2 ds \leq \|\mathbf{v_0}\|^2 + \frac{T_m^2 mt}{\alpha_2}, \tag{2.107}$$

where (2.96) has been used.

Now, employ (2.107) in (2.105) and we may show that

$$\|\mathbf{w}\|^2 - \frac{T_m^2 \mathcal{T}}{2\alpha_1}\int_0^t \|\mathbf{w}\|^2 ds \leq \alpha^2 \left[\frac{\mathbf{v_0}^2}{\alpha_2} + \frac{T_m^2 m\mathcal{T}}{\alpha_2^2}\right].$$

This inequality is integrated by an integrating factor method and we derive

$$\int_0^t \|\mathbf{w}\|^2 ds \le K\alpha^2, \tag{2.108}$$

where

$$K = \frac{2\alpha_1\|\mathbf{v}_0\|^2}{\alpha_2 T_m^2 \mathcal{T}} + \frac{2\alpha_1 m}{\alpha_2^2}.$$

Inequality (2.108) establishes continuous dependence on α in the measure $\int_0^t \|\mathbf{w}\|^2 ds$. We may determine continuous dependence estimates in the measures $\|\theta(t)\|^2$ and $\|\mathbf{w}(t)\|^2$ from (2.103) and (2.105) and (2.107) and these are

$$\|\theta(t)\|^2 \le \frac{KT_m^2}{2}\alpha^2, \tag{2.109}$$

$$\|\mathbf{w}(t)\|^2 \le K_2\alpha^2, \tag{2.110}$$

where

$$K_2 = \frac{KT_m^2 \mathcal{T}}{2\alpha_1} + \frac{\|\mathbf{v}_0\|^2}{\alpha_1\alpha_2} + \frac{mT_m^2 \mathcal{T}}{\alpha_1\alpha_2^2}.$$

2.7 Continuous dependence, Krishnamurti coefficient

A very interesting model to describe a situation of penetrative convection in a viscous fluid was developed by (Krishnamurti, 1997). She also produced an experiment which captured the phenomenon and motivated her model. Linear instability and nonlinear energy stability bounds for a solution to the Krishnamurti model were derived by (Straughan, 2002b). The theoretical model of (Krishnamurti, 1997) relies on a pH indicator called thymol blue being dissolved in water. This gives rise to a double diffusive model with an equation for the temperature of the fluid coupled to an equation for the concentration of thymol blue. The penetrative effect is provided by the heat source depending on the thymol blue concentration. In this section we consider continuous dependence for a Krishnamurti model in a Darcy porous medium. Linear instability and nonlinear energy stability analyses for this model are given by (Hill, 2005a). In his work (Hill, 2005a) also develops stability analyses for a Brinkman theory, a theory where the heat source is nonlinear, and for a theory in which the density in the buoyancy force depends on temperature and concentration.

The partial differential equations governing the Krishnamurti model in a Darcy porous medium are

$$v_i = -p_{,i} + g_i T,$$
$$\frac{\partial v_i}{\partial x_i} = 0,$$
$$\frac{\partial T}{\partial t} + v_i \frac{\partial T}{\partial x_i} = \Delta T + \alpha C, \qquad (2.111)$$
$$\frac{\partial C}{\partial t} + v_i \frac{\partial C}{\partial x_i} = \Delta C.$$

In these equations v_i, p, T, C are the velocity, pressure, temperature and concentration, g_i is the gravity vector ($|\mathbf{g}| \leq 1$), and the Krishnamurti effect is introduced via the αC term in $(2.111)_3$. The Krishnamurti term arises because (Krishnamurti, 1997) takes the heat supply to depend (linearly) on concentration and this gives rise to equations $(2.111)_3$. We here assume (2.111) hold on $\Omega \times (0, T)$ with the boundary conditions

$$v_i n_i = 0, \quad T = h(\mathbf{x}, t), \quad C = r(\mathbf{x}, t), \qquad \text{on } \Gamma \times (0, T]. \qquad (2.112)$$

The initial conditions are

$$T(\mathbf{x}, 0) = T_0(\mathbf{x}), \qquad C(\mathbf{x}, 0) = C_0(\mathbf{x}). \qquad (2.113)$$

The goal of this section is to show that the solution (v_i, p, T, C) depends continuously on changes in the Krishnamurti coefficient α. It is important in analysing a model to know that the addition of a term like the αC Krishnamurti term still retains the well posedness of the original system.

To establish continuous dependence we find it necessary to have an a priori bound for the temperature T. We may invoke the analysis of section 2.6 to see that C is bounded by its initial and boundary values, precisely,

$$|C| \leq C_m = \max \left\{ \sup_{\Omega} |C_0|, \sup_{\Omega \times [0, T]} |r| \right\}.$$

The presence of the αC term in (2.111) prevents us from immediately deducing a maximum principle for T.

2.7.1 An a priori bound for T

We introduce the function H which solves

$$\Delta H = 0 \qquad \text{in } \Omega,$$
$$H = h^{2p-1} \qquad \text{on } \Gamma, \qquad (2.114)$$

where $H = H(\mathbf{x}, t)$ since $h = h(\mathbf{x}, t)$, and p is an integer.

Because of equation $(2.111)_3$ we may write

$$\int_0^t ds \int_\Omega (T^{2p-1} - H)(T_{,t} + v_i T_{,i} - \Delta T - \alpha C) dx = 0. \qquad (2.115)$$

After several integrations by parts we deduce from (2.115)

$$\int_\Omega T^{2p}dx + \frac{2(2p-1)}{p}\int_0^t ds \int_\Omega T_{,i}^p T_{,i}^p dx = \int_\Omega T_0^{2p}dx + 2p(H,T)$$

$$- 2p(H_0,T_0) - 2p\int_0^t (H_{,s},T)ds + 2p\int_0^t ds \int_\Omega Hv_iT_{,i}dx$$

$$+ 2p\int_0^t ds \int_\Gamma \frac{\partial H}{\partial n}h\,dA - \alpha\int_0^t (H,C)ds$$

$$+ \alpha\int_0^t ds \int_\Omega T^{2p-1}Cdx. \qquad (2.116)$$

The second - sixth terms on the right of (2.116) are handled as in (Payne and Straughan, 1998a) and the new terms are the seventh and eighth. The arithmetic-geometric mean inequality is used to see that

$$-\alpha\int_0^t (H,C)ds \le \frac{\alpha}{2}\int_0^t \|H\|^2 ds + \frac{mC_m^2\mathcal{T}}{2}\alpha, \qquad (2.117)$$

where m is the measure of Ω. To handle the last term in (2.116) we employ Young's inequality as follows,

$$\int_0^t ds \int_\Omega T^{2p-1}Cdx \le \left(\frac{2p-1}{p}\right)\int_0^t ds \int_\Omega T^{2p}dx$$

$$+ \frac{1}{2p}\int_0^t ds \int_\Omega C^{2p}dx. \qquad (2.118)$$

From the maximum principle, (Protter and Weinberger, 1967), we know $H \le h_m^{2p-1}$, $h_m = \max_\Gamma |h|$, and then since from $(2.111)_1$ we find $\|\mathbf{v}\| \le \|T\|$, we use (2.117) and (2.118) and follow the analysis of (Payne and Straughan, 1998a) to derive

$$\int_\Omega T^{2p}dx \le \int_\Omega T_0^{2p}dx + 2p(\|H\|\,\|T\| + \|H_0\|\,\|T_0\|)$$

$$+ 2p\sqrt{\int_0^t \|H_{,s}\|^2 ds \int_0^t \|T\|^2 ds}$$

$$+ 2ph_m^{2p-1}\sqrt{\int_0^t \|\nabla T\|^2 ds \int_0^t \|T\|^2 ds}$$

$$+ 2p\sqrt{\int_0^t ds \int_\Gamma h^2 dA \int_0^t ds \int_\Gamma \left(\frac{\partial H}{\partial n}\right)^2 dA}$$

$$+ \frac{\alpha}{2}\int_0^t \|H\|^2 ds + \frac{mC_m^2\mathcal{T}}{2}\alpha + \alpha\left(\frac{2p-1}{p}\right)\int_0^t ds \int_\Omega T^{2p}dx$$

$$+ \frac{mC_m^{2p}\mathcal{T}}{2p}\alpha. \qquad (2.119)$$

The next step is to bound the $\|T\|$ and $\|\nabla T\|$ terms and their integrals. To this end we introduce the function G which satisfies

$$\Delta G = 0 \quad \text{in } \Omega, \qquad G = h(\mathbf{x}, t) \quad \text{on } \Gamma. \tag{2.120}$$

Now form the combination

$$\int_0^t ds \int_\Omega (T - G)(T_{,t} + v_i T_{,i} - \Delta T - \alpha C) dx = 0.$$

After integrations by parts we may derive from this

$$\frac{1}{2}\|T\|^2 + \int_0^t \|\nabla T\|^2 ds = \frac{1}{2}\|T_0\|^2 + \int_0^t ds \int_\Omega T_{,i} G_{,i} dx$$

$$+ (G, T) - (G_0, T_0) - \int_0^t ds \int_\Omega TG_{,s} dx$$

$$+ \int_0^t ds \int_\Omega Gv_i T_{,i} dx + \alpha \int_0^t ds \int_\Omega CT dx$$

$$- \alpha \int_0^t ds \int_\Omega CG dx.$$

We modify the argument of (Payne and Straughan, 1998a), p. 328, to find

$$\int_0^t ds \int_\Omega Gv_i T_{,i} dx \leq G_m \int_0^t ds \int_\Omega |\mathbf{v}|\,|\nabla T|\,dx$$

$$\leq \frac{h_m^2}{2} \int_0^t \|T\|^2 ds + \frac{1}{2} \int_0^t \|\nabla T\|^2 ds.$$

Thus, use of this and the arithmetic-geometric mean inequality in the above allows us to deduce

$$\frac{1}{4}\|T\|^2 + \frac{1}{2} \int_0^t \|\nabla T\|^2 ds \leq \|T_0\|^2 + \int_0^t ds \int_\Gamma h \frac{\partial G}{\partial n} dA$$

$$+ \|G\|^2 + \frac{1}{2}\|G_0\|^2 + \frac{1}{2} \int_0^t \|G_{,s}\|^2 ds + \frac{\alpha}{2} \int_0^t \|G\|^2 ds$$

$$+ \alpha T_m C_m^2 + \left(\frac{1}{2} + \frac{\alpha}{2} + \frac{h_m^2}{4}\right) \int_0^t \|T\|^2 ds. \tag{2.121}$$

(Payne and Straughan, 1998a) show how to use a Rellich identity to bound the G terms in (2.121). The new term here is the $\alpha \int_0^t \|G\|^2 ds/2$ one but this also responds to the (Payne and Straughan, 1998a) treatment. We define the data term $D_1(t)$, for computable constants $h_1, \ldots, h_6$

dependent only on data, by

$$\frac{1}{4}D_1(t) = h_1 \int_\Gamma h^2 dA + h_2 \int_\Gamma |\nabla_s h|^2 dA$$
$$+ h_3 \sqrt{\int_0^t ds \int_\Gamma h^2 dA \int_0^t d\eta \int_\Gamma |\nabla_s h|^2 dA}$$
$$+ h_4 \int_0^t ds \int_\Gamma h_{,s}^2 dA + h_5 \int_0^t ds \int_\Gamma h^2 dA$$
$$+ h_6 \int_0^t d\eta \int_\Omega |\nabla_s h_{,\eta}|^2 dA,$$

where ∇_s is the tangential derivative on Γ. We may show $D_1/4$ is a data bound for all five terms on the right of (2.121) which involve G.

Thus, put $a = 2 + 2\alpha + h_m^2$, then (2.121) leads to

$$\frac{1}{4}\|T\|^2 + \frac{1}{2}\int_0^t \|\nabla T\|^2 ds \le \|T_0\|^2 + \frac{1}{4}D_1 + \alpha T C_m^2 m + \frac{a}{4}\int_0^t \|T\|^2 ds.$$

This inequality may be integrated to find

$$\|T(t)\|^2 \le D_2(t) + a\int_0^t \|T\|^2 ds, \qquad (2.122)$$

where

$$D_2(t) = 4D_1 + 4\|T_0\|^2 + 4m\alpha T C_m^2.$$

Inequality (2.122) may be integrated to obtain the following three bounds,

$$\|T(t)\|^2 \le D_2 + a\int_0^t e^{a(t-s)} D_2(s) ds = D_3(t),$$
$$\int_0^t \|T\|^2 ds \le \int_0^t e^{a(t-s)} D_2(s) ds = D_4(t),$$
$$\int_0^t \|\nabla T\|^2 ds \le \frac{1}{2}D_2 + \frac{a}{2}D_4 = D_5(t).$$

We now return to (2.119). (Payne and Straughan, 1998a) show there are constants $\psi_1, c_1 > 0$ such that

$$\|H\|^2 \le \psi_1 \int_\Gamma h^{4p-2} dA,$$
$$\|H_{,t}\|^2 \le \psi_1 \int_\Gamma |(h^{2p-1})_{,t}|^2 dA,$$
$$\int_\Gamma \left(\frac{\partial H}{\partial n}\right)^2 dA \le c_1 \int_\Gamma |\nabla_s h^{2p-1}|^2 dA.$$

Using these inequalities and the bounds for $\|T\|$ and $\|\nabla T\|$ in (2.119) we may derive

$$\int_\Omega T^{2p} dx \le \int_\Omega T_0^{2p} dx + 2p\left(D_{3\,max}^{1/2} + \|T_0\|\right)\psi_1^{1/2}\sqrt{\int_\Gamma h^{4p-2} dA}$$

$$+ 2pD_4^{1/2}\sqrt{\int_0^t \psi_1 d\eta \int_\Gamma h_{,\eta}^2 h^{4p-4} dA}$$

$$+ 2ph_m^{2p-1}\sqrt{\int_0^t D_3(s)ds \int_0^t D_5(s)ds}$$

$$+ 2pc_1^{1/2}\sqrt{\int_0^t ds \int_\Gamma h^2 dA \int_0^t d\eta \int_\Gamma |\nabla_s h^{2p-1}|^2 dA}$$

$$+ m\alpha T\left(\frac{C_m^{2p}}{2p} + \frac{C_m^2}{2}\right) + \frac{\alpha}{2}\psi_1 \int_0^t ds \int_\Gamma h^{4p-2} dA$$

$$+ \alpha\left(\frac{2p-1}{p}\right)\int_0^t ds \int_\Omega T^{2p} dx. \tag{2.123}$$

The first seven terms on the right of (2.123) are data and we denote these by $F(h)$. With $Q = \int_0^t ds \int_\Omega T^{2p} dx$, (2.123) is

$$Q' - \mu Q \le F,$$

where $\mu = \alpha(2p-1)/p$. This inequality integrates to yield

$$\int_\Omega T^{2p} dx \le \mu \int_0^t F(s)e^{\mu(t-s)} ds + F.$$

We raise both sides of this inequality to the power $1/2p$ to see that

$$\left(\int_\Omega T^{2p} dx\right)^{1/2p} \le \left[F + \mu \int_0^t F(s)e^{\mu(t-s)} ds\right]^{1/2p}. \tag{2.124}$$

Let $p \to \infty$ and since the right hand side of (2.124) is composed of $\int_\Omega T_0^{2p} dx$, h_m^{2p}, C_m^{2p} raised to the power $1/2p$ we arrive at

$$\sup_{\Omega \times [0,T]} |T| \le \max\left\{|T_0|_m, \sup_{[0,T]} h_m, C_m\right\} = T_B. \tag{2.125}$$

This is the a priori bound we sought to achieve.

2.7.2 Continuous dependence

We now let (u_i, T, C_1, p) be a solution to (2.111) – (2.113) for Krishnamurti coefficient α_1 and we let (v_i, S, C_2, q) be another solution for a different Krishnamurti coefficient α_2, but for the same data functions h, r, T_0 and

C_0. Thus (u_i, T, C_1, p) and (v_i, S, C_2, q) satisfy the boundary-initial value problems,

$$
\begin{aligned}
u_i &= -p_{,i} + g_i T, \\
u_{i,i} &= 0, \\
T_{,t} + u_i T_{,i} &= \Delta T + \alpha_1 C_1, \\
C_{1,t} + u_i C_{1,i} &= \Delta C_1,
\end{aligned}
\tag{2.126}
$$

in $\Omega \times (0, \mathcal{T})$,

$$
u_i n_i = 0, \quad T = h, \quad C_1 = r \qquad \text{on } \Gamma \times (0, \mathcal{T}], \tag{2.127}
$$

$$
T(\mathbf{x}, 0) = T_0(\mathbf{x}), \quad C_1(\mathbf{x}, 0) = C_0(\mathbf{x}), \tag{2.128}
$$

and

$$
\begin{aligned}
v_i &= -q_{,i} + g_i S, \\
v_{i,i} &= 0, \\
S_{,t} + v_i S_{,i} &= \Delta S + \alpha_2 C_2, \\
C_{2,t} + v_i C_{2,i} &= \Delta C_2,
\end{aligned}
\tag{2.129}
$$

in $\Omega \times (0, \mathcal{T})$,

$$
v_i n_i = 0, \quad S = h, \quad C_2 = r \qquad \text{on } \Gamma \times (0, \mathcal{T}], \tag{2.130}
$$

$$
S(\mathbf{x}, 0) = S_0(\mathbf{x}), \quad C_2(\mathbf{x}, 0) = C_0(\mathbf{x}). \tag{2.131}
$$

The difference variables w_i, θ, ϕ, π and α are defined by

$$
w_i = u_i - v_i, \ \theta = T - S, \ \phi = C_1 - C_2, \ \pi = p - q, \ \alpha = \alpha_1 - \alpha_2. \tag{2.132}
$$

By direct calculation we see that (w_i, θ, ϕ, π) satisfies the boundary-initial value problem

$$
\begin{aligned}
w_i &= -\pi_{,i} + g_i \theta, \\
w_{i,i} &= 0, \\
\theta_{,t} + w_i T_{,i} + v_i \theta_{,i} &= \Delta \theta + \alpha_1 \phi + \alpha C_2, \\
\phi_{,t} + w_i C_{1,i} + v_i \phi_{,i} &= \Delta \phi,
\end{aligned}
\tag{2.133}
$$

in $\Omega \times (0, \mathcal{T})$,

$$
w_i n_i = 0, \quad \theta = 0, \quad \phi = 0, \qquad \text{on } \Gamma \times (0, \mathcal{T}], \tag{2.134}
$$

$$
\theta(\mathbf{x}, 0) = 0, \quad \phi(\mathbf{x}, 0) = 0. \tag{2.135}
$$

First, observe that multiplying $(2.133)_1$ by w_i, integrating over Ω and using the Cauchy-Schwarz inequality we find

$$
\|\mathbf{w}\| \le \|\theta\|. \tag{2.136}
$$

By multiplying $(2.133)_3$ by θ and integrating over Ω,

$$\frac{d}{dt}\|\theta\|^2 = 2\int_\Omega w_i T\theta_{,i}dx - 2\|\nabla\theta\|^2 + 2\alpha(C_2, \theta) + 2\alpha_1(\phi, \theta).$$

Now use the bound for T and the arithmetic-geometric mean inequality to find

$$\frac{d}{dt}\|\theta\|^2 \le a\|\theta\|^2 + \alpha_1\|\phi\|^2 + k\alpha^2, \qquad (2.137)$$

where we have set

$$a = \frac{T_B^2}{2} + 1 + \alpha_1, \qquad k = mC_m^2.$$

Next, multiply $(2.133)_4$ by ϕ and integrate over Ω to find

$$\begin{aligned}\frac{d}{dt}\|\phi\|^2 &= 2\int_\Omega w_i C_1\phi_{,i}dx - 2\|\nabla\phi\|^2,\\ &\le \frac{C_m^2}{2}\|\mathbf{w}\|^2,\\ &\le \frac{C_m^2}{2}\|\theta\|^2, \qquad (2.138)\end{aligned}$$

where (2.136) has also been employed.

We put $\beta = a + C_m^2/2$ and add (2.137) and (2.138) to deduce

$$\frac{d}{dt}(\|\theta\|^2 + \|\phi\|^2) \le \beta(\|\theta\|^2 + \|\phi\|^2) + k\alpha^2.$$

This inequality is integrated to arrive at

$$\|\theta(t)\|^2 + \|\phi(t)\|^2 \le \zeta(t)\alpha^2, \qquad (2.139)$$

where $\zeta(t) = ke^{\beta t}/\beta$.

Inequality (2.139) is an *a priori* bound and establishes continuous dependence on the Krishnamurti coefficient α for equations (2.111).

2.8 Continuous dependence, Dufour coefficient

This section is devoted to studying the influence the Dufour effect has on double diffusive convective motion in a porous medium of Brinkman type. We focus on the Brinkman equations rather than the Darcy equations. As pointed out in chapter 1, the Brinkman equations of flow in porous media (Brinkman, 1947) have been the subject of intense recent attention. Among recent papers dealing with Brinkman models we cite (Franchi and Straughan, 1996), (Givler and Altobelli, 1994), (Guo and Kaloni, 1995c; Guo and Kaloni, 1995a), (Kladias and Prasad, 1991), (Kwok and Chen, 1987), (Lombardo and Mulone, 2002a; Lombardo and Mulone, 2002b; Lombardo and Mulone, 2003), (Nield and Bejan, 2006),

(Qin and Chadam, 1996), (Qin et al., 1995), (Qin and Kaloni, 1992; Qin and Kaloni, 1994), (Payne and Song, 1997; Payne and Song, 2000), (Payne and Straughan, 1996; Payne and Straughan, 1999a), and the references therein. Double diffusive convective motion is the phenomenon involving the diffusion and convection of two independent fields, such as temperature and a salt field. In section 2.7 we analysed another double diffusive problem. Stability analyses of double diffusive phenomena, in a variety of practical contexts, have occupied much recent attention, cf. (Avramenko and Kuznetsov, 2004), (Bardan et al., 2000; Bardan et al., 2001), (Bardan and Mojtabi, 1998), (Bresch and Sy, 2003), (Budu, 2002), (Carr, 2003a; Carr, 2003b), (Chang, 2004), (Charrier-Mojtabi et al., 1998), (Clark et al., 2002), (Guo and Kaloni, 1995c; Guo and Kaloni, 1995a; Guo and Kaloni, 1995b), (Guo et al., 1994), (Hill, 2005a; Hill, 2003; Hill, 2004b; Hill, 2004a; Hill, 2004c; Hill, 2005b), (Hurle and Jakeman, 1971), (Karimi-Fard et al., 1999), (Knutti and Stocker, 2000), (Lombardo and Mulone, 2002b), (Lombardo et al., 2001), (Malashetty et al., 2006), (Song, 2002), (Stocker, 2001), (Stocker and Schmittner, 1997), (Straughan and Tracey, 1999) and (Ybarra and Velarde, 1979). (Straughan, 2004a), chapter 14 discusses double diffusive and even multi-diffusive convection in detail in a variety of contexts. Further practical studies of double diffusive convection to energy conversion and management via a solar pond occupy the papers by (Rothmeyer, 1980), (Tabor, 1980), and (Zangrando, 1991), the one by Rothmeyer investigating in particular the Soret effect, which is in some sense the mathematical adjoint to the Dufour effect.

To describe the Dufour effect, the equations for convective - diffusive motion in an incompresssible fluid in a Brinkman porous medium may be written as, employing a Boussinesq approximation in the body force term in the momentum equation,

$$
\begin{aligned}
v_i - \lambda\Delta v_i &= -p_{,i} + g_i T + h_i C, \qquad v_{i,i} = 0, \\
T_{,t} + v_i T_{,i} &= -J_{i,i}, \\
C_{,t} + v_i C_{,i} &= -K_{i,i},
\end{aligned}
\tag{2.140}
$$

where v_i, T, C and p represent velocity, temperature, salt concentration and pressure fields, respectively, g_i and h_i are the gravity vector terms arising in the density equation of state, and $\mathbf{J}$ and $\mathbf{K}$ are fluxes of heat and solute, respectively. In equations (2.140) λ is the Brinkman coefficient. The Brinkman equations are discussed at length in (Nield and Bejan, 2006) and in chapter 1, section 1.4 of this book. We observe that in $(2.140)_1$ the T, C terms arise from the body force in a Boussinesq approximation. The v_i term is essentially an interaction force between the fluid and porous matrix. The $\lambda\Delta v_i$ term is an effective viscosity contribution and is believed appropriate when the porosity is not too small. In the Brinkman equations the nonlinear convective terms of Navier-Stokes theory are omitted as is the acceleration, $\partial v_i/\partial t$, term; this is consistent with flow through a porous matrix where the convection and acceleration terms are likely to be negligible. (Hurle and Jakeman, 1971) argue that the general forms for the fluxes $\mathbf{J}$ and $\mathbf{K}$ should be

$$J_i = -\kappa T_{,i} - \rho T C \left(\frac{\partial \mu}{\partial C}\right) D' C_{,i}, \;\; K_i = -\rho D \left[S_T C (1-C) T_{,i} + C_{,i}\right], \;\; (2.141)$$

where κ, D, D', S_T, ρ and μ are, respectively, thermal conductivity, diffusion constant, Dufour coefficient, Soret coefficient, density and chemical potential of the solute. Continuous dependence of the solution on the Soret coefficient is treated in (Straughan and Hutter, 1999). In this section we set the Soret coefficient $S_T = 0$ and concentrate on a Dufour effect. As a first step we treat a linear Dufour effect. This means we treat the $\rho T C D'(\partial \mu / \partial C)$ term in (2.141) as constant. This is in keeping with the approach of (Ybarra and Velarde, 1979). From a mathematical viewpoint we may then, without loss of generality, reduce system (2.140), incorporating the reduced version of (2.141), to the form

$$\begin{aligned}
&v_i - \lambda \Delta v_i = -p_{,i} + g_i T + h_i C, \qquad v_{i,i} = 0, \\
&T_{,t} + v_i T_{,i} = \Delta T + \gamma \Delta C, \qquad\qquad\qquad\quad (2.142)\\
&C_{,t} + v_i C_{,i} = \Delta C,
\end{aligned}$$

where $\gamma > 0$ is a constant and $\gamma \Delta C$ represents the Dufour effect. We now develop a priori bounds to enable us to establish continuous dependence of the solution on changes in the Dufour coefficient (constant) γ.

2.8.1   Continuous dependence on γ.

The continuous dependence result we now establish is truly a priori in that the coefficients appearing in the stability estimate are dependent only on initial and boundary data, and on the geometry of the domain. The proof given here is not identical to that of (Straughan and Hutter, 1999). However, it can be adapted very quickly since the Soret system studied in (Straughan and Hutter, 1999) is obtained by exchanging T and C in (2.142). On the boundary Γ we consider the given data

$$v_i = 0, \quad T = h, \quad C = g, \qquad x \in \Gamma, \qquad (2.143)$$

for prescribed functions h and g. Note that since we are dealing with the Brinkman equations all components of the velocity are prescribed on Γ. The initial data are

$$T(\mathbf{x}, 0) = T_0(\mathbf{x}), \quad C(\mathbf{x}, 0) = C_0(\mathbf{x}), \qquad \mathbf{x} \in \Omega. \qquad (2.144)$$

To study continuous dependence on γ we let (u_i, T, C_1, p) and (v_i, S, C_2, q) be solutions to (2.142) – (2.144) for the same boundary and initial data, but for different Dufour coefficients γ_1 and γ_2. Thus, let (u_i, T, C_1, p) and

(v_i, S, C_2, q) solve the boundary-initial value problems

$$
\begin{aligned}
u_i - \lambda \Delta u_i &= -p_{,i} + g_i T + h_i C_1, \\
u_{i,i} &= 0, \\
T_{,t} + u_i T_{,i} &= \Delta T + \gamma_1 \Delta C_1, \\
C_{1,t} + u_i C_{1,i} &= \Delta C_1,
\end{aligned}
\tag{2.145}
$$

in $\Omega \times (0, \mathcal{T})$,

$$
u_i = 0, \quad T = h, \quad C_1 = g, \qquad \text{on } \Gamma \times (0, \mathcal{T}), \tag{2.146}
$$

$$
T(\mathbf{x}, 0) = T_0(\mathbf{x}), \quad C_1(\mathbf{x}, 0) = C_0(\mathbf{x}), \quad \mathbf{x} \in \Omega, \tag{2.147}
$$

and

$$
\begin{aligned}
v_i - \lambda \Delta v_i &= -q_{,i} + g_i S + h_i C_2, \\
v_{i,i} &= 0, \\
S_{,t} + v_i S_{,i} &= \Delta S + \gamma_2 \Delta C_2, \\
C_{2,t} + v_i C_{2,i} &= \Delta C_2,
\end{aligned}
\tag{2.148}
$$

in $\Omega \times (0, \mathcal{T})$,

$$
v_i = 0, \quad S = h, \quad C_2 = g, \qquad \text{on } \Gamma \times (0, \mathcal{T}), \tag{2.149}
$$

$$
S(\mathbf{x}, 0) = T_0(\mathbf{x}), \quad C_2(\mathbf{x}, 0) = C_0(\mathbf{x}), \quad \mathbf{x} \in \Omega. \tag{2.150}
$$

Define the difference solution (w_i, θ, ϕ, π) and the gamma-difference, γ, by

$$
w_i = u_i - v_i, \quad \theta = T - S, \quad \phi = C_1 - C_2, \quad \pi = p - q, \quad \gamma = \gamma_1 - \gamma_2.
$$

The solution (w_i, θ, ϕ, π) satisfies the partial differential equations

$$
\begin{aligned}
w_i - \lambda \Delta w_i &= -\pi_{,i} + g_i \theta + h_i \phi, \qquad w_{i,i} = 0, \\
\theta_{,t} + w_i T_{,i} + v_i \theta_{,i} &= \Delta \theta + \gamma \Delta C_1 + \gamma_2 \Delta \phi, \\
\phi_{,t} + v_i \phi_{,i} + w_i C_{1,i} &= \Delta \phi,
\end{aligned}
\tag{2.151}
$$

in $\Omega \times (0, \mathcal{T})$, together with the boundary and initial conditions,

$$
w_i = 0, \quad \theta = 0, \quad \phi = 0, \qquad \text{on } \Gamma \times (0, \mathcal{T}), \tag{2.152}
$$

$$
\theta(\mathbf{x}, 0) = 0, \quad \phi(\mathbf{x}, 0) = 0. \tag{2.153}
$$

Our analysis commences by multiplying $(2.151)_1$ by w_i and integrating over Ω to derive

$$
\|\mathbf{w}\|^2 + \lambda \|\nabla \mathbf{w}\|^2 = g_i(\theta, w_i) + h_i(\phi, w_i). \tag{2.154}
$$

Again we suppose, $|\mathbf{g}| \leq 1, \quad |\mathbf{h}| \leq 1$.

Multiply $(2.151)_3$ by θ and integrate over Ω. Multiply $(2.151)_4$ by ϕ and likewise integrate over Ω. In this way one derives

$$\frac{d}{dt}\frac{1}{2}\|\theta\|^2 = -(w_i T_{,i}, \theta) - \|\nabla\theta\|^2 - \gamma(\nabla C_1, \nabla\theta) - \gamma_2(\nabla\theta, \nabla\phi), \quad (2.155)$$

and

$$\frac{d}{dt}\frac{1}{2}\|\phi\|^2 = -(w_i C_{1,i}, \phi) - \|\nabla\phi\|^2. \quad (2.156)$$

We form the combination $(2.155)+\Gamma(2.156)$ for a constant $\Gamma(> 0)$ to be chosen. In this way we obtain

$$\frac{d}{dt}\frac{1}{2}\left(\Gamma\|\phi\|^2 + \|\theta\|^2\right) = -\Gamma(w_i C_{1,i}, \phi) - (w_i T_{,i}, \theta) - \Gamma\|\nabla\phi\|^2$$
$$- \gamma_2(\nabla\theta, \nabla\phi) - \|\nabla\theta\|^2 - \gamma(\nabla C_1, \nabla\theta). \quad (2.157)$$

The first two terms on the right of this expression are cubic. We wish to make a positive - definite form from the next three. So, the idea now is to require Γ so large that

$$\Gamma\|\nabla\phi\|^2 + \gamma_2(\nabla\theta, \nabla\phi) + \|\nabla\theta\|^2 \geq \xi_1\|\nabla\phi\|^2 + \xi_2\|\nabla\theta\|^2,$$

for positive numbers ξ_1, ξ_2. For example by using the arithmetic-geometric mean inequality on the γ_2 term we may deduce

$$\Gamma\|\nabla\phi\|^2 + \gamma_2(\nabla\theta, \nabla\phi) + \|\nabla\theta\|^2 \geq \left(\Gamma - \frac{\gamma_2}{2\alpha}\right)\|\nabla\phi\|^2 + \left(1 - \frac{\alpha\gamma_2}{2}\right)\|\nabla\theta\|^2,$$

for $\alpha > 0$ at our disposal. Let us now choose $\alpha = 1/\gamma_2$ and then select $\Gamma = \gamma_2^2$. Thus, the inequality above becomes

$$\Gamma\|\nabla\phi\|^2 + \gamma_2(\nabla\theta, \nabla\phi) + \|\nabla\theta\|^2 \geq \frac{\gamma_2^2}{2}\|\nabla\phi\|^2 + \frac{1}{2}\|\nabla\theta\|^2. \quad (2.158)$$

Now use the arithmetic-geometric mean inequality on the last term of (2.157). We balance the $\|\nabla\theta\|^2$ term which arises with a piece of the same term from (2.158). Thus, (2.157) together with (2.158) allows us to derive

$$\frac{d}{dt}\frac{1}{2}\left(\Gamma\|\phi\|^2 + \|\theta\|^2\right) \leq -\Gamma(w_i C_{1,i}, \phi) - (w_i T_{,i}, \theta) - \frac{\gamma_2^2}{2}\|\nabla\phi\|^2$$
$$- \frac{1}{4}\|\nabla\theta\|^2 + \gamma^2\|\nabla C_1\|^2. \quad (2.159)$$

Since we have extra dissipation provided by the Brinkman term (as opposed to the Darcy term of section 2.7) we can bound the cubic terms in (2.159) in a different manner. We begin with the following Sobolev inequality

$$\|\mathbf{w}\|_4 \leq c_1\|\nabla\mathbf{w}\|, \quad (2.160)$$

where $\|\cdot\|_4$ is the norm on $L^4(\Omega)$ and $c_1 = c_1(\Omega)$. We also utilise the Poincaré inequality $\lambda_1\|\mathbf{w}\|^2 \leq \|\nabla\mathbf{w}\|^2$. Next, use the Cauchy - Schwarz,

Sobolev and Poincaré inequalities together with the arithmetic-geometric mean inequality to find

$$\begin{aligned}|(w_i C_{1,i}, \phi)| &\le \|\nabla C_1\| \|\mathbf{w}\|_4 \|\phi\|_4 \\ &\le c_1^2 \|\nabla C_1\| \|\nabla \mathbf{w}\| \|\nabla \phi\| \\ &\le \frac{c_1^4}{2} \|\nabla C_1\|^2 \|\nabla \mathbf{w}\|^2 + \frac{1}{2}\|\nabla \phi\|^2.\end{aligned} \tag{2.161}$$

A similar procedure leads to

$$|(w_i T_{,i}, \theta)| \le c_1^4 \|\nabla T\|^2 \|\nabla \mathbf{w}\|^2 + \frac{1}{4}\|\nabla \theta\|^2. \tag{2.162}$$

Now, combine (2.161) and (2.162) in inequality (2.159) to arrive at

$$\begin{aligned}\frac{d}{dt}\frac{1}{2}\left(\gamma_2^2\|\phi\|^2 + \|\theta\|^2\right) &\le \frac{c_1^4 \gamma_2^2}{2}\|\nabla C_1\|^2 \|\nabla \mathbf{w}\|^2 \\ &\quad + c_1^4 \|\nabla T\|^2 \|\nabla \mathbf{w}\|^2 + \gamma^2 \|\nabla C_1\|^2.\end{aligned} \tag{2.163}$$

We need to estimate $\|\nabla \mathbf{w}\|^2$ and then from (2.154) we may find

$$\begin{aligned}\|\mathbf{w}\|^2 + \lambda \|\nabla \mathbf{w}\|^2 &= (g_i \theta, w_i) + (h_i \phi, w_i) \\ &\le \|\theta\|\|\mathbf{w}\| + \|\phi\|\|\mathbf{w}\|\end{aligned}$$

and then we use Poincaré's inequality on a part of $\|\nabla \mathbf{w}\|^2$ to find

$$\|\mathbf{w}\|^2 + \lambda\sqrt{\lambda_1}\|\mathbf{w}\|\|\nabla \mathbf{w}\| \le \|\theta\|\|\mathbf{w}\| + \|\phi\|\|\mathbf{w}\|.$$

From this inequality we derive the estimate

$$\|\mathbf{w}\| + \lambda\sqrt{\lambda_1}\|\nabla \mathbf{w}\| \le \|\theta\| + \|\phi\|. \tag{2.164}$$

What we require in (2.159) is an upper bound for $\|\nabla \mathbf{w}\|^2$ and we may derive this from (2.164), since this inequality shows

$$\|\nabla \mathbf{w}\| \le \frac{\|\theta\| + \|\phi\|}{\lambda\sqrt{\lambda_1}},$$

and squaring

$$\|\nabla \mathbf{w}\|^2 \le \frac{(\|\theta\| + \|\phi\|)^2}{\lambda^2 \lambda_1} \le \frac{2}{\lambda^2 \lambda_1}\left(\|\theta\|^2 + \|\phi\|^2\right). \tag{2.165}$$

Thus, we employ estimate (2.165) in inequality (2.163) to find

$$\begin{aligned}\frac{d}{dt}\frac{1}{2}\left(\gamma_2^2\|\phi\|^2 + \|\theta\|^2\right) &\le \frac{c_1^4}{\lambda^2 \lambda_1}\left(\gamma_2^2 \|\nabla C_1\|^2 + 2\|\nabla T\|^2\right)\left(\|\theta\|^2 + \|\phi\|^2\right) \\ &\quad + \gamma^2 \|\nabla C_1\|^2.\end{aligned} \tag{2.166}$$

We now need a *priori* bounds for $\|\nabla C_1\|$ and $\|\nabla T\|$. To this end we follow analogous steps to section 2.7 and we introduce the harmonic function, H, which adopts the same boundary values as C_1. Thus, define

$$\Delta H = 0, \text{ in } \Omega \times (0, T), \quad H(\mathbf{x}, t) = g(\mathbf{x}, t), \text{ on } \Gamma \times (0, T). \tag{2.167}$$

Form the identity

$$\int_0^t \int_\Omega (C_1 - H)(C_{1,t} + u_i C_{1,i} - \Delta C_1) dx\, d\eta = 0. \tag{2.168}$$

Next perform several integrations in (2.168) and use the boundary values and properties of H to see that

$$\frac{1}{2}\|C_1(t)\|^2 - \frac{1}{2}\|C_0\|^2 - (H, C_1) + (H_0, C_0) + \int_0^t \int_\Omega H_{,\eta} C_1 dx\, d\eta$$
$$- \int_0^t \int_\Omega H u_i C_{1,i} dx\, d\eta + \int_0^t \|\nabla C_1\|^2 d\eta - \int_0^t \oint_\Gamma g \frac{\partial H}{\partial n}\, dA\, d\eta = 0. \tag{2.169}$$

The point of introducing such an H is that we cannot work directly with T or C_1 to form energy-like estimates since they have non-zero boundary values. Instead we work with identities for $T - H$ or $C_1 - H$, functions which are zero on Γ. We may derive *a priori* bounds for H in a straightforward manner. To handle the cubic term in (2.169) we let g_m be the maximum value of g on $\Gamma \times [0, T)$ (g_m is taken positive) and then since H is harmonic we know by the maximum principle that $H \leq g_m$. Upon employing the Cauchy-Schwarz and arithmetic-geometric mean inequalities we derive

$$\int_0^t \int_\Omega H u_i C_{1,i} dx\, d\eta \leq g_m \sqrt{\int_0^t \|u\|^2 d\eta} \sqrt{\int_0^t \|\nabla C_1\|^2 d\eta}$$
$$\leq \frac{1}{2}\int_0^t \|\nabla C_1\|^2 d\eta + \frac{1}{2} g_m^2 \int_0^t \|u\|^2 d\eta, \tag{2.170}$$

where the coefficient of $\int_0^t \|\nabla C_1\|^2 d\eta$ has been deliberately chosen less than 1 so we may dominate it by the equivalent term in (2.169).

From equation $(2.145)_1$ we may show that

$$\|u\|^2 + \lambda \|\nabla u\|^2 = g_i(T, u_i) + h_i(C_1, u_i).$$

We use this equation to derive a bound for $\int_0^t \|u\|^2 d\eta$ to employ in (2.170). We now use the Cauchy - Schwarz inequality and Poincaré's inequality to derive

$$\|u\|^2 + \lambda \|\nabla u\|^2 \leq \|T\|\|u\| + \|C_1\|\|u\|,$$

then

$$\|u\|^2 + \lambda \lambda_1 \|u\|^2 \leq \|T\|\|u\| + \|C_1\|\|u\|.$$

Thus,

$$\|u\| \leq \frac{\|T\| + \|C_1\|}{(1 + \lambda\lambda_1)},$$

from whence,

$$\|u\|^2 \leq \frac{2(\|T\|^2 + \|C_1\|^2)}{(1 + \lambda\lambda_1)}. \tag{2.171}$$

Therefore, from (2.170),

$$\int_0^t \int_\Omega H u_i C_{1,i} dx\, d\eta \le \frac{1}{2} \int_0^t \|\nabla C_1\|^2 d\eta$$
$$+ g_m^2 \left(\int_0^t \|T\|^2 d\eta + \int_0^t \|C_1\|^2 d\eta \right). \qquad (2.172)$$

By using the arithmetic-geometric mean inequality we may now show that

$$(H, C_1) \le \|H\|^2 + \frac{1}{4}\|C_1\|^2, \qquad -(H_0, C_0) \le \frac{1}{2}\|H_0\|^2 + \frac{1}{2}\|C_0\|^2, \quad (2.173)$$

and

$$\int_0^t \int_\Omega H_{,\eta} C_1 \, dx\, d\eta \le \frac{1}{2a} \int_0^t \int_\Omega H_{,\eta}^2 dx\, d\eta + \frac{a}{2} \int_0^t \int_\Omega C_1^2 dx\, d\eta, \quad (2.174)$$

for $a > 0$ to be selected.

We now use the Poincaré inequality on the C_1^2 term on the right, but since $C_1 = g$ on Γ the Poincaré inequality now takes form

$$\lambda_1 \int_\Omega C_1^2 dx \le \int_\Omega |\nabla C_1|^2 dx + k_P \int_\Gamma g^2 dA,$$

where λ_1 and k_P are positive constants depending on Ω. We integrate this inequality over $(0, t)$ to find

$$\int_0^t ds \int_\Omega C_1^2 dx \le \frac{1}{\lambda_1} \int_0^t ds \int_\Omega |\nabla C_1|^2 dx + \frac{k_P}{\lambda_1} \int_0^t ds \int_\Gamma g^2 dA. \quad (2.175)$$

Now use estimate (2.175) on the right of (2.174) to find

$$\int_0^t \int_\Omega H_{,\eta} C_1 \, dx\, d\eta \le \frac{1}{2a} \int_0^t \int_\Omega H_{,\eta}^2 \, dx\, d\eta + \frac{a}{2\lambda_1} \int_0^t ds \int_\Omega |\nabla C_1|^2 dx$$
$$+ \frac{a k_P}{2\lambda_1} \int_0^t ds \int_\Gamma g^2 dA.$$

We choose $a/2\lambda_1 = 1/4$, i.e. $a = \lambda_1/2$, to balance the $\int_0^t ds \int_\Omega |\nabla C_1|^2 dx$ piece with an equivalent piece of the analogous term in (2.169). Thus, the necessary inequality is

$$\int_0^t \int_\Omega H_{,\eta} C_1 \, dx\, d\eta \le \frac{1}{\lambda_1} \int_0^t \int_\Omega H_{,\eta}^2 \, dx\, d\eta + \frac{1}{4} \int_0^t ds \int_\Omega |\nabla C_1|^2 dx$$
$$+ \frac{k_P}{4} \int_0^t ds \int_\Gamma g^2 dA. \qquad (2.176)$$

By use of the Cauchy-Schwarz inequality one finds

$$\int_0^t \int_\Gamma g \frac{\partial H}{\partial n} dA\, d\eta \le \sqrt{\int_0^t \int_\Gamma g^2 dA\, d\eta} \sqrt{\int_0^t \int_\Gamma \left(\frac{\partial H}{\partial n}\right)^2 dA\, d\eta}. \quad (2.177)$$

We next employ (2.173), (2.176) and (2.177) together with (2.170) in equation (2.169) to arrive at

$$\frac{1}{4}\|C_1(t)\|^2 + \frac{1}{4}\int_0^t \|\nabla C_1\|^2 d\eta \leq \|C_0\|^2 + \frac{k_P}{4}\int_0^t ds \int_\Gamma g^2 dA + \|H\|^2$$

$$+ \frac{1}{2}\|H_0\|^2 + \frac{1}{2}\int_0^t \|H_{,\eta}\|^2 d\eta$$

$$+ \sqrt{\int_0^t \int_\Gamma g^2 dA \, d\eta} \sqrt{\int_0^t \int_\Gamma \left(\frac{\partial H}{\partial n}\right)^2 dA \, d\eta}$$

$$+ g_m^2 \int_0^t \|T\|^2 d\eta + g_m^2 \int_0^t \|C_1\|^2 d\eta. \tag{2.178}$$

The next stage involves use of a Rellich identity, cf. (Payne and Weinberger, 1958), to estimate the H terms on the right of (2.178). Details appropriate to the function H are similar to those in (Franchi and Straughan, 1994), p. 449. We now give details.

Recall how the function H is defined in (2.167). Thus we may write

$$0 = \int_\Omega x^i H_{,i} \Delta H \, dx$$

$$= \int_\Omega (x^i H_{,i} H_{,j})_{,j} dx - \int_\Omega x^i_{,j} H_{,i} H_{,j} dx - \int_\Omega x^i H_{,ij} H_{,j} dx$$

$$= \int_\Gamma x^i H_{,i} n_j H_{,j} dA - \int_\Omega \delta^i_j H_{,i} H_{,j} dx - \int_\Omega \frac{x^i}{2}(H_{,j} H_{,j})_{,i} dx$$

$$= \int_\Gamma x_i H_{,i} \frac{\partial H}{\partial n} dA - \int_\Omega H_{,i} H_{,i} dx$$

$$\quad - \frac{1}{2}\int_\Omega (x^i H_{,j} H_{,j})_{,i} dx + \frac{1}{2}\int_\Omega x^i_{,i} H_{,j} H_{,j} dx$$

$$= \int_\Gamma x_i H_{,i} \frac{\partial H}{\partial n} dA - \frac{1}{2}\int_\Gamma x_i n_i H_{,j} H_{,j} dA$$

$$\quad - \int_\Omega H_{,i} H_{,i} dx + \frac{3}{2}\int_\Omega H_{,i} H_{,i} dx,$$

where several integrations by parts and use of the divergence theorem have been performed. Thus, we see that

$$\frac{1}{2}\|\nabla H\|^2 = \frac{1}{2}\int_\Gamma x_i n_i H_{,j} H_{,j} dA - \int_\Gamma x_i H_{,i} \frac{\partial H}{\partial n} dA. \tag{2.179}$$

On Γ we write ∇H as a normal and tangential part, thus

$$H_{,i} = \frac{\partial H}{\partial n} n_i + \nabla_s H \, s_i,$$

where $\nabla_s H \, s_i$ is the tangential derivative, $s^i \nabla_s H = x^i_{;\alpha} a^{\alpha\beta} H_{;\beta}$ where $a^{\alpha\beta}$ is the first fundamental form on Γ and $_{;\alpha}$ denotes surface differentiation.

From this decomposition it follows that $H_{,j}H_{,j} = (\partial H/\partial n)^2 + |\nabla_s H|^2$. Hence, we write the right hand side (RHS) of (2.179) as

$$RHS = \frac{1}{2}\int_\Gamma x_i n_i \left(\frac{\partial H}{\partial n}\right)^2 dA + \frac{1}{2}\int_\Gamma x_i n_i |\nabla_s H|^2 dA$$

$$- \int_\Gamma x_i n_i \left(\frac{\partial H}{\partial n}\right)^2 dA - \int_\Gamma x_i s^i \nabla_s H \frac{\partial H}{\partial n} dA$$

$$= -\frac{1}{2}\int_\Gamma x_i n_i \left(\frac{\partial H}{\partial n}\right)^2 dA - \frac{1}{2}\int_\Gamma x_i n_i |\nabla_s H|^2 dA - \int_\Gamma x_i s^i \nabla_s H \frac{\partial H}{\partial n} dA$$

So (2.179) becomes

$$\frac{1}{2}\|\nabla H\|^2 + \frac{1}{2}\int_\Gamma x_i n_i \left(\frac{\partial H}{\partial n}\right)^2 dA = \frac{1}{2}\int_\Gamma x_i n_i |\nabla_s H|^2 dA$$

$$- \int_\Gamma x_i s_i \frac{\partial H}{\partial n} \nabla_s H \, dA. \qquad (2.180)$$

We suppose now Ω is star shaped and put $m_1 = \min_\Gamma x_i n_i > 0$. Thus, from (2.180) we may determine positive constants c_1 and c_2 depending on Γ such that

$$\|\nabla H\|^2 + c_1 \int_\Gamma \left(\frac{\partial H}{\partial n}\right)^2 dA \leq c_2 \int_\Gamma |\nabla_s H|^2 dA$$

$$= c_2 \int_\Gamma |\nabla_s g|^2 dA. \qquad (2.181)$$

The Poincaré inequality for H has form, since $H \neq 0$ on Γ,

$$\lambda_1 \|H\|^2 \leq \|\nabla H\|^2 + k_P \int_\Gamma H^2 dA,$$

where $k_P = k_P(\Omega) > 0$ and so

$$\|H\|^2 \leq \frac{c_2}{\lambda_1} \int_\Gamma |\nabla_s g|^2 dA + \frac{k_P}{\lambda_1} \int_\Gamma g^2 dA. \qquad (2.182)$$

Furthermore, $\Delta H_{,t} = 0$ in $\Omega \times [0, T]$, with $H_{,t} = g_{,t}$ on Γ. We may apply the above analysis to $\phi = H_{,t}$ to derive an inequality analogous to (2.181) and from this we find

$$\|\nabla H_{,t}\|^2 \leq c_2 \int_\Gamma |\nabla_s g_{,t}|^2 dA. \qquad (2.183)$$

Thus, inequalities (2.180) – (2.183) allow us to obtain estimates for the H terms on the right of (2.178). Clearly, we may determine constants c_α

dependent on Γ such that

$$\|H\|^2 + \frac{1}{2}\|H_0\|^2 \le \frac{3}{2}c_3\int_\Gamma g^2 dA + \frac{3}{2}c_4\int_\Gamma |\nabla_s g|^2 dA, \tag{2.184}$$

$$\int_0^t \|H_{,\eta}\|^2 d\eta \le c_5 \int_0^t\int_\Gamma g_{,\tau}^2 dA\, d\tau + c_6\int_0^t\int_\Gamma |\nabla_s g_{,\tau}|^2 dA\, d\tau, \tag{2.185}$$

$$\int_0^t\int_\Gamma \left(\frac{\partial H}{\partial n}\right)^2 dA\, d\eta \le c_2 \int_0^t\int_\Gamma |\nabla_s g|^2 dA\, d\eta. \tag{2.186}$$

If we now denote by D_1 a data term of form

$$D_1(t) = 4\|T_0\|^2 + k_1\int_\Gamma g^2 dA + k_2\int_\Gamma |\nabla_s g|^2 dA + k_3\int_0^t\int_\Gamma g_{,\tau}^2 dA\, d\tau$$

$$+ k_4\int_0^t\int_\Gamma |\nabla_s g_{,\tau}|^2 dA\, d\tau + k_5\sqrt{\int_0^t\int_\Gamma g^2 dA\, d\eta}\sqrt{\int_0^t\int_\Gamma |\nabla_s g|^2 dA\, d\eta},$$

where k_α may be computed from (2.184) – (2.186), then from (2.178) we may arrive at the inequality

$$\|C_1(t)\|^2 + \int_0^t \|\nabla C_1\|^2 d\eta \le D_1(t) + 4g_m^2\int_0^t (\|T\|^2 + \|C_1\|^2) d\eta. \tag{2.187}$$

We must now carry out a similar procedure for bounding $\|T\|$ and $\|\nabla T\|$ and so we introduce the harmonic function G which assumes the same boundary values as T, i.e. define G to solve

$$\Delta G = 0, \quad \text{in } \Omega \times (0, \mathcal{T}), \qquad G(\mathbf{x}, t) = h(\mathbf{x}, t), \quad \text{on } \Gamma \times (0, \mathcal{T}). \tag{2.188}$$

Since T satisfies (2.145) we may construct the identity

$$\int_0^t\int_\Omega (T - G)(T_{,t} + u_i T_{,i} - \Delta T - \gamma_1 \Delta C_1) dx\, d\eta = 0. \tag{2.189}$$

We now carry out several integrations in (2.189) to arrive at

$$\frac{1}{2}\|T(t)\|^2 - \frac{1}{2}\|T_0\|^2 - (G, T) + (G_0, T_0) + \int_0^t (G_{,\eta}, T) d\eta$$

$$- \int_0^t\int_\Omega Gu_i T_{,i} dx\, d\eta + \int_0^t \|\nabla T\|^2 d\eta + \gamma_1\int_0^t (\nabla C_1, \nabla T) d\eta \tag{2.190}$$

$$- \int_0^t\int_\Gamma h\frac{\partial G}{\partial n} dA\, d\eta - \gamma_1\int_0^t\int_\Gamma g\frac{\partial G}{\partial n} dA\, d\eta = 0.$$

Let h_m denote the maximum value of h on Γ. Then following the procedure leading to (2.170) we estimate the cubic term in (2.190). The arithmetic-geometric mean inequality is used on the γ_1 term and these procedures furnish the bound

$$
\int_0^t \int_\Omega Gu_i T_{,i} dx\, d\eta - \gamma_1 \int_0^t (\nabla C_1, \nabla T)d\eta \le h_m^2 \int_0^t \|\mathbf{u}\|^2 d\eta
$$
$$
+ \frac{1}{2} \int_0^t \|\nabla T\|^2 d\eta + \gamma_1^2 \int_0^t \|\nabla C_1\|^2 d\eta
$$
$$
\le \frac{2h_m^2}{(1+\lambda\lambda_1)} \left(\int_0^t \|T\|^2 d\eta + \int_0^t \|C_1\|^2 d\eta \right)
$$
$$
+ \frac{1}{2} \int_0^t \|\nabla T\|^2 d\eta + \gamma_1^2 \int_0^t \|\nabla C_1\|^2 d\eta, \qquad (2.191)
$$

where in the last step (2.171) has been employed.
   We estimate the $G_{,\eta}$ term as

$$
\int_0^t (G_{,\eta}, T)d\eta \le \frac{1}{2a} \int_0^t \|G_{,\eta}\|^2 d\eta + \frac{a}{2} \int_0^t \|T\|^2 d\eta
$$
$$
\le \frac{1}{2a} \int_0^t \|G_{,\eta}\|^2 d\eta + \frac{a}{2\lambda_1} \int_0^t \|\nabla T\|^2 d\eta
$$
$$
+ \frac{ak_P}{2\lambda_1} \int_0^t d\eta \int_\Gamma h^2 dA
$$

where we have also used the Poincaré inequality for T. Now pick $a/2\lambda_1 = 1/4$, and then

$$
\int_0^t (G_{,\eta}, T)d\eta \le \frac{1}{\lambda_1} \int_0^t \|G_{,\eta}\|^2 d\eta + \frac{1}{4} \int_0^t \|\nabla T\|^2 d\eta
$$
$$
+ \frac{k_P}{4} \int_0^t d\eta \int_\Gamma h^2 dA. \qquad (2.192)
$$

   Upon employing (2.191) and (2.192) in (2.190) we may further use the arithmetic-geometric mean inequality to obtain

$$
\frac{1}{4}\|T(t)\|^2 + \frac{1}{4} \int_0^t \|\nabla T\|^2 d\eta \le \|T_0\|^2 + \frac{1}{2}\|G_0\|^2 + \|G\|^2 + \frac{1}{\lambda_1} \int_0^t \|G_{,\eta}\|^2 d\eta
$$
$$
+ \int_0^t \int_\Gamma h \frac{\partial G}{\partial n} dA\, d\eta + \gamma_1 \int_0^t \int_\Gamma g \frac{\partial G}{\partial n} dA\, d\eta + \gamma_1^2 \int_0^t \|\nabla C_1\|^2 d\eta
$$
$$
+ 2g_m^2 \int_0^t \|C_1\|^2 d\eta + 2g_m^2 \int_0^t \|T\|^2 d\eta. \qquad (2.193)
$$

Next use the Cauchy-Schwarz inequality on the boundary terms,

$$\int_0^t \int_\Gamma h \frac{\partial G}{\partial n}\, dA\, d\eta + \gamma_1 \int_0^t \int_\Gamma g \frac{\partial G}{\partial n}\, dA\, d\eta$$

$$\leq \sqrt{\int_0^t \int_\Gamma h^2 dA\, d\eta} \sqrt{\int_0^t \int_\Gamma \left(\frac{\partial G}{\partial n}\right)^2 dA\, d\eta}$$

$$+ \gamma_1 \sqrt{\int_0^t \int_\Gamma g^2 dA\, d\eta} \sqrt{\int_0^t \int_\Gamma \left(\frac{\partial G}{\partial n}\right)^2 dA\, d\eta} \quad (2.194)$$

By using a Rellich identity argument one may show that analogous inequalities to (2.184) – (2.186) hold for G. We then define the data term D_2 for computable constants $\ell_1, \ldots, \ell_5$ as

$$D_2(t) = 4\|T_0\|^2 + \ell_1 \int_\Gamma h^2 dA + \ell_2 \int_\Gamma |\nabla_s h|^2 dA$$

$$+ \ell_3 \int_0^t \int_\Gamma h_{,\tau}^2 dA\, d\tau + \ell_4 \int_0^t \int_\Gamma |\nabla_s h_{,\tau}|^2 dA\, d\tau$$

$$+ \ell_5 \sqrt{\int_0^t \int_\Gamma h^2 dA\, d\eta} \sqrt{\int_0^t \int_\Gamma |\nabla_s h|^2 dA\, d\eta}$$

$$+ \ell_5 \gamma_1 \sqrt{\int_0^t \int_\Gamma g^2 dA\, d\eta} \sqrt{\int_0^t \int_\Gamma |\nabla_s h|^2 dA\, d\eta}. \quad (2.195)$$

Upon using (2.194) and (2.195) in (2.193) one may produce the inequality

$$\|T(t)\|^2 + \int_0^t \|\nabla T\|^2 d\eta \leq D_2(t) + 8g_m^2 \int_0^t \|C_1\|^2 d\eta$$

$$+ 8g_m^2 \int_0^t \|T\|^2 d\eta + 4\gamma_1^2 \int_0^t \|\nabla C_1\|^2 d\eta. \quad (2.196)$$

We now let α be a constant such that $\alpha > 4\gamma_1^2$ and then form $\alpha(2.187) +$ (2.196). In this manner we obtain the bound

$$\alpha \|C_1(t)\|^2 + (\alpha - 4\gamma_1^2) \int_0^t \|\nabla C_1\|^2 d\eta + \|T(t)\|^2 + \int_0^t \|\nabla T\|^2 d\eta$$

$$\leq \alpha D_1 + D_2 + \left[4\alpha g_m^2 + 8h_m^2\right] \int_0^t \|C_1\|^2 d\eta$$

$$+ (4\alpha g_m^2 + 8h_m^2) \int_0^t \|T\|^2 d\eta. \quad (2.197)$$

Define now $K_1 = 4\alpha g_m^2 + 8h_m^2$, $D(t) = \alpha D_1 + D_2$, and $K = K_1$ if $\alpha > 1$ or $K = K_1/\alpha$ if $\alpha < 1$. Then from (2.197) one may discard the $\|\nabla C_1\|^2$ and $\|\nabla T\|^2$ terms to derive

$$\alpha \|C_1(t)\|^2 + \|T(t)\|^2 \leq D + K\left[\alpha \int_0^t \|C_1\|^2 d\eta + \int_0^t \|T\|^2 d\eta\right].$$

Thus upon integration we see that

$$\alpha \int_0^t \|C_1\|^2 d\eta + \int_0^t \|T\|^2 d\eta \leq P(t), \qquad (2.198)$$

where P is the data term

$$P(t) = \int_0^t e^{K(t-s)} D(s) ds. \qquad (2.199)$$

We still need a priori estimates for $\int_0^t \|\nabla T\|^2 d\eta$ and $\int_0^t \|\nabla C_1\|^2 d\eta$ and these follow by using (2.198) in (2.197) to find

$$\int_0^t \|\nabla T\|^2 d\eta \leq P_2(t), \qquad \int_0^t \|\nabla C_1\|^2 d\eta \leq P_1(t), \qquad (2.200)$$

where P_1 and P_2 are data terms given by

$$P_1(t) = \frac{1}{(\alpha - 4\gamma_1^2)} [D(t) + KP(t)], \qquad P_2(t) = D(t) + KP(t).$$

We are now in a position to complete the continuous dependence estimate on γ. An integration of (2.166) yields

$$\gamma_2^2 \|\phi(t)\|^2 + \|\theta(t)\|^2 \leq \frac{2c_1^4}{\lambda^2 \lambda_1} \int_0^t \left[\gamma_2^2 \|\nabla C_1\|^2 + 2\|\nabla T\|^2 \right] (\|\phi\|^2 + \|\theta\|^2) d\eta$$

$$+ \gamma^2 \int_0^t \|\nabla C_1\|^2 d\eta,$$

$$\leq \frac{2K_1 c_1^4}{\lambda^2 \lambda_1} \int_0^t \left[\gamma_2^2 \|\nabla C_1\|^2 + 2\|\nabla T\|^2 \right] (\gamma_2^2 \|\phi\|^2 + \|\theta\|^2) d\eta$$

$$+ \gamma^2 P_1(t), \qquad (2.201)$$

where $K_2 = \max\{1, \gamma_2^{-2}\}$. Now define $f(t) = 2K_2 c_1^4 [\gamma_2^2 \|\nabla C_1\|^2 + 2\|\nabla T\|^2]/\lambda^2 \lambda_1$. Then an application of Gronwall's inequality to (2.201) furnishes the estimate

$$\gamma_2^2 \|\phi(t)\|^2 + \|\theta(t)\|^2 \leq \gamma^2 P_1(t) + \gamma^2 \int_0^t P_1(s) f(s) \left[\exp \int_s^t f(u) du \right] ds,$$

$$\leq \gamma^2 P_1(t) + \gamma^2 \left[\exp \int_0^t f(s) ds \right] \bar{P}_1(t) \int_0^t f(s) ds, \qquad (2.202)$$

where $\bar{P}_1(t) = \max_{s \in [0,t]} P_1(s)$. Thanks to (2.200) we have $\int_0^t f(s) ds \leq P_3(t)$, where the data term P_3 is given by $P_3(t) = 2K_2 c_1^4 [\gamma_2^2 P_2(t) + 2P_1(t)]/\lambda^2 \lambda_1$. Therefore, from inequality (2.202) we may deduce

$$\gamma_2^2 \|\phi(t)\|^2 + \|\theta(t)\|^2 \leq R(t) \gamma^2, \qquad (2.203)$$

where $R(t)$ is the data term given by $R(t) = P_1(t) + \bar{P}_1(t) P_3(t) \exp [P_3(t)]$.

Inequality (2.203) demonstrates continuous dependence on the Dufour coefficient γ, for the salt concentration C and temperature T.

We may also derive a continuous dependence inequality for the velocity $\mathbf{u}$ by employing (2.154) in combination with (2.203). From (2.154) one easily derives the estimates

$$\|\mathbf{w}\| \leq \frac{\|\theta\| + \|\phi\|}{(1 + \lambda\lambda_1)}, \qquad \text{and} \qquad \|\nabla\mathbf{w}\| \leq \frac{1}{\lambda\sqrt{\lambda_1}}(\|\theta\| + \|\phi\|).$$

These inequalities together with (2.203) yield

$$\|\mathbf{w}(t)\|^2 \leq \frac{2K_2 R(t)}{(1 + \lambda\lambda_1)^2}\gamma^2, \qquad \text{and} \qquad \|\nabla\mathbf{w}(t)\|^2 \leq \frac{2K_2}{\lambda^2\lambda_1} R(t)\gamma^2.$$

$$(2.204)$$

Inequalities (2.204) establish continuous dependence on the Dufour coefficient γ in the L^2 and H^1 measures of $\mathbf{w}$ as indicated.

Very interesting *a priori* bounds and continuous dependence on the Soret coefficient for the system of equations (2.140) are established by (Lin and Payne, 2007a). These writers study equations (2.140) with zero *flux* boundary conditions. The methods they use are very interesting and of necessity different from those described in this section.

2.9 Initial - final value problems

Recently a new class of problem has been shown to be relevant to many applied mathematical situations. This is where the data are not given at time $t = 0$, but instead are prescribed as a linear combination at times $t = 0$ and $t = T$. We shall refer to such situations as initial - final value problems. Specific applications of these ideas are in (Payne and Schaefer, 2002), (Payne et al., 2004), (Ames et al., 2004a; Ames et al., 2004b), (Quintanilla and Straughan, 2005b; Quintanilla and Straughan, 2005a) and the references therein. This class of problem was originally introduced in order to stabilize solutions to the improperly posed problem when the data is given at $t = T$ and one wishes to compute the solution backward in time, see (Ames et al., 1998), (Ames and Payne, 1999) and the references therein. (Ames et al., 2004a) study an initial - final value problem for the first order abstract equation $u_t + Au = f$. (Ames et al., 2004b) investigate an initial - final value problem for the diffusion equation with the spatial domain being an infinite cylinder. (Payne and Schaefer, 2002) study an initial - final value problem for the second order in time abstract equation $u_{tt} + Au = F$. They also investigate a similar initial - final value problem for the equation $u_{tt} + au_t + Au = 0$, for $a > 0$ a constant. (Payne et al., 2004) study an initial - final value problem for some fluid mechanics problems, especially in connection with Stokes flow. Further analyses of initial - final value problems are by (Quintanilla and Straughan, 2005b) who investigate thermoelasticity according to the new developments of (Green and Naghdi, 1991; Green and Naghdi, 1992; Green and Naghdi, 1993).

Further analysis of these theories may be found in (Quintanilla and Racke, 2003), (Quintanilla and Straughan, 2000; Quintanilla and Straughan, 2002; Quintanilla and Straughan, 2004), (Zhang and Zuazua, 2003), (Puri and Jordan, 2004). Another article dealing with initial - final value problems is that of (Quintanilla and Straughan, 2005a) who concentrate on dipolar fluids, see also (Bleustein and Green, 1967), (Green and Naghdi, 1968; Green and Naghdi, 1970), (Green et al., 1965), (Green and Rivlin, 1967), (Akyildiz and Bellout, 2004), (Jordan and Puri, 1999; Jordan and Puri, 2002), (Puri and Jordan, 1999b; Puri and Jordan, 1999a), on the (Green and Naghdi, 1996) extended theory of viscous fluids, and on the Brinkman-Forchheimer model of flow in porous media. The last topic is of interest in this book.

The article of (Quintanilla and Straughan, 2005a) analyses the Brinkman-Forchheimer equations, as used by (Qin and Kaloni, 1998), namely

$$Au_{i,t} = -p_{,i} - u_i + \lambda\Delta u_i - \beta|\mathbf{u}|u_i,$$
$$u_{i,i} = 0.$$
(2.205)

In these equations u_i, p represent the velocity and pressure, and A, λ, β are positive constants.

We take equations (2.205) to be defined on a bounded domain $\Omega \subset \mathbb{R}^3$ on the time interval $(0, T)$ for some $T < \infty$, with the boundary conditions being

$$u_i = 0 \qquad \text{on} \quad \Gamma.$$
(2.206)

The study of (Quintanilla and Straughan, 2005a) uses the initial - final condition

$$u_i(T) + \alpha u_i(0) = f_i,$$
(2.207)

where α is a constant, and $f_i(\mathbf{x})$ is a prescribed function. (The standard initial boundary value problem for (2.205) would replace (2.207) by $u_i(0) = f_i$. The standard final boundary value problem for (2.205) would employ $u_i(T) = f_i$ instead of (2.207).) Here, the objective is to obtain a bound on u_i in terms of f_i and α, employing the relation (2.207).

(Quintanilla and Straughan, 2005a) note that for the final value problem for (2.205), (2.206), i.e. with $\alpha = 0$, a global solution does not exist. By transforming $t \to T - t$ one may show (cf. for example, the arguments in (Straughan, 1998))

$$\|\mathbf{u}(t)\| \geq \frac{\|\mathbf{u}(0)\|}{e^{-\gamma t} - k_2\|\mathbf{u}(0)\|(1 - e^{-\gamma t})/2\gamma}.$$
(2.208)

In this inequality $\gamma = (\lambda\lambda_1 + 1)/A$, $k_2 = 2\beta/Am^{1/2}$, with λ_1 being the first eigenvalue in the membrane problem for Ω and where m is the volume of Ω. The right hand side of (2.208) blows-up at time $\mathcal{T} = [A/(\lambda\lambda_1 + 1)]\log\{1 + [(\lambda\lambda_1 + 1)m^{1/2}/\beta\|\mathbf{u}(0)\|]\}$, and so u_i cannot exist classically beyond this time. (Quintanilla and Straughan, 2005a) then argue that care must be taken with the initial - final value problem defined by (2.205) – (2.207).

(Quintanilla and Straughan, 2005a) derive a bound for u_i by commencing with multiplication of (2.205) by u_i and integration over Ω using the boundary conditions to find

$$\frac{d}{dt}\frac{A}{2}\|\mathbf{u}\|^2 = -\|\mathbf{u}\|^2 - \lambda\|\nabla\mathbf{u}\|^2 - \beta\int_\Omega |\mathbf{u}|^3 dx. \tag{2.209}$$

We employ the Poincaré inequality $-\|\nabla\mathbf{u}\|^2 \leq -\lambda_1\|\mathbf{u}\|^2$ and the Cauchy-Schwarz inequality to find $-\int_\Omega |\mathbf{u}|^3 dx \leq -\|\mathbf{u}\|^{3/2}/m^{1/2}$. Then from (2.209) with $\Phi(t) = \|\mathbf{u}(t)\|^2$ one may show

$$\frac{d\Phi}{dt} \leq -c_1\Phi - c_2\Phi^{3/2}, \tag{2.210}$$

where the constants c_1 and c_2 are given by

$$c_1 = \frac{2(1+\lambda\lambda_1)}{A}, \qquad c_2 = \frac{2\beta}{Am^{1/2}}.$$

Inequality (2.210) is integrated to obtain

$$\|\mathbf{u}(t)\| \leq \frac{\|\mathbf{u}(0)\|e^{-c_1t/2}}{1 + c_2\|\mathbf{u}(0)\|(1 - e^{-\gamma t})/c_1}, \tag{2.211}$$

for t in the interval $0 \leq t \leq T$.

This is a bound for $u_i(t)$ in terms of $u_i(0)$. However, $u_i(0)$ is unknown. We need to remove $\|\mathbf{u}(0)\|$ in (2.211) and convert it to an estimate involving f_i and α. The key is also to retain the c_2 term since this contains the Forchheimer effect (the β term). It is necessary to bound $\|\mathbf{u}(0)\|$ from both above and below.

(Quintanilla and Straughan, 2005a) show that one may demonstrate

$$\|\mathbf{u}(0)\| \geq \frac{\|\mathbf{f}\|}{\sqrt{2(\alpha^2 + e^{-c_1T})}}, \tag{2.212}$$

and provided $|\alpha| > e^{-c_1T/2}$,

$$\|\mathbf{u}(0)\| \leq \frac{1}{(|\alpha| - e^{-c_1T/2})}\|\mathbf{f}\|. \tag{2.213}$$

The lower and upper bounds (2.212) and (2.213) used in (2.211) lead to the estimate

$$\|\mathbf{u}(t)\| \leq e^{-c_1t/2}\frac{\|\mathbf{f}\|}{(|\alpha| - e^{-c_1T/2})}\left[1 + \frac{c_2(1 - e^{-c_1t/2})\|\mathbf{f}\|}{c_1\sqrt{2(\alpha^2 + e^{-c_1T})}}\right]^{-1}, \tag{2.214}$$

provided $|\alpha| > e^{-c_1T/2}$, for t in the interval $0 \leq t \leq T$.

(Quintanilla and Straughan, 2005a) observe that while the bound in (2.214) is not optimal, the system of equations (2.205) is nonlinear, and so an optimal bound would be hard to achieve. If instead one were to consider the equivalent problem for the Brinkman equations, i.e. take $\beta = 0$ in (2.205), we may derive an optimal estimate. We do not include details since they follow very closely the arguments of (Payne et al., 2004) for the Stokes equations. The difference is the addition of the $-u_i$ term in (2.205). The Lagrange identity and non-uniqueness proofs of (Payne et al., 2004) apply here, *mutatis mutandis*.

2.10   The interface problem

In this section we study the problem where a viscous fluid adjoins a porous medium saturated with the same fluid. In thermal convection this was addressed in the fundamental papers by (Nield, 1977) and by (Chen and Chen, 1988). One of the fundamental problems in modelling flow of a fluid over a porous medium is that the conditions at the interface between the fluid and the porous medium are a contentious matter, see e.g. (Beavers and Joseph, 1967), (Caviglia et al., 1992b),(Ciesjko and Kubik, 1999), (Jäger and Mikelic, 1998), (Jäger et al., 1999), (Jones, 1973), (McKay, 2001), (Murdoch and Soliman, 1999), (Nield and Bejan, 2006), pp. 17 – 19, (Ochoa-Tapia and Whitaker, 1995a; Ochoa-Tapia and Whitaker, 1995b; Ochoa-Tapia and Whitaker, 1997), (Saffman, 1971), (Taylor, 1971). Very good agreement with experiment is often achieved by employing the experimentally suggested condition proposed by (Beavers and Joseph, 1967), or its generalization by (Jones, 1973). (Straughan, 2001c; Straughan, 2002a), (Carr, 2004) and (Carr and Straughan, 2003) have investigated various aspects and generalisations of the Nield and Chen-Chen problems. They find that the Beavers-Joseph and Jones boundary conditions give good results over a wide range of parameters. The Beavers-Joseph condition has been successful in the slow flow of a fluid past a porous sphere (Qin and Kaloni, 1993). If one is employing a method based on linearized instability and so is using Stokes' flow, use of a Beavers-Joseph or a Jones condition is probably justified. Numerical schemes are developed for the coupled fluid flow and porous flow problems by (Discacciati et al., 2002), by (Miglio et al., 2003), by (Hoppe et al., 2007), and by (Mu and Xu, 2007). Several computational simulations are reported in these papers. Another interesting numerical contribution to porous/fluid flow is by (Das et al., 2002). This paper presents a finite volume method in three-dimensions. The porous part of the domain is allowed to be anisotropic. It is shown that flow circulation may occur inside the porous medium and the direction of flow may reverse at the interface between the porous medium and fluid. (Layton et al., 2003) prove existence for weak solutions to the problem of Darcy porous media flow coupled to the Stokes equations in a fluid with the Beavers - Joseph interface boundary condition. They also analyse in detail a finite element scheme which formulates the coupled problem as uncoupled steps in the porous and fluid regions thereby allowing a user to employ some of the many existing numerical codes for the separate flow regions. (Das and Lewis, 2007) is another recent very interesting contribution. These writers are interested in the three-dimensional flow pattern and how heterogeneities in the porous medium will affect this. To achieve their aim they interestingly employ two porous layers with different permeabilities.

The purpose of this section is to review work of (Payne and Straughan, 1998a) which studies the manner in which a solution to flow in a fluid which borders a porous medium depends on a coefficient in the Jones boundary conditions. We adopt the notation of (Payne and Straughan, 1998a) and thus, let an appropriate part of the plane $z = x_3 = 0$ denote the boundary between a porous medium occupying a bounded region Ω_2 in $\mathbb{R}^3$, and a linear viscous fluid occupying a bounded region Ω_1 in $\mathbb{R}^3$. The porous region is in $z \geq 0$ while the fluid domain is in $z < 0$, although both Ω_1 and Ω_2 are bounded. The interface between Ω_1 and Ω_2 is denoted by L while the remaining parts of the boundaries of Ω_1 and Ω_2 are denoted, respectively, by Γ_1 and Γ_2. In Ω_1 the fluid velocity is slow such that the governing equations may be taken to be those of Stokes flow. The question of Navier-Stokes flow is addressed in (Payne and Straughan, 1998a). In the porous region Ω_2 the flow is assumed to satisfy the Darcy (1856) equations.

Let (u_i, T, p) denote the velocity, temperature and pressure in Ω_1 while (u_i^m, T^m, p^m) denotes the velocity, temperature and pressure in Ω_2. The Stokes flow equations which hold in the fluid region are

$$\frac{\partial u_i}{\partial t} = -\frac{\partial p}{\partial x_i} + \mu \Delta u_i + g_i T, \qquad \frac{\partial u_i}{\partial x_i} = 0,$$
$$\frac{\partial T}{\partial t} + u_i \frac{\partial T}{\partial x_i} = \kappa \Delta T, \tag{2.215}$$

in $\Omega_1 \times (0, \mathcal{T})$, where μ is the dynamic viscosity, κ is the thermal diffusivity and g_i is the gravity vector which is scaled such that $|\mathbf{g}| \leq 1$.

The relevant Darcy equations which hold in the porous region are,

$$\frac{\mu}{k} u_i^m = -\frac{\partial p^m}{\partial x_i} + g_i T^m, \qquad \frac{\partial u_i^m}{\partial x_i} = 0,$$
$$\frac{\partial T^m}{\partial t} + u_i^m \frac{\partial T^m}{\partial x_i} = \kappa^m \Delta T^m, \tag{2.216}$$

in $\Omega_2 \times (0, \mathcal{T})$. The constant k is the permeability and κ^m is the thermal diffusivity of the porous medium.

The functions u_i, T and T^m satisfy the initial data

$$u_i(x, 0) = f_i(x), \qquad T(x, 0) = T_0(x), \quad x \in \Omega_1,$$
$$T^m(x, 0) = T_0^m(x), \quad x \in \Omega_2. \tag{2.217}$$

On the outer boundary $\Gamma_1 \cup \Gamma_2$ we consider

$$u_i = 0, \quad T = T_U(x,t), \qquad \text{on} \quad \Gamma_1 \times (0,\mathcal{T}),$$
$$u_i^m n_i = 0, \quad T^m = T_L(x,t), \qquad \text{on} \quad \Gamma_2 \times (0,\mathcal{T}), \tag{2.218}$$

for prescribed functions T_U and T_L, with n_i being the unit outward normal. The conditions on the interface L chosen by (Payne and Straughan, 1998a) are

$$u_3 = u_3^m, \qquad T = T^m, \qquad \kappa T_{,3} = \kappa^m T_{,3}^m,$$
$$p^m = p - 2\mu u_{3,3}, \qquad u_{\beta,3} + u_{3,\beta} = \frac{\alpha_1}{\sqrt{k}} u_\beta. \tag{2.219}$$

The coefficient α_1 is determined by experiment for a given fluid and a given porous solid. These boundary conditions are discussed at length in (Nield and Bejan, 2006), see also chapter 6. The condition $u_{\beta,3} + u_{3,\beta} = u_\beta \alpha_1/\sqrt{k}$ essentially derives from the work of (Jones, 1973). The motivation for this arose from (Beavers and Joseph, 1967) who argued on the basis of experimental results that

$$u_{\beta,3} = \frac{\alpha_1}{\sqrt{k}}(u_\beta - u_\beta^m), \qquad \text{on } L \tag{2.220}$$

and (Jones, 1973) generalised this to include the shear stress at the interface, i.e.

$$u_{\beta,3} + u_{3,\beta} = \frac{\alpha_1}{\sqrt{k}}(u_\beta - u_\beta^m). \tag{2.221}$$

(Nield and Bejan, 2006) write that (Saffman, 1971) argues that the last term may essentially be dropped in equation (2.220). This is the justification for $(2.219)_5$.

The object of this section is to describe an *a priori* estimate showing how (u_i, T) and (u_i^m, T^m) depend continuously on the interface coefficient α_1. To do this, let (u_i, p, T) and (u_i^m, p^m, T^m) satisfy (2.215) – (2.219) and let (v_i, q, S) and (v_i^m, q^m, S^m) solve the same boundary initial value problem with identical data functions f_i, T_0, T_0^m, T_U and T_L, but with the Jones coefficient α_1 replaced by a different value α_2. The difference variables (w_i, π, θ) and σ are defined by

$$w_i = u_i - v_i, \quad \pi = p - q, \quad \theta = T - S, \quad \sigma = \alpha_1 - \alpha_2. \tag{2.222}$$

By direct calculation one finds that (w_i, π, θ) satisfy the partial differential equations

$$\frac{\partial w_i}{\partial t} = -\frac{\partial \pi}{\partial x_i} + \mu \Delta w_i + g_i \theta,$$
$$\frac{\partial w_i}{\partial x_i} = 0, \tag{2.223}$$
$$\frac{\partial \theta}{\partial t} + u_i \frac{\partial \theta}{\partial x_i} + w_i \frac{\partial S}{\partial x_i} = \kappa \Delta \theta,$$

in $\Omega_1 \times (0, T)$,

$$\frac{\mu}{k} w_i^m = -\frac{\partial \pi^m}{\partial x_i} + g_i \theta^m,$$

$$\frac{\partial w_i^m}{\partial x_i} = 0, \tag{2.224}$$

$$\frac{\partial \theta^m}{\partial t} + u_i^m \frac{\partial \theta^m}{\partial x_i} + w_i^m \frac{\partial S^m}{\partial x_i} = \kappa^m \Delta \theta^m,$$

in $\Omega_2 \times (0, T)$.

The initial conditions become

$$w_i(x, 0) = 0, \quad \theta(x, 0) = 0, \quad x \in \Omega_1, \quad \theta^m(x, 0) = 0, \quad x \in \Omega_2. \tag{2.225}$$

On the outer boundary the relevant conditions become

$$\begin{aligned} w_i = 0, \quad \theta = 0, \quad &\text{on} \quad \Gamma_1 \times (0, T), \\ w_i^m n_i = 0, \quad \theta^m = 0, \quad &\text{on} \quad \Gamma_2 \times (0, T). \end{aligned} \tag{2.226}$$

The interface boundary conditions may be written

$$\begin{aligned} w_3 = w_3^m, \quad \theta = \theta^m, \quad \kappa \theta_{,3} = \kappa^m \theta_{,3}^m, \\ \pi^m = \pi - 2\mu w_{3,3}, \quad w_{\beta,3} + w_{3,\beta} = \frac{\alpha_1}{\sqrt{k}} w_\beta + \frac{\sigma}{\sqrt{k}} v_\beta, \end{aligned} \tag{2.227}$$

these holding on $L \times (0, T)$.

(Payne and Straughan, 1998a) establish the following theorem which demonstrates continuous dependence of a solution on the interface coefficient α_1.

Theorem 2.10.1 *Suppose $\partial T / \partial n \in L^1(\Gamma_1 \times (0, T))$ and $\partial T^m / \partial n \in L^1(\Gamma_2 \times (0, T))$. Then there exist constants $\gamma (< 2\mu/k)$, B, C and $\hat{A}$, determined in (Payne and Straughan, 1998a) such that*

$$\int_{\Omega_1} w_i w_i \, dx + B \int_0^t \int_{\Omega_1} w_i w_i \, dx \, d\eta + \gamma \int_{\Omega_2} w_i^m w_i^m \, dx$$
$$\leq \frac{C e^{Bt}}{\alpha_1 \alpha_2} \left(\int_{\Omega_1} f_i f_i \, dx + \hat{A} t T_m^2 \right) \sigma^2. \tag{2.228}$$

Furthermore, there is a constant M, depending on t, such that

$$\int_{\Omega_1} \theta^2 \, dx + \int_{\Omega_2} (\theta^m)^2 \, dx \leq \frac{M}{\alpha_1 \alpha_2} \sigma^2. \tag{2.229}$$

The proof of this theorem is technical, care must be taken with the interface terms, and we refer to (Payne and Straughan, 1998a) for full details. Nevertheless we note that the proof is interesting and is based on a combination function $\Phi(t)$ of the form

$$\Phi(t) = \int_{\Omega_1} w_i w_i \, dx + \gamma \int_0^t \int_{\Omega_2} w_i^m w_i^m \, dx \, d\eta.$$

2.11   Lower bounds on the blow-up time

(Payne and Schaefer, 2006; Payne and Schaefer, 2007) and (Payne and Song, 2007a) produce a clever argument to show that one can derive *lower bounds* for the blow-up time for a nonlinear differential equation and for the Navier-Stokes equations with nonlinear forcing terms. Prior to this work there had been many analyses of blow-up which had derived upper bounds on the blow-up time. However, the work of (Payne and Schaefer, 2006; Payne and Schaefer, 2007) and (Payne and Song, 2007a) is novel in that it produces a lower bound for the blow-up time. (Suzuki, 2006) shows how to derive a universal bound, independent of the initial data, which is useful in calculating the initial blow-up rate of a solution, whereas (Hirota and Ozawa, 2006) consider numerical techniques for estimating the blow-up time and the rate of solution increase. (Kirane et al., 2005) investigate critical exponents of Fujita type when fractional derivatives are present. (Fila and Winkler, 2008) demonstrate a solution which blows up in a finite time at a point with the solution remaining bounded elsewhere. Other interesting blow-up results and analysis showing prevention of blow-up are due to (Bhandar et al., 2004), (Boutat et al., 2004), (Tersenov, 2004).

We now consider an analogue of the (Payne and Song, 2007a) problem but for a Brinkman porous medium. The equations for the Brinkman problem with a non-zero inertia and nonlinear forces depending on temperature are, cf. equations (2.76)

$$\alpha \frac{\partial u_i}{\partial t} = -u_i + \lambda \Delta u_i - \frac{\partial p}{\partial x_i} + h_i(T),$$

$$\frac{\partial u_i}{\partial x_i} = 0, \tag{2.230}$$

$$\frac{\partial T}{\partial t} + u_i \frac{\partial T}{\partial x_i} = \Delta T + f(T).$$

In these equations u_i, T, p are velocity, temperature and pressure, α, λ are the inertia and Brinkman coefficients and $h_i(T)$ and $f(T)$ are nonlinear functions of temperature. Equations (2.230) are defined on a bounded spatial region Ω over a time interval $(0, \mathcal{T})$. The boundary conditions considered are

$$u_i = 0, \quad T = 0 \qquad \text{on } \Gamma \times (0, \mathcal{T}), \tag{2.231}$$

while the initial conditions are

$$u_i(\mathbf{x}, 0) = u_i^0(\mathbf{x}), \quad T(\mathbf{x}, 0) = T_0(\mathbf{x}) \geq 0. \tag{2.232}$$

We here only consider the Brinkman model, but one could consider a Darcy model. Also, we only consider Dirichlet conditions on the boundary whereas one could alternatively employ Neumann boundary conditions following (Payne and Schaefer, 2006), (Payne and Song, 2007a). We also note that we could employ $T = \text{constant}$ in (2.231) although care would then need to be taken with the function f.

Since both equation $(2.230)_1$ and equation $(2.230)_3$ are forced by non-linear functions of temperature, one may ask if blow-up occurs, will this be in the first instance via the velocity or the temperature field? We follow (Payne and Song, 2007a) to show this must be via the temperature.

Let t_1 be the blow-up time of the temperature T and t_2 be the blow-up time of the velocity u_i. We wish to show that $t_1 < t_2$. Suppose, therefore, this is false so that $t_2 < t_1$. Then, for $t < t_2$, we multiply equation $(2.230)_1$ by u_i and integrate over Ω to find after integrations by parts and use of the boundary conditions and $(2.230)_2$,

$$\frac{d}{dt}\frac{\alpha}{2}\|\mathbf{u}\|^2 = -\|\mathbf{u}\|^2 - \lambda\|\nabla\mathbf{u}\|^2 + \int_\Omega h_i u_i dx.$$

We employ the Poincaré inequality $\lambda_1\|\mathbf{u}\|^2 \leq \|\nabla\mathbf{u}\|^2$ and the arithmetic-geometric mean inequality for $\gamma > 0$ to now see that

$$\frac{d}{dt}\frac{\alpha}{2}\|\mathbf{u}\|^2 \leq -\left(1 + \lambda\lambda_1 - \frac{\gamma}{2}\right)\|\mathbf{u}\|^2 + \frac{\|\mathbf{h}\|^2}{2\gamma}. \qquad (2.233)$$

Pick $\gamma = (1 + \lambda\lambda_1)$ and then from (2.233) one sees that

$$\frac{d}{dt}\|\mathbf{u}\|^2 \leq -\frac{\gamma}{\alpha}\|\mathbf{u}\|^2 + \frac{\|\mathbf{h}\|^2}{\gamma\alpha}. \qquad (2.234)$$

Since $t < t_2 < t_1$, $h_i(T)$ is bounded and so $\|\mathbf{h}\|^2 \leq M^2$, for some constant M. Employ this bound in (2.234), and integrate with an integrating factor to obtain

$$\|\mathbf{u(t)}\|^2 \leq \|\mathbf{u_0}\|^2 \exp\left[-\left(\frac{1+\lambda\lambda_1}{\alpha}\right)t\right]$$
$$+ \frac{M^2}{(1+\lambda\lambda_1)^2}\left\{1 - \exp\left[-\left(\frac{1+\lambda\lambda_1}{\alpha}\right)t\right]\right\}, \qquad (2.235)$$

where $t \leq t_2$. Now let $t \to t_2$. By assumption $\|\mathbf{u(t)}\|^2$ blows up at $t = t_2$, but inequality (2.235) contradicts this. Thus, $t_1 \leq t_2$, and so t_1 is a lower bound for the blow-up time.

The conditions we now impose on the nonlinear function $f(T)$ are the same as those of (Payne and Schaefer, 2007), namely

$$f(0) = 0, \quad f(s) > 0, \qquad \text{for } s > 0, \qquad (2.236)$$

$$\int_T^\infty \frac{ds}{f(s)} \quad \text{is bounded for } T > 0, \qquad (2.237)$$

and there are constants $n > 2$ and $\beta > 0$ such that

$$f(T)\left(\int_T^\infty \frac{ds}{f(s)}\right)^{n+1} \to \infty \quad \text{as} \quad T \to 0^+, \tag{2.238}$$

$$f'(T)\int_T^\infty \frac{ds}{f(s)} \le n+1-\beta. \tag{2.239}$$

As (Payne and Schaefer, 2007) remark, from the work of (Ball, 1977) and (Kielhöfer, 1975), when the solution does cease to exist globally then the behaviour is that of blow-up.

To now derive a lower bound for the blow-up time t_1 we follow (Payne and Schaefer, 2007), (Payne and Song, 2007a). Put $R = \int_T^\infty ds/f(s)$, $v = 1/R$, and define the function $\phi(t)$ by

$$\phi(t) = \int_\Omega v^n dx.$$

By differentiation

$$\begin{aligned}
\frac{d\phi}{dt} &= n\int_\Omega v^{n-1}v_t dx \\
&= n\int_\Omega \frac{v^{n+1}}{f(T)} T_t \, dx \\
&= n\int_\Omega \frac{v^{n+1}}{f(T)} \left[\Delta T - u_i T_{,i} + f(T)\right] dx. \tag{2.240}
\end{aligned}$$

Using the chain rule one shows

$$\begin{aligned}
\int_\Omega v^{n+1}\frac{u_i T_{,i}}{f(T)}\, dx &= \frac{1}{n}\int_\Omega (v^n)_{,i} u_i dx \\
&= \frac{1}{n}\left[\int_\Omega (v^n u_i)_{,i} dx - \int_\Omega v^n u_{i,i} dx\right] = 0.
\end{aligned}$$

Thus, equation (2.240) reduces to

$$\frac{d\phi}{dt} = n\int_\Omega \frac{v^{n+1}}{f(T)}\left[\Delta T + f(T)\right] dx. \tag{2.241}$$

From this point, the estimate for t_1 effectively follows from the arguments of (Payne and Schaefer, 2007). Integrate the first term on the right of (2.241) by parts to find

$$n\int_\Omega \frac{v^{n+1}\Delta T}{f(T)}dx = -n\int_\Omega \left(\frac{v^{n+1}}{f}\right)_{,i} T_{,i}dx + n\int_\Gamma \frac{v^{n+1}}{f(T)}\frac{\partial T}{\partial v} dS, \tag{2.242}$$

where $\partial/\partial v$ denotes the unit outward normal derivative. Thanks to condition (2.238) the last term in (2.242) is zero. The first term in (2.242) is expanded and then (2.239) is employed to find

$$n \int_\Omega \frac{v^{n+1}\Delta T}{f(T)} dx = n \int_\Omega \frac{v^{n+1} f'(T)}{f^2} T_{,i} T_{,i} dx$$

$$- n(n+1) \int_\Omega \frac{T_{,i}}{f^2} v^{n+2} T_{,i} dx,$$

$$\leq n \int_\Omega \frac{v^{n+2}}{f^2} T_{,i} T_{,i} [n+1-\beta] dx - n(n+1) \int_\Omega \frac{v^{n+2}}{f^2} T_{,i} T_{,i} dx$$

$$= -\beta n \int_\Omega \frac{v^{n+2}}{f^2} T_{,i} T_{,i} dx. \tag{2.243}$$

Inequality (2.243) is now employed in equation (2.241) to find

$$\frac{d\phi}{dt} \leq -\beta n \int_\Omega \frac{v^{n+2}}{f^2} T_{,i} T_{,i} dx + n \int_\Omega v^{n+1} dx.$$

Noting that $v^{(n/2+1)} T_{,i}/f = 2(v^{n/2})_{,i}/n$ this inequality is rearranged as

$$\frac{d\phi}{dt} \leq -\frac{4\beta}{n} \int_\Omega (v^{n/2})_{,i} (v^{n/2})_{,i} dx + n \int_\Omega v^{n+1} dx. \tag{2.244}$$

If m denotes the measure of Ω then from Hölder's inequality and the Cauchy-Schwarz inequality one sees

$$\int_\Omega v^{n+1} dx \leq m^{(n-2)/3n} \left(\int_\Omega v^{3n/2} dx \right)^{2(n+1)/3n}$$

$$\leq m^{(n-2)/3n} \left(\int_\Omega v^{2n} dx \int_\Omega v^n dx \right)^{(n+1)/3n}. \tag{2.245}$$

We next use the Sobolev inequality

$$\int_\Omega \psi^4 dx \leq C \left(\int_\Omega \psi^2 dx \right)^{1/2} \left(\int_\Omega \psi_{,i} \psi_{,i} dx \right)^{3/2},$$

where a value for C is calculated in (Payne, 1964), taking $\psi = v^{n/2}$ to find

$$\int_\Omega v^{2n} dx \leq C \left(\int_\Omega v^n dx \right)^{1/2} \left[\int_\Omega (v^{n/2})_{,i} (v^{n/2})_{,i} dx \right]^{3/2}.$$

This estimate is now used in (2.245) to obtain

$$\int_\Omega v^{n+1} dx \leq m^{(n-2)/3n} C^{(n+1)/3n} \left(\alpha_1 \int_\Omega |\nabla v^{n/2}|^2 dx \right)^{(n+1)/2n}$$

$$\times \left(\frac{1}{\alpha_1} \int_\Omega v^n dx \right)^{(n+1)/2n}, \tag{2.246}$$

where the constant $\alpha_1 > 0$ has been added to allow removal of the $|\nabla v^{n/2}|^2$ term. Next, employ Young's inequality

$$XY \leq \frac{X^p}{p} + \frac{Y^s}{s}, \qquad \frac{1}{p} + \frac{1}{s} = 1,$$

with $X = \alpha_1 \int_\Omega |\nabla v^{n/2}|^2 dx$, $Y = \int_\Omega v^n dx/\alpha_1$, $p = 2n/(n+1) > 1$, and $s = 2n/(n-1)$. Then, from (2.246) we derive

$$\int_\Omega v^{n+1} dx \leq m^{(n-2)/3n} C^{(n+1)/3n} \left(\frac{n+1}{2n}\right) \alpha_1^{2n/(n+1)} \int_\Omega |\nabla v^{n/2}|^2 dx$$

$$+ m^{(n-2)/3n} C^{(n+1)/3n} \left(\frac{n-1}{2n}\right) \frac{1}{\alpha_1^{2n/(n-1)}}$$

$$\times \left(\int_\Omega v^n dx\right)^{(n+1)/(n-1)}. \tag{2.247}$$

Inequality (2.247) is next employed in inequality (2.244) to find

$$\frac{d\phi}{dt} \leq \left[nm^{(n-2)/3n} C^{(n+1)/3n}\left(\frac{n+1}{2n}\right)\alpha_1^{2n/(n+1)} - \frac{4\beta}{n}\right] \int_\Omega |\nabla v^{n/2}|^2 dx$$

$$+ nm^{(n-2)/3n} C^{(n+1)/3n}\left(\frac{n-1}{2n}\right)\frac{1}{\alpha_1^{2n/(n-1)}} \phi^{(n+1)/(n-1)}. \tag{2.248}$$

The constant α_1 is now selected to make the first term on the right of (2.248) zero. Thus, for K computable, from (2.248) we derive

$$\frac{d\phi}{dt} \leq K\phi^{(n+1)/(n-1)}.$$

This inequality is integrated to obtain

$$\frac{1}{[\phi(0)]^{2/(n-1)}} - \frac{1}{[\phi(t)]^{2/(n-1)}} \leq \frac{2Kt}{(n-1)}. \tag{2.249}$$

When $t \to t_1$ (the blow-up time), then (2.249) yields the lower bound $\hat{t}$ for t_1, where

$$t_1 \geq \hat{t} = \left(\frac{n-1}{2K}\right)\frac{1}{[\phi(0)]^{2/(n-1)}}$$

$$= \left(\frac{n-1}{2K}\right)\left(\int_\Omega \left[\int_{T_0(x)}^\infty \frac{ds}{f(s)}\right]^{-n} dx\right)^{-2/(n-1)}. \tag{2.250}$$

The above derivation simply adapts the clever analyses of (Payne and Schaefer, 2007) and (Payne and Song, 2007a) to a Brinkman model.

A lower bound with a more direct derivation may be found by adapting the method of (Payne, 1975), pp. 49, 50. To do this we work with equation (2.230)$_3$. The assumption on the nonlinearity is now inequality (8.31) of (Payne, 1975), namely

$$\int_\Omega T^{2p-1} f(T) dx \leq \int_\Omega |T|^{2p+\gamma}, \tag{2.251}$$

where γ is a positive constant, and (2.251) holds for any positive integer p. Introduce the function

$$\Phi_p(t) = \int_\Omega T^{2p} dx.$$

Then,

$$\frac{d\Phi_p}{dt} = 2p \int_\Omega T^{2p-1} \frac{\partial T}{\partial t} dx$$

$$= 2p \int_\Omega T^{2p-1} \Delta T \, dx + 2p \int_\Omega T^{2p-1} f(T) dx$$

$$- \int_\Omega T^{2p-1} u_i \frac{\partial T}{\partial x_i} dx. \tag{2.252}$$

Integrating by parts and using the boundary conditions,

$$\int_\Omega T^{2p-1} u_i \frac{\partial T}{\partial x_i} \, dx = \frac{1}{2p} \int_\Omega u_i \frac{\partial}{\partial x_i} T^{2p} \, dx$$

$$= \frac{1}{2p} \int_\Gamma u_i n_i T^{2p} \, dS - \frac{1}{2p} \int_\Omega u_{i,i} T^{2p} \, dx$$

$$= 0. \tag{2.253}$$

Further integration by parts and use of the boundary conditions yield

$$2p \int_\Omega T^{2p-1} \Delta T \, dx = 2p \int_\Gamma T^{2p-1} \frac{\partial T}{\partial n} dS - 2p(2p-1) \int_\Omega T^{2p-2} T_{,i} T_{,i} dx$$

$$= -2p(2p-1) \int_\Omega T^{2p-2} T_{,i} T_{,i} dx. \tag{2.254}$$

Now, use (2.253), (2.254) and inequality (2.251) in equation (2.252) to see that

$$\frac{d\Phi_p}{dt} \le -\frac{2(2p-1)}{p} \int_\Omega T_{,i}^p T_{,i}^p \, dx + 2p \int_\Omega |T|^{2p+\gamma} \, dx$$

$$\le 2p \int_\Omega T^{2p+\gamma} dx. \tag{2.255}$$

Next, put

$$T_*(t) = \sup_{x \in \Omega} |T(\mathbf{x}, t)|.$$

Then from (2.255) we may derive

$$\frac{d\Phi_p}{dt} \le 2p T_*^\gamma \Phi_p.$$

An integration of this inequality yields

$$\Phi_p(t) \le \Phi_p(0) \exp\left[2p \int_0^t T_*^\gamma(s) ds \right].$$

Raise both sides of this inequality to the power $1/2p$ and then let $p \to \infty$. In this manner we obtain

$$T_*(t) \leq T_*(0) \exp \left[\int_0^t T_*^\gamma(s)ds \right]. \qquad (2.256)$$

Since t_1 is the blow-up time for T we must have $T_*(t) \to \infty$ as $t \to t_1$, and assuming T is sufficiently regular,

$$\int_0^{t_1} T_*^\gamma(s)ds = \infty. \qquad (2.257)$$

The next step is to raise both sides of inequality (2.256) to the power γ and then, provided $t \leq t_1$, this inequality yields

$$T_*^\gamma(t) \exp \left[-\gamma \int_0^t T_*^\gamma(s)ds \right] \leq T_*^\gamma(0).$$

A further integration of this inequality over $0, t < t_1$, leads to

$$1 - \exp \left[-\gamma \int_0^t T_*^\gamma(s)ds \right] \leq \gamma t T_*^\gamma(0).$$

Let now $t \to t_1$ and employ condition (2.257). In this way we find

$$\frac{1}{\gamma T_*^\gamma(0)} \leq t_1. \qquad (2.258)$$

Inequality (2.258) represents an alternative lower bound for the blow-up time t_1 to the estimate (2.250).

The above proof is a straightforward adaptation of the demonstration of (Payne, 1975), pp. 49, 50.

2.12   Uniqueness in compressible porous flows

So far in this book we have concentrated on fluid flow in a porous medium where the fluid may be treated as incompressible. However, sound propagation through a porous medium is one important example of a situation where flow of a compressible gas in a porous material is necessary. We study in detail wave motion of a compressible fluid in a porous medium in chapter 8 with related material given in chapter 7. Therefore, in this chapter we commence a study of the well posedness of a theory for compressible flow in a porous medium by establishing a uniqueness theorem. Since the wave motion in chapter 8 is typically for sound waves propagating in an infinite medium we here establish a uniqueness theorem for flow in an infinite spatial region. To establish our theorem we appeal to a beautiful result of Dario Graffi, (Graffi, 1960) although Graffi's paper is conveniently found in the selected works, (Graffi, 1999), pages 273 – 280.

The model for compressible flow in a porous material is taken from (De Ville, 1996). It consists of the equations for flow of a barotropic perfect fluid, cf. (Fabrizio, 1994), to which have been added a Darcy term and a Forchheimer term to represent the interaction with the porous matrix. This model is one of equivalent fluid type, and these are discussed in greater detail in section 8.1. The equations we employ are those of (De Ville, 1996), equations (4) and (5), although we assume the fluid is polytropic so that the pressure - density relation is of form $p = a\rho^\gamma$, where p and ρ are pressure and density, a is a positive constant, and γ is a constant with $1 < \gamma < 2$. With v_i being the fluid velocity, k, λ, b_1 positive constants the model of (De Ville, 1996) may be written

$$\frac{\partial \rho}{\partial t} + v_i \frac{\partial \rho}{\partial x_i} + \rho \frac{\partial v_i}{\partial x_i} = 0,$$

$$\frac{\partial v_i}{\partial t} + v_j \frac{\partial v_i}{\partial x_j} + b_1 v v_i + \frac{k}{\rho} v_i = -\frac{a\gamma}{\lambda} \rho^{\gamma-2} \rho_{,i},$$

(2.259)

where we adopt the (Graffi, 1960; Graffi, 1999) notation $v = |\mathbf{v}|$.

(Graffi, 1960; Graffi, 1999) establishes uniqueness for (2.259) when $b_1 = 0, k = 0$. The extension to include these terms is non-trivial and given below. Nevertheless, we extend the (Graffi, 1999) method and employ his notation. Henceforth, we employ the notation (Graffi, 1999) to denote paper 22 of the selected works, pages 273 – 280.

Equations (2.259) are defined on a space - time domain. The time domain is $(0, \mathcal{T})$ and the spatial domain D is either $\mathbb{R}^3$ or the exterior of a bounded domain σ_0 in $\mathbb{R}^3$. In either case D is an unbounded domain. We impose the same hypotheses as (Graffi, 1999), and in particular his hypotheses (a) – (g). However, we have already mentioned hypothesis (b) which states that the pressure is polytropic and we have no need for hypothesis (c) which concerns the body force, since one may regard equation (2.259)$_2$ as defining a particular form for the body force. The remaining hypotheses (a) and (d) – (g) are stated below.

(a) In the domain $D \times (0, \mathcal{T})$ the velocity $\mathbf{v}$ and density ρ are uniformly bounded together with their first derivatives in space and time.

(d) If D has an interior boundary $\partial \sigma_0$, then on $\partial \sigma_0$ we assign $\mathbf{v} \cdot \mathbf{n}$, $\mathbf{n}$ being the unit outward normal to $\partial \sigma_0$, and where the fluid enters so that $\mathbf{v} \cdot \mathbf{n} < 0$ we assign ρ and $\mathbf{v}$.

(e) The values of $\rho(\mathbf{x}, 0)$ and $v_i(\mathbf{x}, 0)$ are assigned.

(f) The density ρ is positive and $|\nabla \rho|/\rho$ is bounded in $D \times (0, \mathcal{T})$.

(g) Let R denote the distance from the origin in D, then $\rho \geq c/R^\beta$, where c is a positive constant and $\beta \geq 0$ is a constant.

Let us observe that the last relation is physically necessary. It allows the density to vanish as $R \to \infty$ although not in an arbitrary way. In fact, it is condition (g) which makes the extension of the (Graffi, 1999) result to system (2.259) non-trivial. One now has to also handle the terms $b_1 v v_i$ and $k v_i/\rho$.

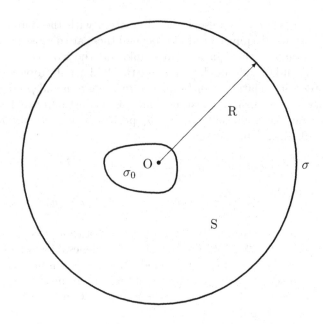

Figure 2.1. Geometry for uniqueness proof

We now denote by S the intersection of the ball of radius R with D. The geometrical configuration is shown in figure 2.1.

The outer boundary of S, i.e. the spherical surface of radius R, is denoted by σ.

To study uniqueness we follow (Graffi, 1999) and let ρ, v_i and $\rho+\rho_1, v_i+v_i^1$ be two solutions to equations (2.259) which both satisfy hypotheses (a) and (d) – (g). By subtraction we find ρ_1 and v_i^1 satisfy the equations

$$\frac{\partial \rho_1}{\partial t} + \frac{\partial}{\partial x_i}\left[(\rho+\rho_1)(v_i+v_i^1) - \rho v_i\right] = 0,$$

$$\frac{\partial v_i^1}{\partial t} + v_j^1 \frac{\partial(v_i+v_i^1)}{\partial x_j} + v_j \frac{\partial v_i^1}{\partial x_j}$$

$$= -b\left\{(\rho+\rho_1)^{\gamma-2}\frac{\partial}{\partial x_i}(\rho+\rho_1) - \rho^{\gamma-2}\frac{\partial \rho}{\partial x_i}\right\} \qquad (2.260)$$

$$- k\left\{\left(\frac{v_i+v_i^1}{\rho+\rho_1}\right) - \frac{v_i}{\rho}\right\}$$

$$- b_1\left\{(v+v_1)(v_i+v_i^1) - vv_i\right\},$$

where we have put $b = a\gamma/\lambda$. The proof of (Graffi, 1999) is very clever and balances the $\mathbf{v}_1 \cdot \nabla \rho_1$ term which arises from $(2.260)_1$ with an equivalent term from $(2.260)_2$. This necessitates the use of a weighted L^2 energy for ρ_1, weighted by both $\rho^{\gamma-3}$ and $(\rho + \rho_1)^{\gamma-3}$.

We begin by multiplying $(2.260)_2$ by v_i^1 and find

$$
\begin{aligned}
\frac{1}{2}\frac{\partial}{\partial t}v_i^2 = & - v_i^1 v_j^1 (v_i + v_i^1)_{,j} - v_i^1 v_j v_{i,j}^1 \\
& - b v_i^1 \{(\rho + \rho_1)^{\gamma-2}(\rho + \rho_1)_{,i} - \rho^{\gamma-2}\rho_{,i}\} \\
& - k v_i^1 \left\{ \left(\frac{v_i + v_i^1}{\rho + \rho_1} \right) - \frac{v_i}{\rho} \right\} \\
& - b_1 v_i^1 \{(v + v_1)(v_i + v_i^1) - v v_i\}.
\end{aligned}
\tag{2.261}
$$

Employ the rearrangement (5) of (Graffi, 1999),

$$
\begin{aligned}
(\rho + \rho_1)^{\gamma-2}&(\rho + \rho_1)_{,i} - \rho^{\gamma-2}\rho_{,i} \\
& = \frac{1}{2}\{(\rho + \rho_1)^{\gamma-2} - \rho^{\gamma-2}\}\{(\rho + \rho_1)_{,i} + \rho_{,i}\} \\
& \quad + \frac{1}{2}\rho^{\gamma-2}\rho_{1,i} + (\rho + \rho_1)^{\gamma-2}\rho_{1,i}.
\end{aligned}
\tag{2.262}
$$

The first term on the right of (2.262) is handled by firstly noting that from hypothesis (f) there is a positive constant n such that

$$
|(\rho + \rho_1)_{,i}| \le n(\rho + \rho_1), \qquad |\rho_{,i}| \le n\rho,
$$

then

$$
\begin{aligned}
\big|\{(\rho + \rho_1)^{\gamma-2} &- \rho^{\gamma-2}\}\{(\rho + \rho_1)_{,i} + \rho_{,i}\}\big| \\
& \le n(\rho + \rho_1 + \rho)|(\rho + \rho_1)^{\gamma-2} - \rho^{\gamma-2}| \\
& = n|(\rho + \rho_1)^{\gamma-1} - \rho^{\gamma-1} + \rho(\rho + \rho_1)^{\gamma-2} - (\rho + \rho_1)\rho^{\gamma-2}|.
\end{aligned}
\tag{2.263}
$$

To bound the terms on the right of (2.263) one uses the intermediate value theorem, and for $0 < \theta < 1$, and $0 < \theta' < 1$, one finds

$$
|(\rho + \rho_1)^{\gamma-1} - \rho^{\gamma-1}| \le |(\gamma - 1)(\rho + \theta\rho_1)^{\gamma-2}\rho_1|,
\tag{2.264}
$$

and

$$
\begin{aligned}
|\rho(\rho + \rho_1)^{\gamma-2} - (\rho + \rho_1)\rho^{\gamma-2}| & = \rho(\rho + \rho_1)\left| \frac{1}{(\rho + \rho_1)^{3-\gamma}} - \frac{1}{\rho^{3-\gamma}} \right| \\
& = \rho(\rho + \rho_1)\left| \frac{(\rho + \rho_1)^{3-\gamma} - \rho^{3-\gamma}}{(\rho + \rho_1)^{3-\gamma}\rho^{3-\gamma}} \right| \\
& = \left| \frac{(3 - \gamma)(\rho + \theta'\rho_1)^{2-\gamma}\rho_1}{(\rho + \rho_1)^{2-\gamma}\rho^{2-\gamma}} \right|.
\end{aligned}
\tag{2.265}
$$

Combining (2.264) and (2.265) in (2.263) one then obtains

$$
\begin{aligned}
&\left|\{(\rho + \rho_1)^{\gamma-2} - \rho^{\gamma-2}\}\{(\rho + \rho_1)_{,i} + \rho_{,i}\}\right| \\
&\leq n\left|(\gamma - 1)(\rho + \theta\rho_1)^{\gamma-2} + \frac{(3 - \gamma)(\rho + \theta'\rho_1)^{2-\gamma}}{(\rho + \rho_1)^{2-\gamma}\rho^{2-\gamma}}\right||\rho_1|.
\end{aligned} \tag{2.266}
$$

If ρ_1 is positive then the greater value of the right of (2.266) is achieved with $\theta = 0, \theta' = 1$ (since $\gamma - 2 < 0$) whereas if ρ_1 is negative we select $\theta = 1, \theta' = 0$, which in turn yield,

$$
\begin{aligned}
\left|\{(\rho + \rho_1)^{\gamma-2} - \rho^{\gamma-2}\}\{(\rho + \rho_1)_{,i} + \rho_{,i}\}\right| &\leq 2n\rho^{\gamma-2}|\rho_1|, \\
\left|\{(\rho + \rho_1)^{\gamma-2} - \rho^{\gamma-2}\}\{(\rho + \rho_1)_{,i} + \rho_{,i}\}\right| &\leq 2n(\rho + \rho_1)^{\gamma-2}|\rho_1|.
\end{aligned}
$$

These results together in (2.266) lead to

$$
\begin{aligned}
&\left|\{(\rho + \rho_1)^{\gamma-2} - \rho^{\gamma-2}\}\{(\rho + \rho_1)_{,i} + \rho_{,i}\}\right| \\
&\leq 2n\{(\rho + \rho_1)^{\gamma-2} + \rho^{\gamma-2}\}|\rho_1|.
\end{aligned} \tag{2.267}
$$

We now see that from (2.267)

$$
\begin{aligned}
&\left|-\frac{b}{2}v_i^1\{(\rho + \rho_1)^{\gamma-2}(\rho + \rho_1)_{,i} - \rho^{\gamma-2}\rho_{,i}\}\right| \\
&\leq bn\{(\rho + \rho_1)^{\gamma-2} + \rho^{\gamma-2}\}|\rho_1 \mathbf{v}_1| \\
&\leq \frac{bn}{2}\{(\rho + \rho_1)^{2\gamma-4}\rho_1^2 + \rho^{2\gamma-4}\rho_1^2 + 2v_1^2\},
\end{aligned} \tag{2.268}
$$

where in the last line the arithmetic-geometric mean inequality has been employed. Thus, from (2.268), (2.262) and (2.261) we obtain

$$
\begin{aligned}
\frac{1}{2}\frac{\partial}{\partial t}v_1^2 \leq &-v_i^1 v_j^1(v_i + v_i^1)_{,j} - v_i^1 v_j v_{i,j}^1 \\
&+ \frac{bn}{2}\{(\rho + \rho_1)^{2\gamma-4}\rho_1^2 + \rho^{2\gamma-4}\rho_1^2 + 2v_1^2\} \\
&- \frac{b}{2}v_i^1\{\rho^{\gamma-2}\rho_{1,i} + (\rho + \rho_1)^{\gamma-2}\rho_{1,i}\} \\
&- kv_i^1\left\{\left(\frac{v_i + v_i^1}{\rho + \rho_1}\right) - \frac{v_i}{\rho}\right\} \\
&- b_1 v_i^1\{(v + v_1)(v_i + v_i^1) - vv_i\}.
\end{aligned} \tag{2.269}
$$

The first two terms on the right of (2.269) are written as

$$
-v_i^1 v_j^1(v_i + v_i^1)_{,j} - \frac{1}{2}\frac{\partial}{\partial x_j}(v_j v_1^2) + \frac{1}{2}v_{j,j}v_1^2. \tag{2.270}
$$

The final term of (2.269) is handled with the identity, (Payne and Straughan, 1999a),

$$[(v+v_1)(v_i+v_i^1)-vv_i]v_i^1 = \frac{1}{2}(v+v_1+v)v_1^2$$
$$+ \frac{1}{2}(|\mathbf{v}+\mathbf{v_1}|-|\mathbf{v}|)^2(v+v_1+v). \qquad (2.271)$$

Thus, recalling hypothesis (a) the gradients of $\mathbf{v}$ and $\mathbf{v}+\mathbf{v}^1$ are bounded in (2.270) and then from employment of (2.271) and (2.270) in inequality (2.269) we deduce that, after integration over S,

$$\frac{d}{dt}\frac{1}{2}\int_S v_1^2 dx \leq \frac{3}{2}N_1\int_S v_1^2 dx - \frac{1}{2}\int_S \frac{\partial}{\partial x_j}(v_j v_1^2)dx$$
$$+ \frac{bn}{2}\int_S \left[(\rho+\rho_1)^{2\gamma-4}\rho_1^2 + \rho^{2\gamma-4}\rho_1^2 + 2v_1^2\right]dx$$
$$- \frac{b}{2}\int_S v_i^1\left[\rho^{\gamma-2}\rho_{1,i} + (\rho+\rho_1)^{\gamma-2}\rho_{1,i}\right]dx$$
$$- k\int_S v_i^1\left[\left(\frac{v_i+v_i^1}{\rho+\rho_1}\right) - \frac{v_i}{\rho}\right]dx, \qquad (2.272)$$

where N_1 is a bound for $|\nabla\mathbf{v}|$ and $|\nabla(\mathbf{v}+\mathbf{v}^1)|$ and we have discarded the last term of (2.269) thanks to (2.271). (In studying continuous dependence one may desire to retain the right hand side of (2.271) and then use the effect of v_1 in L^3, cf. section 4.6.2.)

To handle the last term in (2.272) we note

$$- k\int_S v_i^1\left[\left(\frac{v_i+v_i^1}{\rho+\rho_1}\right) - \frac{v_i}{\rho}\right]dx$$
$$= -k\int_S \frac{v_1^2}{(\rho+\rho_1)}dx + k\int_S \frac{\rho_1 v_i v_i^1}{\rho(\rho+\rho_1)}dx. \qquad (2.273)$$

The arithmetic-geometric mean inequality is used on the last term in the form

$$k\int_S \frac{\rho_1 v_i v_i^1}{\rho(\rho+\rho_1)}dx \leq k\int_S \frac{v_1^2}{(\rho+\rho_1)}dx + \frac{k}{4}\int_S \frac{v^2\rho_1^2}{\rho^2(\rho+\rho_1)}dx.$$

This inequality is inserted in (2.273) and the result employed in (2.272) to find

$$\frac{d}{dt}\frac{1}{2}\int_S v_i^2 dx \le \frac{3}{2}N_1\int_S v_i^2 dx - \frac{1}{2}\int_{\sigma\cup\partial\sigma_0} n_i v_i^2 v_i dS$$
$$+ \frac{bn}{2}\int_S \left[(\rho+\rho_1)^{2\gamma-4}\rho_1^2 + \rho^{2\gamma-4}\rho_1^2 + 2v_i^2\right]dx$$
$$-\frac{b}{2}\int_S v_i^1\left[\rho^{\gamma-2}\rho_{1,i} + (\rho+\rho_1)^{\gamma-2}\rho_{1,i}\right]dx$$
$$+\frac{k}{4}\int_S \frac{v^2\rho_1^2}{\rho^2(\rho+\rho_1)} dx .$$

(2.274)

To continue we note that from $(2.259)_2$,

$$kv_i = -\rho v_{i,t} - \rho v_j v_{i,j} - b_1\rho v v_i - b\rho^{\gamma-1}\rho_{,i} .$$

Thus, recollecting hypotheses (a) and (f) we see there are constants n_1, n_2 such that

$$v \le n_1\rho + n_2\rho^\gamma .$$

Thus, there are further constants $m_1, m_2, m_3, \ell_1, \ell_2$ and ℓ_3 such that

$$v^2 \le m_1\rho^2 + m_2\rho^{\gamma+1} + m_3\rho^{2\gamma},$$

and

$$\frac{kv^2}{4\rho^2(\rho+\rho_1)} \le \frac{\ell_1 + \ell_2\rho^{\gamma-1} + \ell_3\rho^{2\gamma-2}}{(\rho+\rho_1)} .$$

(2.275)

To incorporate the (Graffi, 1999) weights $\rho^{\gamma-3}$, $(\rho+\rho_1)^{\gamma-3}$ we write

$$\frac{\ell_1}{\rho+\rho_1} = (\rho+\rho_1)^{\gamma-3} \cdot \ell_1(\rho+\rho_1)^{2-\gamma} .$$

(2.276)

Then we use Young's inequality for arbitrary $\alpha > 0$,

$$\frac{\ell_3\rho^{2\gamma-2}}{\rho+\rho_1} \le \ell_3\left[\frac{(\alpha\rho^{2\gamma-2})^p}{p} + \frac{[\alpha^{-1}(\rho+\rho_1)^{-1}]^q}{q}\right]$$

$p^{-1} + q^{-1} = 1$. Pick $q = 3 - \gamma > 1$, then $p = (3-\gamma)/(2-\gamma) > 1$. Thus,

$$\frac{\ell_3\rho^{2\gamma-2}}{\rho+\rho_1} \le \frac{\ell_3}{(3-\gamma)\alpha^{(3-\gamma)}}(\rho+\rho_1)^{\gamma-3}$$
$$+ \frac{\ell_3(2-\gamma)\alpha^{(3-\gamma)/(2-\gamma)}}{(3-\gamma)}\rho^{\gamma(3-\gamma)/(2-\gamma)} \cdot \rho^{\gamma-3} .$$

(2.277)

A similar calculation utilizing Young's inequality shows

$$\frac{\ell_2\rho^{\gamma-1}}{\rho+\rho_1} \le \frac{\ell_2}{(3-\gamma)\beta^{(3-\gamma)}}(\rho+\rho_1)^{\gamma-3}$$
$$+ \frac{\ell_2(2-\gamma)\beta^{(3-\gamma)/(2-\gamma)}}{(3-\gamma)}\rho^{(3-\gamma)/(2-\gamma)} \cdot \rho^{\gamma-3} .$$

(2.278)

Thus, (2.275) – (2.278) in inequality (2.274) give

$$
\frac{d}{dt}\frac{1}{2}\int_S v_i^2 dx \le \frac{3}{2}N_1\int_S v_1^2 dx - \frac{1}{2}\int_{\sigma\cup\partial\sigma_0} n_i v_1^2 v_i dS
$$

$$
+ \frac{bn}{2}\int_S \left[(\rho+\rho_1)^{2\gamma-4}\rho_1^2 + \rho^{2\gamma-4}\rho_1^2 + 2v_1^2\right]dx
$$

$$
- \frac{b}{2}\int_S v_i^1\left[\rho^{\gamma-2}\rho_{1,i} + (\rho+\rho_1)^{\gamma-2}\rho_{1,i}\right]dx
$$

$$
+ \int_S (\rho+\rho_1)^{\gamma-3}\left[\ell_1(\rho+\rho_1)^{2-\gamma}\right. \tag{2.279}
$$

$$
+ \frac{\ell_2}{(3-\gamma)\beta^{3-\gamma}} + \frac{\ell_3}{(3-\gamma)\alpha^{3-\gamma}}\left.\right]\rho_1^2 dx
$$

$$
\int_S \rho^{\gamma-3}\frac{(2-\gamma)}{(3-\gamma)}\left[\ell_2\beta^{(3-\gamma)/(2-\gamma)}\rho^{(3-\gamma)/(2-\gamma)}\right.
$$

$$
+ \ell_3\alpha^{(3-\gamma)/(2-\gamma)}\rho^{\gamma(3-\gamma)/(2-\gamma)}\left.\right]\rho_1^2 dx .
$$

The next step is to multiply equation $(2.260)_1$ by $\rho^{\gamma-3}\rho_1$ and then by $(\rho+\rho_1)^{\gamma-3}\rho_1$ and integrate over S. This part of the calculation follows that of (Graffi, 1999).

Upon multiplying $(2.260)_1$ by $\rho^{\gamma-3}\rho_1$ one may show that

$$
\frac{1}{2}\frac{\partial}{\partial t}(\rho^{\gamma-3}\rho_1^2) = \rho^{\gamma-2}\rho_{1,i}v_i^1
$$

$$
- \frac{\partial}{\partial x_i}\left[\rho^{\gamma-2}\rho_1 v_i^1 + \frac{1}{2}\rho^{\gamma-3}\rho_1^2(v_i + v_i^1)\right]
$$

$$
+ \frac{\rho_1^2}{2}\frac{\partial}{\partial t}\rho^{\gamma-3} + \left[(\rho^{\gamma-2})_{,i} - \rho^{\gamma-3}\rho_{,i}\right]\rho_1 v_i^1 \tag{2.280}
$$

$$
- \frac{1}{2}\rho^{\gamma-3}\rho_1^2(v_i + v_i^1)_{,i} + \frac{1}{2}(\rho^{\gamma-3})_{,i}(v_i + v_i^1)\rho_1^2 .
$$

For the fourth term on the right,

$$
\left|\left[(\rho^{\gamma-2})_{,i} - \rho^{\gamma-3}\rho_{,i}\right]\rho_1 v_i^1\right| = |(3-\gamma)\rho^{\gamma-3}\rho_{,i}v_i\rho_1|
$$

$$
\le n(3-\gamma)\rho^{\gamma-2}|\rho_1 v_1|
$$

$$
\le \frac{n}{2}(3-\gamma)\left[\rho^{2\gamma-4}\rho_1^2 + v_1^2\right], \tag{2.281}
$$

where hypothesis (f) has been employed. Further, using hypotheses (a) and (f), we have for constants q_1, q_2,

$$
\left|-\frac{1}{2}\rho^{\gamma-3}\rho_1^2(v_i + v_i^1)_{,i}\right| \le q_1\rho^{\gamma-3}\rho_1^2, \tag{2.282}
$$

$$
\left|\frac{1}{2}(\rho^{\gamma-3})_{,i}(v_i + v_i^1)\rho_1^2\right| \le q_2\rho^{\gamma-3}\rho_1^2. \tag{2.283}
$$

The third term on the right of (2.280) is handled by noting

$$\frac{\rho_1^2}{2}\frac{\partial}{\partial t}\rho^{\gamma-3} \le \frac{(3-\gamma)}{2}\rho_1^2\left|\frac{\rho_t}{\rho}\right|\rho^{\gamma-3}.$$

Then, from (2.259)$_1$,

$$\frac{\rho_t}{\rho} = -\frac{v_i}{\rho}\rho_{,i} - v_{i,i}.$$

Using hypotheses (a) and (f),

$$\left|\frac{\rho_t}{\rho}\right| \le q_3,$$

for a constant q_3. Then, for a further constant $q_4 > 0$,

$$\frac{\rho_1^2}{2}\frac{\partial}{\partial t}\rho^{\gamma-3} \le q_4\rho^{\gamma-3}\rho_1^2. \tag{2.284}$$

Thus, combining (2.281) – (2.284) in equation (2.280) we find

$$\frac{1}{2}\frac{\partial}{\partial t}(\rho^{\gamma-3}\rho_1^2) = \rho^{\gamma-2}\rho_{1,i}v_i^1$$
$$-\frac{\partial}{\partial x_i}\left[\rho^{\gamma-2}\rho_1 v_i^1 + \frac{1}{2}\rho^{\gamma-3}\rho_1^2(v_i + v_i^1)\right] \tag{2.285}$$
$$+ m_2\rho^{2\gamma-4}\rho_1^2 + m_3v_1^2 + m_4\rho^{\gamma-3}\rho_1^2.$$

Similarly, we multiply equation (2.260)$_1$ by $(\rho+\rho_1)^{\gamma-3}\rho_1$ and obtain

$$\frac{1}{2}\frac{\partial}{\partial t}\left[(\rho+\rho_1)^{\gamma-3}\rho_1^2\right] = -\frac{\rho_1^2}{2}\frac{\partial}{\partial t}(\rho+\rho_1)^{\gamma-3}$$
$$+ (\rho+\rho_1)^{\gamma-3}\rho_1\left[\rho v_i - (\rho+\rho_1)(v_i + v_i^1)\right]_{,i}.$$

The last term of this expression may be rewritten

$$(\rho+\rho_1)^{\gamma-3}\rho_1\left\{-\left[(\rho+\rho_1)v_i^1\right]_{,i} - (\rho_1 v_i)_{,i}\right\}$$
$$= -\left[(\rho+\rho_1)^{\gamma-2}\rho_1 v_i^1\right]_{,i} + (\rho+\rho_1)^{\gamma-2}v_i^1\rho_{1,i}$$
$$- (\rho+\rho_1)^{\gamma-3}\rho_1^2 v_{i,i} - (\rho+\rho_1)^{\gamma-3}\frac{v_i}{2}(\rho_1^2)_{,i}$$
$$= -\frac{\partial}{\partial x_i}\left[(\rho+\rho_1)^{\gamma-2}\rho_1 v_i^1 + \frac{1}{2}(\rho+\rho_1)^{\gamma-3}v_i\rho_1^2\right]$$
$$- \frac{1}{2}(\rho+\rho_1)^{\gamma-3}v_{i,i}\rho_1^2$$
$$+ \frac{(\gamma-3)}{2}v_i\frac{(\rho+\rho_1)_{,i}}{(\rho+\rho_1)}(\rho+\rho_1)^{\gamma-3}\rho_1^2 + (\rho+\rho_1)^{\gamma-2}v_i^1\rho_{1,i}.$$

Hence, we find

$$
\begin{aligned}
\frac{1}{2}\frac{\partial}{\partial t}\big[(\rho+\rho_1)^{\gamma-3}\rho_1^2\big] &= -\frac{\rho_1^2}{2}\frac{\partial}{\partial t}(\rho+\rho_1)^{\gamma-3}\\
&\quad -\frac{\partial}{\partial x_i}\big[(\rho+\rho_1)^{\gamma-2}\rho_1 v_i^1 + \frac{1}{2}(\rho+\rho_1)^{\gamma-3}v_i\rho_1^2\big]\\
&\quad -\frac{1}{2}(\rho+\rho_1)^{\gamma-3}v_{i,i}\rho_1^2\\
&\quad +\frac{(\gamma-3)}{2}v_i\frac{(\rho+\rho_1)_{,i}}{(\rho+\rho_1)}(\rho+\rho_1)^{\gamma-3}\rho_1^2 + (\rho+\rho_1)^{\gamma-2}v_i^1\rho_{1,i}.
\end{aligned}
\tag{2.286}
$$

From equation $(2.259)_1$

$$
\frac{(\rho+\rho_1)_t}{(\rho+\rho_1)} = -\frac{(v_i+v_i^1)}{(\rho+\rho_1)}(\rho+\rho_1)_{,i} - (v_i+v_i^1)_{,i}
$$

and hence recollecting hypotheses (a) and (f) we find from (2.286) that there is a constant m_5 such that

$$
\begin{aligned}
\frac{1}{2}\frac{\partial}{\partial t}\big[(\rho+\rho_1)^{\gamma-3}\rho_1^2\big] &\le m_5(\rho+\rho_1)^{\gamma-3}\rho_1^2\\
&\quad -\frac{\partial}{\partial x_i}\big[(\rho+\rho_1)^{\gamma-2}\rho_1 v_i^1 + \frac{1}{2}(\rho+\rho_1)^{\gamma-3}v_i\rho_1^2\big]\\
&\quad +(\rho+\rho_1)^{\gamma-2}v_i^1\rho_{1,i}.
\end{aligned}
\tag{2.287}
$$

Upon adding (2.285) and (2.287) and integrating over S we may derive,

$$
\begin{aligned}
\frac{1}{2}\frac{d}{dt}\int_S \big\{[\rho^{\gamma-3}+(\rho+\rho_1)^{\gamma-3}]\rho_1^2\big\}dx &\le \int_S [\rho^{\gamma-2}+(\rho+\rho_1)^{\gamma-2}]v_i^1\rho_{1,i}dx\\
&\quad -\int_{\sigma\cup\partial\sigma_0} n_i\Big[\{\rho^{\gamma-2}+(\rho+\rho_1)^{\gamma-2}\}\rho_1 v_i^1\\
&\qquad +\frac{1}{2}\rho^{\gamma-3}\rho_1^2(v_i+v_i^1) + \frac{1}{2}(\rho+\rho_1)^{\gamma-3}v_i\rho_1^2\Big]dS\\
&\quad +m_2\int_S \rho^{2\gamma-4}\rho_1^2 dx + m_3\int_S v_1^2 dx\\
&\quad +\int_S [m_4\rho^{\gamma-3}+m_5(\rho+\rho_1)^{\gamma-3}]\rho_1^2 dx.
\end{aligned}
\tag{2.288}
$$

The idea is now to add (2.279) and (2.288) together in such a way that the terms involving $v_i^1 \rho_{1,i}$ add to zero. So, we add $(2.279) + (b/2)(2.288)$ to derive

$$\frac{d}{dt}\left[\frac{1}{2}\int_S v_1^2 dx + \frac{b}{4}\int_S \rho^{\gamma-3}\rho_1^2 dx + \frac{b}{4}\int_S (\rho+\rho_1)^{\gamma-3}\rho_1^2 dx\right]$$
$$\leq \int_\sigma \frac{1}{2}[\rho^{2\gamma-4} + (\rho+\rho_1)^{2\gamma-4}]\rho_1^2 dS + \frac{1}{2}\int_\sigma (1+n_1)v_1^2 dS$$
$$+ \int_\sigma \frac{n_1}{2}[\rho^{\gamma-3} + (\rho+\rho_1)^{\gamma-3}]\rho_1^2 dS$$
$$+ \int_S \rho_1^2 \rho^{\gamma-3}[m_4 + r_1\rho^{\gamma-1} + r_2\rho^{(3-\gamma)/(2-\gamma)} + r_3\rho^{\gamma(3-\gamma)/(2-\gamma)}]dx$$
$$+ \int_S \rho_1^2 (\rho+\rho_1)^{\gamma-3}\left[r_4 + \frac{bn}{2}(\rho+\rho_1)^{\gamma-1}\right]dx$$
$$+ r_4\int_S v_1^2 dx\,, \tag{2.289}$$

where

$$r_1 = m_2 + \frac{bn}{2}, \qquad r_2 = \left(\frac{2-\gamma}{3-\gamma}\right)\ell_2\beta^{(3-\gamma)/(2-\gamma)},$$
$$r_3 = \left(\frac{2-\gamma}{3-\gamma}\right)\ell_3\alpha^{(3-\gamma)/(2-\gamma)},$$
$$r_4 = m_5 + \frac{\ell_2}{(3-\gamma)\beta^{3-\gamma}} + \frac{\ell_3}{(3-\gamma)\alpha^{3-\gamma}}, \qquad r_5 = m_3 + \frac{3N_1}{2} + bn.$$

Now invoke hypothesis (a), let n_1 be a bound for $\rho, \rho + \rho_1$, and integrate (2.289) twice over the time interval $(0, h)$ to see that

$$\int_0^h dt \int_S \left[\frac{1}{2}v_1^2 + \frac{b}{4}\rho^{\gamma-3}\rho_1^2 + \frac{b}{4}(\rho+\rho_1)^{\gamma-3}\rho_1^2\right]dx$$
$$\leq \int_0^h dt \int_\sigma h(n_1^{\gamma-1} + n_1)[\rho^{\gamma-3} + (\rho+\rho_1)^{\gamma-3}]\rho_1^2 dS$$
$$+ \int_0^h dt \int_\sigma h\frac{(1+n_1)}{2}v_1^2 dS + \int_0^h dt \int_S hk_1[\rho^{\gamma-3} + (\rho+\rho_1)^{\gamma-3}]\rho_1^2 dx$$
$$+ r_4 h\int_0^h dt \int_S v_1^2 dx\,, \tag{2.290}$$

for a constant k_1 independent of h. Let $n_1^{\gamma-1} + n_1, (1 + n_1)/2$ be denoted by constants r_5, r_6. Then we rewrite (2.290) as

$$\int_0^h dt \int_S v_1^2(1 - 2r_4h)dx + \int_0^h dt \int_S \rho_1^2\rho^{\gamma-3}\left(\frac{b}{2} - 2k_1h\right)dx$$

$$+ \int_0^h dt \int_S \rho_1^2(\rho + \rho_1)^{\gamma-3}\left(\frac{b}{2} - 2k_1h\right)dx$$

$$\leq 2r_6h \int_0^h dt \int_\sigma v_1^2 dS + 2r_5h \int_0^h dt \int_\sigma \rho^{\gamma-3}\rho_1^2 dS$$

$$+ 2r_5h \int_0^h dt \int_\sigma (\rho + \rho_1)^{\gamma-3}\rho_1^2 dS . \tag{2.291}$$

Now suppose h is such that

$$1 - 2r_4h \geq \frac{1}{2}, \qquad \frac{b}{2} - 2k_1h \geq \frac{b}{4},$$

then define the Graffi function $G(R)$ by

$$G(R) = \int_0^h dt \int_S \left(v_1^2 + \frac{b}{2}[\rho^{\gamma-3} + (\rho + \rho_1)^{\gamma-3}]\rho_1^2\right)dx . \tag{2.292}$$

Then from (2.291) we see that for a constant $A = \max\{8r_6h, 8r_5h/b\}$,

$$G(R) \leq AG'(R).$$

This inequality integrates to see that for $R \geq R_0 > 0$,

$$G(R) \geq G(R_0) \exp\left(\frac{R - R_0}{A}\right) . \tag{2.293}$$

Now, $|\rho_1| = |\rho + \rho_1 - \rho| \leq |\rho + \rho_1| + |\rho|$ and so by hypothesis (a), $|\rho_1|$ and v_1 are bounded then $G(R)$ has maximum growth in R like $R^{\beta(3-\gamma)+3}$ using also hypothesis (g). Thus,

$$\lim_{R \to 0} \frac{G(R)}{R^{\beta(3-\gamma)+3+\epsilon}} = 0.$$

This contradicts (2.293) and so $v_i^1 \equiv 0$, $\rho_1 \equiv 0$ on $S \times (0, h)$. Since the bounds in hypotheses (a), (d)-(g) are independent of h we may reapply the argument on $(h, 2h)$ etc., to conclude uniqueness on $S \times (0, T)$.

3
Spatial Decay

3.1 Spatial decay for the Darcy equations

A special class of stability problems are those which investigate how the
solution to a problem decays in space given data on certain boundaries.
Within porous media such studies are relatively recent. In studies of fluid
mechanics such spatial decay estimates have a longer history. For example, for the steady Navier-Stokes equations estimates have been provided
by (Horgan, 1978), (Horgan and Wheeler, 1978), (Ames and Payne, 1989).
Estimates for the time dependent diffusion equation are given by (Horgan et al., 1984), (Lin and Payne, 1994), (Payne and Philippin, 1995),
for such diffusion equations backward in time by (Lin and Payne, 1993),
(Franchi and Straughan, 1994), and for the Stokes equations by (Ames
et al., 1993), (Chirita et al., 2001), (Chirita and Ciarletta, 2003), (Song,
2003). Other areas of recent interest for spatial decay estimates in Continuum Mechanics are in swelling porous elastic soils, (Bofill and Quintanilla,
2003), in anisotropic elasticity, (Chirita and Ciarletta, 2006), the dual-
phase-lag heat equation, (Horgan and Quintanilla, 2005), in generalized
heat transmission, (Lin and Payne, 2004), (Payne and Song, 2004a; Payne
and Song, 2004b), and in Maxwell's equations associated with electromagnetism of continuous media, see (Fabrizio and Morro, 2003), pp. 366 – 373.
The spatial decay estimates of interest in this book are of the type derived
in porous media by (Payne and Song, 1997) for the equations of Darcy and
Brinkman type, by (Qin and Kaloni, 1998) for the Brinkman-Forchheimer
model, by (Payne and Song, 2002) for the Forchheimer equations, and then

B. Straughan, *Stability and Wave Motion in Porous Media*,
DOI: 10.1007/978-0-387-76543-3_3, © Springer Science+Business Media, LLC 2008

by (Payne and Song, 2000), (Song, 2002) for double diffusive convection in a porous medium incorporating the Soret effect. The work we describe in the first four sections of this chapter is all new, although we employ methods similar to those of (Payne and Song, 1997; Payne and Song, 2002). We also describe the interesting spatial decay estimate derived by (Ames et al., 2001) for the situation where a porous medium is in contact with a clear fluid.

There are estimates of spatial decay type for other theories of porous media. For example, (Iesan and Quintanilla, 1995) derive bounds for a porous type theory where the material is an elastic body which contains voids.

3.1.1 Nonlinear temperature dependent density.

We now study the following system of Darcy equations for non-isothermal flow in a saturated porous medium, cf. chapter 1, sections 1.2, 1.6.1,

$$u_i = -p_{,i} + g_i T + \hat{g}_i T^2,$$
$$u_{i,i} = 0, \tag{3.1}$$
$$T_{,t} + u_i T_{,i} = \Delta T,$$

where we have taken the density in equation (1.15) to be quadratic in T. Here u_i, p, T are velocity, pressure, and temperature, and $g_i, \hat{g}_i$ are gravity vectors which we assume (without loss of generality) satisfy the constraint

$$|\mathbf{g}|, |\hat{\mathbf{g}}| \leq 1.$$

We have taken the thermal diffusivity to be equal to 1 without any loss for the analysis contained herein.

Let D be a domain in $\mathbb{R}^2$ and then we consider the semi-infinite cylinder $R \subset \mathbb{R}^3$ which is formed by the domain D running from $z = 0$ to $z = \infty$. The domain $D \times \{z\}$ we denote by D_z and R_z is the domain $D \times (z, \infty)$, as shown in figure 3.1

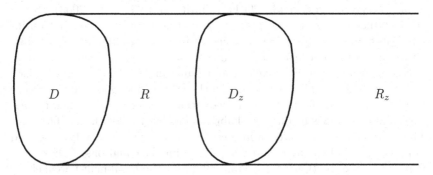

Figure 3.1. Spatial cylinder domain

We denote the boundary of R by ∂R. Observe that ∂R is composed of $D(z = 0)$ together with the curved boundary of the cylinder which we denote by ∂R_c, and the limit boundary of D as $z \to \infty$.

Equations (3.1) are defined on $R \times \{t > 0\}$, with the following conditions,

$$T(x_1, x_2, x_3, 0) = 0 \text{ in } R,$$
$$u_\alpha n_\alpha = 0 \text{ on } \partial D \times \{z > 0\} \times \{t \geq 0\},$$
$$T(x_1, x_2, x_3, t) = 0 \text{ on } \partial D \times \{z > 0\} \times \{t \geq 0\}, \tag{3.2}$$
$$u_3(x_1, x_2, 0, t) = f(x_1, x_2, t), \text{ in } \bar{D} \times \{t > 0\},$$
$$T(x_1, x_2, 0, t) = h(x_1, x_2, t), \text{ in } \bar{D} \times \{t > 0\}.$$

In addition, the following decay-like bounds are imposed on the solution,

$$u_i, T, T_{,i} = o(1), \ p = O(1) \text{ as } z(= x_3) \to \infty, \ \forall x_1, x_2, t. \tag{3.3}$$

(Payne and Song, 1997) derive spatial decay estimates for a solution to (3.1) in R together with conditions (3.2), (3.3), when $\hat{g}_i = 0$. We here allow $\hat{g}_i \neq 0$ and our presentation is different from that of (Payne and Song, 1997).

Next, note that for different cross section z−places, z_1 and z_2 we show, as in (Payne and Song, 1997),

$$\int_{D_{z_2}} u_3 dA = \int_{D_{z_1}} u_3 dA - \int_{z_1}^{z_2} dz \int_{D_z} u_{3,3} dA$$

but since $u_{i,i} = 0$, with repeated Greek indices denoting summation over 1 and 2, $u_{3,3} = -u_{\alpha,\alpha}$, whence

$$\int_{D_{z_2}} u_3 dA = \int_{D_{z_1}} u_3 dA + \int_{z_1}^{z_2} dz \int_{D_z} u_{\alpha,\alpha} dA$$
$$= \int_{D_{z_1}} u_3 dA + \int_{z_1}^{z_2} dz \int_{\partial D_z} u_\alpha n_\alpha d\ell$$

by the divergence theorem, where $d\ell$ denotes integration around ∂D_z. But, since $u_\alpha n_\alpha = 0$ on ∂D_z the second term on the right vanishes. Thus, for arbitrary $0 < z_1 \leq z_2 < \infty$,

$$\int_{D_{z_1}} u_3 dA = \int_{D_{z_2}} u_3 dA. \tag{3.4}$$

If we apply (3.4) to $z = 0$ and $z \to \infty$, then

$$\int_{D_0} f \, dA = \int_{D_0} u_3 dA = \lim_{z \to \infty} \int_{D_z} u_3 dA.$$

Upon use of the asymptotic condition $u_3 = o(1)$ as $z \to \infty$ we find

$$\int_D f \, dA = 0 \ \forall t \geq 0. \tag{3.5}$$

Note that we write $D \equiv D_0$ and $R \equiv R_0$. Condition (3.5) means that there is zero net flow through the cross section D, i.e. for $z = 0$. (Payne and Song, 2002) show how to deal with the situation in which non-zero net flow is allowed and we briefly mention this in section 3.3.

3.1.2 An appropriate "energy" function.

The idea is to derive an estimate for a positive-definite energy function $E(z,t)$. This function has form

$$E(z,t) = k \int_0^t \int_{R_z} T_{,i}T_{,i}dx\,ds + \int_0^t \int_{R_z} u_i u_i dx\,ds, \qquad (3.6)$$

where $k(> 0)$ is a constant to be selected to our advantage. The estimate follows by estimating each of the integrals in (3.6) in turn.

We begin with $\int_0^t \int_{R_z} T_{,i}T_{,i}dx\,ds$. We integrate by parts and use the divergence theorem to see that

$$\int_0^t \int_{R_z} T_{,i}T_{,i}dx\,ds = \int_0^t \int_{R_z} \frac{\partial}{\partial x_i}(TT_{,i})dx\,ds - \int_0^t \int_{R_z} T\Delta T\,dx\,ds$$

$$= \int_0^t ds \int_z^\infty d\xi \int_{D_\xi} (TT_{,i})_{,i}dA - \int_0^t \int_{R_z} T\Delta T\,dx\,ds$$

$$= \int_0^t \lim_{z \to \infty} \int_{D_z} TT_{,3}dA\,ds - \int_0^t \int_{D_z} TT_{,3}dA\,ds$$

$$+ \int_0^t ds \int_z^\infty d\xi \int_{\partial D_\xi} TT_{,\alpha}n_\alpha\,d\ell - \int_0^t \int_{R_z} T\Delta T\,dx\,ds.$$

The first and third terms on the right are zero since $T, T_{,3} = o(1)$ as $z \to \infty$, and $T = 0$ on ∂D_z. Thus,

$$\int_0^t \int_{R_z} T_{,i}T_{,i}dx\,ds = -\int_0^t \int_{D_z} TT_{,3}dA\,ds - \int_0^t \int_{R_z} T\Delta T dx\,ds$$

and we now substitute from the differential equation $(3.1)_3$ for ΔT to find

$$\int_0^t \int_{R_z} T_{,i}T_{,i}dx\,ds = -\int_0^t \int_{D_z} TT_{,3}dA\,ds$$
$$-\int_0^t \int_{R_z} T(T_{,s} + u_i T_{,i})dx\,ds \qquad (3.7)$$

By integration,

$$-\int_0^t ds \int_{R_z} TT_{,s}dx = -\frac{1}{2}\int_0^t ds \frac{\partial}{\partial s}\int_{R_z} T^2 dx$$
$$= -\frac{1}{2}\int_{R_z} T^2(\mathbf{x},t)dx \qquad (3.8)$$

since $T(\mathbf{x}, 0) = 0$. Further,

$$
\begin{aligned}
-\int_0^t ds \int_{R_z} Tu_i T_{,i} dx &= -\frac{1}{2}\int_0^t ds \int_{R_z} (u_i T^2)_{,i} dx \\
&= -\frac{1}{2}\int_0^t ds \lim_{z\to\infty}\int_{D_z} u_3 T^2 dA + \frac{1}{2}\int_0^t ds \int_{D_z} u_3 T^2 dA \\
&\quad - \frac{1}{2}\int_0^t ds \int_{\partial R_c} u_\alpha n_\alpha T^2 d\ell\, dz.
\end{aligned}
$$

The first and third terms on the right vanish due to the asymptotic conditions as $z \to \infty$ and the boundary conditions and so,

$$
-\int_0^t ds \int_{R_z} Tu_i T_{,i} dx = \frac{1}{2}\int_0^t ds \int_{D_z} u_3 T^2 dA. \tag{3.9}
$$

Use of (3.9) in (3.7) yields

$$
\begin{aligned}
\int_0^t ds \int_{R_z} T_{,i} T_{,i} dx = &-\int_0^t ds \int_{D_z} TT_{,3} dA - \frac{1}{2}\int_{R_z} T^2(t) dx \\
&+ \frac{1}{2}\int_0^t ds \int_{D_z} u_3 T^2 dA.
\end{aligned}
$$

The term in $T^2(t)$ is now discarded to derive

$$
\int_0^t ds \int_{R_z} T_{,i} T_{,i} dx \leq -\int_0^t ds \int_{D_z} TT_{,3} dA + \frac{1}{2}\int_0^t ds \int_{D_z} u_3 T^2 dA. \tag{3.10}
$$

The next step is to appeal to the maximum principle for T. The maximum principle for differential equations is discussed in detail in (Protter and Weinberger, 1967). For Darcy and Brinkman equations of porous media the maximum principle is proved in e.g. (Payne and Straughan, 1998a) or (Payne et al., 2001). Further details may be found in chapter 2 where the maximum principle is discussed in connection with structural stability. Because of the boundary conditions, we may assert that

$$
T(\mathbf{x}, t) \leq T_M \equiv \sup_{D\times[0,t]} h(x_\alpha, t), \tag{3.11}
$$

where x_α denotes x_1, x_2 and T_M is the maximum value of the function T. Use of this in (3.10) together with the Cauchy-Schwarz inequality yields

$$
\begin{aligned}
\int_0^t ds \int_{R_z} T_{,i} T_{,i} dx \leq &\sqrt{\int_0^t \int_{D_z} T^2 dA\, ds \int_0^t \int_{D_z} T_{,3}^2 dA\, ds} \\
&+ \frac{T_M}{2}\sqrt{\int_0^t \int_{\partial R_z} u_3^2 dA\, ds \int_0^t \int_{D_z} T^2 dA\, ds}. \tag{3.12}
\end{aligned}
$$

We now employ Poincaré's inequality in the form

$$\lambda_1 \int_{D_z} T^2 dA \leq \int_{D_z} T_{,\alpha} T_{,\alpha} dA \tag{3.13}$$

to see that

$$\int_0^t ds \int_{R_z} T_{,i} T_{,i} dx \leq \frac{1}{\sqrt{\lambda_1}} \sqrt{\int_0^t ds \int_{D_z} T_{,\alpha} T_{,\alpha} dA \int_0^t ds \int_{D_z} T_{,3}^2 dA}$$

$$+ \frac{T_M}{2\sqrt{\lambda_1}} \sqrt{\int_0^t ds \int_{D_z} u_3^2 dA \int_0^t ds \int_{D_z} T_{,\alpha} T_{,\alpha} dA}$$

$$\leq \beta_2 \int_0^t ds \int_{D_z} T_{,i} T_{,i} dA$$

$$+ \alpha_2 \int_0^t ds \int_{D_z} u_i u_i dA, \tag{3.14}$$

where

$$\beta_2 = \frac{1}{\sqrt{\lambda_1}} + \frac{T_M \mu}{4\sqrt{\lambda_1}} \quad \text{and} \quad \alpha_2 = \frac{T_M \mu}{4\sqrt{\lambda_1}},$$

and where $\mu(> 0)$ is a constant at our disposal.

We next turn to estimate $\int_0^t ds \int_{R_z} u_i u_i dx$. Using the differential equation $(3.1)_1$,

$$\int_0^t ds \int_{R_z} u_i u_i dx = \int_0^t ds \int_{R_z} u_i(-p_{,i} + g_i T + \hat{g}_i T^2) dx$$

$$\leq \left(\frac{1}{2\epsilon_3} + \frac{T_M}{2\epsilon_4} \right) \int_0^t ds \int_{R_z} u_i u_i dx$$

$$+ \left(\frac{\epsilon_3}{2\lambda_1} + \frac{\epsilon_4 T_M}{2\lambda_1} \right) \int_0^t ds \int_{R_z} T_{,i} T_{,i} dx$$

$$- \int_0^t ds \int_{R_z} u_i p_{,i} dx. \tag{3.15}$$

In deriving this inequality we have employed the maximum principle (3.11), the arithmetic - geometric mean inequality, and Poincaré's inequality (3.13).

To handle the pressure term we follow the argument in (Payne and Song, 2002).

$$-\int_0^t ds \int_{R_z} u_i p_{,i} dx = -\int_0^t ds \int_{R_z} (u_i p)_{,i} dx$$

$$= -\int_0^t ds \int_z^\infty d\xi \int_{D_\xi} (u_i p)_{,i} dA$$

$$= \int_0^t ds \int_{D_z} p u_3 dA, \qquad (3.16)$$

since the D term vanishes as $z \to \infty$ as does the term on ∂D vanish. Next, introduce the function w_α which solves

$$\begin{aligned} w_{\alpha,\alpha} &= u_3 \text{ in } D_\xi \\ w_\alpha &= 0 \text{ on } \partial D_\xi. \end{aligned} \qquad (3.17)$$

Then,

$$-\int_0^t ds \int_{R_z} u_i p_{,i} dx = \int_0^t ds \int_{D_z} p u_3 dA$$

$$= \int_0^t ds \int_{D_z} p w_{\alpha,\alpha} dA$$

$$= \int_0^t ds \int_{D_z} (p w_\alpha)_{,\alpha} dA - \int_0^t ds \int_{D_z} p_{,\alpha} w_\alpha dA$$

$$= \int_0^t ds \int_{\partial D_z} p w_\alpha n_\alpha dA - \int_0^t ds \int_{D_z} p_{,\alpha} w_\alpha dA,$$

where we have integrated by parts and used the divergence theorem. Since $w_\alpha = 0$ on ∂D_z, we see that

$$-\int_0^t ds \int_{R_z} u_i p_{,i} dx = -\int_0^t ds \int_{D_z} p_{,\alpha} w_\alpha dA$$

$$= -\int_0^t ds \int_{D_z} w_\alpha(-u_\alpha + g_\alpha T + \hat{g}_\alpha T^2) dA$$

substituting for $p_{,\alpha}$ from the differential equation. We now use the maximum principle and the Cauchy-Schwarz inequality to find

$$-\int_0^t ds \int_{R_z} u_i p_{,i} dx \leq \int_0^t ds \int_{D_z} u_\alpha w_\alpha dA$$

$$+ (1 + T_M) \sqrt{\int_0^t ds \int_{D_z} w_\alpha w_\alpha dA \int_0^t ds \int_{D_z} T^2 dA}. \qquad (3.18)$$

The next step is to employ Poincaré's inequality on the $w_\alpha w_\alpha$ term to find

$$\int_{D_z} w_\alpha w_\alpha dA \leq \frac{1}{\lambda_1} \int_{D_z} w_{\alpha,\beta} w_{\alpha,\beta} dA.$$

After this we employ the Babuska - Aziz inequality, which holds for functions ω_α which solve a boundary value problem like (3.17) (this inequality is discussed in (Horgan and Payne, 1983) where they also show how to estimate the constant C)

$$\int_{D_z} \omega_{\alpha,\beta}\omega_{\alpha,\beta}dA \le C \int_{D_z} (\omega_{\alpha,\alpha})^2 dA. \tag{3.19}$$

Noting that $\omega_{\alpha,\alpha} = u_3$ in D_z, we use the last two inequalities in (3.18) to see that

$$-\int_0^t ds \int_{R_z} u_i p_{,i} dx \le \int_0^t ds \int_{D_z} u_\alpha \omega_\alpha dA$$
$$+ (1+T_M)\sqrt{\frac{C}{\lambda_1}}\sqrt{\int_0^t ds \int_{D_z} u_3^2 dA \int_0^t ds \int_{D_z} T^2 dA}. \tag{3.20}$$

Next, employ the Cauchy-Schwarz inequality and then a similar procedure to that above involving Poincaré's inequality and the Babuska-Aziz inequality to show that

$$\int_0^t ds \int_{D_z} u_\alpha \omega_\alpha dA \le \sqrt{\int_0^t ds \int_{D_z} u_\alpha u_\alpha dA \int_0^t ds \int_{D_z} \omega_\alpha \omega_\alpha dA}$$
$$\le \sqrt{\frac{C}{\lambda_1}}\sqrt{\int_0^t ds \int_{D_z} u_\alpha u_\alpha dA \int_0^t ds \int_{D_z} u_3^2 dA}$$
$$\le \sqrt{\frac{C}{\lambda_1}} \int_0^t ds \int_{D_z} u_i u_i dA. \tag{3.21}$$

Furthermore, using the Cauchy-Schwarz, Poincaré and arithmetic-geometric mean inequalities we may show, for $\zeta > 0$ a constant to be selected,

$$\sqrt{\int_0^t ds \int_{D_z} u_3^2 dA \int_0^t ds \int_{D_z} T^2 dA}$$
$$\le \frac{1}{\sqrt{\lambda_1}}\sqrt{\int_0^t ds \int_{D_z} u_3^2 dA \int_0^t ds \int_{D_z} T_{,\alpha}T_{,\alpha}dA}$$
$$\le \frac{1}{2\zeta\sqrt{\lambda_1}} \int_0^t ds \int_{D_z} u_i u_i dA + \frac{\zeta}{2\sqrt{\lambda_1}} \int_0^t ds \int_{D_z} T_{,i}T_{,i}dA. \tag{3.22}$$

We employ (3.22) and (3.21) in (3.20) to arrive at

$$-\int_0^t ds \int_{R_z} u_i p_{,i} dx \le \left(\sqrt{\frac{C}{\lambda_1}} + \frac{\sqrt{C}}{2\zeta\lambda_1}\right) \int_0^t ds \int_{D_z} u_i u_i dA$$
$$+ \frac{\zeta\sqrt{C}}{2\lambda_1} \int_0^t ds \int_{D_z} T_{,i}T_{,i}dA. \tag{3.23}$$

Now use (3.23) in (3.15) to find

$$\int_0^t ds \int_{R_z} u_i u_i dx \le \alpha_1 \int_0^t ds \int_{D_z} u_i u_i dA + \beta_1 \int_0^t ds \int_{D_z} T_{,i} T_{,i} dA$$
$$+ \gamma_1 \int_0^t ds \int_{R_z} u_i u_i dx + \gamma_2 \int_0^t ds \int_{R_z} T_{,i} T_{,i} dx \tag{3.24}$$

where the constants $\alpha_1, \beta_1, \gamma_1$ and γ_2 are given by

$$\alpha_1 = \sqrt{\frac{C}{\lambda_1}} + \frac{\sqrt{C}}{2\zeta\lambda_1}, \qquad \beta_1 = \frac{\zeta\sqrt{C}}{2\lambda_1},$$

$$\gamma_1 = \frac{1}{2\epsilon_3} + \frac{T_M}{2\epsilon_4}, \qquad \gamma_2 = \frac{\epsilon_3}{2\lambda_1} + \frac{\epsilon_4 T_M}{2\lambda_1}.$$

Inequalities (3.14) and (3.24) are our fundamental inequalities to allow us to estimate $E(z,t)$. With $E(z,t)$ defined as in (3.6) we find using (3.14), (3.24),

$$E(z,t) = k \int_0^t ds \int_{R_z} T_{,i} T_{,i} dx + \int_0^t ds \int_{R_z} u_i u_i dx$$

$$\le (k\alpha_2 + \alpha_1) \int_0^t ds \int_{D_z} u_i u_i dA + (k\beta_2 + \beta_1) \int_0^t ds \int_{D_z} T_{,i} T_{,i} dA$$

$$+ \gamma_1 \int_0^t ds \int_{R_z} u_i u_i dx + \gamma_2 \int_0^t ds \int_{R_z} T_{,i} T_{,i} dx. \tag{3.25}$$

To remove the R_z terms from the right we must now select $k - \gamma_2 > 0$ and $1 - \gamma_1 > 0$. To ensure this we pick ϵ_3, ϵ_4 such that

$$1 - \frac{1}{2\epsilon_3} - \frac{T_M}{2\epsilon_4} > 0$$

and then choose k so large that

$$k - \frac{\epsilon_3}{2\lambda_1} - \frac{\epsilon_4 T_M}{2\lambda_1} > 0.$$

For example, we may select $\epsilon_3 = 2, \epsilon_4 = 2T_M$ and then $k > 1/\lambda_1 + T_M^2/\lambda_1$. The choice $k = 2/\lambda_1 + 2T_M^2/\lambda_1$ leads to

$$E(z,t) \le 2(k\alpha_2 + \alpha_1) \int_0^t ds \int_{D_z} u_i u_i dA + 2(k\beta_2 + \beta_1) \int_0^t ds \int_{D_z} T_{,i} T_{,i} dA.$$

Let

$$\delta = \max\left\{ \frac{2k\beta_2 + 2\beta_1}{k}, 2k\alpha_2 + 2\alpha_1 \right\}$$

and then we derive

$$E(z,t) \le \delta\left(k \int_0^t ds \int_{D_z} T_{,i} T_{,i} dA + \int_0^t ds \int_{D_z} u_i u_i dA \right). \tag{3.26}$$

Since

$$E = k \int_0^t ds \int_z^\infty d\xi \int_{D_\xi} T_{,i}T_{,i}dA + \int_0^t ds \int_z^\infty d\xi \int_{D_\xi} u_i u_i dA$$

we see that

$$\frac{\partial E}{\partial z} = -\left[k \int_0^t ds \int_{D_z} T_{,i}T_{,i}dA + \int_0^t ds \int_{D_z} u_i u_i dA \right].$$

Thus, inequality (3.26) is

$$E \le -\delta \frac{\partial E}{\partial z}. \tag{3.27}$$

This inequality is now integrated to find

$$E(z,t) \le E(0,t)e^{-z/\delta}. \tag{3.28}$$

Inequality (3.28) is a decay bound for $E(z,t)$ in terms of $E(0,t)$. However, $E(0,t)$ is not in terms of the data f and h, and so we now turn to estimate $E(0,t)$ in terms of data.

3.1.3 A data bound for $E(0,t)$.

We derive a bound for $E(0,t)$ in a similar manner to that prescribed by (Payne and Song, 1997).

Let now $S(\mathbf{x},t)$ be defined by

$$S(\mathbf{x},t) = h(x_1,x_2,t)e^{-\sigma z} \tag{3.29}$$

where $\sigma > 0$ is a constant to be chosen. Then, recalling the form of E, we put $z = 0$ and estimate the two relevant terms, $\int_0^t ds \int_R T_{,i}T_{,i}dx$ and $\int_0^t ds \int_R u_i u_i dx$.

$$\int_0^t ds \int_R T_{,i}T_{,i}dx = \int_0^t ds \int_R \frac{\partial}{\partial x_i}(TT_{,i})dx - \int_0^t ds \int_R T\Delta T dx$$

$$= \int_0^t ds \lim_{z\to\infty} \int_{D_z} TT_{,3}dA - \int_0^t ds \int_{D_0} TT_{,3}dA$$

$$+ \int_0^t ds \int_0^\infty dz \int_{\partial D_z} TT_{,\alpha}n_\alpha d\ell - \int_0^t ds \int_R T\Delta T dx,$$

where we have integrated by parts and employed the divergence theorem. Due to the boundary conditions the first and thrid terms vanish and since $S \equiv T$ on $D_0(= D)$ one finds

$$\int_0^t ds \int_R T_{,i}T_{,i}dx = -\int_0^t ds \int_D ST_{,3}dA - \int_0^t ds \int_R T\Delta T dx. \tag{3.30}$$

Furthermore, by differentiation and use of the divergence theorem,

$$\int_0^t ds \int_R (ST_{,i})_{,i}dx = \int_0^t ds \int_R S_{,i}T_{,i}dx + \int_0^t ds \int_R S\Delta T dx$$

and

$$\int_0^t ds \int_R (ST_{,i})_{,i} dx = \int_0^t ds \int_0^\infty dz \int_{\partial D_z} ST_{,\alpha} n_\alpha d\ell$$
$$+ \int_0^t ds \lim_{z \to \infty} \int_{D_z} ST_{,3} dA - \int_0^t ds \int_{D_0} ST_{,3} dA.$$

The first two terms on the right vanish because of the boundary conditions, and then we combine these two identities to deduce

$$-\int_0^t ds \int_D ST_{,3} dA = \int_0^t ds \int_R S_{,i} T_{,i} dx + \int_0^t ds \int_R S\Delta T dx. \qquad (3.31)$$

Use of (3.31) in (3.30) leads to

$$\int_0^t ds \int_R T_{,i} T_{,i} dx = \int_0^t ds \int_R S_{,i} T_{,i} dx + \int_0^t ds \int_R (S - T)\Delta T dx$$
$$= \int_0^t ds \int_R S_{,i} T_{,i} dx$$
$$+ \int_0^t ds \int_R (S - T)(T_{,s} + u_i T_{,i}) dx \qquad (3.32)$$

where in the last line the differential equation has been used to substitute for ΔT.

The last term in (3.32) has four components which we expand as

$$\int_0^t ds \int_R ST_s dx - \int_0^t ds \int_R TT_s dx$$
$$+ \int_0^t ds \int_R Su_i T_{,i} dx - \int_0^t ds \int_R u_i TT_{,i} dx.$$

We now rearrange each of these terms.

$$\int_0^t ds \int_R ST_s dx = \int_0^t ds \frac{\partial}{\partial s} \int_R ST dx - \int_0^t ds \int_R T \frac{\partial S}{\partial s} dx$$
$$= \int_R S(t)T(t) dx - \int_0^t d\eta \int_R TS_\eta dx,$$

where we note the term in $S(0)T(0)$ is zero. Also,

$$-\int_0^t ds \int_R TT_s dx = -\frac{1}{2} \int_0^t \frac{\partial}{\partial s} \int_R T^2 dx = -\frac{1}{2} \int_R T^2(t) dx,$$

since $T(0) = 0$.

$$\int_0^t ds \int_R Su_i T_{,i} dx = \int_0^t ds \int_R \frac{\partial}{\partial x_i}(Su_i T) dx - \int_0^t ds \int_R u_i TS_{,i} dx$$

$$= \int_0^t ds \lim_{z \to \infty} \int_{D_z} Su_3 T dA ds - \int_0^t ds \int_{D_0} u_3 ST dA$$

$$+ \int_0^t ds \int_0^\infty dz \int_{\partial D_z} n_\alpha u_\alpha ST d\ell - \int_0^t ds \int_R u_i TS_{,i} dx.$$

Since $T = 0$ on ∂D_z, using the asymptotic behaviour of S, T, u_3, and noting that on $D_0(= D)$, $u_3 = f$, $T = S = h$, this expression reduces to

$$\int_0^t ds \int_R Su_i T_{,i} dx = -\int_0^t ds \int_D fh^2 dA - \int_0^t ds \int_R u_i TS_{,i} dx.$$

The remaining term of the four in question is handled as follows,

$$-\int_0^t ds \int_R u_i TT_{,i} dx = -\frac{1}{2} \int_0^t ds \int_R (u_i T^2)_{,i} dx$$

$$= -\frac{1}{2} \int_0^t ds \lim_{z \to \infty} \int_{D_z} u_3 T^2 dA + \frac{1}{2} \int_0^t ds \int_{D_0} u_3 T^2 dA$$

$$-\frac{1}{2} \int_0^t ds \int_0^\infty dz \int_{\partial D_z} u_\alpha n_\alpha T^2 d\ell$$

$$= \frac{1}{2} \int_0^t ds \int_D fh^2 dA,$$

since $u_3 = f$ and $T = h$ on D_0.

Using the above rearrangements (3.32) becomes

$$\int_0^t ds \int_R T_{,i} T_{,i} dx = \int_0^t ds \int_R S_{,i} T_{,i} dx - \int_0^t d\eta \int_R TS_\eta dx$$

$$+ \int_R S(t) T(t) dx - \frac{1}{2} \int_R T^2(t) dx$$

$$- \frac{1}{2} \int_0^t ds \int_D fh^2 dA - \int_0^t ds \int_R u_i TS_{,i} dx. \quad (3.33)$$

We discard the $T^2(t)$ term. Also, the maximum principle is invoked on the term in $S(t)T(t)$ to see that

$$\int_R S(t) T(t) dx \le \sup_{D \times [0,t]} h(x_\alpha, t) \int_0^\infty dz\, e^{-\sigma z} \int_D h(x_\alpha, t) dA$$

$$= \Phi(t) \quad (3.34)$$

where the data term $\Phi(t)$ is defined by (3.34).

Hence, from (3.33) we obtain

$$\int_0^t ds \int_R T_{,i}T_{,i}dx \le \Phi(t) - \frac{1}{2}\int_0^t ds \int_D fh^2 dA$$

$$+ \int_0^t ds \int_R S_{,i}T_{,i}dx - \int_0^t d\eta \int_R TS_\eta dx \qquad (3.35)$$

$$- \int_0^t ds \int_R u_i S_{,i}T dx. \qquad (3.36)$$

The arithmetic-geometric mean inequality is now used on the last three terms on the right to find for positive constants $\alpha_1, \alpha_2, \alpha_3$, to be selected

$$\int_0^t ds \int_R S_{,i}T_{,i}dx \le \frac{1}{2\alpha_1}\int_0^t ds \int_R S_{,i}S_{,i}dx + \frac{\alpha_1}{2}\int_0^t ds \int_R T_{,i}T_{,i}dx$$

and

$$-\int_0^t d\eta \int_R TS_\eta dx \le \frac{1}{2\alpha_2}\int_0^t d\eta \int_R S_\eta^2 dx + \frac{\alpha_2}{2}\int_0^t ds \int_R T^2 dx$$

$$\le \frac{1}{2\alpha_2}\int_0^t d\eta \int_R S_\eta^2 dx + \frac{\alpha_2}{2\lambda_1}\int_0^t ds \int_R T_{,i}T_{,i}dx,$$

where Poincaré's inequality has also been used. With the aid of the maximum principle,

$$-\int_0^t ds \int_R u_i S_{,i}T dx \le T_M \int_0^t ds \int_R |u_i S_{,i}|dx$$

$$\le \frac{T_M^2}{2\alpha_3}\int_0^t ds \int_R S_{,i}S_{,i}dx + \frac{\alpha_3}{2}\int_0^t ds \int_R u_i u_i dx.$$

The above three inequalities are now used in (3.36) to arrive at

$$\int_0^t ds \int_R T_{,i}T_{,i}dx \le \Phi(t) - \frac{1}{2}\int_0^t ds \int_D fh^2 dA$$

$$+ \Big(\frac{T_M^2}{2\alpha_3} + \frac{1}{2\alpha_1}\Big)\int_0^t ds \int_R S_{,i}S_{,i}dx$$

$$+ \frac{1}{2\alpha_2}\int_0^t d\eta \int_R S_\eta^2 dx$$

$$+ \frac{1}{2}(\alpha_1 + \alpha_2/\lambda_1)\int_0^t ds \int_R T_{,i}T_{,i}dx$$

$$+ \frac{1}{2}\alpha_3\int_0^t ds \int_R u_i u_i dx. \qquad (3.37)$$

We now derive a bound for the u_i term. To this end use the differential equation to find

$$\int_0^t ds \int_R u_i u_i dx = \int_0^t ds \int_R u_i(-p_{,i} + g_i T + \hat{g}_i T^2)dx. \qquad (3.38)$$

The pressure term is handled as follows,

$$-\int_0^t ds \int_R u_i p_{,i} dx = -\int_0^t ds \int_0^\infty dz \int_{D_z} \frac{\partial}{\partial x_i}(u_i p) dA$$

$$= \int_0^t \int_{D_0} u_3 p \, dA \, d\eta,$$

since the other terms which arise after use of the divergence theorem vanish. The function ϕ_i of (Payne and Song, 1997) is now introduced as a function satisfying

$$\phi_{i,i} = 0 \text{ in } R \times \{t > 0\}$$
$$\phi_i n_i = u_i n_i \text{ on } \partial R \times \{t > 0\}. \tag{3.39}$$

In fact, they select

$$\phi_3(\mathbf{x}, t) = f(x_1, x_2, t) e^{-\sigma z}. \tag{3.40}$$

Observe that $\phi_3 = f = u_3$ on D.

Then,

$$\int_0^t ds \int_{D_0} u_3 p \, dA = \int_0^t ds \int_{D_0} \phi_3 p \, dA.$$

But,

$$\int_0^t ds \int_R (p\phi_i)_{,i} dx = -\int_0^t ds \int_{D_0} \phi_3 p \, dA$$

since the other boundary terms vanish after use of the divergence theorem. Then, because $\phi_{i,i} = 0$,

$$\int_0^t ds \int_{D_0} u_3 p \, dA = -\int_0^t ds \int_R p_{,i} \phi_i dx$$

$$= \int_0^t ds \int_R \phi_i (u_i - g_i T - \hat{g}_i T^2) dx$$

where in the last line the differential equation has been employed. Putting this into (3.38),

$$\int_0^t ds \int_R u_i u_i dx = \int_0^t ds \int_R u_i g_i T dx + \int_0^t ds \int_R u_i \hat{g}_i T^2 dx$$

$$+ \int_0^t ds \int_R \phi_i u_i dx - \int_0^t ds \int_R g_i \phi_i T dx$$

$$- \int_0^t ds \int_R \hat{g}_i \phi_i T^2 dx. \tag{3.41}$$

We next use the arithmetic-geometric mean inequality on all five terms on the right of (3.41) and bound one of the T's in the T^2 term by the maximum principle. Poincaré's inequality is also employed on the resulting T^2 terms, so we derive from (3.41)

$$\int_0^t ds \int_R u_i u_i dx \leq \frac{1}{2}\left(\frac{1}{\beta_1} + \frac{1}{\beta_2} + \frac{1}{\beta_3}\right) \int_0^t ds \int_R u_i u_i dx$$

$$+ \frac{1}{2\lambda_1}\left(\beta_1 + \beta_2 T_M^2 + \frac{1}{\beta_4} + \frac{T_M^2}{\beta_5}\right) \int_0^t ds \int_R T_{,i} T_{,i} dx$$

$$+ \frac{1}{2}(\beta_3 + \beta_4 + \beta_5) \int_0^t ds \int_R \phi_i \phi_i dx, \tag{3.42}$$

where $\beta_i > 0$ are constants. In fact, we now pick $\beta_1 = \beta_2 = \beta_3 = 3$. Then from (3.42) we find

$$\int_0^t ds \int_R u_i u_i dx \leq \frac{1}{\lambda_1}\left(3 + 3T_M^2 + \frac{1}{\beta_4} + \frac{T_M^2}{\beta_5}\right) \int_0^t ds \int_R T_{,i} T_{,i} dx$$

$$+ (3 + \beta_4 + \beta_5) \int_0^t ds \int_R \phi_i \phi_i dx. \tag{3.43}$$

We now return to inequality (3.37) and use (3.43) to bound the u_i term. This leads to

$$\int_0^t ds \int_R T_{,i} T_{,i} dx \leq \Phi(t) - \frac{1}{2}\int_0^t ds \int_D f h^2 dA$$

$$+ \frac{1}{2}(\alpha_1^{-1} + T_M^2 \alpha_3^{-1}) \int_0^t ds \int_R S_{,i} S_{,i} dx$$

$$+ \frac{1}{2\alpha_2} \int_0^t d\eta \int_R S_\eta^2 dx$$

$$+ \frac{1}{2}\left\{\alpha_1 + \alpha_2 + \frac{\alpha_3}{\lambda_1}\left[3 + 3T_M^2 + \frac{1}{\beta_4} + \frac{T_M^2}{\beta_5}\right]\right\} \int_0^t ds \int_R T_{,i} T_{,i} dx$$

$$+ \frac{1}{2}\alpha_3(3 + \beta_4 + \beta_5) \int_0^t ds \int_R \phi_i \phi_i dx.$$

Pick now $\beta_4 = \beta_5 = 1$, then $\alpha_1 = 1/3$, $\alpha_2 = \lambda_1/3$, $\alpha_3 = \lambda_1/12$. The above inequality may now be reduced to

$$\frac{1}{2}\int_0^t ds \int_R T_{,i} T_{,i} dx \leq \Phi(t) - \frac{1}{2}\int_0^t ds \int_D f h^2 dA$$

$$+ \left(\frac{3}{2} + \frac{6T_M^2}{\lambda_1}\right) \int_0^t ds \int_R S_{,i} S_{,i} dx$$

$$+ \frac{3}{2\lambda_1} \int_0^t d\eta \int_R S_\eta^2 dx$$

$$+ \frac{5\lambda_1}{24} \int_0^t ds \int_R \phi_i \phi_i dx. \tag{3.44}$$

Inequality (3.44) is effectively a bound for $\int_0^t ds \int_R T_{,i}T_{,i}dx$, but it remains to show the ϕ_i term can be bounded by data. Since $\phi_3 = f(x_\alpha, t)e^{-\sigma z}$, $\phi_{i,i} = 0$ yields

$$\phi_{\alpha,\alpha} = \sigma f e^{-\sigma z}. \tag{3.45}$$

Then,

$$\int_0^t ds \int_R \phi_i\phi_i dx = \int_0^t ds \int_0^\infty dz \int_{D_z} \phi_\alpha\phi_\alpha dA$$
$$+ \int_0^t ds \int_R f^2 e^{-2\sigma z} dx.$$

The first term on the right is bounded using the Poincaré and Babuska-Aziz inequalities as follows,

$$\int_0^t ds \int_0^\infty dz \int_{D_z} \phi_\alpha\phi_\alpha dA \leq \frac{1}{\lambda_1} \int_0^t ds \int_0^\infty dz \int_{D_z} \phi_{\alpha,\beta}\phi_{\alpha,\beta} dA$$
$$\leq \frac{C}{\lambda_1} \int_0^t ds \int_0^\infty dz \int_{D_z} \phi_{\alpha,\alpha}^2 dA$$
$$\leq \frac{C\sigma^2}{\lambda_1} \int_0^t ds \int_R f^2 e^{-2\sigma z} dx.$$

This then leads to the data bound

$$\int_0^t ds \int_R \phi_i\phi_i dx \leq \left(1 + \frac{C\sigma^2}{\lambda_1}\right) \int_0^t ds \int_R f^2 e^{-2\sigma z} dx. \tag{3.46}$$

Insertion of this into (3.44), together with the definition of S gives us the data bound

$$\int_0^t ds \int_R T_{,i}T_{,i}dx \leq 2\Phi(t) - \int_0^t ds \int_D fh^2 dA$$
$$+ \int_0^t d\eta \int_R \left\{\left[3 + \frac{12T_M^2}{\lambda_1}\right](\sigma^2 h^2 + h_{,\alpha}h_{,\alpha}) + \frac{3}{\lambda_1}h_\eta^2\right\} e^{-2\sigma z} dx$$
$$+ \frac{5\lambda_1}{12}\left(1 + \frac{C\sigma^2}{\lambda_1}\right) \int_0^t ds \int_R e^{-2\sigma z} f^2 dx$$
$$\equiv \Psi(t), \tag{3.47}$$

where $\Psi(t)$ is defined as indicated. We then find a u_i bound using (3.47) in (3.43),

$$\int_0^t ds \int_R u_i u_i dx \leq \frac{4}{\lambda_1}(1 + T_M^2)\Psi(t) + 5\left(1 + \frac{C\sigma^2}{\lambda_1}\right) \int_0^t ds \int_R e^{-2\sigma z} f^2 dx$$
$$\equiv \chi(t), \tag{3.48}$$

with $\chi(t)$ being defined in (3.48).

Thus,

$$E(0, t) \leq k\Psi(t) + \chi(t) = \text{data}.$$

Our *a priori* spatial decay estimate now follows with the aid of (3.28) and is

$$E(z, t) \leq \left[k\Psi(t) + \chi(t) \right] e^{-z/\delta}. \tag{3.49}$$

A spatial decay bound for equations (3.1) when the permeability is anisotropic is established by (Song, 2006), while (Payne and Song, 2007b) produce spatial decay results for a doubly diffusive convection Darcy system.

3.2   Spatial decay for the Brinkman equations

We now study the following system of Brinkman equations for non-isothermal flow in a saturated porous medium, cf. (3.1),

$$
\begin{aligned}
-\lambda \Delta u_i + u_i &= -p_{,i} + g_i T + \hat{g}_i T^2, \\
u_{i,i} &= 0, \\
T_{,t} + u_i T_{,i} &= \Delta T.
\end{aligned}
\tag{3.50}
$$

Here u_i, p, T are velocity, pressure, and temperature, and $g_i, \hat{g}_i$ are gravity vectors which we again assume (without loss of generality) satisfy the constraint

$$|\mathbf{g}|, |\hat{\mathbf{g}}| \leq 1.$$

We have taken the thermal diffusivity to be equal to 1 without any loss for the analysis contained herein, and $\lambda > 0$ is the Brinkman coefficient.

R and D are again as in 3.1.1, so let D be a domain in $\mathbb{R}^2$ and then we consider the semi-infinite cylinder $R \subset \mathbb{R}^3$ which is formed by the domain D running from $z = 0$ to $z = \infty$. The domain $D \times \{z\}$ we denote by D_z and R_z is the domain $D \times (z, \infty)$.

The boundary of R is ∂R. Again, ∂R is composed of $D(z = 0)$ together with the curved boundary of the cylinder which we denote by ∂R_c, and the limit boundary of D as $z \to \infty$.

Equations (3.50) are defined on $R \times \{t > 0\}$, with the following conditions,

$$
\begin{aligned}
T(x_1, x_2, x_3, 0) &= 0 \text{ in } R, \\
u_i &= 0 \text{ on } \partial D \times \{z > 0\} \times \{t \geq 0\}, \\
T(x_1, x_2, x_3, t) &= 0 \text{ on } \partial D \times \{z > 0\} \times \{t \geq 0\}, \\
u_i(x_1, x_2, 0, t) &= f_i(x_1, x_2, t), \text{ in } \bar{D} \times \{t > 0\}, \\
T(x_1, x_2, 0, t) &= h(x_1, x_2, t), \text{ in } \bar{D} \times \{t > 0\}.
\end{aligned}
\tag{3.51}
$$

In addition, the following decay-like bounds are imposed on the solution,

$$u_i, T = o(1), \quad p, u_{i,j}, T_{,i} = O(1/z) \text{ as } z(= x_3) \to \infty, \quad \forall x_1, x_2, t. \quad (3.52)$$

(Payne and Song, 1997) derive spatial decay estimates for a solution to
(3.50) in R together with conditions (3.51), (3.52), when $\hat{g}_i = 0$. We here
allow $\hat{g}_i \neq 0$. Our presentation is different from that of (Payne and Song,
1997), although we are following their analysis.

The argument given leading to (3.4) is valid here and so if f_3 has a zero
integral over D we see that

$$\int_{D_z} u_3 dA = \int_D f_3 dA = 0, \quad \forall z \geq 0, \ \forall t > 0.$$

Due to the presence of the Δu_i term in equations (3.50) the argument
in section 3.1 is not sufficient. Instead, for the Brinkman system, (Payne
and Song, 1997) show that one must introduce an energy function which
integrates a further time in z, not just over R_z. This leads to a second
order differential inequality rather than a first order one. To see this, we
commence with estimates for the functionals $\int_0^t ds \int_{R_z} (\xi - z) T_{,i} T_{,i} dx$ and
$\int_0^t ds \int_{R_z} (\xi - z) u_{i,j} u_{i,j} dx$. Note that

$$\int_0^t ds \int_{R_z} (\xi - z) f \, dx \equiv \int_0^t ds \int_z^\infty d\xi \int_\xi^\infty d\mu \int_{D_\mu} f \, dA$$

where $dx \equiv dA \, d\xi = dx_1 dx_2 d\xi$.

3.2.1 An estimate for grad$T$.

We begin with an estimate for $\int_0^t \int_{R_z} (\xi - z) T_{,i} T_{,i} dx \, ds$.

First, by integration by parts

$$\int_0^t ds \int_{R_z} (\xi - z) T_{,i} T_{,i} dx = \int_0^t ds \int_z^\infty d\xi \int_{D_\xi} [(\xi - z) T_{,i} T]_{,i} dA$$

$$- \int_0^t ds \int_{R_z} T[(\xi - z) T_{,i}]_{,i} dx.$$

The first term on the right is zero as is seen by using the divergence theorem
and the boundary conditions. Carrying out the differentiation on the second
term we then find

$$\int_0^t ds \int_{R_z} (\xi - z) T_{,i} T_{,i} dx = - \int_0^t ds \int_{R_z} T T_{,3} dx - \int_0^t ds \int_{R_z} (\xi - z) T \Delta T \, dx$$

$$= - \int_0^t ds \int_{R_z} T T_{,3} dx$$

$$- \int_0^t ds \int_{R_z} (\xi - z) T (T_{,s} + u_i T_{,i}) dx \quad (3.53)$$

where the differential equation $(3.50)_3$ has been used to substitute for ΔT. The last term in (3.53) may be rearranged as

$$-\frac{1}{2}\int_0^t ds \int_{R_z} (\xi - z)\frac{\partial T^2}{\partial s}\,dx - \frac{1}{2}\int_0^t ds \int_{R_z} (\xi - z)(u_i T^2)_{,i}dx$$

$$= -\frac{1}{2}\int_{R_z} (\xi - z)T^2(t)dx + \frac{1}{2}\int_0^t ds \int_{R_z} u_3 T^2(\xi - z)_{,\xi}dx$$

where we have used the divergence theorem, the boundary conditions, and integration by parts. The above expression is employed in (3.53) to find

$$\int_0^t ds \int_{R_z} (\xi - z)T_{,i}T_{,i}dx = -\int_0^t ds \int_{R_z} TT_{,3}dx - \frac{1}{2}\int_{R_z} (\xi - z)T^2(t)dx$$

$$+ \frac{1}{2}\int_0^t ds \int_{R_z} u_3 T^2 dx. \tag{3.54}$$

The arithmetic-geometric mean and Poincaré inequalities are now used on the first term on the right to derive, for arbitrary $\gamma_1 > 0$,

$$-\int_0^t ds \int_{R_z} TT_{,3}dx \leq \frac{1}{2}\gamma_1 \int_0^t ds \int_{R_z} T^2 dx + \frac{1}{2\gamma_1}\int_0^t ds \int_{R_z} T_{,3}^2 dx$$

$$\leq \frac{\gamma_1}{2\lambda_1}\int_0^t ds \int_{R_z} T_{,\alpha}T_{,\alpha}dx$$

$$+ \frac{1}{2\gamma_1}\int_0^t ds \int_{R_z} T_{,3}^2 dx. \tag{3.55}$$

To bound the last term in (3.54) we use the maximum principle, followed by the arithmetic-geometric mean inequality and then Poincaré's inequality to obtain, for arbitrary $\gamma_2 > 0$,

$$\frac{1}{2}\int_0^t ds \int_{R_z} u_3 T^2 dx \leq \frac{T_M}{2}\left(\frac{1}{2\gamma_2}\int_0^t ds \int_{R_z} u_3^2 dx + \frac{1}{2}\gamma_2 \int_0^t ds \int_{R_z} T^2 dx\right)$$

$$\leq \frac{T_M}{4\gamma_2}\int_0^t ds \int_{R_z} u_3^2 dx$$

$$+ \frac{\gamma_2 T_M}{4\lambda_1}\int_0^t ds \int_{R_z} T_{,\alpha}T_{,\alpha}dx. \tag{3.56}$$

Next, the second term on the right of (3.54) is non-positive and discarded. Then, with the aid of (3.55) and (3.56) we find from (3.54),

$$\int_0^t ds \int_{R_z} (\xi - z)T_{,i}T_{,i}dx \leq \frac{T_M}{4\gamma_2}\int_0^t ds \int_{R_z} u_3^2 dx + \frac{1}{2\gamma_1}\int_0^t ds \int_{R_z} T_{,3}^2 dx$$

$$+ \left(\frac{\gamma_1}{2\lambda_1} + \frac{\gamma_2 T_M}{4\lambda_1}\right)\int_0^t ds \int_{R_z} T_{,\alpha}T_{,\alpha}dx. \tag{3.57}$$

3.2.2 An estimate for grad **u**.

We now integrate by parts, use the divergence theorem and the boundary conditions, and then the differential equation to see that

$$\int_0^t ds \int_{R_z} (\xi - z) u_{i,j} u_{i,j} dx = \int_0^t ds \int_{R_z} [(\xi - z) u_{i,j} u_i]_{,j} dx$$

$$- \int_0^t ds \int_{R_z} u_i [(\xi - z) u_{i,j}]_{,j} dx$$

$$= - \int_0^t ds \int_{R_z} u_i u_{i,3} dx - \int_0^t ds \int_{R_z} (\xi - z) u_i \Delta u_i dx$$

$$= - \int_0^t ds \int_{R_z} u_i u_{i,3} dx$$

$$- \int_0^t ds \int_{R_z} (\xi - z) \frac{u_i}{\lambda} \left[u_i + p_{,i} - g_i T - \hat{g}_i T^2 \right] dx. \quad (3.58)$$

The arithmetic-geometric mean inequality is used on the $u_i u_{i,3}$ term and then Poincaré's inequality is used on the resulting $u_i u_i$ term. The arithmetic-geometric mean inequality is used on the $u_i T$ term, and again on the $u_i T^2$ term after use of the maximum principle. For positive constants γ_0, γ_3 and μ_1 we then deduce from (3.58)

$$\int_0^t ds \int_{R_z} (\xi - z) u_{i,j} u_{i,j} dx \le \frac{1}{2} \gamma_3 \int_0^t ds \int_{R_z} u_{i,3} u_{i,3} dx$$

$$+ \frac{1}{2\gamma_3 \lambda_1} \int_0^t ds \int_{R_z} u_{i,\alpha} u_{i,\alpha} dx$$

$$+ \left(\frac{\gamma_0}{2\lambda} + \frac{T_M^2 \mu_1}{2\lambda} - \frac{1}{\lambda} \right) \int_0^t ds \int_{R_z} (\xi - z) u_i u_i dx$$

$$+ \frac{1}{2\lambda \lambda_1} \left(\frac{1}{\gamma_0} + \frac{1}{\mu_1} \right) \int_0^t ds \int_{R_z} (\xi - z) T_{,\alpha} T_{,\alpha} dx$$

$$- \frac{1}{\lambda} \int_0^t ds \int_{R_z} (\xi - z) u_i p_{,i} dx. \quad (3.59)$$

The difficult term to bound in (3.59) is the pressure term. This is handled by (Payne and Song, 1997) by introducing the function ω_α which solves

$$\omega_{\alpha,\alpha} = u_3 \quad \text{in} \quad D_\xi,$$
$$\omega_\alpha = 0 \quad \text{on} \quad \partial D_\xi. \quad (3.60)$$

A series of integrations by parts, use of the divergence theorem and boundary conditions, employment of (3.60) and then use of the differential equation $(3.50)_1$, leads to

$$-\frac{1}{\lambda}\int_0^t ds \int_{R_z}(\xi - z)u_i p_{,i}dx = -\frac{1}{\lambda}\int_0^t ds \int_{R_z}[(\xi - z)u_i p]_{,i}dx$$

$$+\frac{1}{\lambda}\int_0^t ds \int_{R_z}(\xi - z)_{,\xi}u_3 p\,dx$$

$$=\frac{1}{\lambda}\int_0^t ds \int_{R_z}u_3 p\,dx$$

$$=\frac{1}{\lambda}\int_0^t ds \int_{R_z}\omega_{\alpha,\alpha}p\,dx$$

$$=\frac{1}{\lambda}\int_0^t ds \int_{R_z}(\omega_\alpha p)_{,\alpha}dx - \frac{1}{\lambda}\int_0^t ds \int_{R_z}\omega_\alpha p_{,\alpha}dx$$

$$=-\frac{1}{\lambda}\int_0^t ds \int_{R_z}\omega_\alpha\big[g_\alpha T + \hat{g}_\alpha T^2 - u_\alpha + \lambda\Delta u_\alpha\big]dx. \quad (3.61)$$

The last term, in Δu_α integrates by parts to find

$$-\int_0^t ds \int_{R_z}\omega_\alpha\Delta u_\alpha dx = -\int_0^t ds \int_{R_z}(\omega_\alpha u_{\alpha,i})_{,i}dx + \int_0^t ds \int_{R_z}\omega_{\alpha,i}u_{\alpha,i}dx$$

$$=\int_0^t ds \int_{D_z}\omega_\alpha u_{\alpha,3}dA$$

$$+\int_0^t ds \int_{R_z}\omega_{\alpha,i}u_{\alpha,i}dx. \quad (3.62)$$

Thus, use of (3.62) in (3.61) shows

$$-\frac{1}{\lambda}\int_0^t ds \int_{R_z}(\xi - z)u_i p_{,i}dx = -\frac{1}{\lambda}\int_0^t ds \int_{R_z}\omega_\alpha T g_\alpha dx$$

$$-\frac{1}{\lambda}\int_0^t ds \int_{R_z}\omega_\alpha \hat{g}_\alpha T^2 dx + \frac{1}{\lambda}\int_0^t ds \int_{R_z}u_\alpha\omega_\alpha dx$$

$$+\int_0^t ds \int_{D_z}\omega_\alpha u_{\alpha,3}dA$$

$$+\int_0^t ds \int_{R_z}\omega_{\alpha,i}u_{\alpha,i}dx. \quad (3.63)$$

The maximum principle is used on the T^2 term and then the arithmetic-geometric mean inequality is employed on each term on the right of (3.63). Poincaré's inequality is used on the resulting terms in $\omega_\alpha\omega_\alpha$ and $T_{,\alpha}T_{,\alpha}$ integrated over R_z. In this manner we derive for arbitrary positive constants $\gamma_4 - \gamma_8$,

$$-\frac{1}{\lambda}\int_0^t ds \int_{R_z} (\xi - z)u_i p_{,i} dx$$

$$\leq \left[\frac{1}{2\lambda\lambda_1}(1 + T_M)\gamma_8 + \gamma_7 + \frac{1}{2}\gamma_5\right]\int_0^t ds \int_{R_z} \omega_{\alpha,\beta}\omega_{\alpha,\beta} dx$$

$$+ \frac{1}{2}\gamma_6 \int_0^t ds \int_{R_z} \omega_{\alpha,3}\omega_{\alpha,3} dx + \frac{1}{2}\gamma_4 \int_0^t ds \int_{D_z} \omega_\alpha\omega_\alpha dA$$

$$+ \frac{1}{2\lambda\gamma_7}\int_0^t ds \int_{R_z} u_\alpha u_\alpha dx + \frac{1}{2\gamma_5}\int_0^t ds \int_{R_z} u_{\alpha,\beta}u_{\alpha,\beta} dx$$

$$+ \frac{1}{2\gamma_6}\int_0^t ds \int_{R_z} u_{\alpha,3}u_{\alpha,3} dx + \frac{2}{\gamma_4}\int_0^t ds \int_{D_z} u_{\alpha,3}u_{\alpha,3} dA$$

$$+ \left(\frac{1 + T_M}{2\gamma_8\lambda\lambda_1}\right)\int_0^t ds \int_{R_z} T_{,\alpha}T_{,\alpha} dx. \tag{3.64}$$

We now need a series of inequalities derived from the Babuska-Aziz and Poincaré inequalities,

$$\int_0^t ds \int_{R_z} \omega_{\alpha,\beta}\omega_{\alpha,\beta} dx \leq C \int_0^t ds \int_{R_z} (\omega_{\alpha,\alpha})^2 dx$$

$$= C \int_0^t ds \int_{R_z} u_3^2 dx$$

$$\leq \frac{C}{\lambda_1}\int_0^t ds \int_{R_z} u_{3,\alpha}u_{3,\alpha} dx, \tag{3.65}$$

$$\int_0^t ds \int_{R_z} \omega_{\alpha,3}\omega_{\alpha,3} dx \leq \frac{1}{\lambda_1}\int_0^t ds \int_{R_z} \omega_{\alpha,3\beta}\omega_{\alpha,3\beta} dx$$

$$\leq \frac{C}{\lambda_1}\int_0^t ds \int_{R_z} (\omega_{\alpha,\alpha3})^2 dx$$

$$= \frac{C}{\lambda_1}\int_0^t ds \int_{R_z} u_{3,3}^2 dx, \tag{3.66}$$

$$\int_0^t ds \int_{D_z} \omega_\alpha\omega_\alpha dA \leq \frac{1}{\lambda_1}\int_0^t ds \int_{D_z} \omega_{\alpha,\beta}\omega_{\alpha,\beta} dA$$

$$\leq \frac{C}{\lambda_1}\int_0^t ds \int_{D_z} (\omega_{\alpha,\alpha})^2 dA$$

$$= \frac{C}{\lambda_1}\int_0^t ds \int_{D_z} u_3^2 dA$$

$$\leq \frac{C}{\lambda_1^2}\int_0^t ds \int_{D_z} u_{3,\alpha}u_{3,\alpha} dA. \tag{3.67}$$

Upon use of (3.65) – (3.67) in (3.64) we derive

$$-\frac{1}{\lambda}\int_0^t ds \int_{R_z}(\xi-z)u_i p_{,i}dx$$

$$\leq \frac{C}{\lambda_1}\Big[\frac{1}{2\lambda\lambda_1}(1+T_M)\gamma_8 + \gamma_7 + \frac{1}{2}\gamma_5\Big]\int_0^t ds \int_{R_z} u_{3,\alpha}u_{3,\alpha}dx$$

$$+\frac{C\gamma_6}{2\lambda_1}\int_0^t ds \int_{R_z} u_{3,3}^2 dx + \Big(\frac{2}{\gamma_4}+\frac{C\gamma_4^2}{4\lambda_1^2}\Big)\int_0^t ds \int_{D_z} u_{3,\alpha}u_{3,\alpha}dA$$

$$+\frac{1}{2\lambda\gamma_7}\int_0^t ds \int_{R_z} u_\alpha u_\alpha dx + \frac{1}{2\gamma_5}\int_0^t ds \int_{R_z} u_{\alpha,\beta}u_{\alpha,\beta}dx$$

$$+\frac{1}{2\gamma_6}\int_0^t ds \int_{R_z} u_{\alpha,3}u_{\alpha,3}dx + \frac{2}{\gamma_4}\int_0^t ds \int_{D_z} u_{\alpha,3}u_{\alpha,3}dA$$

$$+\Big(\frac{1+T_M}{2\gamma_8\lambda\lambda_1}\Big)\int_0^t ds \int_{R_z} T_{,\alpha}T_{,\alpha}dx. \tag{3.68}$$

Finally use of (3.68) in (3.59) leads to the estimate for $u_{i,j}u_{i,j}$,

$$\int_0^t ds \int_{R_z}(\xi-z)u_{i,j}u_{i,j}dx$$

$$\leq \frac{1}{2}\gamma_3 \int_0^t ds \int_{R_z} u_{i,3}u_{i,3}dx + \frac{1}{2\gamma_3\lambda_1}\int_0^t ds \int_{R_z} u_{i,\alpha}u_{i,\alpha}dx$$

$$+\Big(\frac{\gamma_0}{2\lambda}+\frac{T_M^2\mu_1}{2\lambda}-\frac{1}{\lambda}\Big)\int_0^t ds \int_{R_z}(\xi-z)u_i u_i dx$$

$$+\frac{C}{\lambda_1}\Big[\frac{1}{2\lambda\lambda_1}(1+T_M)\gamma_8 + \gamma_7 + \frac{1}{2}\gamma_5\Big]\int_0^t ds \int_{R_z} u_{3,\alpha}u_{3,\alpha}dx$$

$$+\frac{C\gamma_6}{2\lambda_1}\int_0^t ds \int_{R_z} u_{3,3}^2 dx + \Big(\frac{2}{\gamma_4}+\frac{C\gamma_4^2}{4\lambda_1^2}\Big)\int_0^t ds \int_{D_z} u_{3,\alpha}u_{3,\alpha}dA$$

$$+\frac{1}{2\lambda\gamma_7}\int_0^t ds \int_{R_z} u_\alpha u_\alpha dx + \frac{1}{2\gamma_5}\int_0^t ds \int_{R_z} u_{\alpha,\beta}u_{\alpha,\beta}dx$$

$$+\frac{1}{2\gamma_6}\int_0^t ds \int_{R_z} u_{\alpha,3}u_{\alpha,3}dx + \frac{2}{\gamma_4}\int_0^t ds \int_{D_z} u_{\alpha,3}u_{\alpha,3}dA$$

$$+\frac{1}{2\lambda\lambda_1}\Big(\frac{1}{\gamma_0}+\frac{1}{\mu_1}\Big)\int_0^t ds \int_{R_z}(\xi-z)T_{,\alpha}T_{,\alpha}dx$$

$$+\Big(\frac{1+T_M}{2\gamma_8\lambda\lambda_1}\Big)\int_0^t ds \int_{R_z} T_{,\alpha}T_{,\alpha}dx. \tag{3.69}$$

The idea is now to combine (3.57) and (3.69) opportunely. To this end, for constants k_1 and k_2 at our disposal, form $k_1(3.57)$ added to (3.69) and then add $k_2 \int_0^t ds \int_{R_z} (\xi - z)u_i u_i dx$ to the result. This leads to the inequality

$$
\begin{aligned}
&\left[k_1 - \frac{1}{2\lambda\lambda_1} \left(\frac{1}{\gamma_0} + \frac{1}{\mu_1} \right) \right] \int_0^t ds \int_{R_z} (\xi - z)T_{,i}T_{,i} dx \\
&+ \int_0^t ds \int_{R_z} (\xi - z)u_{i,j}u_{i,j} dx \\
&+ \left[k_2 - \frac{\gamma_0}{2\lambda} - \frac{T_M^2 \mu_1}{2\lambda} + \frac{1}{\lambda} \right] \int_0^t ds \int_{R_z} (\xi - z)u_i u_i dx \\
&\leq \frac{k_1}{2\gamma_1} \int_0^t ds \int_{R_z} T_{,3}^2 dx + \left[\frac{k_1}{2\lambda_1} \left(\gamma_1 + \frac{\gamma_2 T_M}{2} \right) + \frac{1 + T_M}{2\gamma_8 \lambda\lambda_1} \right] \int_0^t ds \int_{R_z} T_{,\alpha}T_{,\alpha} dx \\
&+ \frac{k_1 T_M}{4\gamma_2} \int_0^t ds \int_{R_z} u_3^2 dx + \frac{1}{2}\gamma_3 \int_0^t ds \int_{R_z} u_{i,3}u_{i,3} dx \\
&+ \frac{1}{2\gamma_3 \lambda_1} \int_0^t ds \int_{R_z} u_{i,\alpha}u_{i,\alpha} dx \\
&+ \frac{C}{\lambda_1} \left[\frac{1}{2\lambda\lambda_1}(1 + T_M)\gamma_8 + \gamma_7 + \frac{1}{2}\gamma_5 \right] \int_0^t ds \int_{R_z} u_{3,\alpha}u_{3,\alpha} dx \\
&+ \frac{C\gamma_6}{2\lambda_1} \int_0^t ds \int_{R_z} u_{3,3}^2 dx + \frac{1}{2\lambda\gamma_7} \int_0^t ds \int_{R_z} u_\alpha u_\alpha dx \\
&+ \frac{1}{2\gamma_5} \int_0^t ds \int_{R_z} u_{\alpha,\beta}u_{\alpha,\beta} dx + \frac{1}{2\gamma_6} \int_0^t ds \int_{R_z} u_{\alpha,3}u_{\alpha,3} dx \\
&+ \left(\frac{2}{\gamma_4} + \frac{C\gamma_4^2}{4\lambda_1^2} \right) \int_0^t ds \int_{D_z} u_{3,\alpha}u_{3,\alpha} dA \\
&+ \frac{2}{\gamma_4} \int_0^t ds \int_{D_z} u_{\alpha,3}u_{\alpha,3} dA.
\end{aligned} \tag{3.70}
$$

We pick

$$
k_1 > \frac{1}{2\lambda\lambda_1} \left(\frac{1}{\gamma_0} + \frac{1}{\mu_1} \right) \qquad \text{and} \qquad k_2 > \frac{\gamma_0}{2\lambda} + \frac{\mu_1 T_M^2}{2\lambda} - \frac{1}{\lambda}.
$$

For example, pick

$$
\frac{1}{\gamma_0} + \frac{1}{\mu_1} = 2\lambda\lambda_1, \quad k_1 = 2
$$

and pick k_2 such that

$$
k_2 - \frac{\gamma_0}{2\lambda} - \frac{\mu_1 T_M^2}{2\lambda} + \frac{1}{\lambda} \equiv K > 0.
$$

Then, if we define a positive-definite function $E(z,t)$ by

$$E(z,t) = \int_0^t ds \int_{R_z} (\xi - z) T_{,i} T_{,i} dx + \int_0^t ds \int_{R_z} (\xi - z) u_{i,j} u_{i,j} dx$$

$$+ K \int_0^t ds \int_{R_z} (\xi - z) u_i u_i dx. \tag{3.71}$$

From (3.70) we may then determine constants $\Gamma_1 - \Gamma_4$ such that

$$E(z,t) \leq \Gamma_1 \int_0^t ds \int_{R_z} u_i u_i dx + \Gamma_2 \int_0^t ds \int_{R_z} T_{,i} T_{,i} dx$$

$$+ \Gamma_3 \int_0^t ds \int_{R_z} u_{i,j} u_{i,j} dx + \Gamma_4 \int_0^t ds \int_{D_z} u_{i,j} u_{i,j} dA. \tag{3.72}$$

Observe now that

$$\frac{\partial E}{\partial z} = E_z = - \int_0^t ds \int_{R_z} T_{,i} T_{,i} dx - \int_0^t ds \int_{R_z} u_{i,j} u_{i,j} dx$$

$$- K \int_0^t ds \int_{R_z} u_i u_i dx, \tag{3.73}$$

$$\frac{\partial^2 E}{\partial z^2} = E_{zz} = \int_0^t ds \int_{D_z} T_{,i} T_{,i} dA + \int_0^t ds \int_{D_z} u_{i,j} u_{i,j} dA$$

$$+ K \int_0^t ds \int_{D_z} u_i u_i dA. \tag{3.74}$$

Then, from (3.72) we derive

$$E(z,t) \leq -\Gamma E_z + \Gamma_4 E_{zz}, \tag{3.75}$$

where

$$\Gamma = \max\{\Gamma_2, \Gamma_3, \Gamma_1/K\}.$$

Inequality (3.75) may be integrated as in (Payne and Song, 1997) who note that the inequality may be rewritten

$$E_{zz} - m_1 E_z - m_2 E \geq 0, \tag{3.76}$$

with $m_1 = \Gamma/\Gamma_4$, $m_2 = \Gamma_4^{-1}$. They then set

$$a = \frac{1}{2} m_1 + \frac{1}{2} \sqrt{m_1^2 + 4m_2}, \qquad b = -\frac{1}{2} m_1 + \frac{1}{2} \sqrt{m_1^2 + 4m_2},$$

and (3.76) is

$$E_{zz} - (a - b) E_z - ab E \geq 0. \tag{3.77}$$

This inequality may be factorized as

$$\left(\frac{\partial}{\partial z} - a \right) \left(\frac{\partial E}{\partial z} + bE \right) \geq 0 \qquad \text{or} \qquad \left(\frac{\partial}{\partial z} + b \right) \left(\frac{\partial E}{\partial z} - aE \right) \geq 0. \tag{3.78}$$

(Payne and Song, 1997) show that (3.78) integrates to derive either

$$E(z,t) \le E(0,t)e^{-bz}, \tag{3.79}$$

or

$$-E_z(z,t) + aE(z,t) \le \left[-E_z(0,t) + aE(0,t)\right]e^{-bz}. \tag{3.80}$$

Neither (3.79) nor (3.80) is a decay bound in terms of data, since $E(0,t)$ and $-E_z(0,t)$ are not directly functions of f_i and h. It remains to bound these functions in terms of the data f_i and h. This may be done as in (Payne and Song, 1997), with the details not dissimilar to those of section 3.1.3. We do not include details here since this would be long and the technicalities would obscure the content. The key thing in this section is to show how the Brinkman equations lead to a second order differential inequality for a suitable function E and how this may be bounded above by a decreasing exponential function of z.

3.3   Spatial decay for the Forchheimer equations

In this section we study the following system of Forchheimer equations for non-isothermal flow in a saturated porous medium, cf. chapter 1,

$$\begin{aligned}
b|\mathbf{u}|u_i + c|\mathbf{u}|^2 u_i + u_i &= -p_{,i} + g_i T, \\
u_{i,i} &= 0, \\
T_{,t} + u_i T_{,i} &= \Delta T.
\end{aligned} \tag{3.81}$$

Again u_i, p, T are velocity, pressure, and temperature, and g_i is a gravity vector which we assume (without loss of generality) satisfies the constraint

$$|\mathbf{g}| \le 1.$$

The thermal diffusivity is chosen to be equal to 1 without any loss for the analysis contained herein. The coefficients $b > 0$ and $c > 0$ are Forchheimer coefficients and $|\mathbf{u}| = \sqrt{u_i u_i}$.

The domains D, R, D_z, R_z, and ∂R and ∂R_c are as in sections 3.1 and 3.2.

Equations (3.81) are defined on $R \times \{t > 0\}$, with the following conditions,

$$\begin{aligned}
T(x_1, x_2, x_3, 0) &= 0 \text{ in } R, \\
u_\alpha n_\alpha &= 0 \text{ on } \partial D \times \{z > 0\} \times \{t \ge 0\}, \\
T(x_1, x_2, x_3, t) &= 0 \text{ on } \partial D \times \{z > 0\} \times \{t \ge 0\}, \\
u_3(x_1, x_2, 0, t) &= f(x_1, x_2, t), \text{ in } \bar{D} \times \{t > 0\}, \\
T(x_1, x_2, 0, t) &= h(x_1, x_2, t), \text{ in } \bar{D} \times \{t > 0\}.
\end{aligned} \tag{3.82}$$

In addition, the following decay-like bounds are imposed on the solution,

$$u_i, T = O(1), \quad u_3, T_{,i}, p = o(1/z) \text{ as } z(= x_3) \to \infty, \quad \forall x_1, x_2, t. \quad (3.83)$$

The first analysis of spatial decay for a Forchheimer system is due to (Payne and Song, 2002) and they study (3.81) – (3.83) but with $c = 0$. The analysis contained herein is a natural extension of their work.

The nonlinear b, c terms in $(3.81)_1$ necessitate that we do not work immediately with a space time integral. Thus, we commence by taking the inner product of $(3.81)_1$ with u_i and integrate over R_z,

$$b \int_{R_z} |\mathbf{u}|^3 dx + c \int_{R_z} |\mathbf{u}|^4 dx + \int_{R_z} u_i u_i dx$$
$$= \int_{R_z} g_i T u_i dx - \int_{R_z} u_i p_{,i} dx. \quad (3.84)$$

Again, the pressure term is the tricky one. To handle this, we proceed as follows,

$$-\int_{R_z} u_i p_{,i} dx = -\int_{R_z} (u_i p)_{,i} dx$$
$$= -\oint_{\partial R_z} u_i n_i p \, dS$$
$$= \int_{D_z} u_3 p \, dA \quad (3.85)$$

where the divergence theorem and boundary conditions have been employed. The function w_α of (3.60) is now utilized.

$$\int_{D_z} u_3 p dA = \int_{D_z} w_{\alpha,\alpha} p \, dA$$
$$= \int_{D_z} (w_\alpha p)_{,\alpha} dA - \int_{D_z} w_\alpha p_{,\alpha} dA$$
$$= -\int_{D_z} w_\alpha p_{,\alpha} dA$$
$$= \int_{D_z} w_\alpha [b|\mathbf{u}|u_\alpha + c|\mathbf{u}|^2 u_\alpha + u_\alpha - g_\alpha T] dA \quad (3.86)$$

where we have integrated by parts, used the divergence theorem, boundary conditions, and the differential equation.

Since D is a two-dimensional domain the Sobolev inequality holds for w_α in the form

$$\int_{D_z} (w_\alpha w_\alpha)^2 dA \leq \Lambda^2 \int_{D_z} w_\alpha w_\alpha dA \int_{D_z} w_{\alpha,\beta} w_{\alpha,\beta} dA, \quad (3.87)$$

where $\Lambda^2 > 0$ is a constant. Also, with the aid of the Babuska-Aziz and Poincaré inequalities we know that

$$\int_{D_z} \omega_{\alpha,\beta}\omega_{\alpha,\beta}dA \leq C\int_{D_z} \omega_{\alpha,\alpha}^2 dA = C\int_{D_z} u_3^2 dA, \qquad (3.88)$$

and

$$\int_{D_z} \omega_\alpha \omega_\alpha dA \leq \frac{1}{\lambda_1}\int_{D_z} \omega_{\alpha,\beta}\omega_{\alpha,\beta}dA \leq \frac{C}{\lambda_1}\int_{D_z} u_3^2 dA. \qquad (3.89)$$

These inequalities are needed to estimate the right hand side of (3.86).

The first three terms on the right of (3.86) are manipulated using Hölder's inequality, the Cauchy-Schwarz inequality, and inequalities (3.87) – (3.89), as now shown.

$$
\begin{aligned}
\int_{D_z} \omega_\alpha |\mathbf{u}| u_\alpha dA &\leq \left(\int_{D_z} (\omega_\alpha \omega_\alpha)^{3/2} dA\right)^{1/3}\left(\int_{D_z} |\mathbf{u}|^3 dA\right)^{2/3} \\
&\leq \left(\int_{D_z} (\omega_\alpha \omega_\alpha)^2 dA \int_{D_z} \omega_\alpha \omega_\alpha dA\right)^{1/6}\left(\int_{D_z} |\mathbf{u}|^3 dA\right)^{2/3} \\
&\leq \left(\frac{\Lambda^2 C^2}{\lambda_1}\left[\int_{D_z} u_3^2 dA\right]^2\right)^{1/6}\left(\frac{C}{\lambda_1}\int_{D_z} u_3^2 dA\right)^{1/6}\left(\int_{D_z} |\mathbf{u}|^3 dA\right)^{2/3} \\
&= \frac{\Lambda^{1/3} C^{1/2}}{\lambda_1^{1/3}}\left(\int_{D_z} u_3^2 dA\right)^{1/2}\left(\int_{D_z} |\mathbf{u}|^3 dA\right)^{2/3} \\
&\leq \frac{\Lambda^{1/3} C^{1/2}}{\lambda_1^{1/3}}\left[\left(\int_{D_z} dA\right)^{1/3}\left(\int_{D_z} |u_3|^3 dA\right)^{2/3}\right]^{1/2}\left(\int_{D_z} |\mathbf{u}|^3 dA\right)^{2/3} \\
&\leq k_1 \int_{D_z} |\mathbf{u}|^3 dA, \qquad (3.90)
\end{aligned}
$$

where $m = m(D)$ is the measure of D and

$$k_1 = \frac{m^{1/6}\Lambda^{1/3}C^{1/2}}{\lambda_1^{1/3}}.$$

The second term on the right of (3.86) is bounded by,

$$
\begin{aligned}
\int_{D_z} \omega_\alpha |\mathbf{u}|^2 u_\alpha dA &\leq \left(\int_{D_z} (\omega_\alpha \omega_\alpha)^2 dA\right)^{1/4}\left(\int_{D_z} |\mathbf{u}|^4 dA\right)^{3/4} \\
&\leq \frac{\Lambda^{1/2} C^{1/2}}{\lambda_1^{1/4}}\left(\int_{D_z} u_3^2 dA\right)^{1/2}\left(\int_{D_z} |\mathbf{u}|^4 dA\right)^{3/4} \\
&\leq \frac{\Lambda^{1/2} C^{1/2} m^{1/4}}{\lambda_1^{1/4}}\left(\int_{D_z} u_3^4 dA\right)^{1/4}\left(\int_{D_z} |\mathbf{u}|^4 dA\right)^{3/4} \\
&\leq k_2 \int_{D_z} |\mathbf{u}|^4 dA, \qquad (3.91)
\end{aligned}
$$

where the constant k_2 has the form

$$k_2 = \frac{m^{1/4}\Lambda^{1/2}C^{1/2}}{\lambda_1^{1/4}}.$$

Also, the third term on the right of (3.86) is estimated via

$$\int_{D_z} \omega_\alpha u_\alpha dA \leq \sqrt{\int_{D_z} u_\alpha u_\alpha dA \int_{D_z} \omega_\alpha \omega_\alpha dA}$$

$$\leq \sqrt{\frac{C}{\lambda_1}} \sqrt{\int_{D_z} u_\alpha u_\alpha dA \int_{D_z} u_3^2 dA}$$

$$\leq \sqrt{\frac{C}{\lambda_1}} \int_{D_z} u_i u_i dA. \tag{3.92}$$

The final term in (3.86) is manipulated with the arithmetic-geometric mean inequality for a constant $\beta > 0$ to be chosen, and then with the Poincaré and Babuska-Aziz inequalities,

$$-\int_{D_z} \omega_\alpha g_\alpha T dA \leq \frac{1}{2\beta} \int_{D_z} T^2 dA + \frac{\beta}{2} \int_{D_z} \omega_\alpha \omega_\alpha dA$$

$$\leq \frac{1}{2\beta} \int_{D_z} T^2 dA + \frac{\beta C}{2\lambda_1} \int_{D_z} u_3^2 dA. \tag{3.93}$$

We may now employ (3.90) – (3.93) in (3.86) to arrive at

$$\int_{D_z} u_3 p\, dA \leq \frac{1}{2\beta} \int_{D_z} T^2 dA + \frac{\beta C}{2\lambda_1} \int_{D_z} u_3^2 dA$$

$$+ \sqrt{\frac{C}{\lambda_1}} \int_{D_z} u_i u_i dA + k_2 c \int_{D_z} |\mathbf{u}|^4 dA \tag{3.94}$$

$$+ k_1 b \int_{D_z} |\mathbf{u}|^3 dA.$$

The next step is to use (3.94) together with (3.85) in equation (3.84). We also use the arithmetic-geometric mean inequality on the first term on the right of (3.84) for $\gamma > 0$ a constant to be selected. In this way we may obtain

$$b \int_{R_z} |\mathbf{u}|^3 dx + c \int_{R_z} |\mathbf{u}|^4 dx + \int_{R_z} u_i u_i dx \leq \frac{\gamma}{2} \int_{R_z} u_i u_i dx$$

$$+ \frac{1}{2\gamma} \int_{R_z} T^2 dx + \frac{1}{2\beta} \int_{D_z} T^2 dA + \frac{\beta C}{2\lambda_1} \int_{D_z} u_3^2 dA$$

$$+ \sqrt{\frac{C}{\lambda_1}} \int_{D_z} u_i u_i dA + k_2 c \int_{D_z} |\mathbf{u}|^4 dA \tag{3.95}$$

$$+ k_1 b \int_{D_z} |\mathbf{u}|^3 dA.$$

Next, select $\gamma = 1$. We treat the T^2 term over D_z by integration, the Cauchy-Schwarz and Poincaré inequalities,

$$
\begin{aligned}
\int_{D_z} T^2 dA &= -\int_{R_z} (T^2)_{,\xi} dx \\
&= -2 \int_z^\infty d\xi \int_{D_\xi} TT_{,3} dA \\
&\leq 2 \sqrt{\int_{R_z} T^2 dx \int_{R_z} T_{,3}^2 dx} \\
&\leq \frac{2}{\sqrt{\lambda_1}} \int_{R_z} T_{,i} T_{,i} dx.
\end{aligned}
\tag{3.96}
$$

The T^2 term over R_z in (3.95) is also treated by the Poincaré inequality and then from (3.95) we may obtain

$$
\begin{aligned}
b \int_{R_z} |\mathbf{u}|^3 dx &+ c \int_{R_z} |\mathbf{u}|^4 dx + \frac{1}{2} \int_{R_z} u_i u_i dx \\
&\leq \left(\frac{1}{2\lambda_1} + \frac{1}{\beta\sqrt{\lambda_1}} \right) \int_{R_z} T_{,i} T_{,i} dx \\
&+ \left(\frac{\beta C}{2\lambda_1} + \sqrt{\frac{C}{\lambda_1}} \right) \int_{D_z} u_i u_i dA \\
&+ k_1 b \int_{D_z} |\mathbf{u}|^3 dA + k_2 c \int_{D_z} |\mathbf{u}|^4 dA.
\end{aligned}
\tag{3.97}
$$

Inequality (3.97) is now integrated over $(0, t)$ to yield an estimate for a functional of u_i

$$
\begin{aligned}
b \int_0^t ds \int_{R_z} |\mathbf{u}|^3 dx &+ c \int_0^t ds \int_{R_z} |\mathbf{u}|^4 dx + \frac{1}{2} \int_0^t ds \int_{R_z} u_i u_i dx \\
&\leq \left(\frac{1}{2\lambda_1} + \frac{1}{\beta\sqrt{\lambda_1}} \right) \int_0^t ds \int_{R_z} T_{,i} T_{,i} dx \\
&+ \left(\frac{\beta C}{2\lambda_1} + \sqrt{\frac{C}{\lambda_1}} \right) \int_0^t ds \int_{D_z} u_i u_i dA \\
&+ k_1 b \int_0^t ds \int_{D_z} |\mathbf{u}|^3 dA + k_2 c \int_0^t ds \int_{D_z} |\mathbf{u}|^4 dA.
\end{aligned}
\tag{3.98}
$$

3.3.1   An estimate for $\operatorname{grad} T$

We begin with the same T function as in section 3.2. Using integration by parts,

$$\int_0^t ds \int_{R_z} (\xi - z) T_{,i} T_{,i} dx = \int_0^t ds \int_{R_z} \left[(\xi - z) T_{,i} T \right]_{,i} dx$$
$$- \int_0^t ds \int_{R_z} T \left[(\xi - z) T_{,i} \right]_{,i} dx.$$

The first term on the right is zero as may be seen by employing the divergence theorem and boundary conditions. The differentiation is performed on the second term and then the resulting expression involving ΔT is transformed by using the differential equation $(3.81)_3$, to find

$$\int_0^t ds \int_{R_z} (\xi - z) T_{,i} T_{,i} dx = - \int_0^t ds \int_{R_z} T T_{,3} dx$$
$$- \frac{1}{2} \int_0^t ds \frac{\partial}{\partial s} \int_{R_z} (\xi - z) T^2 dx$$
$$- \frac{1}{2} \int_0^t ds \int_{R_z} (\xi - z)(u_i T^2)_{,i} dx.$$

We perform the differentiation on the second term on the right, then integrate by parts and use the divergence theorem on the last term to deduce that

$$\int_0^t ds \int_{R_z} (\xi - z) T_{,i} T_{,i} dx = - \int_0^t ds \int_{R_z} T T_{,3} dx$$
$$- \frac{1}{2} \int_{R_z} (\xi - z) T^2(t) dx$$
$$+ \frac{1}{2} \int_0^t ds \int_{R_z} u_3 T^2 dx. \qquad (3.99)$$

Since the second term on the right of (3.99) is non-positive we discard this. The Cauchy-Schwarz, arithmetic-geometric mean and Poincaré inequalities together with the maximum principle are then used on the remaining two terms on the right of (3.99) to find

$$\int_0^t ds \int_{R_z} (\xi - z) T_{,i} T_{,i} dx \leq \sqrt{\int_0^t ds \int_{R_z} T^2 dx} \sqrt{\int_0^t ds \int_{R_z} T_{,3}^2 dx}$$

$$+ \frac{T_M}{2} \sqrt{\int_0^t ds \int_{R_z} u_3^2 dx} \sqrt{\int_0^t ds \int_{R_z} T^2 dx}$$

$$\leq \frac{1}{\sqrt{\lambda_1}} \int_0^t ds \int_{R_z} T_{,i} T_{,i} dx$$

$$+ \frac{T_M}{2\mu} \int_0^t ds \int_{R_z} u_3^2 dx$$

$$+ \frac{T_M \mu}{2\lambda_1} \int_0^t ds \int_{R_z} T_{,i} T_{,i} dx, \qquad (3.100)$$

where $\mu > 0$ is a constant at our disposal. Thus, we obtain our required estimate for $\mathrm{grad}\, T$, namely

$$\int_0^t ds \int_{R_z} (\xi - z) T_{,i} T_{,i} dx \leq \left(\frac{1}{\sqrt{\lambda_1}} + \frac{T_M \mu}{2\lambda_1} \right) \int_0^t ds \int_{R_z} T_{,i} T_{,i} dx$$

$$+ \frac{T_M}{2\mu} \int_0^t ds \int_{R_z} u_i u_i dx. \qquad (3.101)$$

We now form the combination (3.101)+Γ(3.98) to arrive at

$$\int_0^t ds \int_{R_z} (\xi - z) T_{,i} T_{,i} dx + \frac{1}{2} \Gamma \int_0^t ds \int_{R_z} u_i u_i dx + b\Gamma \int_0^t ds \int_{R_z} |\mathbf{u}|^3 dx$$

$$+ c\Gamma \int_0^t ds \int_{R_z} |\mathbf{u}|^4 dx$$

$$\leq \left(\frac{1}{\sqrt{\lambda_1}} + \frac{T_M \mu}{2\lambda_1} + \frac{\Gamma}{2\lambda_1} + \frac{\Gamma}{\beta \sqrt{\lambda_1}} \right) \int_0^t ds \int_{R_z} T_{,i} T_{,i} dx$$

$$+ \frac{T_M}{2\mu} \int_0^t ds \int_{R_z} u_i u_i dx$$

$$+ \Gamma \left(\frac{\beta C}{2\lambda_1} + \sqrt{\frac{C}{\lambda_1}} \right) \int_0^t ds \int_{D_z} u_i u_i dA$$

$$+ \Gamma k_1 b \int_0^t ds \int_{D_z} |\mathbf{u}|^3 dA$$

$$+ \Gamma k_2 c \int_0^t ds \int_{D_z} |\mathbf{u}|^4 dA. \qquad (3.102)$$

To remove the $u_i u_i$ term integrated over R_z from the right hand side of (3.102) we now select $\Gamma > T_M / \mu$, e.g. pick $\Gamma = 2 T_M / \mu$. Then, we may arrive at

$$\int_0^t ds \int_{R_z} (\xi - z) T_{,i} T_{,i} dx + \alpha_1 \int_0^t ds \int_{R_z} u_i u_i dx + \alpha_2 \int_0^t ds \int_{R_z} |\mathbf{u}|^3 dx$$

$$+ \alpha_3 \int_0^t ds \int_{R_z} |\mathbf{u}|^4 dx \leq a_1 \int_0^t ds \int_{R_z} T_{,i} T_{,i} dx$$

$$+ a_2 \int_0^t ds \int_{D_z} u_i u_i dA + a_3 \int_0^t ds \int_{D_z} |\mathbf{u}|^3 dA$$

$$+ a_4 \int_0^t ds \int_{D_z} |\mathbf{u}|^4 dA. \tag{3.103}$$

Now define the function $E(z,t)$ by

$$E(z,t) = \int_0^t ds \int_{R_z} (\xi - z) T_{,i} T_{,i} dx + \alpha_1 \int_0^t ds \int_{R_z} u_i u_i dx$$

$$+ \alpha_2 \int_0^t ds \int_{R_z} |\mathbf{u}|^3 dx + \alpha_3 \int_0^t ds \int_{R_z} |\mathbf{u}|^4 dx. \tag{3.104}$$

Note that

$$\frac{\partial E}{\partial z} = E_z = - \int_0^t ds \int_{R_z} T_{,i} T_{,i} dx - \alpha_1 \int_0^t ds \int_{D_z} u_i u_i dA$$

$$- \alpha_2 \int_0^t ds \int_{D_z} |\mathbf{u}|^3 dA - \alpha_3 \int_0^t ds \int_{D_z} |\mathbf{u}|^4 dA.$$

Let now ζ be the number

$$\zeta = \max \left\{ 1, \frac{a_2}{a_1 \alpha_1}, \frac{a_3}{a_1 \alpha_2}, \frac{a_4}{a_1 \alpha_3} \right\}. \tag{3.105}$$

Then, from (3.103) we may derive the inequality

$$E \leq -a_1 \zeta E_z. \tag{3.106}$$

This inequality is integrated to see that

$$E(z,t) \leq E(0,t) e^{-z/a_1 \zeta}. \tag{3.107}$$

As it stands, (3.107) is not an *a priori* estimate because $E(0,t)$ is not given directly in terms of the data functions f and h.

3.3.2 An estimate for $E(0,t)$

In order to find a data bound for $E(0,t)$ we put $k_1 = \lambda_1^{-1/2} + T_M \mu / 2\lambda_1$, $k_2 = T_M / 2\mu$ and note (3.101) becomes

$$\int_0^t ds \int_{R_z} (\xi - z) T_{,i} T_{,i} dx \leq k_1 \int_0^t ds \int_{R_z} T_{,i} T_{,i} dx + k_2 \int_0^t ds \int_{R_z} u_i u_i dx.$$

Upon using this with $z = 0$ in the definition for E, (3.104), we find

$$E(0,t) \leq k_1 \int_0^t ds \int_R T_{,i}T_{,i}dx + (\alpha_1 + k_2) \int_0^t ds \int_R u_i u_i dx$$

$$+ \alpha_2 \int_0^t ds \int_R |\mathbf{u}|^3 dx + \alpha_3 \int_0^t ds \int_R |\mathbf{u}|^4 dx. \qquad (3.108)$$

Define now S as in (Payne and Song, 2002), namely S solves

$$\frac{\partial S}{\partial t} = \Delta S \quad \text{in} \quad R \times \{t > 0\}, \qquad (3.109)$$

where S satisfies the same boundary and initial conditions as T. From the triangle inequality

$$\left(\int_0^t ds \int_R T_{,i}T_{,i}dx \right)^{1/2} \leq \left(\int_0^t ds \int_R (T-S)_{,i}(T-S)_{,i}dx \right)^{1/2}$$

$$+ \left(\int_0^t ds \int_R S_{,i}S_{,i}dx \right)^{1/2}$$

and squaring this we may deduce

$$\int_0^t ds \int_R T_{,i}T_{,i}dx \leq 2 \Bigg[\int_0^t ds \int_R (T-S)_{,i}(T-S)_{,i}dx$$

$$+ \int_0^t ds \int_R S_{,i}S_{,i}dx \Bigg]. \qquad (3.110)$$

Next, integrating by parts and using the equations $S_{,t} = \Delta S$, $\Delta T = u_i T_{,i} + T_{,t}$, recalling $T = S$ on the boundary and initially, we may show

$$\int_0^t ds \int_R (T-S)_{,i}(T-S)_{,i}dx = \int_0^t ds \int_R [(T-S)_{,i}(T-S)]_{,i}dx$$

$$- \int_0^t ds \int_R (T-S)\Delta(T-S)dx$$

$$= - \int_0^t ds \int_R (T-S)(T-S)_{,t}dx$$

$$- \int_0^t ds \int_R (T-S)u_i T_{,i}dx$$

$$= - \frac{1}{2} \int_R (T-S)^2 dx \Big|_{\text{time } t}$$

$$+ \int_0^t ds \int_R (T-S)_{,i}u_i T dx.$$

Drop the first term on the right and split the second into two terms to see that

$$\int_0^t ds \int_R (T-S)_{,i}(T-S)_{,i}dx \le \frac{1}{2}\int_0^t ds \int_R (u_i T^2)_{,i}dx - \int_0^t ds \int_R S_{,i}u_i T dx$$

$$\le -\frac{1}{2}\int_0^t ds \int_D u_i n_i T^2 dA$$

$$+ T_M \sqrt{\int_0^t ds \int_R S_{,i}S_{,i}dx \int_0^t ds \int_R u_i u_i dx}$$

$$\le -\frac{1}{2}\int_0^t ds \int_D h^2 f dA + \frac{T_M^2}{2\epsilon}\int_0^t ds \int_R S_{,i}S_{,i}dx$$

$$+ \frac{\epsilon}{2}\int_0^t ds \int_R u_i u_i dx, \tag{3.111}$$

where $\epsilon > 0$ is to be chosen. (Lin and Payne, 1994) analyse the spatial decay problem for S in a semi-infinite cylinder in detail. They obtain true spatial decay estimates for a functional of form $\int_0^t ds \int_{R_z} S_{,i}S_{,i}dx + \int_0^t ds \int_{R_z} S^2 dx$. In particular, they show how to bound the $S_{,i}S_{,i}$ piece in (3.111) in terms of data, namely in terms of integrals over D of h^2, $(\partial h/\partial t)^2$ and $|\text{grad}_s h|^2$, where $\text{grad}_s h$ denotes the tangential gradient of h. Thus, we employ (3.111) in (3.110) and let Q_1 be the data bound

$$\left(2 + \frac{T_M^2}{\epsilon}\right)\int_0^t ds \int_R S_{,i}S_{,i}dx - \int_0^t ds \int_D h^2 f \, dA \le Q_1. \tag{3.112}$$

In this manner we find

$$\int_0^t ds \int_R T_{,i}T_{,i}dx \le Q_1 + \epsilon \int_0^t ds \int_R u_i u_i dx. \tag{3.113}$$

3.3.3 An estimate for $u_i u_i$

We multiply the differential equation $(3.81)_1$ by u_i and integrate over $R \times (0,t)$ to obtain

$$\int_0^t ds \int_R u_i u_i dx + b \int_0^t ds \int_R |\mathbf{u}|^3 dx + c \int_0^t ds \int_R |\mathbf{u}|^4 dx =$$

$$= -\int_0^t ds \int_R p_{,i}u_i dx + \int_0^t ds \int_R g_i u_i T dx. \tag{3.114}$$

We now utilize the function ϕ_i defined in (3.39) in order to manipulate the pressure term,

$$
-\int_0^t ds \int_R p_{,i} u_i dx = -\int_0^t ds \int_0^\infty dz \int_{D_z} \frac{\partial}{\partial x_i}(pu_i)dA
$$

$$
= \int_0^t ds \int_D pu_i n_i dA
$$

$$
= \int_0^t ds \int_D p\phi_i n_i dA
$$

$$
= -\int_0^t ds \int_R p_{,i}\phi_i dx
$$

$$
= \int_0^t ds \int_R \phi_i(-g_i T + u_i + b|\mathbf{u}|u_i + c|\mathbf{u}|^2 u_i)dx. \qquad (3.115)
$$

Upon use of (3.115) in (3.114) we find

$$
\int_0^t ds \int_R u_i u_i dx + b\int_0^t ds \int_R |\mathbf{u}|^3 dx + c\int_0^t ds \int_R |\mathbf{u}|^4 dx
$$

$$
= \int_0^t ds \int_R Tg_i(u_i - \phi_i)dx + \int_0^t ds \int_R \phi_i u_i dx
$$

$$
+ b\int_0^t ds \int_R \phi_i|\mathbf{u}|u_i dx + c\int_0^t ds \int_R \phi_i|\mathbf{u}|^2 u_i dx.
$$

We next use the arithmetic-geometric mean inequality on the terms on the right to derive

$$
\int_0^t ds \int_R u_i u_i dx + b\int_0^t ds \int_R |\mathbf{u}|^3 dx + c\int_0^t ds \int_R |\mathbf{u}|^4 dx
$$

$$
\leq \left(\frac{1}{2\delta_1} + \frac{1}{2\delta_2}\right)\int_0^t ds \int_R T^2 dx + \frac{\delta_1}{2}\int_0^t ds \int_R u_i u_i dx
$$

$$
+ \left(\frac{\delta_2}{2} + \frac{b}{2\delta_3}\right)\int_0^t ds \int_R \phi_i\phi_i dx + \left(\frac{b\delta_3}{2} + \frac{\delta_4 c}{2}\right)\int_0^t ds \int_R |\mathbf{u}|^4 dx
$$

$$
+ \frac{c}{2\delta_4}\int_0^t ds \int_R |\mathbf{u}|^2|\phi|^2 dx
$$

$$
\leq \left(\frac{1}{2\delta_1} + \frac{1}{2\delta_2}\right)\int_0^t ds \int_R T^2 dx + \frac{\delta_1}{2}\int_0^t ds \int_R u_i u_i dx
$$

$$
+ \left(\frac{\delta_2}{2} + \frac{b}{2\delta_3}\right)\int_0^t ds \int_R \phi_i\phi_i dx
$$

$$
+ \left(\frac{b\delta_3}{2} + \frac{c\delta_4}{2} + \frac{c\delta_5}{4\delta_4}\right)\int_0^t ds \int_R |\mathbf{u}|^4 dx
$$

$$
+ \frac{c}{4\delta_4\delta_5}\int_0^t ds \int_R |\phi|^4 dx, \qquad (3.116)
$$

where $\delta_1, \ldots, \delta_5 > 0$ are at our disposal. Select now, for example, $\delta_1 = 1$, $\delta_2 = 1$, $\delta_3 = c/3b$, $\delta_4 = 1/3$, and $\delta_5 = 2/9$. Then, bounding the T^2 term in (3.116) using Poincaré's inequality we derive from (3.116)

$$\int_0^t ds \int_R u_i u_i dx + b \int_0^t ds \int_R |\mathbf{u}|^3 dx + c \int_0^t ds \int_R |\mathbf{u}|^4 dx$$
$$\leq \frac{1}{\lambda_1} \int_0^t ds \int_R T_{,i} T_{,i} dx + \left(\frac{1}{2} + \frac{3b^2}{2c}\right) \int_0^t ds \int_R \phi_i \phi_i dx$$
$$+ \frac{27c}{8} \int_0^t ds \int_R |\phi|^4 dx. \tag{3.117}$$

3.3.4   Bounding ϕ_i

The next step is to bound the ϕ_i contributions in terms of data. The function ϕ_i is chosen as just before (3.45) so that $\phi_3 = f(x_1, x_2, t)e^{-\sigma z}$ and then ϕ_1, ϕ_2 may be chosen to vanish on ∂D and satisfy (3.45) namely

$$\phi_{\alpha,\alpha} = \sigma \phi_3. \tag{3.118}$$

To bound the $|\phi|^4$ term note

$$\int_0^t ds \int_R |\phi|^4 dx = \int_0^t ds \int_R (\phi_3^2 + \phi_\alpha \phi_\alpha)^2 dx$$
$$\leq 2 \int_0^t ds \int_R \phi_3^4 dx + 2 \int_0^t ds \int_R (\phi_\alpha \phi_\alpha)^2 dx,$$
$$\leq 2 \int_0^t ds \int_R \phi_3^4 dx$$
$$+ 2\Lambda^2 \int_0^t ds \int_0^\infty dz \left(\int_{D_z} \phi_\alpha \phi_\alpha dA \int_{D_z} \phi_{\alpha,\beta} \phi_{\alpha,\beta} dA\right),$$

where the Sobolev inequality (3.87) has been employed. The Poincaré inequality is employed on $\int_{D_z} \phi_\alpha \phi_\alpha dA$ and then the Babuska-Aziz inequality is utilized to obtain

$$\int_0^t ds \int_R |\phi|^4 dx \leq 2 \int_0^t ds \int_R \phi_3^4 dx + \frac{2\Lambda^2 C^2}{\lambda_1} \int_0^t ds \int_0^\infty dz \left[\int_{D_z} (\phi_{\alpha,\alpha})^2 dx\right]^2.$$

Recalling the definition of ϕ_3 and (3.118) we find

$$\int_0^t ds \int_R |\phi|^4 dx \leq \frac{1}{2\sigma} \int_0^t ds \int_D f^4 dA + \frac{\sigma \Lambda^2 C^2}{2\lambda_1} \int_0^t ds \left[\int_D f^2 dA\right]^2$$
$$= \hat{Q}_2 \text{ (data)}, \tag{3.119}$$

where $\hat{Q}_2$ is the indicated data term. Similarly, one shows

$$\int_0^t ds \int_R \phi_i \phi_i dx \leq \left(\frac{1}{2\sigma} + \frac{C\sigma}{2\lambda_1}\right) \int_0^t ds \int_D f^2 dA \equiv \hat{Q}_3 \text{ (data)}. \tag{3.120}$$

Upon employment of (3.119) and (3.120) in (3.117) we derive

$$\frac{1}{2}\int_0^t ds \int_R u_i u_i dx + b \int_0^t ds \int_R |\mathbf{u}|^3 dx + \frac{c}{2}\int_0^t ds \int_R |\mathbf{u}|^4 dx$$
$$\leq \frac{1}{\lambda_1}\int_0^t ds \int_R T_{,i}T_{,i}dx + Q_2 + Q_3, \qquad (3.121)$$

where $Q_2 = 27c\hat{Q}_2/8$ and $Q_3 = \hat{Q}_3(1 + 3b^2/c)/2$.

If we now utilize (3.121) in (3.113) and pick $\epsilon = \lambda_1/2$ we may obtain

$$\int_0^t ds \int_R T_{,i}T_{,i}dx \leq 2Q_1 + \lambda_1(Q_2 + Q_3). \qquad (3.122)$$

This inequality is now used in (3.121) to find

$$\int_0^t ds \int_R u_i u_i dx + b \int_0^t ds \int_R |\mathbf{u}|^3 dx + c \int_0^t ds \int_R |\mathbf{u}|^4 dx$$
$$\leq \frac{4Q_1}{\lambda_1} + 3(Q_2 + Q_3). \qquad (3.123)$$

Finally, we may use (3.122) and (3.123) in inequality (3.108) to find a data bound for $E(0,t)$,

$$E(0,t) \leq k_1 \big[2Q_1 + \lambda_1(Q_2 + Q_3)\big] + A\Big[\frac{4Q_1}{\lambda_1} + 3(Q_2 + Q_3)\Big]$$
$$= Q_4 \text{ (data)}. \qquad (3.124)$$

The required spatial decay estimate follows by utilizing (3.124) in inequality (3.107),

$$E(z,t) \leq Q_4 e^{-z/a_1\zeta}. \qquad (3.125)$$

3.4   Spatial decay for a Krishnamurti model

(Krishnamurti, 1997) produced a very interesting model for studying penetrative convection in a fluid. A complete linear instability and nonlinear energy stability analysis for her model was provided by (Straughan, 2002b). Her model relies on a pH indicator called thymol blue being dissolved in water and as such is a double diffusive model with an equation for the temperature of the fluid coupled to an equation for the concentration of thymol blue. The penetrative effect is provided by the heat source depending on the thymol blue concentration. In this section we consider spatial decay for a Krishnamurti model in a Darcy porous medium. Stability and instability studies for this model are given by (Hill, 2005a), who also considers a Brinkman theory, a theory where the heat source is nonlinear, and when the density in the buoyancy force depends on temperature and concentration, see also (Hill, 2003; Hill, 2004b; Hill, 2004a; Hill, 2004c; Hill, 2005b).

The model we consider is a direct adaption of the (Krishnamurti, 1997) one in which the buoyancy force term depends only on temperature in a linear manner, and the heat soure depends linearly on thymol blue concentration. Thus, the equations are

$$
\begin{aligned}
u_i &= -p_{,i} + g_i T, \\
u_{i,i} &= 0, \\
T_{,t} + u_i T_{,i} &= \Delta T + \gamma C, \\
C_{,t} + u_i C_{,i} &= \Delta C,
\end{aligned}
\tag{3.126}
$$

where the notation is as in section 3.1, C represents a concentration, and $\gamma > 0$ is a constant.

(Payne and Song, 2002) develop spatial decay estimates for a double diffusive model in either a Darcy or Brinkman porous material. They allow for a Soret effect, thus, have a linear term in ΔT added to the right hand side of $(3.126)_4$, although they do not have the γC term in $(3.126)_3$. They also have a $g_i^1 C$ term in $(3.126)_1$. In fact, (Payne and Song, 2002) study their double diffusive model in the steady case in which $C_{,t} = 0$, $T_{,t} = 0$. The time dependent analogue is considered for a two-dimensional spatial domain by (Song, 2002). We here also develop spatial decay estimates for (3.126) in the steady case. Thus, we study the system

$$
\begin{aligned}
u_i &= -p_{,i} + g_i T, \\
u_{i,i} &= 0, \\
u_i T_{,i} &= \Delta T + \gamma C, \\
u_i C_{,i} &= \Delta C.
\end{aligned}
\tag{3.127}
$$

The spatial domains R, D, R_z, D_z are as in section 3.1. On the boundary we assume

$$
u_i n_i = 0, \quad T = 0, \quad C = 0 \quad \text{on } \partial D,
\tag{3.128}
$$

with flow conditions at $z = 0$,

$$
u_3 = F_1, \quad T = F_2, \quad C = F_3 \quad \text{on } D_0, \text{ i.e. } z = 0.
\tag{3.129}
$$

The asymptotic conditions at infinity are as in (Payne and Song, 2002)

$$
u_i, p, T, C, \nabla T, \nabla C \to 0 \text{ uniformly in } (x_1, x_2) \text{ as } x_3 = z \to \infty.
\tag{3.130}
$$

Since $u_{i,i} = 0$ we also require (cf. the argument leading to (3.5)),

$$
\int_{D_0} u_3 \, dA = \int_{D_0} F_1 \, dA = 0.
\tag{3.131}
$$

Our goal is to establish an exponential decay estimate in z for a function of form

$$
E = K_1 \int_{R_z} T_{,i} T_{,i} \, dx + K_2 \int_{R_z} C_{,i} C_{,i} \, dx + \int_{R_z} u_i u_i \, dx,
$$

for K_1, K_2 suitable constants.

3.4.1 Estimates for $T_{,i}T_{,i}$ and $C_{,i}C_{,i}$

We observe that we have a maximum principle for C but not for T. To bound C we integrate by parts to find

$$\int_{R_z} C_{,i}C_{,i}dx = \int_{R_z} (C_{,i}C)_{,i}dx - \int_{R_z} C\Delta C dx$$

$$= -\int_{D_z} CC_{,3}dA - \int_{R_z} \frac{1}{2}(u_iC^2)_{,i}dx$$

$$= -\int_{D_z} CC_{,3}dA + \frac{1}{2}\int_{D_z} u_3C^2 dA, \qquad (3.132)$$

where we use the divergence theorem and $(3.127)_4$. From the maximum principle $C \leq C_M = \max F_3$ in D. Hence, using the maximum principle, the Cauchy-Schwarz inequality, and the arithmetic-geometric mean inequality in (3.132) we may derive

$$\int_{R_z} C_{,i}C_{,i}dx \leq \frac{1}{\sqrt{\lambda_1}}\left(\int_{D_z} C_{,\alpha}C_{,\alpha}dA\right)^{1/2}\left(\int_{D_z} C_{,3}^2 dA\right)^{1/2}$$

$$+ \frac{C_M}{2\sqrt{\lambda_1}}\left(\int_{D_z} u_3^2 dA\right)^{1/2}\left(\int_{D_z} C_{,\alpha}C_{,\alpha}dA\right)^{1/2}$$

$$\leq \frac{1}{\sqrt{\lambda_1}}\left(1 + \frac{a_2C_M}{4}\right)\int_{D_z} C_{,i}C_{,i}dA$$

$$+ \frac{C_M}{4a_2\sqrt{\lambda_1}}\int_{D_z} u_3^2 dA, \qquad (3.133)$$

where $a_2 > 0$ is a constant at our disposal and λ_1 is the constant in Poincaré's inequality for D.

To bound $T_{,i}T_{,i}$ we commence in similar fashion. Thus, we obtain

$$\int_{R_z} T_{,i}T_{,i}dx = \int_{R_z} (T_{,i}T)_{,i}dx - \int_{R_z} T\Delta T dx$$

$$= -\int_{D_z} TT_{,3}dA - \int_{R_z} T(u_iT_{,i} - \gamma C)dx$$

$$= -\int_{D_z} TT_{,3}dA + \frac{1}{2}\int_{D_z} u_3T^2 dA + \gamma\int_{R_z} TC dx. \qquad (3.134)$$

Now,

$$\int_{R_z} TC dx = \int_z^\infty d\xi \int_{D_\xi} TC \, dA.$$

From the Cauchy-Schwarz and Poincaré inequalities we have

$$
\begin{aligned}
\int_{D_\xi} TC\, dA &\leq \sqrt{\int_{D_\xi} T^2 dA} \sqrt{\int_{D_\xi} C^2 dA} \\
&\leq \frac{1}{\lambda_1} \sqrt{\int_{D_\xi} T_{,\alpha} T_{,\alpha} dA} \sqrt{\int_{D_\xi} C_{,\alpha} C_{,\alpha} dA} \\
&\leq \frac{1}{2\lambda_1 a} \int_{D_\xi} T_{,\alpha} T_{,\alpha} dA + \frac{a}{2\lambda_1} \int_{D_\xi} C_{,\alpha} C_{,\alpha} dA,
\end{aligned}
$$

for $a > 0$ to be selected. This inequality may be integrated to see that

$$
\begin{aligned}
\int_{R_z} TC\, dx &\leq \frac{1}{2\lambda_1 a} \int_{R_z} T_{,\alpha} T_{,\alpha} dx + \frac{a}{2\lambda_1} \int_{R_z} C_{,\alpha} C_{,\alpha} dx, \\
&\leq \frac{1}{2\lambda_1 a} \int_{R_z} T_{,i} T_{,i} dx + \frac{a}{2\lambda_1} \int_{R_z} C_{,i} C_{,i} dx. \quad (3.135)
\end{aligned}
$$

Next,

$$
\begin{aligned}
-\int_{D_z} TT_{,3}\, dA &\leq \frac{1}{\sqrt{\lambda_1}} \sqrt{\int_{D_z} T_{,\alpha} T_{,\alpha} dA} \sqrt{\int_{D_z} T_{,3}^2 dA} \\
&\leq \frac{1}{\sqrt{\lambda_1}} \int_{D_z} T_{,i} T_{,i} dA. \quad (3.136)
\end{aligned}
$$

In addition, we use the Cauchy-Schwarz inequality followed by the Sobolev inequality (3.87) to find

$$
\begin{aligned}
\frac{1}{2} \int_{D_z} u_3 T^2 dA &\leq \frac{1}{2} \sqrt{\int_{D_z} u_3^2 dA} \sqrt{\int_{D_z} T^4 dA} \\
&\leq \frac{\Lambda}{2} \left(\int_{D_z} u_3^2 dA \right)^{1/2} \left(\int_{D_z} T^2 dA \int_{D_z} T_{,\alpha} T_{,\alpha} dA \right)^{1/2} \\
&\leq \frac{\Lambda}{2} \left(\int_{D_z} u_3^2 dA \right)^{1/2} \int_{D_z} T_{,\alpha} T_{,\alpha} dA \\
&\leq \frac{\Lambda}{2\lambda_1} \left[a_2 \int_{D_z} u_3^2 dA + \frac{1}{a_2} \left(\int_{D_z} T_{,\alpha} T_{,\alpha} dA \right)^2 \right], \quad (3.137)
\end{aligned}
$$

where $a_2 > 0$ is a constant at our disposal.

Estimates (3.135), (3.136) and (3.137) together in (3.134) lead to the bound

$$\int_{R_z} T_{,i}T_{,i}dx \leq \frac{1}{\sqrt{\lambda_1}}\int_{D_z} T_{,i}T_{,i}dA + \frac{\gamma}{2\lambda_1 a}\int_{R_z} T_{,i}T_{,i}dx$$

$$+ \frac{\gamma a}{2\lambda_1}\int_{R_z} C_{,i}C_{,i}dx + \frac{\Lambda a_2}{2\lambda_1}\int_{D_z} u_3^2 dA$$

$$+ \frac{\Lambda}{2\lambda_1 a_2}\left(\int_{D_z} T_{,\alpha}T_{,\alpha}dA\right)^2. \tag{3.138}$$

We now form the combination (3.138)+ξ(3.133) for $\xi > 0$ a constant to be selected. In fact, we choose $a = \gamma/\lambda_1$ and $\xi = 1/2 + \gamma^2/2\lambda_1^2$ and then derive the inequality

$$\int_{R_z} T_{,i}T_{,i}dx + \int_{R_z} C_{,i}C_{,i}dx \leq \frac{2}{\sqrt{\lambda_1}}\int_{D_z} T_{,i}T_{,i}dA$$

$$+ \left(1 + \frac{\gamma^2}{\lambda_1^2}\right)\frac{1}{\sqrt{\lambda_1}}\left(1 + \frac{a_2 C_M}{4}\right)\int_{D_z} C_{,i}C_{,i}dA$$

$$+ \left[\frac{\Lambda a_2}{\lambda_1} + \left(1 + \frac{\gamma^2}{\lambda_1^2}\right)\frac{C_M}{4a_2\sqrt{\lambda_1}}\right]\int_{D_z} u_3^2 dA$$

$$+ \frac{\Lambda}{\lambda_1 a_2}\left(\int_{D_z} T_{,\alpha}T_{,\alpha}dA\right)^2. \tag{3.139}$$

3.4.2   An estimate for the $u_i u_i$ term

We must now derive an estimate for the term $\int_{R_z} u_i u_i dx$. From the differential equation (3.127)$_1$ we find

$$\int_{R_z} u_i u_i dx = \int_{R_z} u_i(-p_{,i} + g_i T)dx. \tag{3.140}$$

For the pressure term we proceed via the introduction of the function ω_α of (3.17). In this way we see that

$$-\int_{R_z} u_i p_{,i} dx = \int_{D_z} u_3 p\, dA$$

$$= \int_{D_z} \omega_{\alpha,\alpha} p\, dA$$

$$= -\int_{D_z} \omega_\alpha p_{,\alpha} dA$$

$$= \int_{D_z} \omega_\alpha(u_\alpha - g_\alpha T)dA$$

where $\alpha = 1, 2$ and $(3.127)_1$ have been employed. The Cauchy-Schwarz, Poincaré and Babuska-Aziz inequalities are now used to see that

$$-\int_{R_z} u_i p_{,i} dx \leq \sqrt{\frac{C}{\lambda_1}} \sqrt{\int_{D_z} u_3^2 dA \int_{D_z} u_\alpha u_\alpha dA}$$

$$+ \frac{\sqrt{C}}{\lambda_1} \sqrt{\int_{D_z} u_3^2 dA \int_{D_z} T_{,\alpha} T_{,\alpha} dA}. \qquad (3.141)$$

Hence, utilizing (3.141) in (3.140) with further use of the arithmetic-geometric mean and Poincaré inequalities we may show that for $a_3, a_4 > 0$ to be chosen,

$$\int_{R_z} u_i u_i dx \leq \frac{a_3}{2} \int_{R_z} u_i u_i dx + \frac{1}{2a_3 \lambda_1} \int_{R_z} T_{,i} T_{,i} dx$$

$$+ \sqrt{C} \Big(\frac{1}{\sqrt{\lambda_1}} + \frac{1}{2\lambda_1 a_4} \Big) \int_{D_z} u_i u_i dA$$

$$+ \frac{a_4 \sqrt{C}}{2\lambda_1} \int_{D_z} T_{,i} T_{,i} dA. \qquad (3.142)$$

We now choose $a_3 = 1, a_2 = 4, a_4 = 2$, and add $\lambda_1(3.142)$ to (3.139) to derive

$$\frac{1}{2} \int_{R_z} u_i u_i dx + \frac{1}{2} \int_{R_z} T_{,i} T_{,i} dx + \int_{R_z} C_{,i} C_{,i} dx$$

$$\leq \Big(\frac{2}{\sqrt{\lambda_1}} + \sqrt{C} \Big) \int_{D_z} T_{,i} T_{,i} dA$$

$$+ \frac{1}{\sqrt{\lambda_1}} \Big(1 + \frac{\gamma^2}{\lambda_1^2} \Big)(1 + C_M) \int_{D_z} C_{,i} C_{,i} dA$$

$$+ \Big[\frac{5\sqrt{C}}{4} + \frac{4\Lambda}{\lambda_1} + \frac{C_M}{16\sqrt{\lambda_1}} \Big(1 + \frac{\gamma^2}{\lambda_1^2} \Big) \Big] \int_{D_z} u_i u_i dA$$

$$+ \frac{\Lambda}{4\lambda_1} \Big(\int_{D_z} T_{,i} T_{,i} dA \Big)^2. \qquad (3.143)$$

Next, define the constants A and B by

$$A = \max \Big\{ \frac{4}{\sqrt{\lambda_1}} + 2\sqrt{C}, \frac{(1 + C_M)}{\sqrt{\lambda_1}} \Big(1 + \frac{\gamma^2}{\lambda_1^2} \Big), \frac{5\sqrt{C}}{2} + \frac{8\Lambda}{\lambda_1} + \frac{C_M}{8\sqrt{\lambda_1}} \Big(1 + \frac{\gamma^2}{\lambda_1^2} \Big) \Big\},$$

$B = \Lambda/\lambda_1$, and define the function $H(z)$ by

$$H(z) = \frac{1}{2} \int_{R_z} u_i u_i dx + \frac{1}{2} \int_{R_z} T_{,i} T_{,i} dx + \int_{R_z} C_{,i} C_{,i} dx.$$

From inequality (3.143) we may now show that H satisfies the inequality

$$H(z) \leq -AH'(z) + B \big[-H'(z) \big]^2, \qquad (3.144)$$

where $H' = dH/dz$.

3.4.3 Integration of the H inequality

The inequality (3.144) is integrated by (Horgan and Payne, 1992), p. 656. For completeness we sketch the steps.

Complete the square in (3.144) to find

$$\left(-H' + \frac{A}{2B}\right)^2 \geq \frac{H}{B} + \frac{A^2}{4B^2},$$

from which one finds

$$-\frac{dH}{dz} \geq \frac{1}{\sqrt{B}}\left(\sqrt{H + d^2} - d\right),$$

where $d = A/2\sqrt{B}$. Separating variables leads to

$$-\int_{H(0)}^{H(z)} \frac{ds}{\left(\sqrt{s + d^2} - d\right)} \geq \frac{z}{\sqrt{B}},$$

which integrates to yield

$$2\left\{\left[H(0) + d^2\right]^{1/2} - \left[H(z) + d^2\right]^{1/2}\right\}$$
$$+ 2d \log\left[\frac{\sqrt{H(0) + d^2} - d}{\sqrt{H(z) + d^2} - d}\right] \geq \frac{z}{\sqrt{B}}.$$

Now drop the term $-2\left[H(z) + d^2\right]^{1/2}$ and rearrange to find

$$\log\left[\frac{\sqrt{H(0) + d^2} - d}{\sqrt{H(z) + d^2} - d}\right] \geq \left(\frac{z}{2d\sqrt{B}} - \frac{\sqrt{H(0) + d^2}}{d}\right).$$

After taking the exponential one obtains

$$H(z) \leq H(0) \exp\left(\frac{\sqrt{H(0) + d^2}}{d}\right) e^{-z/2d\sqrt{B}}. \tag{3.145}$$

While this is an exponential spatial decay estimate the term $H(0)$ is not in terms of the data F_1, F_2 and F_3. It remains to find a suitable bound for $H(0)$.

3.4.4 A bound for H(0)

This subsection follows the analogous analysis of (Payne and Song, 2002). Let S be a solution of the problem

$$\Delta S = 0 \quad \text{in } R$$
$$S = C \quad \text{on } \partial R. \tag{3.146}$$

Then by using the triangle inequality and squaring,

$$\int_R C_{,i} C_{,i} dx \leq 2 \int_R (C - S)_{,i} (C - S)_{,i} dx + 2 \int_R S_{,i} S_{,i} dx. \tag{3.147}$$

Now, recall C satisfies the equation $\Delta C = u_i C_{,i}$ and $C = F_3$ at $z = 0$, so by integrating by parts and use of the divergence theorem we find

$$\int_R (C-S)_{,i}(C-S)_{,i}dx = -\int_R (C-S)\Delta C dx \tag{3.148}$$

$$= -\int_R (C-S)u_i C_{,i}dx \tag{3.149}$$

$$= -\frac{1}{2}\int_R (u_i C^2)_{,i}dx + \int_R u_i C_{,i}S dx \tag{3.150}$$

$$= \frac{1}{2}\int_{D_0} u_3 F_3^2 dA - \int_R u_i S_{,i}C dx + \oint_{\partial R} u_i n_i C S\, dS \tag{3.151}$$

$$= -\frac{1}{2}\int_{D_0} u_3 F_3^2 dA - \int_R u_i S_{,i}C dx \tag{3.152}$$

$$\leq -\frac{1}{2}\int_{D_0} F_1 F_3^2 dA + C_M\sqrt{\int_R u_i u_i dx \int_R S_{,i}S_{,i}dx}, \tag{3.153}$$

where $C_M = \max_{\bar{D}} F_3$. Upon using (3.153) in (3.147) and using the arithmetic-geometric mean inequality for $\alpha_1 > 0$ to be selected we may show that

$$\int_R C_{,i}C_{,i}dx \leq (2+\alpha_1 C_M^2)\int_R S_{,i}S_{,i}dx + \frac{1}{\alpha_1}\int_R u_i u_i dx$$
$$- \int_{D_0} F_1 F_3^2 dA. \tag{3.154}$$

To bound the $S_{,i}S_{,i}$ term we recollect S satisfies (3.146) and then

$$\int_R S_{,i}S_{,i}dx = \int_R (S_{,i}S)_{,i}dx - \int_R S\Delta S dx$$
$$= -\int_{D_0} F_3 S_{,3}dA \tag{3.155}$$

because $S = C = F_3$ on D_0. Furthermore,

$$0 = \int_R S_{,3}\Delta S dx$$
$$= \frac{1}{2}\int_R (S_{,3}^2)_{,3}dx + \int_R (S_{,3}S_{,\alpha})_{,\alpha}dx - \frac{1}{2}\int_R \frac{\partial}{\partial z}(S_{,\alpha}S_{,\alpha})dx$$
$$= -\frac{1}{2}\int_{D_0} S_{,3}^2 dA + \frac{1}{2}\int_{D_0} S_{,\alpha}S_{,\alpha}dA$$

and so we see that

$$\int_{D_0} S_{,3}^2 dA = \int_{D_0} S_{,\alpha}S_{,\alpha}dA = \int_{D_0} F_{3,\alpha}F_{3,\alpha}dA. \tag{3.156}$$

Now, combining (3.155) with (3.156) together with the Cauchy-Schwarz inequality we derive

$$\int_R S_{,i}S_{,i}dx \le \sqrt{\int_{D_0} F_3^2 dA \int_{D_0} S_{,3}^2 dA}$$

$$= \sqrt{\int_{D_0} F_3^2 dA \int_{D_0} F_{3,\alpha}F_{3,\alpha}dA}. \tag{3.157}$$

Upon employing (3.157) in (3.154) we arrive at

$$\int_R C_{,i}C_{,i}dx \le L_1 + \frac{1}{\alpha_1}\int_R u_i u_i dx, \tag{3.158}$$

where L_1 is the data term

$$L_1 = -\int_{D_0} F_1 F_3^2 dA + (2 + \alpha_1 C_M^2)\sqrt{\int_{D_0} F_3^2 dA \int_{D_0} F_{3,\alpha}F_{3,\alpha}dA}.$$

Since there is no maximum principle for T we now introduce the function S which solves

$$\begin{aligned} \Delta S &= u_i S_{,i} &&\text{in } R \\ S &= T &&\text{on } \partial R. \end{aligned} \tag{3.159}$$

Again from the triangle inequality

$$\int_R T_{,i}T_{,i}dx \le 2\int_R (T-S)_{,i}(T-S)_{,i}dx + 2\int_R S_{,i}S_{,i}dx. \tag{3.160}$$

The second term is

$$\int_R (T-S)_{,i}(T-S)_{,i}dx = -\int_R (T-S)\Delta(T-S)dx$$

$$= \gamma\int_R (T-S)Cdx$$

$$\le \frac{\gamma}{2a\lambda_1}\int_R (T-S)_{,i}(T-S)_{,i}dx + \frac{\gamma a}{2\lambda_1}\int_R C_{,i}C_{,i}dx \tag{3.161}$$

where we now select $a = \gamma/\lambda_1$. We next use (3.161) in (3.160) to find

$$\int_R T_{,i}T_{,i}dx \le \frac{2\gamma^2}{\lambda_1^2}\int_R C_{,i}C_{,i}dx + 2\int_R S_{,i}S_{,i}dx. \tag{3.162}$$

Note now that S satisfies the same equation as C and if we replace F_3 by F_2 it satisfies the same boundary conditions. Thus, S will satisfy a bound like (3.158). In fact, if we define the data term L_2 by

$$L_2 = -\int_{D_0} F_1 F_2^2 dA + (2 + \alpha_1 S_M^2)\sqrt{\int_{D_0} F_2^2 dA \int_{D_0} F_{2,\alpha}F_{2,\alpha}dA}$$

where $\mathcal{S}_M = \max_{\bar{D}} F_2$ then $\mathcal{S}$ satisfies the following estimate,

$$\int_R \mathcal{S}_{,i} \mathcal{S}_{,i} dx \leq L_2 + \frac{1}{\alpha_1} \int_R u_i u_i dx. \tag{3.163}$$

Next, (3.158) and (3.163) are used in (3.162) to determine the following bound for a functional of T,

$$\int_R T_{,i} T_{,i} dx \leq L_3 + \frac{2}{\alpha_1} \left(1 + \frac{\gamma^2}{\lambda_1^2}\right) \int_R u_i u_i dx, \tag{3.164}$$

where $L_3 = 2\gamma^2 L_1 / \lambda_1^2 + 2L_2$.

3.4.5 Bound for $u_i u_i$ at $z = 0$

From the differential equation for u_i we show

$$\int_R u_i u_i dx = \int_R u_i(-p_{,i} + g_i T) dx. \tag{3.165}$$

Now, introduce the function ϕ_i defined in (3.39), where $\phi_3 = F_1(x_\alpha) e^{-\sigma z}$ and $\phi_{\alpha,\alpha} = -\phi_{3,3}$. Then,

$$-\int_R p_{,i} u_i dx = \int_{D_0} u_3 p \, dA$$

$$= -\int_R p_{,i} \phi_i dx$$

$$= \int_R \phi_i(u_i - g_i T) dx.$$

This expression is used in (3.165) and then we use the arithmetic-geometric mean inequality to find

$$\int_R u_i u_i dx = \int_R u_i g_i T dx - \int_R g_i \phi_i T dx + \int_R \phi_i u_i dx$$

$$\leq \frac{1}{2} \int_R u_i u_i dx + 2 \int_R T^2 dx + \frac{5}{4} \int_R \phi_i \phi_i dx.$$

Rearranging and using Poincaré's inequality one finds

$$\frac{1}{2} \int_R u_i u_i dx \leq \frac{2}{\lambda_1} \int_R T_{,i} T_{,i} dx + \frac{5}{4} \int_R \phi_i \phi_i dx. \tag{3.166}$$

The $\phi_i \phi_i$ integral is bounded as in (3.46) to obtain

$$\int_R \phi_i \phi_i dx \leq \hat{C} \int_D F_1^2 dA \tag{3.167}$$

where

$$\hat{C} = \frac{1}{2\sigma} + \frac{C\sigma}{2\lambda_1}.$$

Thus, from (3.166)

$$\int_R u_i u_i dx \leq \frac{4}{\lambda_1} \int_R T_{,i} T_{,i} dx + L_4, \qquad (3.168)$$

where L_4 is the data term

$$L_4 = \frac{5\hat{C}}{2} \int_D F_1^2 dA.$$

The next step is to employ (3.164) in inequality (3.168) to derive, with the data term $L_5 = 8L_3/\lambda_1 + 2L_4$, after selecting $\alpha_1 = 16(1 + \gamma^2/\lambda_1^2)/\lambda_1$,

$$\int_R u_i u_i dx \leq L_5. \qquad (3.169)$$

The data bound (3.169) leads in turn to bounds for $T_{,i} T_{,i}$ and $C_{,i} C_{,i}$ from (3.164) and (3.158), namely

$$\int_R T_{,i} T_{,i} dx \leq L_3 + \frac{\lambda_1 L_5}{8}, \qquad (3.170)$$

$$\int_R C_{,i} C_{,i} dx \leq L_1 + \frac{\lambda_1}{16} \frac{L_5}{(1 + \gamma^2/\lambda_1^2)}. \qquad (3.171)$$

Hence, since

$$H(0) = \frac{1}{2} \int_R u_i u_i dx + \frac{1}{2} \int_R T_{,i} T_{,i} dx + \int_R C_{,i} C_{,i} dx,$$

we find the following bound for $H(0)$ in terms of the data functions F_1, F_2 and F_3,

$$H(0) \leq L_1 + \frac{L_3}{2} + L_5 \left[\frac{1}{2} + \frac{\lambda_1}{16} \left(\frac{2 + \gamma^2/\lambda_1^2}{1 + \gamma^2/\lambda_1^2} \right) \right]. \qquad (3.172)$$

This is a data bound for $H(0)$ and so inequality (3.145) represents a true spatial decay estimate.

3.5   Spatial decay for a fluid-porous model

In the last section of chapter 3 we describe work of (Ames et al., 2001). These writers tackled a very interesting but highly technical problem. They studied spatial decay in a semi-infinite cylinder which is partly composed of a saturated porous medium and partly filled with a viscous fluid. This is thus an extension of the work in sections 3.1 – 3.4 where the semi-infinite cylinder was always filled with a saturated porous material. One of the difficulties facing (Ames et al., 2001) is the boundary conditions on the interface between the porous medium and the fluid. The boundary conditions for this situation have been discussed already in section 2.10 in connection with a structural stability question and they are analysed in depth in chapter 6 in a variety of situations.

(Ames et al., 2001) consider a semi-infinite cylinder Ω which is divided in two parts. The configuration is as shown in figure 3.2

The axis of the cylinder Ω is in the $x_3 = z$ direction and the fluid occupies the region Ω_1 which has $x_2 > 0$, whereas the porous medium is in the domain Ω_2 which has $x_2 < 0$. The interface between the two media is at $x_2 = 0$ and is denoted by L. The cross sections in Ω_1 and Ω_2 are denoted by D_1 and D_2, respectively, and Ω_1 and Ω_2 have lateral boundaries Γ_1 and Γ_2.

The equations which (Ames et al., 2001) employ are Stokes' equations in the fluid, with the acceleration term omitted and a Boussinesq approximation for the density. Thus, in $\Omega_1 \times \{t > 0\}$ they have

$$\mu \Delta u_i - \frac{\partial p}{\partial x_i} + g_i T = 0,$$

$$\frac{\partial u_i}{\partial x_i} = 0, \tag{3.173}$$

$$\frac{\partial T}{\partial t} + u_i \frac{\partial T}{\partial x_i} = \kappa \Delta T,$$

where u_i, p, T, μ, κ and g_i represent velocity, pressure, temperature, dynamic viscosity, thermal diffusivity and the gravity vector, respectively. In the porous domain $\Omega_2 \times \{t > 0\}$ the equations are governed by Darcy's law and are

$$\frac{\mu}{k} v_i = -\frac{\partial \pi}{\partial x_i} + g_i \theta,$$

$$\frac{\partial v_i}{\partial x_i} = 0, \tag{3.174}$$

$$\frac{\partial \theta}{\partial t} + v_i \frac{\partial \theta}{\partial x_i} = \kappa_m \Delta \theta.$$

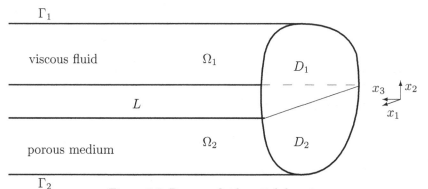

Figure 3.2. Porous - fluid spatial domain

Here v_i, π, θ, k and κ_m are the velocity, pressure, temperature, permeability and thermal diffusivity, respectively.

The initial conditions are that

$$u_i = 0, T = 0 \quad \text{in } \Omega_1 \quad \text{for } t = 0,$$
$$\theta = 0 \quad \text{in } \Omega_2 \quad \text{for } t = 0. \tag{3.175}$$

The boundary conditions on the lateral and end walls are for Ω_1,

$$u_i = 0, T = 0 \quad \text{on } \Gamma_1 \times \{t > 0\},$$
$$u_3 = f, T = h \quad \text{on } D_1 \times \{x_3 = 0\} \times \{t > 0\}, \tag{3.176}$$

while for Ω_2,

$$v_i n_i = 0, \theta = 0 \quad \text{on } \Gamma_2 \times \{t > 0\},$$
$$v_3 = f^m, \theta = h^m \quad \text{on } D_2 \times \{x_3 = 0\} \times \{t > 0\}. \tag{3.177}$$

The asymptotic boundary conditions are

$$|\mathbf{u}|, |\mathbf{v}|, |T|, |\theta| = O(1), \qquad |u_3|, |v_3|, |\nabla \mathbf{u}|, |\nabla T|, |p|, |\pi| = o(1/z), \tag{3.178}$$

uniformly in x_1, x_2 and t, as $z \to \infty$. The boundary conditions on the interface L are taken to be

$$u_2 = v_2, \quad T = \theta, \quad \kappa \frac{\partial T}{\partial x_2} = \kappa_m \frac{\partial \theta}{\partial x_2},$$
$$\pi = p - 2\mu \frac{\partial u_2}{\partial x_2}, \quad \frac{\partial u_\gamma}{\partial x_2} + \frac{\partial u_2}{\partial x_\gamma} = \frac{\alpha}{\sqrt{k}} u_\gamma, \quad \gamma = 1, 3. \tag{3.179}$$

Thus, the velocity, temperature and heat flux are continuous across L as is the normal pressure. The last boundary condition represents a (Jones, 1973) contribution to the (Beavers and Joseph, 1967) boundary condition but with the porous velocity neglected as discussed by (Nield and Bejan, 2006) and used by (Payne and Straughan, 1998a). (For further analysis of the interface conditions see chapter 6.)

(Ames et al., 2001) let $\Omega_1(z)$ and $\Omega_2(z)$ denote the domains

$$\Omega_1(z) = \Omega_1 \cap \{x_3 > z\}, \qquad \Omega_2(z) = \Omega_2 \cap \{x_3 > z\}.$$

Their work establishes an exponential spatial decay bound for a function of form

$$E(z, t) = A \int_0^t ds \int_{\Omega_1(z)} (\xi - z)(u_{i,j} + u_{j,i})u_{i,j} dx$$
$$+ B \int_0^t ds \int_{\Omega_2(z)} (\xi - z)v_i v_i dx$$
$$+ C \int_0^t ds \int_{\Omega_1(z)} (\xi - z)T_{,i} T_{,i} dx$$
$$+ D \int_0^t ds \int_{\Omega_2(z)} (\xi - z)\theta_{,i}\theta_{,i} dx.$$

We do not describe the proof because it is technical and lengthy. However, we note that they require a zero net flow into the pipe via the condition

$$\int_{D_1} f \, dA + \int_{D_2} f^m \, dA = 0.$$

Their proof hinges on establishing a second order differential inequality for E of form

$$\frac{\partial^2 E}{\partial z^2} - k_1 \frac{\partial E}{\partial z} - k_2 E \geq 0,$$

where $k_1, k_2 > 0$ are constants.

4
Convection in Porous Media

To commence this chapter we investigate the stability of convective flow in a layer of saturated porous material which is subject to a vertical temperature gradient. The appropriate equations are now presented according to the theories of Darcy, Forchheimer, Darcy with anisotropic permeability, and Brinkman.

While my previous book, (Straughan, 2004a), concentrates on energy stability applications in fluid mechanics, there are inevitably, some investigations there of convection flows in porous media. The object of this book is not to cover the same material, apart from the basic analysis which for reasons of clarity is contained in sections 4.1 and 4.2. However, the rest of this chapter concentrates on novel flow situations in porous media, many very recent, and some given here for the first time. Some of this work deals with nonlinear stability by means of the energy method, but not exclusively so.

We concentrate entirely on stability of flow in porous media. It is worth pointing out, however, that several of the stability ideas discussed here have already found useful application in other areas and will continue to do so. For example, we mention the important mathematical biology / medical area, cf. (Avramenko and Kuznetsov, 2004), (Ghorai and Hill, 2005), (Mulone et al., 2007), (Pieters, 2004), chapter 5, (Quinlan and Straughan, 2005), (Rappoldt et al., 2003), (Rionero, 2006b; Rionero, 2006a), (van Duijn et al., 2001), when interfaces are involved, (Jovanovic and Vulkov, 2005), and the important area of control theory, cf. (Alvarez-Ramirez et al., 2001), (Khadra et al., 2005), (Ruszkowski et al., 2005), (Ydstie, 2002).

B. Straughan, *Stability and Wave Motion in Porous Media*,
DOI: 10.1007/978-0-387-76543-3_4, © Springer Science+Business Media, LLC 2008

4.1   Equations for thermal convection in a porous medium

4.1.1   The Darcy equations

The derivation of the Darcy equations for thermal convection is discussed in chapter 1, sections 1.2 and 1.6.1. We here simply present the relevant system of equations.

The complete system of equations for thermal convection in a porous medium according to Darcy's law may be taken to be

$$0 = -p_{,i} - \frac{\mu}{k} v_i - k_i g \rho(T)$$
$$v_{i,i} = 0 \tag{4.1}$$
$$T_{,t} + v_i T_{,i} = \kappa \Delta T.$$

In equations (4.1), v_i, T and p are the variables to be solved for, i.e. the velocity, temperature and pressure. The quantities μ, k, g and κ are constants and represent viscosity, permeability, gravity and thermal diffusivity, respectively. The vector $\mathbf{k} = (0, 0, 1)$ and $\rho(T)$ is the density - temperature relationship. In this section we assume a linear one, namely,

$$\rho = \rho_0(1 - \alpha[T - T_0]), \tag{4.2}$$

where ρ_0 is the density at temperature $T = T_0$ and α is the coefficient of thermal expansion. We can always redefine the pressure p by putting

$$\tilde{p} = p + \rho_0 g[1 + \alpha T_0]z, \tag{4.3}$$

and then (4.1) may be rewritten as

$$0 = -\tilde{p}_{,i} - \frac{\mu}{k} v_i + k_i g \rho_0 \alpha T. \tag{4.4}$$

It is worth observing that we have replaced the body force term ρf_i in equation (1.15) by the appropriate representation for thermal convection, namely $-k_i g \rho(T)$, in equation (4.1).

4.1.2   The Forchheimer equations

As stated in chapter 1, section 1.3, when the flow rate is large the Forchheimer equations may be more appropriate than the Darcy ones, and for the problem of thermal convection in a saturated porous material these are usually taken to be

$$0 = -p_{,i} - \frac{\mu}{k} v_i - b|\mathbf{v}|v_i - k_i g \rho(T)$$
$$v_{i,i} = 0 \tag{4.5}$$
$$T_{,t} + v_i T_{,i} = \kappa \Delta T.$$

The Forchheimer equations are believed more appropriate when the velocity is not small, the idea being that the pressure gradient is no longer proportional to the velocity itself.

4.1.3   The Darcy equations with anisotropic permeability

For many practical situations the permeability k is not isotropic. (Just imagine rock strata. Many times this has a preferred direction and so anisotropic permeability is a realistic thing to consider.) When the permeability is not isotropic we can generalise the Darcy equations in a straightforward manner. This then yields a system of equations with rich mathematical properties which is very amenable to linear and energy stability techniques but which also applies to many real life mechanisms. For an anisotropic permeability k in $(4.1)_1$ is no longer the same in all directions. We must replace k by a tensor and so we generalise the velocity equation $(4.1)_1$ to have form, cf. section 1.5,

$$\mu v_i = -K_{ij}p_{,j} - K_{ij}k_j g\rho(T),$$

where K_{ij} is the permeability tensor. We shall require K_{ij} to be invertible so that the inverse tensor M_{ij} satisfies $M_{ij}K_{jk} = C\delta_{ik}$ where C is an appropriate constant. Then the system of equations for convective motion in an anisotropic porous medium of Darcy type is given by

$$\frac{\mu}{C} M_{ij}v_j = -p_{,i} - k_i g\rho(T)$$
$$v_{i,i} = 0 \tag{4.6}$$
$$T_{,t} + v_i T_{,i} = \kappa\Delta T.$$

We concentrate on the case of a transversely isotropic material where the axis of isotropy is at an angle β to the horizontal. This model was first studied by (Tyvand and Storesletten, 1991), who also adopt the linear density - temperature relationship (4.2). The permeability tensor $\mathbf{K}^*$ is selected as in (Tyvand and Storesletten, 1991), namely,

$$\mathbf{K}^* = K_\parallel \mathbf{i}'\mathbf{i}' + K_\perp(\mathbf{j}'\mathbf{j}' + \mathbf{k}'\mathbf{k}') \tag{4.7}$$

where $K_\parallel$ and $K_\perp$ are the longitudinal and transverse components of permeability. The plane of the porous layer is that for which $\mathbf{k}$ is orthogonal to the layer and $\mathbf{i}$ is aligned with the projection of $\mathbf{i}'$ on the (x, y) plane. The angle β is the angle between the vectors $\mathbf{i}$ and $\mathbf{i}'$. The inverse permeability tensor $\mathbf{M}$ satisfies

$$\mathbf{M}.\mathbf{K}^* = K_\perp\mathbf{I}, \tag{4.8}$$

$\mathbf{I}$ being the identity. Then, see (Tyvand and Storesletten, 1991)

$$\begin{aligned}\mathbf{M} &= \xi\mathbf{i}'\mathbf{i}' + \mathbf{j}'\mathbf{j}' + \mathbf{k}'\mathbf{k}' \\ &= M_{11}\mathbf{ii} + M_{13}(\mathbf{ik} + \mathbf{ki}) + M_{33}\mathbf{kk} + \mathbf{jj}\end{aligned} \tag{4.9}$$

where

$$M_{11} = \xi\cos^2\beta + \sin^2\beta, \qquad M_{13} = (\xi - 1)\cos\beta\sin\beta,$$
$$M_{33} = \cos^2\beta + \xi\sin^2\beta,$$

(4.10)

with $\xi = K_\perp/K_\parallel$ being the anisotropy parameter.

We here adopt the equation of state (4.2) and employ a modified pressure $\tilde{p}$ as in (4.3) (although we discard the tilde). Then, equations (4.6) may be written as

$$p_{,i} = -\frac{\mu}{K_\perp} M_{ij}v_j + k_i g\rho_0\alpha T,$$
$$v_{i,i} = 0,$$
$$T_{,t} + v_i T_{,i} = \kappa\Delta T,$$

(4.11)

where M_{ij} is the symmetric tensor with components as given by (4.9).

There are situations in which one also needs the thermal diffusivity to be anisotropic. We do not study this explicitly, but we draw attention to work of (Storesletten, 1993) where this effect is investigated at length.

4.1.4   The Brinkman equations

As is pointed out in chapter 1, section 1.4, it is often argued that (4.1) are insufficient to describe porous flow situations near a solid wall, or when the porosity is close to one. For such a scenario we may replace (4.1) by the Brinkman equations (4.12)

$$0 = -p_{,i} - \frac{\mu}{k} v_i + \lambda\Delta v_i - k_i g\rho(T)$$
$$v_{i,i} = 0$$
$$T_{,t} + v_i T_{,i} = \kappa\Delta T.$$

(4.12)

Again, we at first focus on the density relation (4.2). Frequently the argument is advanced that the Brinkman equations, due to the inclusion of the viscosity term λ, may be more relevant for flows involving a solid boundary, cf. e.g. (Nield and Bejan, 2006).

4.2   Stability of thermal convection

In this section we investigate the so called Bénard problem in a porous medium. This is the problem where a layer of porous material is saturated with fluid and the bottom of the layer is hotter than the top. If the temperature difference is large enough a cellular fluid motion ensues and this is known as Bénard convection. In fact, we deal with each of the porous systems introduced in section 4.1, namely the equations of Darcy, Forchheimer, the equations for Darcy convection with anisotropic permeability, and those of Brinkman. We shall show that one can establish optimal stability results. Namely, that the linear instability and nonlinear stability Rayleigh numbers are the same, i.e. $R_L^2 \equiv R_E^2$. By linearising the perturbation equations for a stability problem we derive a theory which can give information on when a flow becomes unstable. This critical Rayleigh number we call R_L^2. However, this gives no information on what the full nonlinear system dictates for stability. If we can obtain useful estimates from the nonlinear theory these will give a true vision on the global nonlinear stability picture. We stress this aspect throughout this chapter.

The physical picture is one of a saturated porous medium of infinite extent in the x and y directions bounded by the planes $z = 0$ and $z = d(> 0)$ with gravity, g, in the negative $z-$direction. The upper boundary is held at fixed temperature $T = T_U$ while the lower is held at constant temperature $T = T_L$, with $T_L > T_U$. The problem is to determine under what conditions heating from below will lead to convective (cellular) fluid motion in the porous medium. We begin with the equations of Darcy.

4.2.1   The Bénard problem for the Darcy equations

We commence with the steady solution to equations (4.1) when the density is given by the relation (4.2), namely $\rho(T) = \rho_0(1 - \alpha[T - T_0])$. The steady solution with zero velocity in accordance with a conduction only state is

$$\bar{v}_i = 0, \qquad \bar{T} = -\beta z + T_L, \tag{4.13}$$

with the steady pressure $\bar{p}$ determined from the differential equation $d\bar{p}/dz = g\rho_0(1 - \alpha[\bar{T}(z) - T_0])$. This represents the situation where no convective motion is occurring and the temperature gradient is constant throughout the layer.

The nonlinear perturbation equations which arise from (4.1) using (4.13) as steady solution, with $v_i = \bar{v}_i + u_i$, $T = \bar{T} + \theta$, $p = \bar{p} + \pi$, are

$$0 = -\pi_{,i} - \frac{\mu}{k}\, u_i + g\rho_0\alpha k_i\theta,$$

$$u_{i,i} = 0,$$

$$\theta_{,t} + u_i\theta_{,i} = \beta w + \kappa\Delta\theta,$$

where $w = u_3$. It is convenient to non-dimensionalize these equations with the scalings $t = t^*T$, $u_i = u_i^*U$, $x_i = x_i^*d$, $\theta = T^\sharp\theta^*$, $T = d^2/\kappa$, $P = \mu Ud/k$, $U = \kappa/d$, $T^\sharp = dU\sqrt{\mu\beta/\kappa g\rho_0\alpha k}$, and the Rayleigh number R^2 defined by

$$R^2 = \frac{d^2 g\rho_0\alpha k\beta}{\mu\kappa}. \tag{4.14}$$

Next, the stars are omitted and the non-dimensional (fully nonlinear) perturbation equations for the thermal convection problem arising from

Darcy's equations are

$$0 = -\pi_{,i} - u_i + Rk_i\theta,$$
$$u_{i,i} = 0, \qquad\qquad\qquad (4.15)$$
$$\theta_{,t} + u_i\theta_{,i} = Rw + \Delta\theta.$$

The boundary conditions are that

$$\theta = w = 0, \qquad z = 0, 1, \qquad\qquad (4.16)$$

with (u_i, θ, π) satisfying a plane tiling periodicity in the x, y directions. Since Darcy's law only contains u_i in the momentum equation $(4.15)_1$ we only prescribe the normal component of velocity on the boundary. In addition, we assume that $\mathbf{u}, \theta, p$ have an (x, y)-dependence consistent with one that has a repetitive shape that tiles the plane, such as two-dimensional rolls or hexagons. The hexagon solution was originally given by (Christopherson, 1940) namely,

$$u(x, y) = \cos\frac{1}{2}a(\sqrt{3}\,x + y) + \cos\frac{1}{2}a(\sqrt{3}\,x - y) + \cos ay. \qquad (4.17)$$

In particular, the (x, y)-dependence is consistent with a wavenumber, a, for which with

$$\Delta^* = \frac{\partial^2}{\partial x^2} + \frac{\partial^2}{\partial y^2},$$

u satisfies the relation

$$\Delta^* u = -a^2 u.$$

Whatever shape the cell has in the (x, y)-plane, its Cartesian product with $(0, 1)$ is the period cell V. (In this book we do not discuss the problem of which cell shape is actually taken up when convection commences in the fluid. This requires an analysis of the possible patterns which may occur in the nonlinear theory once convective motion ensues. We refer the reader to the concise mathematical analysis of (Mielke, 1997) for a very clear account of this.)

Throughout this chapter $\|\cdot\|$ and $(\cdot, \cdot)$ denote the norm and inner product on the Hilbert space $L^2(V)$.

4.2.2 Linear instability

To determine linear instability from (4.15) we drop the $u_i\theta_{,i}$ term and write $u_i = u_i(\mathbf{x})e^{\sigma t}$, $\theta = \theta(\mathbf{x})e^{\sigma t}$, $\pi = \pi(\mathbf{x})e^{\sigma t}$, to find

$$0 = -\pi_{,i} - u_i + Rk_i\theta,$$
$$u_{i,i} = 0, \qquad\qquad\qquad (4.18)$$
$$\sigma\theta = Rw + \Delta\theta.$$

Equations (4.18) together with the boundary conditions (4.16) define an eigenvalue problem for σ. In general $\sigma = \sigma_r + i\sigma_i$, $\sigma_r, \sigma_i \in \mathbb{R}$. If $\sigma_i \neq 0 \implies \sigma_r < 0$ it is said that the principle of *Exchange of Stabilities* holds. When $\sigma_i \equiv 0$ then $\sigma \in \mathbb{R}$ and so exchange of stabilities holds automatically.

We now show $\sigma \in \mathbb{R}$ (for each σ_i in the spectrum). To do this let V be the period cell for the solution (u_i, θ, π) and then multiply $(4.18)_1$ by the complex conjugate of u_i, u_i^*, and integrate over V. The result is, after using the boundary conditions

$$\|\mathbf{u}\|^2 = R(\theta, w^*) \tag{4.19}$$

where $w^* = u_3^*$. Now multiply $(4.18)_3$ by the complex conjugate of θ, θ^*, and integrate over V. One may now show that

$$\sigma\|\theta\|^2 = R(w, \theta^*) - \|\nabla\theta\|^2. \tag{4.20}$$

In (4.19) and (4.20) it is understood that u_i, θ are complex and so $\|\cdot\|$ is to be interpreted accordingly, e.g.

$$\|\theta\|^2 = \int_V \theta\theta^* dx.$$

Next, add (4.19) to (4.20) to find

$$\sigma\|\theta\|^2 = R\big[(\theta, w^*) + (w, \theta^*)\big] - \|\mathbf{u}\|^2 - \|\nabla\theta\|^2.$$

Since $\sigma = \sigma_r + i\sigma_i$ then the imaginary part of this equation shows that

$$\sigma_i\|\theta\|^2 = 0.$$

Thus, $\sigma_i = 0$ and $\sigma \in \mathbb{R}$. Exchange of stabilities holds and it is sufficient to take $\sigma = 0$ in (4.18) to have the equations which govern the boundary for linearised instability, i.e.

$$\begin{aligned} 0 &= -\pi_{,i} - u_i + Rk_i\theta, \\ u_{i,i} &= 0, \\ 0 &= Rw + \Delta\theta. \end{aligned} \tag{4.21}$$

Let us denote the lowest eigenvalue for (4.21) together with the associated boundary conditions by R_L. It must be emphasized that the linear theory only yields a boundary for *instability*, i.e. whenever $R > R_L$ the solution to (4.18) has $\sigma = \sigma_r > 0$, thus it grows in time and is unstable. In particular, the linearized equations do not yield any information on *nonlinear* stability. It is, in general, possible for the solution to the full nonlinear equations (4.15) to become unstable at a value of R lower than R_L, and in this case subcritical instabilities occur. Details of the calculation of R_L are given later in section 4.2.4.

4.2.3 Nonlinear stability

To investigate nonlinear energy stability we multiply $(4.15)_1$ by u_i, $(4.15)_3$ by θ and integrate each over V. This yields the following two equations

$$0 = R(\theta, w) - \|\mathbf{u}\|^2, \tag{4.22}$$

$$\frac{d}{dt}\frac{1}{2}\|\theta\|^2 = R(w, \theta) - \|\nabla\theta\|^2. \tag{4.23}$$

We employ Joseph's coupling parameter method (Joseph, 1965; Joseph, 1966). Hence multiply one of (4.22) or (4.23) by a positive parameter, add the equations and then select this parameter optimally. For example, add λ times (4.23) to (4.22). The result is

$$\frac{dE}{dt} = RI - D \tag{4.24}$$

where now

$$E(t) = \frac{\lambda}{2}\|\theta(t)\|^2, \tag{4.25}$$

$$I(t) = (1+\lambda)(w, \theta), \tag{4.26}$$

$$D(t) = \|\mathbf{u}\|^2 + \lambda\|\nabla\theta\|^2. \tag{4.27}$$

The idea is now to optimize an inequality involving the right hand side of (4.24). Hence, define R_E by

$$\frac{1}{R_E} = \max_H \frac{I}{D} \tag{4.28}$$

where now H is such that $u_i \in L^2(V)$, $\theta \in H^1(V)$, and $w = 0, \theta = 0$, on $z = 0, 1$. In this way we find from (4.24)

$$\frac{dE}{dt} = RD\frac{I}{D} - D$$

$$\leq RD\left(\max_H \frac{I}{D}\right) - D = -D\left(\frac{R_E - R}{R_E}\right). \tag{4.29}$$

Since $D \geq \lambda\pi^2\|\theta\|^2 + \|\mathbf{u}\|^2$ by using the Poincaré inequality, we see that $D \geq 2\pi^2 E$. Thus our nonlinear stability criterion is now

$$R < R_E \tag{4.30}$$

for then $\omega = 2\pi^2(R_E - R)/R_E > 0$ and (4.29) leads to

$$\frac{dE}{dt} \leq -\omega E. \tag{4.31}$$

Then by integration and rearranging

$$E(t) \leq \exp(-\omega t)E(0). \tag{4.32}$$

Thus, (4.28), (4.30) yield *unconditional* nonlinear stabilty in the porous Bénard problem when the equations for the porous medium are those of

Darcy. By unconditional stability we mean for all initial perturbations, and so for all $E(0)$. This is sometimes called global stability. Of course, for the Darcy problem, all (4.32) shows is that $\|\theta(t)\|$ decays exponentially. However, from (4.22) we may use the arithmetic-geometric mean inequality to deduce

$$\|\mathbf{u}\|^2 = R(\theta, w) \leq \frac{R^2}{2}\|\theta\|^2 + \frac{1}{2}\|w\|^2 \leq \frac{R^2}{2}\|\theta\|^2 + \frac{1}{2}\|\mathbf{u}\|^2,$$

and this leads to

$$\|\mathbf{u}\|^2 \leq R^2\|\theta\|^2.$$

Hence, (4.30) leads also to exponential decay of $\|\mathbf{u}(t)\|$.

4.2.4   Variational solution to (4.28)

The nonlinear stability threshold is now given by the variational problem (4.28). The approach with nonlinear energy stability calculations is to find a variational problem like (4.28), determine the Euler-Lagrange equations and maximize in the coupling parameter λ to obtain the best value of R_E. The maximum problem (4.28) is

$$\frac{1}{R_E} = \max_H \frac{(1+\lambda)(\theta, w)}{\|\mathbf{u}\|^2 + \lambda\|\nabla\theta\|^2}. \tag{4.33}$$

It is convenient (but not necessary) to rescale θ by putting $\hat{\theta} = \sqrt{\lambda}\,\theta$. Then, since we are seeking the maximum over a linear space H we find that (4.33) is equivalent to

$$\frac{1}{R_E} = \max_H \frac{f(\lambda)\,(\theta, w)}{\|\mathbf{u}\|^2 + \|\nabla\theta\|^2}$$

where $f(\lambda) = (1+\lambda)/\sqrt{\lambda}$. Hence, the Euler-Lagrange equations arising from (4.28) are determined from

$$R_E\delta I - \delta D = 0$$

where now (we incorporate the constraint $u_{i,i} = 0$ in I)

$$\delta I = \frac{d}{d\epsilon} f\left(\theta + \epsilon\eta, w + \epsilon h_3\right)\Big|_{\epsilon=0} - 2\frac{d}{d\epsilon}\left(\pi, u_{i,i} + \epsilon h_{i,i}\right)\Big|_{\epsilon=0}$$
$$= f(\lambda)(w, \eta) + f(\lambda)(h_3, \theta) + 2(\pi_{,i}, h_i)$$

and

$$\delta D = \frac{d}{d\epsilon}\left(\|\mathbf{u} + \epsilon\mathbf{h}\|^2 + \|\nabla(\theta + \epsilon\eta)\|^2\right)\Big|_{\epsilon=0}$$
$$= 2\left[(u_i, h_i) - (\Delta\theta, \eta)\right].$$

Thus, the Euler-Lagrange equations which arise from (4.28) are

$$0 = -\pi_{,i} - u_i + \frac{1}{2} R_E f \theta k_i,$$

$$u_{i,i} = 0, \qquad (4.34)$$

$$0 = \frac{1}{2} f R_E w + \Delta \theta.$$

One now uses the parameteric differentiation method to show $\lambda = 1$ is the optimal value, i.e. that value which maximises R_E. The details are now given.

Let $R_E^1, u_i^1, \theta^1, \pi^1$ denote the solution to the eigenvalue problem arising from (4.34) with $\lambda = \lambda^1 > 0$, and let $R_E^2, u_i^2, \theta^2, \pi^2$ denote the analogous solution when λ has another value, $\lambda = \lambda^2 > 0$, say. Multiply $(4.34)_1$ holding for $\lambda = \lambda^1$ by u_i^2 and integrate over V to obtain using the boundary conditions and $(4.34)_2$,

$$\frac{1}{2} R_E^1 (f^1 \theta^1, w^2) - (u_i^1, u_i^2) = 0, \qquad (4.35)$$

where $f^1 = f(\lambda^1)$. Likewise, multiply $(4.34)_1$ holding for $\lambda = \lambda^2$ by u_i^1 and integrate over V to obtain

$$\frac{1}{2} R_E^2 (f^2 \theta^2, w^1) - (u_i^2, u_i^1) = 0. \qquad (4.36)$$

Now multiply $(4.34)_3$ holding for $\lambda = \lambda^1$ by θ^2, and holding for $\lambda = \lambda^2$ by θ^1, to obtain after integration by parts and use of the boundary conditions

$$\frac{1}{2} R_E^1 (f^1 w^1, \theta^2) - (\nabla \theta^1, \nabla \theta^2) = 0, \qquad (4.37)$$

$$\frac{1}{2} R_E^2 (f^2 w^2, \theta^1) - (\nabla \theta^2, \nabla \theta^1) = 0. \qquad (4.38)$$

Form the combination (4.36) + (4.38) − (4.35) − (4.37) to find

$$R_E^2 (f^2 w^2, \theta^1) + R_E^2 (f^2 \theta^2, w^1) - R_E^1 (f^1 \theta^1, w^2) - R_E^1 (f^1 w^1, \theta^2) = 0.$$

Now add in the (zero-total) contributions as follows

$$R_E^2 ([f^2 - f^1] w^2, \theta^1) + (R_E^2 - R_E^1)(f^1 \theta^1, w^2)$$
$$+ R_E^2 ([f^2 - f^1] w^1, \theta^2) + (R_E^2 - R_E^1)(f^1 w^1, \theta^2) = 0.$$

Divide by $\lambda_2 - \lambda_1 \neq 0$ and take the limit $\lambda_2 \to \lambda_1$ to derive (with $R_E^1, f^1, w^1, \theta^1$ replaced by R_E, f, w and θ),

$$R_E \left(\frac{\partial f}{\partial \lambda} w, \theta \right) + \frac{\partial R_E}{\partial \lambda} (fw, \theta) = 0. \qquad (4.39)$$

To make use of this result we take $u_i^2 \equiv u_i^1$ in (4.35) and $\theta^2 \equiv \theta^1$ in (4.37) to see, once the outcomes are added together,

$$R_E (fw, \theta) = \|\mathbf{u}\|^2 + \|\nabla \theta\|^2. \qquad (4.40)$$

Next, substitute for (fw, θ) in (4.39) to find

$$R_E^2 \left(\frac{\partial f}{\partial \lambda} w, \theta \right) + \frac{\partial R_E}{\partial \lambda} \left(\|\mathbf{u}\|^2 + \|\nabla\theta\|^2 \right) = 0. \tag{4.41}$$

At the optimal value of R_E (i.e. largest as a function of λ) $\partial R_E / \partial \lambda = 0$ and then from (4.41) we see that this equation is satisfied if $\partial f / \partial \lambda = 0$. This then yields $\lambda = 1$ as the best value.

With $\lambda = 1$ equations (4.34) become

$$\begin{aligned}
0 &= -\pi_{,i} - u_i + R_E \theta k_i, \\
u_{i,i} &= 0, \\
0 &= R_E w + \Delta\theta.
\end{aligned} \tag{4.42}$$

These are the same as the linear instability equations (4.21). Thus, we have the optimal result that $R_L = R_E$. This means that the linear instability critical Rayleigh number is the same as the nonlinear stability Rayleigh number. This result holds for all initial data. The basic solution (4.13) is unstable for $R > R_L$ and globally stable for $R < R_E$. Thus, since $R_E = R_L$ this is thus an optimal result. It is to be stressed that this is for the Darcy equations of porous convection.

We now calculate the critical Rayleigh number $R_L = R_E$ arising from (4.42), or equivalently from (4.21). To do this we first remove the pressure by taking curlcurl of (4.42)$_1$. This gives

$$0 = \Delta u_i + R_E(k_j \theta_{,ij} - k_i \Delta\theta), \tag{4.43}$$

since $u_{i,i} = 0$. Now, pick $i = 3$ in this equation. Thus, we instead of (4.42) find a coupled system in w and θ of form

$$\begin{aligned}
0 &= \Delta w - R_E \Delta^* \theta, \\
0 &= R_E w + \Delta\theta,
\end{aligned} \tag{4.44}$$

where $\Delta^* = \partial^2/\partial x^2 + \partial^2/\partial y^2$, and $w = \theta = 0$ at $z = 0, 1$. Since $R_E = R_L$ we omit the E and simply solve (4.44) for R. We seek a solution of form $w = W(z)f(x, y)$, $\theta = \Theta(z)f(x, y)$ where $f(x, y)$ is a planform which tiles the plane. Typically f is the hexagonal solution given by (4.17). So,

$$\Delta^* f = -a^2 f, \tag{4.45}$$

where a is the horizontal wavenumber.

Let $D = d/dz$ then from (4.44) we eliminate θ to find

$$\Delta^2 w = -R^2 \Delta^* w.$$

Thus,

$$(D^2 - a^2)^2 W = R^2 a^2 W. \tag{4.46}$$

Since $w = \theta = 0$ on $z = 0, 1$, we may show from (4.46) that $D^{(2n)}W = 0$ on $z = 0, 1$. Hence, in (4.46) we may select $W = \sin n\pi z$. Thus (4.46) leads to

$$\frac{(n^2\pi^2 + a^2)^2}{a^2} = R^2.$$

We want the smallest value of R^2 as a function of n and so take $n = 1$. Then we must minimise $R^2(a^2)$ in a^2, i.e. minimise $R^2 = (\pi^2 + a^2)^2/a^2$. This leads to $a_c^2 = \pi^2$, where c denotes the critical value. Then, $R_c^2 = 4\pi^2$. Thus,

$$R_E^2 = R_L^2 = 4\pi^2.$$

For a Rayleigh number Ra less than $4\pi^2$ we cannot have any instability, i.e. all perturbations decay rapidly to zero. This is a nonlinear result which holds for all initial data. On the other hand if the Rayleigh number Ra is greater than $4\pi^2$ instability occurs and cellular convective motion is witnessed.

4.2.5   The Bénard problem for the Forchheimer equations

Suppose now that the porous medium is governed by the Forchheimer equations (4.5). We assume the porous material fills the three-dimensional region $\{(x, y) \in \mathbb{R}^2\} \times \{z \in (0, d)\}$ as in section 4.2.1. The boundary conditions are the same, the steady state solution is (4.13), and the same non-dimensionalisation leading to (4.15) is employed except we additionally need to account for the Forchheimer term in $(4.5)_1$, $b|\mathbf{v}|v_i$. Thus, the new non-dimensional variable $F = kb\kappa/\mu d$ arises. Instead of the dimensionless perturbation equations (4.15), when we employ the Forchheimer equations (4.5) we arrive at the dimensionless perturbation equations

$$0 = -\pi_{,i} - u_i + R\theta k_i - F|\mathbf{u}|u_i,$$
$$u_{i,i} = 0, \qquad\qquad\qquad\qquad (4.47)$$
$$\theta_{,t} + u_i\theta_{,i} = Rw + \Delta\theta,$$

where $w = u_3$. These equations hold on $\{(x, y) \in \mathbb{R}^2\} \times \{z \in (0, 1)\} \times \{t > 0\}$. The boundary conditions are again (4.16), i.e.

$$\theta = w = 0, \qquad z = 0, 1,$$

with (u_i, θ, π) satisfying a plane tiling periodicity in the (x, y) directions.

To determine the linear instability boundary from (4.5) we now discard the $u_i\theta_{,i}$ term and also the Forchheimer term $F|\mathbf{u}|u_i$. After writing $u_i = u_i(\mathbf{x})e^{\sigma t}$ with a similar representation for θ and π the linearized equations are derived. These are exactly the same as those of Darcy theory, i.e. (4.18). Thus, the linear instability boundary is exactly the same as that found by employing Darcy's law.

To develop an energy theory from (4.47) we proceed exactly as in the Darcy case, i.e. equations (4.24) – (4.28), the only difference is that we must now include a term of form $F < |\mathbf{u}|^3 >$. Thus, with E, I and D defined by (4.25) – (4.27) we find instead of (4.24) the equation

$$\frac{dE}{dt} = RI - D - F < |\mathbf{u}|^3 > . \tag{4.48}$$

Next, we discard the non-positive Forchheimer term from the right of (4.48) and find E satisfies the differential inequality

$$\frac{dE}{dt} \leq RI - D.$$

The development from this point is exactly the same as for the Darcy equations. Thus, we again find the optimal result that the nonlinear critical Rayleigh number $Ra_E = R_E^2$ is the same as the linear critical Rayleigh number $Ra_L = R_L^2$ and indeed, we have $Ra_E = Ra_L = 4\pi^2$.

Hence, in the current situation the Forchheimer term plays no role in the instability or stability threshold since the optimum result is achieved which is the same as that found in the Darcy case. However, the Forchheimer term does in general make a difference. For example, in sections 2.3 and 2.9, it is shown how the Forchheimer theory leads to sharper estimates than the Darcy theory for continuous depedence, and in an initial - final value problem.

4.2.6 The Bénard problem for the Darcy equations with anisotropic permeability

In this subsection we treat the problem of stability of a layer of saturated porous medium heated from below when the permeability is anisotropic. The anisotropy we allow is the one discussed in section 4.1.3 where the permeability in the direction at an angle of $\beta°$ to the horizontal is different from that in the orthogonal directions. The relevant equations governing convection are then (4.6).

Again, the steady solution whose stability is under investigation is

$$\bar{v}_i \equiv 0, \quad \bar{T} = -\beta z + T_L ,$$

where the porous medium occupies the infinite plane layer $\mathbb{R}^2 \times \{z \in (0, d)\}$ with upper and lower boundary conditions as given in (4.16). Let (u_i, θ, π) be perturbations to $(\bar{v}_i, \bar{T}, \bar{p})$ and then the perturbation equations are

$$\frac{\mu}{K_\perp} M_{ij} u_j = -\pi_{,i} + \rho_0 g \alpha k_i \theta, \quad u_{i,i} = 0,$$
$$\theta_{,t} + u_i \theta_{,i} = \beta w + \kappa \Delta \theta. \tag{4.49}$$

Equations (4.49) are non-dimensionalized with the scalings $\mathbf{x} = \mathbf{x}^* d$, $t = t^* T$, $T = d^2/\kappa$, $u_i = u_i^* U$, $U = \kappa/d$, $\pi = \pi^* P$, $P = \mu U d/K_\perp$, $\theta = \theta^* T^\sharp$, $T^\sharp = U d \sqrt{\mu \beta / \kappa g \rho_0 \alpha K_\perp}$, with the Rayleigh number $Ra = R^2$

being defined by

$$R^2 = \frac{d^2 g \rho_0 \alpha K_\perp \beta}{\mu \kappa}.$$

The perturbation equations in non-dimensional form are

$$M_{ij} u_j = -\pi_{,i} + R k_i \theta,$$
$$u_{i,i} = 0, \tag{4.50}$$
$$\theta_{,t} + u_i \theta_{,i} = Rw + \Delta\theta.$$

We note that the linearised instability equations become

$$M_{ij} u_j = -\pi_{,i} + R k_i \theta,$$
$$u_{i,i} = 0, \tag{4.51}$$
$$\sigma\theta = Rw + \Delta\theta,$$

where u_i has been written as $u_i(\mathbf{x})e^{\sigma t}$, with similar forms for θ and π.

To develop a nonlinear energy stability analysis we multiply $(4.50)_1$ by u_i and integrate over a period cell V to obtain

$$\int_V M_{ij} u_j u_i \, dV = R(\theta, w). \tag{4.52}$$

Likewise, we multiply $(4.50)_3$ by θ and integrate over V to find

$$\frac{d}{dt} \frac{1}{2} \|\theta\|^2 = R(\theta, w) - \|\nabla\theta\|^2. \tag{4.53}$$

By adding $\lambda(4.52)$ to (4.53) we derive an energy equation of form

$$\frac{dE}{dt} = RI - D \tag{4.54}$$

where now

$$E = \frac{1}{2} \|\theta\|^2, \qquad I = (1 + \lambda)(\theta, w),$$

$$D = \|\nabla\theta\|^2 + \lambda \int_V M_{ij} u_i u_j \, dV.$$

We now use the form of M_{ij} given by (4.10) to see that

$$M_{ij} u_i u_j = v^2 + \xi(u \cos\beta + w \sin\beta)^2 + (u \sin\beta - w \cos\beta)^2,$$

where $(u, v, w) = \mathbf{u}$. If $0 < \xi < 1$, then we rewrite this as

$$M_{ij} u_i u_j = v^2 + \left[\xi + (1 - \xi)\sin^2\beta\right] u^2$$
$$+ \left[\xi + (1 - \xi)\cos^2\beta\right] w^2 + 2(\xi - 1)\sin\beta \cos\beta \, uw,$$
$$\geq v^2 + \xi(u^2 + w^2),$$

and so $M_{ij}u_iu_j$ is clearly positive definite. If $\xi \geq 1$ then we arrive at the same conclusion from the rearrangement

$$M_{ij}u_iu_j = v^2 + \left[(\xi - 1)\cos^2\beta + 1\right]u^2$$
$$+ \left[(\xi - 1)\sin^2\beta + 1\right]w^2 + 2(\xi - 1)\sin\beta\cos\beta\, uw\,.$$

Thus $M_{ij}u_iu_j \geq \min\{1,\xi\}\, u_iu_i$. Hence, $\int_V M_{ij}u_iu_j\, dV \geq k_0\|\mathbf{u}\|^2$ where $k_0 = \min\{\xi,1\}$, and from Poincaré's inequality $\|\nabla\theta\|^2 \geq \pi^2\|\theta\|^2$. Thus I/D is bounded and one can show a maximising solution exists to the problem

$$\frac{1}{R_E} = \max_H \frac{I}{D} \tag{4.55}$$

where H is the same space of admissible solutions as in the Darcy Bénard problem. Then from (4.54) we derive

$$\frac{dE}{dt} \leq -D\left(\frac{R_E - R}{R_E}\right) \leq -2\pi^2\left(\frac{R_E - R}{R_E}\right)E$$

provided $R < R_E$. From this inequality we find

$$E(t) \leq E(0)\,\exp\left(-\frac{2\pi^2}{R_E}(R_E - R)t\right)$$

and nonlinear stability follows, for all initial data. It remains to find R_E to solve the nonlinear stability problem. The Euler-Lagrange equations from (4.55) are (replacing u_i by $\hat{u}_i = \sqrt{\lambda}\,u_i$ and then dropping the hat),

$$R_E\frac{f}{2}\theta k_i - M_{ij}u_j = \pi_{,i}, \qquad u_{i,i} = 0,$$
$$R_E\frac{f}{2}w + \Delta\theta = 0 \tag{4.56}$$

where $f(\lambda) = (1 + \lambda)/\sqrt{\lambda}$.

One now uses variation of parameters. Thus, let $f/2 = h$, and multiply $(4.56)_1$ evaluated at λ^2 by u_i^1, $(4.56)_1$ evaluated at λ^1 by u_i^2, $(4.56)_3$ evaluated at λ^2 by θ^1, and $(4.56)_3$ evaluated at λ^1 by θ^2. After integration over V the results are

$$R_E^2 h^2(\theta^2, w^1) - \int_V M_{ij}u_j^2 u_i^1\, dV = 0, \tag{4.57}$$

$$R_E^1 h^1(\theta^1, w^2) - \int_V M_{ij}u_j^1 u_i^2\, dV = 0, \tag{4.58}$$

$$R_E^2 h^2(w^2, \theta^1) - (\nabla\theta^1, \nabla\theta^2) = 0, \tag{4.59}$$

$$R_E^1 h^1(w^1, \theta^2) - (\nabla\theta^1, \nabla\theta^2) = 0. \tag{4.60}$$

We now form the combination $(4.57) - (4.58) + (4.59) - (4.60)$ to find

$$\left(R_E^2 h^2 - R_E^1 h^1\right)\left[(\theta^2, w^1) + (w^2, \theta^1)\right] = 0.$$

Next, recall $\lambda_1 \neq \lambda_2$, divide by $\lambda_2 - \lambda_1$ to obtain

$$\left[\frac{R_E^2(h^2 - h^1) + (R_E^2 - R_E^1)h^1}{\lambda_2 - \lambda_1}\right]\left[(\theta^2, w^1) + (w^2, \theta^1)\right] = 0.$$

Take the limit $\lambda_2 \to \lambda_1$ to find

$$\left[R_E\frac{\partial h}{\partial \lambda} + h\frac{\partial R_E}{\partial \lambda}\right](\theta, w) = 0. \tag{4.61}$$

Directly from (4.56) one shows

$$2R_E h(\theta, w) = \|\nabla\theta\|^2 + \int_V M_{ij}u_i u_j\, dV. \tag{4.62}$$

Thus, rearranging between (4.61) and (4.62) we see that

$$\left[\frac{\|\nabla\theta\|^2 + \int_V M_{ij}u_i u_j\, dV}{R_E f}\right]\left[f\frac{\partial R_E}{\partial \lambda} + R_E\frac{\partial f}{\partial \lambda}\right] = 0$$

where we note $f = 2h$. At the maximum value of R_E, $\partial R_E/\partial\lambda = 0$, and so $\partial f/\partial\lambda = 0$ gives the best value of λ. It is easily seen that $\lambda = 1$. Hence, for the maximum value of R_E as a function of λ we set $\lambda = 1$ in (4.56). Thus the Euler-Lagrange equations become

$$\begin{aligned} R_E\theta k_i - M_{ij}u_j = \pi_{,i}, \qquad u_{i,i} = 0, \\ R_E w + \Delta\theta = 0. \end{aligned} \tag{4.63}$$

Observe that equations (4.63) are the same as the linear instability equations (4.51) if $\sigma = 0$. In fact, by multiplying (4.51)$_1$ by u_i^*, (4.51)$_3$ by θ^* and integrating, thanks to the symmetry of M_{ij} one finds as in the Darcy case that $\sigma \in \mathbb{R}$. Hence, the equations for linear instability are exactly the same as those for nonlinear energy stability. Therefore, the linear critical Rayleigh number Ra_L is the same as the nonlinear critical Rayleigh number Ra_E, even when the permeability is transversely isotropic in the direction along the angle $\beta°$. Since this is an unconditional nonlinear stability result this means no subcritical instabilities can arise even in this anisotropic case.

Recall that equations (4.63) hold on $\mathbb{R}^2 \times (0, 1)$ and are to be solved subject to the boundary conditions

$$w = \theta = 0, \qquad z = 0, 1,$$

together with periodicity in x, y.

Despite the fact that equations (4.63) appear uncomplicated their solution is non-trivial. Details may be found in (Tyvand and Storesletten, 1991), who deal with the linear instability aspect. The details are interesting and show the inclined transverse isotropy has a pronounced effect.

4.2.7 The Bénard problem for the Brinkman equations

The equations for thermal convection according to the Brinkman model are given in (4.12). These are rewritten for clarity

$$0 = -p_{,i} - \frac{\mu}{k} v_i + \tilde{\lambda}\Delta v_i - k_i g\rho(T),$$

$$v_{i,i} = 0, \tag{4.64}$$

$$T_{,t} + v_i T_{,i} = \kappa\Delta T,$$

where again we assume $\rho(T) = \rho_0(1 - \alpha(T - T_0))$. These equations are fundamentally different from the three other porous systems we have analysed in that the order of (4.64) is two higher due to the $\tilde{\lambda}\Delta v_i$ (Brinkman) term. Thus, in prescribing boundary conditions on the planar boundaries $z = 0, d$, we must specify all components of $\mathbf{v}$, not just v_3.

For the thermal convection problem in hand (4.64) hold on $\mathbb{R}^2 \times (0, d) \times \{t > 0\}$ and the boundary conditions are $v_i = 0$, $z = 0, d$, $T = T_U$, $z = d$, $T = T_L$, $z = 0$, with $T_L > T_U$. The steady solution whose stability we investigate is

$$\bar{T} = -\beta z + T_L, \qquad \bar{v}_i = 0,$$

where $\beta = (T_L - T_U)/d$. Letting $v_i = \bar{v}_i + u_i$, $T = \bar{T} + \theta$, $p = \bar{p} + \pi$, the perturbation equations arising from (4.64) are

$$0 = -\pi_{,i} - \frac{\mu}{k} u_i + \tilde{\lambda}\Delta u_i + k_i g\rho_0\alpha\theta,$$

$$u_{i,i} = 0, \tag{4.65}$$

$$\theta_{,t} + u_i\theta_{,i} = \beta w + \kappa\Delta\theta.$$

These equations are non-dimensionalized with the scalings $\mathbf{x} = \mathbf{x}^* d$, $t = t^*\mathcal{T}$, $\pi = \pi^* P$, $u_i = u_i^* U$, $P = Ud\mu/k$, $\mathcal{T} = d^2/\kappa$, $U = \kappa/d$, $\lambda = \tilde{\lambda}k/d^2\mu$, and the Rayleigh number $Ra = R^2$ is defined as

$$Ra = \frac{d^2 g\rho_0\alpha k\beta}{\mu\kappa}.$$

The non-dimensional perturbation equations arising from (4.65) are (dropping *'s)

$$\pi_{,i} = -u_i + \lambda\Delta u_i + R\theta k_i, \tag{4.66}$$

$$u_{i,i} = 0, \tag{4.67}$$

$$\theta_{,t} + u_i\theta_{,i} = Rw + \Delta\theta, \tag{4.68}$$

which hold on $\mathbb{R}^2 \times \{z \in (0, 1)\} \times \{t > 0\}$. The boundary conditions are

$$u_i = 0, \theta = 0, \qquad \text{on } z = 0, 1,$$

and u_i, θ, π satisfy a plane tiling periodicity.

Firstly we note that the linearized equations which follow from (4.66) – (4.68) are after putting $u_i = u_i(\mathbf{x})e^{\sigma t}$, with similar forms for θ and π,

$$\pi_{,i} = -u_i + \lambda \Delta u_i + R\theta k_i, \tag{4.69}$$

$$u_{i,i} = 0, \tag{4.70}$$

$$\sigma\theta = Rw + \Delta\theta. \tag{4.71}$$

One may show $\sigma \in \mathbb{R}$. To do this one multiplies (4.69) by u_i^*, (4.71) by θ^*, integrates each over a period cell V and adds. Taking the imaginary part of the result leads to the stated conclusion.

A nonlinear energy analysis may be developed by multiplying (4.66) by u_i and integrating over V to find

$$0 = -\|\mathbf{u}\|^2 - \lambda\|\nabla\mathbf{u}\|^2 + R(\theta, w). \tag{4.72}$$

In a similar manner we multiply (4.68) by θ and integrate over V to obtain

$$\frac{d}{dt}\frac{1}{2}\|\theta\|^2 = R(\theta, w) - \|\nabla\theta\|^2. \tag{4.73}$$

For a positive coupling parameter ξ, form $\xi(4.73) + (4.72)$ to find

$$\frac{dE}{dt} = RI - D, \tag{4.74}$$

where now

$$E = \frac{1}{2}\xi\|\theta\|^2, \qquad I = (1 + \xi)(w, \theta),$$

$$D = \|\mathbf{u}\|^2 + \lambda\|\nabla\mathbf{u}\|^2 + \xi\|\nabla\theta\|^2.$$

One sets

$$\frac{1}{R_E} = \max_H \frac{I}{D} \tag{4.75}$$

where $H = \{u_i, \theta \in H^1(V) | u_{i,i} = 0\}$ and the solutions satisfy a horizontal plane tiling periodicity. Then from (4.74)

$$\frac{dE}{dt} \leq -D\left(1 - \frac{R}{R_E}\right).$$

If $R < R_E$ then put $a = (R_E - R)/R_E(> 0)$ and note that $D \geq \xi\pi^2\|\theta\|^2 = 2\pi^2 E$. Thus, one derives

$$\frac{dE}{dt} \leq -2\pi^2 aE.$$

This yields

$$E(t) \leq \exp(-2\pi^2 at)\, E(0)$$

from which global nonlinear energy stability follows (i.e. for all initial data). The only condition imposed is $R < R_E$.

We next put $f(\xi) = (1 + \xi)/2\sqrt{\xi}$ and let $\sqrt{\xi}\theta \to \theta$ in (4.75) to scale out the ξ from the denominator. The Euler-Lagrange equations which then arise from (4.75) are

$$R_E f(\xi)\theta k_i - u_i + \lambda\Delta u_i = \pi_{,i}, \tag{4.76}$$
$$u_{i,i} = 0, \tag{4.77}$$
$$R_E f(\xi)w + \Delta\theta = 0, \tag{4.78}$$

where in (4.76), π is a Lagrange multiplier.

The steps in the variation of parameters proof of the previous subsections may be followed to find

$$R_E^2 f^2(\theta^2, w^1) - (u_i^2, u_i^1) - \lambda(\nabla u_i^2, \nabla u_i^1) = 0,$$
$$R_E^1 f^1(\theta^1, w^2) - (u_i^1, u_i^2) - \lambda(\nabla u_i^1, \nabla u_i^2) = 0,$$
$$R_E^2 f^2(w^2, \theta^1) - (\nabla\theta^1, \nabla\theta^2) = 0,$$
$$R_E^1 f^1(w^1, \theta^2) - (\nabla\theta^2, \nabla\theta^1) = 0,$$

where now λ is constant and f^i denotes $f(\xi^i)$, $i = 1, 2$. From the above equations one arrives at, for $\partial R_E/\partial\xi = 0$,

$$f^{-1}\frac{df}{d\xi}\left(\|\mathbf{u}\|^2 + \lambda\|\nabla\mathbf{u}\|^2 + \|\nabla\theta\|^2\right) = 0.$$

Thus, the optimal value of ξ is $\xi = 1$.

With $\xi = 1$ equations (4.76) – (4.78) reduce to

$$R_E\theta k_i - u_i + \lambda\Delta u_i = \pi_{,i},$$
$$u_{i,i} = 0, \tag{4.79}$$
$$R_E w + \Delta\theta = 0.$$

Notice that (4.79) are the same as (4.69) – (4.71) with $\sigma = 0$, which we adopt since $\sigma \in \mathbb{R}$. Thus we may again conclude that $Ra_L = Ra_E$, i.e. the linear instability boundary coincides with the nonlinear stability one. Since this result is unconditional, i.e. it holds for all initial data, this precludes any subcritical instabilities. Note that this optimal result holds for the Darcy theory, Forchheimer theory, the transversely isotropic theory covered earlier, and also for the Brinkman theory.

To find $R_E(= R_L)$ we set $R_E = R$ and take curlcurl $(4.79)_1$ and then (4.79) reduce to the system

$$-R\Delta^*\theta + \Delta w - \lambda\Delta^2 w = 0,$$
$$Rw + \Delta\theta = 0, \tag{4.80}$$

with

$$w = \theta = 0 \quad \text{at} \quad z = 0, 1. \tag{4.81}$$

Two more boundary conditions are needed on w at $z = 0, 1$ and these depend on whether the surfaces are fixed or free of tangential stress. If either one is fixed then numerical solution of (4.80) is recommended.

For purposes of illustration we here consider two stress free surfaces and then in addition to (4.81) we eliminate θ from (4.80) to derive

$$(\lambda \Delta^3 - \Delta^2)w = R^2 \Delta^* w$$

and then $w = f(x, y) \sin n\pi z$ leads to

$$R^2 = \frac{\lambda(n^2\pi^2 + a^2)^3 + (n^2\pi^2 + a^2)^2}{a^2}.$$

It is worth observing that as $\lambda \to 0$ we obtain the equivalent expression for a Darcy porous material, whereas if we let $\lambda \to \infty$ we approach that for a fluid. To minimize R^2 in n we take $n = 1$ and then $dR^2/da^2 = 0$ yields the critical value of a^2 as

$$a_c^2 = \frac{-(\lambda\pi^2 + 1) + (\lambda\pi^2 + 1)\sqrt{1 + 8\pi^2\lambda/(\lambda\pi^2 + 1)}}{4\lambda}.$$

When $\lambda \to \infty$, $a_c^2 \to \pi^2/2$ as in the fluid case, whereas when $\lambda \to 0$, $a_c^2 \to \pi^2$ which is the Darcy case.

A detailed analysis of the linear instability problem for the Brinkman model is provided by (Rees, 2002). He presents numerical solutions and a detailed asymptotic analysis for the Rayleigh number as a function of the Darcy number (which is a non-dimensional form of the ratio of the Brinkman coefficient to the fluid viscosity, i.e. the term λ in (4.66)).

4.3   Stability and symmetry

4.3.1   Symmetric operators

In general, the equations governing problems in hydrodynamic stability (including those in porous media) are typically of the form

$$Au_t = L_S u + L_A u + N(u), \tag{4.82}$$

where u is a Hilbert space valued function, u_t is its time derivative, A is a bounded linear operator (typically a matrix with constant entries), $L = L_S + L_A$ is an unbounded, sectorial linear operator, and $N(u)$ represents the nonlinear terms. The operator L_S is the symmetric part of L while L_A denotes the anti-symmetric part. Such abstract equations and examples in fluid dynamics are discussed in e.g. (Doering and Gibbon, 1995), (Flavin and Rionero, 1995), and (Straughan, 1998; Straughan, 2004a).

The classical theory of linear instability writes

$$u = e^{\sigma t}\phi$$

and discards the $N(u)$ term in (4.82). One is then faced with solving the eigenvalue problem

$$\sigma A\phi = L_S\phi + L_A\phi, \tag{4.83}$$

where σ is the eigenvalue and ϕ the eigenfunction.

It is important to note that equation (4.82) involves both the skew-symmetric operator L_A and the symmetric operator L_S. In general, σ is complex, and one looks for the eigenvalue with largest real part to become positive for instability.

A classical nonlinear energy stability analysis, on the other hand, commences by forming the inner product of u with (4.82). If $(\cdot, \cdot)$ denotes the inner product on the Hilbert space in question then one finds

$$\frac{d}{dt}\frac{1}{2}(u, Au) = (u, L_S u) + (u, N(u)) \tag{4.84}$$

since $(u, L_A u) = 0$. Nonlinear energy stability follows from (4.84) and it is very important to note that in this way the nonlinear stability boundary does not involve the skew part of L, L_A. Thus, one may expect, in general, that the linear instability and nonlinear stability boundaries are very different. Details of how nonlinear stability follows from (4.84) may be found in section 4.3 of (Straughan, 2004a), or from the paper of (Galdi and Straughan, 1985).

In fact, the reason why the nonlinear energy stability analyses of section 4.2 give optimal results is due to the fact that the associated operator L is symmetric.

There are two fundamental problems arising from (4.84) when one is faced with deriving *unconditional* nonlinear stability results. These are

(a) the effect of L_A on the nonlinear stability boundary;

(b) what does one do when $(u, N(u)) \not\geq 0$?

When the operator L is far from symmetric traditional energy stability arguments can break down completely, or yield very poor results for certain classes of problem. For example, in parallel shear flows progress is very difficult, as explained in chapter 8 of (Straughan, 1998). In this regard though, an interested reader may wish to consider the recent articles of (Doering et al., 2000), (Kaiser and Mulone, 2005) and (Kaiser and von Wahl, 2005). Certain classes of viscoelastic flows prove severely problematic to tackle via energy methods, as is shown in the interesting paper of (Doering et al., 2006a).

Due to the failure of the classical energy method to yield sharp, or at least useful, nonlinear stability thresholds in problems such as shear flows, much research effort has recently been directed toward this area and a variety of novel approaches involving clever choices of Lyapunov functional have been suggested, cf. (Kaiser and Mulone, 2005), (Kaiser and von Wahl, 2005), (Lombardo and Mulone, 2005), (Mulone, 2004), (Nerli et al., 2007), (Pieters, 2004), (Pieters and van Duijn, 2006), (Rionero, 2004;

Rionero, 2005; Rionero, 2006c; Rionero, 2006b; Rionero, 2006a), (van Duijn et al., 2002), and further applications of Mulone's method may be found in (Mulone and Straughan, 2006), (Mulone et al., 2007).

We illustrate some of these new techiques now by application to the problem of double diffusive convection in a porous layer.

4.3.2 Heated and salted below

For this problem we have a layer of porous material saturated with water which contains salt (NaCl). The layer is heated from below but the salt gradient is arranged so that the greater salt concentration is toward the bottom of the layer. In this way there are two competing effects, namely that of temperature which has a tendency to destabilize and convectively overturn the fluid whereas the salt gradient opposes this and acts as a stabilizing agent. This competition leads to a non-symmetric operator in the problem and historically is important in nonlinear energy stability theory since it was the first where a generalized energy was employed to try and obtain a sharp nonlinear stability threshold, see (Joseph, 1970).

We commence with a layer of saturated porous material contained between the planes $z = \pm d/2$, and the equations are those for a Darcy porous medium coupled with the equations for temperature T and salt concentration C, as derived in chapter 1. Thus, with $\mathbf{v}$ denoting the velocity field the equations are

$$
\begin{aligned}
0 &= -p_{,i} - \rho g k_i - \frac{\mu}{K} v_i, \\
v_{i,i} &= 0, \\
\frac{1}{M} \frac{\partial T}{\partial t} + v_i T_{,i} &= k \Delta T, \\
\phi \frac{\partial C}{\partial t} + v_i C_{,i} &= k_C \Delta C,
\end{aligned}
\tag{4.85}
$$

where the density is linear in T and C, viz.,

$$
\rho = \rho_0 \big(1 - \alpha[T - T_0] + \alpha_C[C - C_0] \big),
$$

with ρ_0, T_0 and C_0 being reference values. The constant ϕ is porosity and $M = (\rho_0 c_p)_f / (\rho_0 c)_m$, where c_p is the specific heat of the fluid at constant pressure, and

$$
(\rho_0 c)_m = (1 - \phi)(\rho_0 c)_s + \phi(\rho_0 c_p)_f,
$$

with s and f denoting solid and fluid values, respectively. Let $\mathbf{v} = (u, v, w)$ and then to reflect the fact that the fluid is heated and salted from below the boundary conditions are

$$
w = 0, \quad T = T_0 \pm \frac{1}{2}(T_L - T_U), \quad C = C_0 \pm \frac{1}{2}(C_L - C_U), \qquad \text{at } z = \mp d/2.
$$

In these equations $C_L > C_U$, $T_L > T_U$, L denoting the values on the lower plane $z = -d/2$ while U denotes the values on the upper plane $z = d/2$, with $T_0 = (T_L + T_U)/2$, $C_0 = (C_L + C_U)/2$.

One non-dimensionalizes equations (4.85) via the scalings

$$d\mathbf{x}^* = \mathbf{x}, \quad t^* = \frac{kM}{d^2}t, \quad \mathbf{v}^* = \frac{d}{k}\mathbf{v}, \quad \epsilon = \phi M,$$

$$T^* = \frac{T - T_0}{T_L - T_U}, \quad C^* = \frac{C - C_0}{C_L - C_U}, \quad Le = \frac{k}{k_C},$$

where Le is the Lewis number. The Rayleigh and salt Rayleigh numbers are introduced as

$$\mathcal{R} = \frac{\alpha g(T_L - T_U)dK}{\nu k}, \qquad \mathcal{C} = \frac{\alpha_C g(C_L - C_U)dK}{\nu k}.$$

One obtains a steady state

$$\bar{\mathbf{v}} \equiv 0, \qquad \bar{T} = -z, \qquad \bar{C} = -z,$$

cf. (Mulone and Straughan, 2006). Denoting by u_i, θ and γ perturbations to $\bar{\mathbf{v}}, \bar{T}$ and $\bar{C}$ one may then show that $(\mathbf{u}, \theta, \gamma)$ satisfy the non-dimensional partial differential equation system

$$
\begin{aligned}
\pi_{,i} &= -u_i + (\mathcal{R}\theta - Le\mathcal{C}\gamma)k_i, \\
u_{i,i} &= 0, \\
\frac{\partial\theta}{\partial t} + u_i\theta_{,i} &= w + \Delta\theta, \\
\epsilon Le\frac{\partial\gamma}{\partial t} + Le\, u_i\gamma_{,i} &= w + \Delta\gamma,
\end{aligned}
\tag{4.86}
$$

where the boundary conditions are

$$w = \theta = \gamma = 0 \qquad \text{at} \qquad z = \pm\frac{1}{2},$$

together with the fact that u_i, θ, γ satisfy a plane tiling periodicity with planform Γ. The period cell $\Gamma \times (-1/2, 1/2)$ is denoted by V.

The key thing is to observe that a standard L^2 energy stability analysis multiplies (4.86)$_1$ by u_i, (4.86)$_3$ by θ, and (4.86)$_4$ by γ and integrates each equation over V. However, the point is that the $-Le\mathcal{C}(\gamma, w)$ term which arises from (4.86)$_1$ effectively cancels out the (w, γ) term arising from (4.86)$_4$ and the stabilizing effect of the salt field is lost. This has been a major problem in energy stability theory since it was first raised by (Joseph, 1970). To my knowledge it still has not been fully resolved in that a sharp *global* nonlinear stability threshold has not been achieved (by global we mean for all initial data, or at least for a class of finite initial data). Nevertheless, the technique introduced by (Mulone, 2004) was employed by (Mulone and Straughan, 2006) on the heated - salted below problem to achieve a very sharp nonlinear stability threshold, albeit at the expense of only establishing conditional nonlinear stability (i.e. for a restricted class of initial data). Since Mulone's technique is interesting we briefly describe it here.

4.3.3 Symmetrization

The technique of (Mulone, 2004) is somewhat akin to that used in linear algebra whereby if an $n \times n$ matrix A has n linearly independent eigenvectors one chooses these eigenvectors to define the columns of a matrix S and then $S^{-1}AS$ is a diagonal matrix Λ, see e.g. (Strang, 1988), p. 254.

First, we observe that if in (4.86) we have $\epsilon = 1, Le = 1$, we may set $\varphi = \mathcal{R}\theta - \mathcal{C}\gamma$, and then provided $\mathcal{R} - \mathcal{C} = F^2 > 0$, one puts $\psi = \varphi F^{-1}$ and equations (4.86) may be arranged in the form

$$\pi_{,i} = -u_i + F\psi k_i,$$
$$u_{i,i} = 0, \tag{4.87}$$
$$\frac{\partial \psi}{\partial t} + u_i \psi_{,i} = Fw + \Delta\psi.$$

This system is the same as (4.15) in section 4.2.1 and so the linear operator L is symmetric. Thus, the nonlinear energy stability boundary is equal to the linear instability one and subcritical instabilities are not possible.

For the more realistic case $\epsilon \neq 1, Le \neq 1$ we may use the method of Mulone, full details being given in (Mulone and Straughan, 2006).

Mulone's method involves the following sequence of ideas.

1. One starts with the linearized version of (4.86) and replaces the Laplacian operator by its principal eigenvalue. Let this matrix be L_1, say.

2. Compute the eigenvalues of L_1.

3. Introduce a matrix Q of eigenvectors of L_1 (or generalized eigenvectors in the case of a multiple eigenvalue with different geometric and algebraic multiplicity) and its inverse Q^{-1}. (The matrix Q has as jth column the jth eigenvector, but if the jth eigenvalue is complex then the jth and $(j+1)$th columns are the real and imaginary parts of the jth eigenvector).

4. Introduce a variable $\mathbf{Y} = Q^{-1}\mathbf{X}$, where $\mathbf{X} = (u_i, \theta, \gamma)$.

5. Write the linear system $\partial \mathbf{Y}/\partial t = Q^{-1}L_1 Q\mathbf{Y}$.

6. Transform (4.86) into the equivalent *nonlinear* system for Y_i and define a "natural" energy functional $E_1(t) = \|\mathbf{Y}\|^2/2$.

7. Demonstrate coincidence of the linear instability and nonlinear energy stability boundaries in the new measure E_1.

8. Control the nonlinear terms by an extra functional E_2 so that the Lyapunov functional employed is $E(t) = E_1 + bE_2$ for a suitable constant $b > 0$.

We do not go into the technical details here since they are quite involved and depend on the relative values of $\epsilon, Le, \mathcal{R}, \mathcal{C}$. Full details are given in

(Mulone and Straughan, 2006). However, we stress that we have only transformed the original nonlinear system into an equivalent nonlinear one by employing a technique very similar to to that used to diagonalize a matrix. In this way one may obtain an optimal Lyapunov functional which yields coincidence of the linear instability and nonlinear stability boundaries. The drawback of the method is that in the general case the stability obtained is only conditional. To the best of my knowledge the question of obtaining global nonlinear stability bounds is still open.

It is worth pointing out that another way of achieving equality of the linear instability and nonlinear stability boundaries for many classes of problem has been developed by (Rionero, 2004; Rionero, 2005; Rionero, 2006c; Rionero, 2006b; Rionero, 2006a). His interesting idea is to introduce a Lyapunov functional based directly on the eigenvalues of linearized instability theory. The papers of (Rionero, 2004; Rionero, 2005; Rionero, 2006c; Rionero, 2006b; Rionero, 2006a) concentrate on systems of two equations and establish principally conditional stability results. However, I understand from Professor Rionero that he has extended his technique to systems of three equations, and also to the situation where the coefficients in the equations may depend on the spatial coordinate $\mathbf{x}$. These extensions will be particularly useful.

4.3.4 Pointwise constraint

(Pieters, 2004), (Pieters and van Duijn, 2006), and (van Duijn et al., 2002) have made a very valuable contribution to nonlinear energy stability theory in porous media. They effectively noted that everyone before them did not use the momentum equation as a pointwise constraint. In the context of section 4.3.3, this refers to using $(4.86)_1$ as a constraint. Before these writers, the standard analysis multiplied equation $(4.86)_1$ by u_i and integrated over V to yield

$$0 = -\|\mathbf{u}\|^2 + \mathcal{R}(\theta, w) - Le\mathcal{C}(\gamma, w). \tag{4.88}$$

Equation (4.88) is used in the integrated form in an energy stability analysis. However, (Pieters, 2004), (Pieters and van Duijn, 2006), and (van Duijn et al., 2002) observe that, in many cases, much sharper results may be obtained by using energy identities arising from $(4.86)_3$ and $(4.86)_4$, but keeping $(4.86)_1$ as a pointwise constraint in the energy maximization problem.

From $(4.86)_3$ and $(4.86)_4$, the θ and γ energy equations become

$$\frac{d}{dt}\frac{1}{2}\|\theta\|^2 = (w, \theta) - \|\nabla\theta\|^2, \tag{4.89}$$

$$\frac{d}{dt}\frac{\epsilon Le}{2}\|\gamma\|^2 = (w, \gamma) - \|\nabla\gamma\|^2. \tag{4.90}$$

Upon eliminating π from $(4.86)_1$ one obtains the equation

$$\Delta w = \mathcal{R}\Delta^*\theta - Le\mathcal{C}\Delta^*\gamma \qquad (4.91)$$

where $\Delta^* = \partial^2/\partial x^2 + \partial^2/\partial y^2$ is the horizontal Laplacian.

We now define

$$E = \frac{1}{2}\|\theta\|^2 + \frac{\epsilon Le}{2}\|\gamma\|^2,$$

and

$$I = (w,\theta) + (w,\gamma) \qquad \text{and} \qquad D = \|\nabla\theta\|^2 + \|\nabla\gamma\|^2$$

and then from (4.89) and (4.90) form the energy identity

$$\frac{dE}{dt} = I - D.$$

The classical approach would be to derive the Euler-Lagrange equations for the maximum

$$\max \frac{I}{D} = \frac{1}{R_E}$$

where R_E is the energy stability threshold. This calculation is still performed, but instead of involving the integrated form of equation $(4.86)_1$, equation (4.88), one adds (4.91) as a constraint and studies the maximization problem

$$\frac{1}{R_E} = \max \frac{\lambda_1(w,\theta) + \lambda_2(w,\gamma) + \int_V w(\Delta\ell - \mathcal{R}\theta\Delta^*\ell + \mathcal{C}Le\Delta^*\ell)dV}{\lambda_1\|\nabla\theta\|^2 + \lambda_2\|\nabla\gamma\|^2}$$

where ℓ is a Lagrange multiplier and λ_1, λ_2 are coupling parameters. For many problems, solving the Euler-Lagrange equations arising via this approach proves superior to that involving the integrated form (4.88) and yields a sharper nonlinear energy stability threshold. We return to this point later in section 5.4.

4.4   Thermal non-equilibrium

4.4.1   Thermal non-equilibrium model

(Straughan, 2006) shows that the global nonlinear stability threshold for convection with a thermal non-equilibrium model is exactly the same as the linear instability boundary. This result is shown to hold for the porous medium equations of Darcy, Forchheimer, or Brinkman. This optimal result is important because it shows that linearised instability theory has captured completely the physics of the onset of convection.

It may be that in some applications of porous media flow the temperature of the fluid may be different from that of the temperature of the porous matrix. Such an area may be in drying or freezing of foods and other mundane materials which frequently need to be processed extremely quickly, see e.g. (Zorrilla and Rubiolo, 2005a; Zorrilla and Rubiolo, 2005b), (Martins and Silva, 2004), (Sanjuán et al., 1999), (Gigler et al., 2000b; Gigler et al., 2000a), or to applications in everyday food technology such as microwave heating, e.g. (Dincov et al., 2004). Certainly another area where local thermal non-equilibrium theory is likely to feature strongly is in rapid heat transfer, from e.g. computer chips via use of porous metal foams, e.g. (Calmidi and Mahajan, 2000), (Zhao et al., 2004), and their use in heat pipes, e.g. (Nield and Bejan, 2006), pp. 472 – 474.

(Banu and Rees, 2002) and (Malashetty et al., 2005) have investigated the linear instability problem for convection in a porous medium when the temperature of the fluid may differ from that of the solid pores. These are important papers which utilize the local thermal non-equilibrium theory given by (Nield and Bejan, 2006), pp. 204, 205, and by (Nield and Kuznetsov, 2001). (Rees et al., 2008) is an important contribution to the thermal non-equilibrium theory which uses this model to analyse the situation where a hot fluid is injected into a relatively cold porous medium. These writers show that for a sufficiently large injection velocity the mathematical equations may well become hyperbolic and a thermal shock wave can form.

(Banu and Rees, 2002) analyses the onset of thermal convection when the porous medium is modelled using Darcy's law whereas (Malashetty et al., 2005) provide a similar analysis utilizing a Brinkman model. (Straughan, 2006) shows that the results of (Banu and Rees, 2002) and those of (Malashetty et al., 2005) are very strong in the sense that they are optimal in that the global nonlinear stability boundary one obtains from using local thermal non-equilibrium theory is exactly the same as the linear instability ones found by (Banu and Rees, 2002) and by (Malashetty et al., 2005). In this way the work of (Banu and Rees, 2002) and that of (Malashetty et al., 2005) is complete in that their results completely capture the physics of the onset of thermal convection and no subcritical instabilities are possible. (Straughan, 2006) also demonstrates the equivalence between the nonlinear stability and linear instability boundaries for local thermal non-equilibrium convection in a Darcy porous medium when the layer is undergoing a constant angular rotation about an axis in the same direction as gravity. The paper by (Sheu, 2006) is also an interesting contribution which uses a thermal non-equilibrium model to study chaotic convection, while (Malashetty et al., 2007) also include the effect of rotation.

We describe the thermal non-equilibrium model analysis. Consider a layer of porous material saturated with fluid and contained between the planes $z = 0$ and $z = d$. The temperatures of the solid, T_s, and fluid, T_f, are maintained at constant values T_L, T_U, on the planes $z = 0$ and $z = d$, viz.

$$T_s = T_f = T_L, \quad z = 0; \qquad T_s = T_f = T_U, \quad z = d. \qquad (4.92)$$

We suppose $T_L > T_U$, because when $T_U \geq T_L$ one may demonstrate global nonlinear stability always holds. The equations for thermal convection in a porous material allowing for different solid and fluid temperatures are, cf. (Banu and Rees, 2002), (Malashetty et al., 2005), (Nield and Bejan, 2006),

$$v_i = -\frac{K}{\mu}p_{,i} + \frac{\rho_f g \alpha K}{\mu}T_f k_i - \gamma_1|\mathbf{v}|v_i + \hat{\lambda}\Delta v_i, \tag{4.93}$$

$$v_{i,i} = 0, \tag{4.94}$$

$$\epsilon(\rho c)_f T_{,t}^f + (\rho c)_f v_i T_{,i}^f = \epsilon k_f \Delta T_f + h(T_s - T_f), \tag{4.95}$$

$$(1-\epsilon)(\rho c)_s T_{,t}^s = (1-\epsilon)k_s \Delta T_s - h(T_s - T_f). \tag{4.96}$$

We suppose these equations hold in the domain $\mathbb{R}^2 \times \{z \in (0,d)\} \times \{t > 0\}$, $\mathbf{k} = (0,0,1)$, and Δ is the three-dimensional Laplacian. The variables v_i, p, T_f and T_s are the velocity, pressure and fluid and solid temperatures, respectively. The constants $K, \mu, g,\ \alpha, \gamma_1, \hat{\lambda},\ \epsilon, \rho_\alpha, c_\alpha, k_\alpha$ ($\alpha = f,s$), are permeability, dynamic viscosity, gravity, thermal expansion coefficient, the Forchheimer coefficient, the Brinkman coefficient, porosity, density, specific heat, thermal diffusion coefficient (where $\alpha = f,s$, denotes fluid or solid), $(\rho c)_\alpha = \rho_\alpha c_\alpha$, $\alpha = f,s$, and h is a coefficient representing heat transfer between the fluid and solid matrix.

The steady solution whose stability is under investigation is

$$\bar{\mathbf{v}} \equiv 0, \qquad \bar{T}_f = \bar{T}_s = -\beta z + T_L, \tag{4.97}$$

where

$$\beta = \frac{T_L - T_U}{d} \tag{4.98}$$

is the temperature gradient. The steady pressure $\bar{p}(z)$ is a quadratic function which may be found from (4.93).

4.4.2 Stability analysis

Our stability analysis begins by introducing perturbations u_i, π, θ, ϕ to $\bar{v}_i, \bar{p}, \bar{T}_f$ and $\bar{T}_s$ by

$$v_i = u_i + \bar{v}_i, \quad p = \pi + \bar{p}, \quad T_f = \theta + \bar{T}_f, \quad T_s = \phi + \bar{T}_s. \tag{4.99}$$

Perturbation equations are derived from (4.93) – (4.96) and are then non-dimensionalized with velocity, pressure, temperature, time and length scales

$$U = \epsilon k_f/(\rho c)_f d, \qquad P = \mu d U/K, \qquad T^\sharp = U d\sqrt{\mu \beta c_f/\epsilon k_f g \alpha K},$$

$$T = (\rho c)_f d^2/k_f, \qquad L = d.$$

The Rayleigh number Ra is given by

$$Ra = R^2 = d^2 \rho_f^2 \sqrt{\beta c_f g \alpha K/\epsilon k_f \mu},$$

and non-dimensional Forchheimer and Brinkman coefficients, F, λ, are

$$F = \gamma_1 U, \qquad \lambda = \hat{\lambda}/d^2.$$

(Banu and Rees, 2002) introduced the following non-dimensional coefficients, H and γ

$$H = hd^2/\epsilon k_f \qquad \gamma = \epsilon k_f/(1 - \epsilon)k_s.$$

The non-dimensional perturbation equations are

$$u_i = -\pi_{,i} + R\theta k_i - F|\mathbf{u}|u_i + \lambda \Delta u_i, \qquad (4.100)$$
$$u_{i,i} = 0, \qquad (4.101)$$
$$\theta_{,t} + u_i\theta_{,i} = Rw + \Delta\theta + H(\phi - \theta), \qquad (4.102)$$
$$A\phi_{,t} = \Delta\phi - H\gamma(\phi - \theta), \qquad (4.103)$$

where now these equations hold on $\mathbb{R}^2 \times \{z \in (0,1)\} \times \{t > 0\}$. The variable $w = u_3$, and $A = \rho_s c_s k_f/k_s \rho_f c_f$ is a non-dimensional thermal inertia coefficient.

The boundary conditions are

$$u_i n_i = 0, \quad \theta = 0, \quad \phi = 0, \qquad \text{on } z = 0, 1, \qquad (4.104)$$

when $\lambda = 0$ (i.e. Darcy or Forchheimer flow), where n_i denotes the unit outward normal, whereas the boundary conditions are

$$u_i = 0, \quad \theta = 0, \quad \phi = 0, \qquad \text{on } z = 0, 1, \qquad (4.105)$$

if $\lambda \neq 0$ (i.e. Brinkman flow). In addition u_i, π, θ, ϕ satisfy a plane tiling periodicity in x, y.

One may deduce the equivalence between the linear instability boundary and the nonlinear stability one by writing (4.100) – (4.105) as an abstract system of partial differential equations in a Hilbert space and then verifying that appropriate conditions hold. In other words, show that the linear operator L of section 4.3.1 is in this case symmetric. One may alternatively proceed as in (Straughan, 2006) as we do here.

Derive energy identities by multiplying (4.100) by u_i, (4.102) by θ, and (4.103) by ϕ/γ to obtain after integration by parts and using the fact that u_i is solenoidal,

$$0 = -\|\mathbf{u}\|^2 - F\|\mathbf{u}\|_3^3 - \lambda\|\nabla\mathbf{u}\|^2 + R(\theta, w), \qquad (4.106)$$

$$\frac{d}{dt}\frac{1}{2}\|\theta\|^2 = R(w, \theta) - \|\nabla\theta\|^2 - H(\theta, \theta - \phi), \qquad (4.107)$$

$$\frac{d}{dt}\frac{A}{2\gamma}\|\phi\|^2 = -\frac{1}{\gamma}\|\nabla\phi\|^2 - H(\phi, \phi - \theta). \qquad (4.108)$$

Define E, I, D by

$$E(t) = \frac{1}{2}\|\theta\|^2 + \frac{A}{2\gamma}\|\phi\|^2,$$

$$I = 2(\theta, w), \tag{4.109}$$

$$D = \|\mathbf{u}\|^2 + \lambda\|\nabla\mathbf{u}\|^2 + \|\nabla\theta\|^2 + \frac{1}{\gamma}\|\nabla\phi\|^2 + H\|\theta - \phi\|^2.$$

Adding (4.106) – (4.108) one deduces

$$\frac{dE}{dt} = RI - D - F\|\mathbf{u}\|_3^3 \leq RI - D = DR\frac{I}{D} - D.$$

From this one may show

$$\frac{dE}{dt} \leq -D\left(1 - \frac{R}{R_E}\right), \tag{4.110}$$

where

$$R_E^{-1} = \max_{\mathcal{H}} \frac{I}{D} \tag{4.111}$$

where $\mathcal{H} = \{(\mathbf{u}, \theta, \phi)|\, u_i \in L^2(V), \theta, \phi \in H^1(V), u_{i,i} = 0, u_i, \theta, \phi, \pi$ are periodic over a plane tiling domain in x and $y\}$. If $R < R_E$ then put $a = 1 - R/R_E > 0$. Next, employ Poincaré's inequality to find $D \geq BE$ where $B = \min\{2\pi^2, 2\pi^2 A^{-1}\}$. Thus, from (4.110) we deduce $dE/dt \leq -aBE$ from which it follows that $E \to 0$ exponentially in time.

The exponential decay of E guarantees exponential decay of θ and ϕ (in $L^2(V)$ norm). To obtain decay of $\mathbf{u}$ we note from (4.100) that one may show

$$\|\mathbf{u}\|^2 + F\|\mathbf{u}\|_3^3 + \lambda\|\nabla\mathbf{u}\|^2 = R(\theta, w)$$

$$\leq \frac{R^2}{2}\|\theta\|^2 + \frac{1}{2}\|w\|^2,$$

and so

$$\|\mathbf{u}\|^2 + 2F\|\mathbf{u}\|_3^3 + 2\lambda\|\nabla\mathbf{u}\|^2 \leq R^2\|\theta\|^2. \tag{4.112}$$

Thus, $R < R_E$ also guarantees exponential decay of $\|\mathbf{u}\|$ in Darcy theory, of $\|\mathbf{u}\|_3$ in the Forchheimer case, and of $\|\nabla\mathbf{u}\|$ when the Brinkman model is employed.

Thus, R_E represents a global (i.e. for all initial data) nonlinear stability threshold. The quantity R_E is calculated from the Euler-Lagrange equations which follow from (4.111), namely

$$R_E\theta k_i - u_i + \lambda\Delta u_i = \omega_{,i},$$

$$u_{i,i} = 0,$$

$$R_E w + \Delta\theta - H\theta + H\phi = 0, \tag{4.113}$$

$$\frac{1}{\gamma}\Delta\phi + H\theta - H\phi = 0,$$

where ω is a Lagrange multiplier.

(Banu and Rees, 2002) and (Malashetty et al., 2005) show that the strong form of the principle of exchange of stabilities holds for the linearised version of (4.100) – (4.105), i.e. they show one may take the growth rate equal to zero. The point now is to note that (4.113) is identically the same eigenvalue problem as the linearised one from (4.100) – (4.105) with the growth rate σ (which arises from a time dependence like $e^{\sigma t}$) equal to zero. Thus, the linear instability eigenvalues of (Banu and Rees, 2002) and of (Malashetty et al., 2005), R_L^2 are exactly the same as the ones for global nonlinear stability, R_E^2. What this means is that if $R^2 > R_L^2$ there is instability of solution (4.97); this is true also for the nonlinear equations due to Sattinger's instability theory, (Sattinger, 1970) p. 813. If, however, $R^2 < R_E^2 \equiv R_L^2$ there is definitely nonlinear asymptotic stability of solution (4.97). If $R^2 = R_E^2$ there is stability since $\dot{E} \leq 0$. Since $R_L \equiv R_E$ this means no subcritical instabilities can arise. Such a statement implying nonexistence of subcritical instabilities is not true for all convection problems, cf. (Proctor, 1981).

(Straughan, 2006) further studies the problem of convection for a thermal non-equilibrium porous medium model when the layer is undergoing a rotation. The Boussinesq approximation in a rotating frame of reference is addressed by (Ramos and Vargas, 2005).

The theory studied in this section is not the same as that for a bidisperse porous medium. In that theory two temperatures are involved, but also two velocities corresponding to the macro-pores and to the remainder of the structure. Uses of bidispersive porous media in catalytic systems in chemistry have been known for some time, cf. (Szczygiel, 1999). A suitable thermomechanical theory is more recent and convection problems for this class of porous media are studied in detail by (Nield and Kuznetsov, 2006; Nield and Kuznetsov, 2007; Nield and Kuznetsov, 2008).

4.5   Resonant penetrative convection

4.5.1   Nonlinear density, heat source model

(Straughan, 2004b) developed an interesting convection situation in a porous medium. He showed that there is a range of parameters in which the convection may switch from the lower part of the layer to being predominantly in the upper part of the layer. His work develops linear instability and nonlinear energy stability thresholds. (Normand and Azouni, 1992) employed a linear instability analysis to produce an extremely interesting study in penetrative convection in a *fluid* layer. Penetrative convection is described in detail in chapter 17 of (Straughan, 2004a). This phenomenon refers to the physical situation where one part of a fluid layer has a tendency to become convectively unstable while the rest of the layer wishes to remain stable. However, if instability prevails then the convective motion which follows may penetrate into the stable layer and induce a secondary motion there. Penetrative convection in a fluid or porous layer, especially involving salt, is a topic of much recent activity, cf. (Carr, 2003a; Carr, 2003c; Carr, 2004; Carr, 2003b), (Carr and de Putter, 2003), (Carr and Straughan, 2003), (Chasnov and Tse, 2001), (Hill, 2005a; Hill, 2003; Hill, 2004b; Hill, 2004a; Hill, 2004c; Hill, 2005b), (Kato et al., 2003), (Krishnamurti, 1997), (Larson, 2000; Larson, 2001), (Mahidjiba et al., 2003), (Nield and Bejan, 2006), (Normand and Azouni, 1992), (Payne and Straughan, 1987) (Straughan, 2002b; Straughan, 2004a), (Tse and Chasnov, 1998), (Vaidya and Wulandana, 2006), and (Zhang and Schubert, 2000; Zhang and Schubert, 2002). Two mechanisms for producing penetrative convection are that of a nonlinear density-temperature relationship which accounts for the maximum density of water, and that involving a heat source or sink. For penetrative convection modelled by either a nonlinear density, or by a heat supply term, the growth rates involved in a linear instability analysis have been found to be real. When both effects are combined, however, (Normand and Azouni, 1992) demonstrated the striking effect that the growth rate can be complex. They basically found this by adjusting the heat source - nonlinear density situation to produce a stable layer of fluid bounded above and below by potentially unstable layers. Instability could occur in one or other layer and if the parameters are in a certain range a resonance-like effect occurs where the convection could essentially oscillate between one layer and the other. This is a very striking result and shows that if competing effects are present in a very simple Bénard convection situation then very complex behaviour may arise. Other resonant and similarly complex behaviour in different fluid systems has also been observed, cf. (Johnson and Narayanan, 1996), (Jordan and Puri, 2002), (Chen, 1993), (Naulin et al., 2005), and (Chen and Chang, 1992).

(Straughan, 2004b) presents a linearised analysis and develops a nonlinear stability analysis for an analogue of the (Normand and Azouni, 1992) problem, but in a saturated porous medium rather than in a fluid layer. This problem is briefly described here.

4.5.2 Basic equations

Consider a layer of saturated porous material bounded by the horizontal planes $z = 0$ and $z = h(> 0)$, which is assumed infinite in horizontal extent. The upper boundary $z = h$ is held at the constant temperature $4°C$, which is consistent with the density of water at its maximum, i.e. $T = T_m^\circ C = 4°C$. The lower boundary $z = 0$ is held at a fixed temperature T_0 which is either in the range $0 \leq T_0 < 4$, or $T_0 > 4$. In either case the density of water at $z = 0$ is smaller than that at $z = h$. The equation governing the temperature field in the porous medium is the standard one, cf. section 1.6.1, although we allow for a heat source or sink Q, so that

$$\frac{\partial T}{\partial t} + v_i \frac{\partial T}{\partial x_i} = \kappa \Delta T + Q. \tag{4.114}$$

Here T, v_i and κ are temperature, fluid (seepage) velocity, and the effective thermal diffusivity.

In the steady state the velocity field $\bar{v}_i \equiv 0$. We look for a steady temperature field of form $\bar{T} = \bar{T}(z)$, z being x_3. The presence of the heat source or heat sink means $\bar{T}(z)$ is a quadratic function when Q is constant. The constant Q is chosen such that the temperature field has a maximum greater than 4°C in $(0, h)$ if $0 \le T_0 < 4$, and has a minimum less than 4°C in $(0, h)$ if $T_0 > 4$. This leads to a situation in which there are effectively three layers in $(0, h)$, one stable with two potentially unstable. The maximum or minimum temperature, T_{ex}, in $(0, h)$ is given when $d\bar{T}/dz = 0$. One may then show, cf. (Straughan, 2004b), that this three layer situation depends on parameters γ and μ given by

$$\gamma = \frac{\Delta T_1}{\Delta T_2} > 0 \quad \text{and} \quad \mu = \frac{2}{\gamma}\left[1 + \gamma + \sqrt{1+\gamma}\right], \tag{4.115}$$

where

$$\Delta T_1 = T_m - T_0, \qquad \Delta T_2 = T_{ex} - T_m.$$

The steady temperature field in $(0, h)$ may then be written as, cf. (Straughan, 2004b),

$$\bar{T}(z) - T_m = \frac{\Delta T_1}{h^2(1-\mu)} z^2 - \frac{\mu \Delta T_1}{h(1-\mu)} z - \Delta T_1. \tag{4.116}$$

In addition to the equation for the temperature field we add the Forchheimer or Darcy equations together with the incompressibility condition, i.e. the momentum and continuity equations, viz.

$$\frac{\hat{\mu}}{K} v_i + \hat{\lambda}|\mathbf{v}|v_i = -p_{,i} - \rho(T)gk_i, \tag{4.117}$$

$$\frac{\partial v_i}{\partial x_i} = 0. \tag{4.118}$$

The quantities p, ρ, g are pressure, density, and acceleration due to gravity, and $\mathbf{k} = (0, 0, 1)$. The density is quadratic, so

$$\rho = \rho_0\left(1 - \alpha[T - T_m]^2\right), \tag{4.119}$$

where α is a suitable expansion coefficient and ρ_0 is the density of water at $T = T_m = 4$°C. The nonlinear relation is necessary since we have a porous layer saturated with water whose temperature is in the 4°C range, i.e. the maximum temperature range.

Perturbation variables (u_i, θ, π) are introduced via

$$v_i = \bar{v}_i + u_i, \qquad T = \bar{T} + \theta, \qquad p = \bar{p} + \pi,$$

and then perturbation equations are derived. Upon using h as a length scale, the time, velocity, pressure and temperature scales are chosen as

$$T = h^2/\kappa, \qquad U = \kappa/h, \qquad P = \hat{\mu}Uh/K, \qquad T^\sharp = U\sqrt{\hat{\mu}h/\kappa g\alpha K\rho_0}.$$

The Rayleigh number Ra and its square root R are then defined as

$$Ra = R^2 = (\Delta T_1)^2 h\left(\frac{g\alpha k\rho_0}{\hat{\mu}\kappa}\right). \tag{4.120}$$

For a non-dimensional steady temperature field $F(z)$ given by

$$F(z) = (1 - \mu)z^2 + \mu z - 1, \tag{4.121}$$

the non-dimensional perturbation equations have form

$$\begin{aligned}
u_i + \lambda u_i|\mathbf{u}| &= -\pi_{,i} + 2k_i RF\theta + k_i\theta^2, \\
u_{i,i} &= 0, \\
\theta_{,t} + u_i\theta_{,i} &= \Delta\theta - RF'w.
\end{aligned} \tag{4.122}$$

Here $F' = dF/dz$, λ is a non-dimensional Forchheimer coefficient, $w = u_3$, and equations (4.122) are defined on the domain $\mathbb{R}^2 \times (0,1) \times \{t > 0\}$.

The associated boundary conditions are

$$w = 0, \quad \theta = 0, \qquad z = 0, 1, \tag{4.123}$$

with u_i, θ, π satisfying a plane tiling periodicity in the x, y−plane,

4.5.3 Linear instability analysis

Equations (4.122) are linearized and a time dependence like $u_i = e^{\sigma t}u_i(\mathbf{x})$, $\theta = e^{\sigma t}\theta(\mathbf{x})$, $\pi = e^{\sigma t}\pi(\mathbf{x})$, is assumed. Upon removing the pressure perturbation the linearized instability equations are found in the form

$$\begin{aligned}
\Delta w &= 2RF\Delta^*\theta, \\
\sigma\theta &= -RF'w + \Delta\theta,
\end{aligned} \tag{4.124}$$

where $\Delta^* = \partial^2/\partial x^2 + \partial^2/\partial y^2$ is the horizontal Laplacian.

If we denote the depths of the "fictitious" layers as d_1 (lowest) with the two layers of depth d_2 above, one shows that with $n = d_1/d_2$,

$$\mu = \frac{2(n+1)}{n} \qquad \text{and} \qquad \gamma = n(n+2). \tag{4.125}$$

One might expect resonant-like behaviour when $d_1 = d_2$ and this corresponds to $\mu = 4$. In the porous medium context $Ra_1 \propto (\Delta T_1)^2 d_1$ and $Ra_2 \propto (\Delta T_2)^2 d_2$ and these will be equal when $d_1/d_2 = (\Delta T_2/\Delta T_1)^2$, i.e. when $n = 1/\gamma^2$. Since $\gamma = n(n+2)$ this gives a value of n solving

$$n^5 + 4n^4 + 4n^3 = 1. \tag{4.126}$$

Thus, $n \approx 0.54$ for which $\mu \approx 5.6$. Hence, we might expect complex growth rates in the region $\mu \in [4, 5.7]$ and possible oscillatory convection there. Computations do bear this out.

Upon introducing a plane tiling form f, one puts $w = W(z)f(x,y)$, $\theta = \Theta(z)f(x,y)$, and introduces the wavenumber a by $\Delta^* f = -a^2 f$. The linear instability equations (4.124) then reduce to

$$(D^2 - a^2)W = -2Ra^2 F\Theta,$$
$$(D^2 - a^2)\Theta - RF'W = \sigma\Theta, \tag{4.127}$$

where $D = d/dz$, $z \in (0,1)$. The boundary conditions are

$$W = \Theta = 0, \qquad z = 0, 1. \tag{4.128}$$

System (4.127), (4.128) is solved numerically in (Straughan, 2004b).

4.5.4 Nonlinear stability analysis

The goal of (Straughan, 2004b), in addition to finding resonance in the linearized problem, is to develop an unconditional nonlinear energy stability theory for system (4.122), (4.123).

Let V be a period cell for a disturbance to (4.122), and let $\| \cdot \|$ and $(\cdot, \cdot)$ be the norm and inner product on $L^2(V)$. Energy identities are derived by multiplying (4.122)$_1$ by u_i and integrating over V, and (4.122)$_3$ by θ and integrating over V, to find

$$\|\mathbf{u}\|^2 + \lambda\|\mathbf{u}\|_3^3 = 2R(F\theta, w) + (\theta^2, w), \tag{4.129}$$

$$\frac{d}{dt}\frac{1}{2}\|\theta\|^2 = -R(F'\theta, w) - \|\nabla\theta\|^2, \tag{4.130}$$

where $\| \cdot \|_3$ is the norm on $L^3(V)$. To control the cubic term on the right of (4.129) we need an energy functional which contains more than the L^2 norm $\|\theta\|$ and so derive an equation for $\|\theta\|_3$. One shows

$$\frac{d}{dt}\frac{1}{3}\|\theta\|_3^3 = -R(F'w, \theta^2 \operatorname{sgn}(\theta)) - \frac{8}{9}\|\nabla|\theta|^{3/2}\|^2. \tag{4.131}$$

Define our Lyapunov functional $E(t)$ as

$$E(t) = \frac{1}{2}\|\theta\|^2 + \frac{b}{3}\|\theta\|_3^3, \tag{4.132}$$

where $b > 0$ is a coupling parameter at our disposal.

One may then show, see (Straughan, 2004b), that provided

$$\lambda \geq \frac{81R^2 F_m^2}{64\pi^4}, \tag{4.133}$$

$$\frac{dE}{dt} \leq -D\left(\frac{R_E - R}{R_E}\right) - \epsilon\|\theta\|_3^3, \tag{4.134}$$

for some small ϵ. In (4.133) F_m is the maximum value of $|F'(z)|$ in $[0, 1]$. For $R < R_E$ one employs Poincaré's inequality on D to derive the inequality

$$\frac{dE}{dt} \leq -cE, \qquad (4.135)$$

from which one obtains exponential decay of $E(t)$ and hence global non-linear stability. The conditions which must hold are that $R < R_E$ and λ satisfies the restriction (4.133).

Details of the solution of the eigenvalue problem to determine R_E are given in (Straughan, 2004b). In fact, the numerical calculations of (Straughan, 2004b) fix μ and determine

$$Ra_E = \max_{\zeta > 0} \min_{a^2} R_E^2(a^2; \zeta, \mu). \qquad (4.136)$$

4.5.5 Behaviour observed

The key finding of (Straughan, 2004b) is that in the neighbourhood of $\mu = \mu_c = 5.2311$ there is a switch of convection from one arising in the lower part of the layer for $\mu < \mu_c$ to one commencing in the upper part of the layer for $\mu > \mu_c$. At μ_c a resonant-like behaviour may occur and convection may arise in the lower or upper layer. In the neighbourhood of μ_c the linear instability Rayleigh number, Ra_L, against the wave number, a, curve consists of three branches, two on which σ is real at criticality while the intermediate branch is complex, as shown in figure 4.1, where curves 1, 2, 3 are for μ values of 5.1, 5.2311, 5.3, respectively. In curve 1 the minimum is on the right hand branch, in curve 3 it is on the left, whereas curve 2 represents the critical situation $\mu_c = 5.2311$ in which the minimum occurs simultaneously on the right and left branches. Away from μ_c the Ra_L versus a curve behaves not dissimilarly to that of the standard Bénard problem.

(Straughan, 2004b) reports interesting findings where his linear analysis differs from that of (Normand and Azouni, 1992) in that unlike (Normand and Azouni, 1992) he never finds convection commences as oscillatory con-

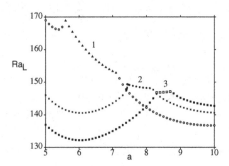

Figure 4.1. Ra_L vs. a. Curve 1, $\circ$=real, $\triangle$=complex, $\mu = 5.1$. Curve 2, $\times$=real, $\bullet$=complex, μ_c=5.2311. Curve 3, $\blacksquare$=real, $\square$=complex, μ=5.3

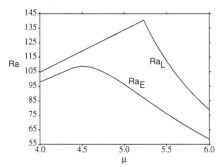

Figure 4.2. Critical Rayleigh numbers, Ra_L linear, Ra_E nonlinear. Unstable above Ra_L curve, globally stable below Ra_E curve

vection even though the Ra_L versus a curves have a complex branch. The linear curves near μ_c are shown in figure 4.1.

The nonlinear energy stability results show that the nonlinear critical Rayleigh number Ra_E is very close to the linear instability one Ra_L for $\mu \leq 4$. Figure 4.2 demonstrates the variation for μ larger than this. The numerical findings, including Ra_L versus a curves, linear eigenfunctions and nonlinear energy stability thresholds are discussed in detail in (Straughan, 2004b).

4.6   Throughflow

4.6.1   Penetrative convection with throughflow

(Hill et al., 2007) investigate penetrative convection in a layer of porous material saturated with water when there is superposed vertical through-flow of the water. To incorporate penetrative convection they employ a density quadratic in temperature. A linearised instability analysis is given and the critical Rayleigh numbers obtained are compared with those derived by a weighted nonlinear energy stability analysis. The basic class of problem considered by (Hill et al., 2007), i.e. where there is an underlying vertical background flow superimposed, is an important one due to applications in everyday situations. In fluid dynamics an important application is to cloud physics, as studied by (Krishnamurti, 1975), (Somerville and Gal-Chen, 1979). Industrial processes employ throughflow in porous media, as described by e.g. (Nield, 1987a; Nield, 1987b; Nield, 1998), (Nield and Bejan, 2006). Indeed, (Rosensweig, 1985), p. 289, describes where fluidized beds may be employed and these are created by throughflow through solid particles. Throughflow may be important for industrial application in that it yields a potential method to control the onset and behaviour of convection. In order to control convection we need to be able to accurately determine the convection threshold from a nonlinear analysis. Since many practical problems involve more than a Boussinesq approximation in the

system of equations governing convection, the aim of (Hill et al., 2007) was
to develop a nonlinear energy stability analysis which will yield unconditional (for all initial data) nonlinear stability bounds which are close to
those of linearised instability and hence will be useful in a practical situation. These writers focussed attention on a saturated porous medium,
when the density is a nonlinear function of temperature, thereby incorporating penetrative convection. They constructed a suitable weighted energy
function which allowed derivation of a global stability bound close to that
of the instablity one. However, their analysis hinges on the fact that the
density is a nonlinear function of T of form

$$\rho(T) = \rho_0\left(1 - \alpha(T - 4)^2\right), \tag{4.137}$$

where T is the temperature and α is the coefficient of thermal expansion.

Since a constant vertical throughflow is imposed, a steady state solution
for the velocity is $\bar{\mathbf{v}} = (0, 0, T_f)$, where T_f is constant. This leads to the
steady state temperature field of form

$$\bar{T}(z) = \xi(e^{cz} - 1)$$

where $\xi = T_U/(e^{cd}-1)$ and $c = T_f/\kappa$. It is important to note that the steady
temperature field is not linear in z as in the classical Bénard problem.
This is entirely due to the throughflow, and is present even if the density
is linear in T. This aspect of the problem leads to non-symmetry, and
a straightforward energy analysis is only partly successful in yielding a
nonlinear stability boundary, cf. (Qiao and Kaloni, 1998).

The non-dimensional perturbation equations of (Hill et al., 2007) are

$$u_i = -\pi_{,i} + b_i\theta^2 - 2b_i RM(z)\theta,$$
$$u_{i,i} = 0, \tag{4.138}$$
$$\theta_{,t} + u_i\theta_{,i} + RF(z)w + Q\theta_{,z} = \Delta\theta,$$

where $M(z) = 4/T_U - (e^{Qz} - 1)/(e^Q - 1)$ and $F(z) = Qe^{Qz}/(e^Q - 1)$. If
one simply multiplies (4.138)$_3$ by θ and integrates over the period cell V
then the term $Q(\theta_{,z}, \theta)$ disappears and an important stabilizing effect of
throughflow is lost. Hence, (Hill et al., 2007) must of necessity employ a
weighted functional $\int_V \mu(z)\theta^2 dV$. Their choice is dictated by the fact that
they must also remove the $b_i\theta^2$ term which enters the analysis through
equation (4.138)$_1$. However, their *global* analysis *only works* when the $b_i\theta^2$
term is present and so works *only* for the nonlinear density (4.137).

4.6.2 Forchheimer model with throughflow

In this section we present an entirely different method of analysis to that
of (Hill et al., 2007) whereby one may derive a global nonlinear stability
result when (4.137) is replaced by a linear relationship in T. The linear

relationship will feature in many practical situations and so our analysis is important.

Hence, we replace (4.137) by the equation

$$\rho(T) = \rho_0\left(1 - \alpha(T - T_0)\right), \tag{4.139}$$

where α is the coefficient of thermal expansion in the fluid. If we absorb the constant terms on the right of (4.139) in the pressure, then the equations governing convection in a porous medium may be written

$$\frac{\mu}{K}v_i + \lambda_F|\mathbf{v}|v_i = -p_{,i} + k_i g\rho_0\alpha T,$$
$$v_{i,i} = 0, \tag{4.140}$$
$$T_{,t} + \epsilon v_j T_{,j} = \kappa\Delta T,$$

where μ, K, λ_F, g and κ are, respectively, dynamic viscosity, permeability, Forchheimer coefficient, gravity, and thermal diffusivity. Furthermore, $\epsilon = \rho^f c_p^f/\phi\rho^m c_p^m$, with f denoting fluid, m denoting average over the porous medium, and ϕ is the porosity. These equations hold in the layer $\{x, y\} \in \mathbb{R}^2$, $z \in (0, d)$, for $t > 0$.

The temperature field satisfies the boundary conditions

$$T = T_L, \quad z = 0, \qquad T = T_U, \quad z = d, \tag{4.141}$$

T_L, T_U constants with $T_L > T_U$. We assume there is a basic flow W in the vertical direction (W may be positive or negative) and then the steady velocity field is $\mathbf{v} = (0, 0, W)$. Equation $(4.140)_3$ then yields that $T(z)$ is a solution to

$$\epsilon W T' = \kappa T''.$$

Using the boundary conditions one finds the basic temperature field

$$T(z) = T_L \exp\left(\frac{\epsilon W z}{\kappa}\right) - \frac{[T_U - T_L \exp(\epsilon W d/\kappa)]}{[\exp(\epsilon W d/\kappa) - 1]}\left[1 - \exp\left(\frac{\epsilon W z}{\kappa}\right)\right].$$

The quantity which arises in the perturbation equations is $T'(z)$ and it is convenient to introduce Q and S as

$$\pm Q = \frac{\epsilon|W|}{\kappa}, \qquad S = Qd,$$

where S is dimensionless. Then, $T'(z)$ may be written

$$T'(z) = \frac{\mp Qd}{[\exp(\pm Qd) - 1]}\left(\frac{T_L - T_U}{d}\right)\exp(\pm Qz).$$

We introduce perturbation velocity and temperature fields to (v_i, T) by (w_i, θ), so that $u_i = v_i + w_i$ is the total velocity field in the perturbed state.

Then, we non-dimensionalize with the scalings

$$\lambda = \frac{\lambda_F K U}{\mu}, \qquad \Omega = \frac{\epsilon U d}{\kappa}, \qquad T^\sharp = \frac{U}{\kappa}\sqrt{\frac{\epsilon d \mu (T_L - T_U)}{g\rho_0 \alpha}},$$

$$\mathcal{T} = \frac{d^2}{\kappa}, \qquad \theta = T^\sharp \hat{\theta}, \qquad w_i = U\hat{w}_i, \qquad t = \hat{t}\mathcal{T}, \qquad \mathbf{x} = \hat{\mathbf{x}}d,$$

where U is a fixed scaling for the velocity field. (We select U different from W since we wish to examine the effect of $W \to 0$.) The Rayleigh number $Ra = R^2$ is introduced as

$$R = \frac{K}{\kappa}\sqrt{\frac{\epsilon d(T_L - T_U)g\rho_0\alpha}{\mu}},$$

and the function F defined by

$$F(z; S) = \pm \frac{S}{(e^{\pm S} - 1)}\exp(\pm Sz).$$

In terms of these variables the non-dimensional equations for (w_i, θ) are found to be (where we omit the $\wedge$'s and π is the non-dimensional pressure perturbation),

$$w_i + \lambda(|\mathbf{u}|u_i - |\mathbf{v}|v_i) = -\pi_{,i} + Rk_i\theta,$$
$$w_{i,i} = 0, \tag{4.142}$$
$$\theta_{,t} + \Omega w_i\theta_{,i} = \pm FRw \mp S\theta_{,z} + \Delta\theta.$$

4.6.3   Global nonlinear stability analysis

Our goal is to derive a *global* nonlinear energy stability analysis from which we may determine a nonlinear critical Rayleigh number which we may then compare to the analogous linear instability critical Rayleigh number. To this end we introduce firstly a weighted energy functional

$$\mathcal{F}(t) = \frac{1}{2}\int_V \xi(z)\theta^2 dV,$$

where $\xi(z)$ is a positive linear function of z (although other ξ may be considered) and V is a period cell for the perturbation. By direct calculation we find using $(4.142)_3$ and $(4.142)_2$,

$$\begin{aligned}
\frac{d\mathcal{F}}{dt} ={}& \frac{\Omega}{2}\int_V \xi'w\theta^2 dV \pm R\int_V F\xi w\theta dV \\
& \pm \frac{S}{2}\int_V \xi'\theta^2 dV - \int_V \xi|\nabla\theta|^2 dV.
\end{aligned} \tag{4.143}$$

In order to control the cubic terms which arise we also introduce another functional

$$\mathcal{G}(t) = \frac{1}{3}\int_V |\theta|^3 dV,$$

cf. (Payne and Straughan, 2000b). Using $(4.142)_3$ and Poincaré's inequality one may establish that

$$\frac{d\mathcal{G}}{dt} \leq \pm R \int_V F w \theta^2 dV - \frac{8\lambda_1}{9} \|\theta\|_3^3, \tag{4.144}$$

where $\lambda_1 = \pi^2$ is the constant in Poincaré's inequality for V.

By multiplying the momentum perturbation equation $(4.142)_1$ by w_i and integrating over V we see that

$$0 = -\|\mathbf{w}\|^2 - \lambda \int_V (|\mathbf{u}|u_i - |\mathbf{v}|v_i) w_i dV + R(\theta, w), \tag{4.145}$$

where $(\cdot, \cdot)$ is the inner product on $L^2(V)$. Next, use the fact that

$$(|\mathbf{u}|u_i - |\mathbf{v}|v_i) w_i = \frac{1}{2}(|\mathbf{u}| + |\mathbf{v}|) w_i w_i + \frac{1}{2}(|\mathbf{u}| - |\mathbf{v}|)^2 (|\mathbf{u}| + |\mathbf{v}|), \quad (4.146)$$

(Payne and Straughan, 1999a), and from the triangle inequality note that

$$|\mathbf{w}| = |\mathbf{u} - \mathbf{v}| \leq |\mathbf{u}| + |\mathbf{v}|. \tag{4.147}$$

Thus, using (4.146) and (4.147) we deduce

$$\frac{1}{2} \int_V |\mathbf{w}|^3 dV \leq \frac{1}{2} \int_V (|\mathbf{u}| + |\mathbf{v}|) w_i w_i dV$$

$$\leq \int_V (|\mathbf{u}|u_i - |\mathbf{v}|v_i) w_i dV. \tag{4.148}$$

Upon employing (4.148) in (4.145) we see that

$$0 \leq -\|\mathbf{w}\|^2 - \frac{\lambda}{2}\|\mathbf{w}\|_3^3 + R(\theta, w). \tag{4.149}$$

To develop a global nonlinear energy stability analysis now let $\zeta, a > 0$ be coupling parameters and form $(4.149) + \zeta(4.143) + a(4.144)$. Thus, we find

$$\frac{d}{dt}\left(a\mathcal{G} + \zeta\mathcal{F}\right) \leq R(\theta, w) \pm R\zeta \int_V F\xi w\theta dV \pm \frac{S\zeta}{2}\int_V \xi'\theta^2 dV - \|\mathbf{w}\|^2$$

$$- \zeta \int_V \xi|\nabla\theta|^2 dV + \frac{\zeta\Omega}{2}\int_V \xi'w\theta^2 dV \pm aR \int_V F w\theta^2 dV$$

$$- \frac{8a\lambda_1}{9}\|\theta\|_3^3 - \frac{\lambda}{2}\|\mathbf{w}\|_3^3. \tag{4.150}$$

From this point we restrict attention to the upper signs in (4.142), which corresponds to a vertical flow from bottom to top of the layer. We select $\xi = \xi_1 - \xi_2 z$, $\xi_1 > \xi_2 > 0$ (constants). We stress, however, that progress is possible with flow in the downward direction (lower signs in (4.142)) and with other choices of $\xi(z)$, not linear in z.

Then, from inequality (4.150) one may find that

$$\frac{d}{dt}\left(a\mathcal{G} + \zeta\mathcal{F}\right) \leq R(\theta, w) + R\zeta \int_V F\xi w\theta dV - \frac{S\zeta}{2}\xi_2\|\theta\|^2 - \|\mathbf{w}\|^2$$
$$- \zeta \int_V \xi|\nabla\theta|^2 dV - \frac{\zeta\Omega\xi_2}{2} \int_V w\theta^2 dV + aR \int_V F|w| |\theta|^2 dV$$
$$- \frac{8a\lambda_1}{9}\|\theta\|_3^3 - \frac{\lambda}{2}\|\mathbf{w}\|_3^3. \tag{4.151}$$

For α, β positive constants we use Young's inequality as follows

$$-\frac{\zeta\Omega\xi_2}{2} \int_V w\theta^2 dV \leq \frac{\zeta\Omega\xi_2\alpha}{3} \int_V |\mathbf{w}|^3 dV + \frac{\zeta\Omega\xi_2}{6\alpha^2} \int_V |\theta|^3 dV,$$

$$aR \int_V F|w| |\theta|^2 dV \leq \frac{aRF_m\beta^3}{3} \int_V |\mathbf{w}|^3 dV + \frac{2aRF_m}{3\beta} \int_V |\theta|^3 dV,$$

where

$$F_m = \max_{\{z\in[0,1]\}} |F(z)|.$$

These two inequalities are employed in (4.151) and we find

$$\frac{d}{dt}\left(a\mathcal{G} + \zeta\mathcal{F}\right) \leq R(\theta, w) + R\zeta \int_V F\xi w\theta dV - \frac{S\zeta}{2}\xi_2\|\theta\|^2 - \|\mathbf{w}\|^2$$
$$- \zeta \int_V \xi|\nabla\theta|^2 dV + \left(\frac{\zeta\Omega\xi_2\alpha}{2} + \frac{a\beta^2 F_m R}{3} - \frac{\lambda}{2}\right)\|\mathbf{w}\|_3^3$$
$$+ \left(\frac{\zeta\Omega\xi_2}{6\alpha^2} + \frac{2aRF_m}{3\beta} - \frac{8a\pi^2}{9}\right)\|\theta\|_3^3. \tag{4.152}$$

To handle the cubic terms in (4.152) we put $a = a' + k\epsilon$ for some $\epsilon > 0$ and require that

$$\frac{8a'\pi^2}{9} - \frac{2a'RF_m}{3\beta} - \frac{\zeta\Omega\xi_2}{6\alpha^2} = 0. \tag{4.153}$$

We then minimize

$$\frac{\zeta\Omega\xi_2\alpha}{3} + \frac{a'\beta^2 F_m R}{3}. \tag{4.154}$$

Solve (4.153) for a' and then utilize this in (4.154) so we now minimize

$$M = \frac{\zeta\Omega\xi_2\alpha}{3} + \frac{F_m R\zeta\Omega\xi_2\beta^3}{2\alpha^2(8\pi^2\beta - 6RF_m)}$$

as a function of α to obtain

$$M = \frac{3^{1/3}\zeta\Omega\xi_2(F_m R)^{1/3}\beta}{2(8\pi^2\beta - 6RF_m)}.$$

This is minimized in β and we obtain

$$\beta = \frac{9RF_m}{8\pi^2}, \qquad \alpha = \frac{9RF_m}{8\pi^2}, \qquad a' = \frac{4\zeta\Omega\xi_2\pi^2}{9R^2F_m^2}.$$

In this manner from (4.152) we obtain

$$\frac{d}{dt}\left(a\mathcal{G} + \zeta\mathcal{F}\right) \leq R(\theta, w) + R\zeta\int_V F\xi w\theta dV - \frac{S\zeta}{2}\xi_2\|\theta\|^2 - \|\mathbf{w}\|^2$$
$$- \zeta\int_V \xi|\nabla\theta|^2 dV - \frac{8\pi^2 k\epsilon}{9}\|\theta\|_3^3 - \hat{b}\|\mathbf{w}\|_3^3, \qquad (4.155)$$

where

$$\hat{b} = \frac{\lambda}{2} - \frac{9\zeta\Omega\xi_2 RF_m}{16\pi^2} - \frac{243R^3 F_m^3\epsilon}{512\pi^6}.$$

We now require λ to be such that

$$\lambda > \frac{9\zeta\Omega\xi_2 RF_m}{8\pi^2}. \qquad (4.156)$$

By selecting ϵ arbitrarily small we can ensure $\hat{b} \geq 0$. Thus, (4.155) reduces to

$$\frac{d}{dt}\left(a\mathcal{G} + \zeta\mathcal{F}\right) \leq R(\theta, w) + R\zeta\int_V F\xi w\theta dV - \frac{S\zeta}{2}\xi_2\|\theta\|^2 - \|\mathbf{w}\|^2$$
$$- \zeta\int_V \xi|\nabla\theta|^2 dV - \epsilon\|\theta\|_3^3 - \hat{b}\|\mathbf{w}\|_3^3, \qquad (4.157)$$

with

$$\mathcal{G} = \frac{1}{3}\int_V |\theta|^3 dV, \qquad \mathcal{F} = \frac{1}{2}\int_V (\xi_1 - \xi_2 z)\theta^2 dV.$$

It is necessary to retain the term in ϵ in order to ensure exponential decay of the function $E = a\mathcal{G} + \zeta\mathcal{F}$.

To proceed from (4.157) we define I and D by

$$I = \int_V \theta w(1 + \zeta F\xi)dV, \qquad D = \frac{S\zeta\xi_2}{2}\|\theta\|^2 + \zeta\int_V \xi|\nabla\theta|^2 dV + \|\mathbf{w}\|^2.$$

Then define R_E by

$$R_E = \max_H \frac{I}{D}, \qquad (4.158)$$

where H is the space of admissible solutions to the problem. From (4.157) we then derive

$$\frac{dE}{dt} \leq RI - D - \epsilon\|\theta\|_3^3$$
$$\leq -D\left(\frac{R_E - R}{R_E}\right) - \epsilon\|\theta\|_3^3. \qquad (4.159)$$

If $R < R_E$ it is then easy to show from (4.159) that $E(t)$ decays exponentially and *global* nonlinear stability is established. Thus, the conditions for global stability are restriction (4.156) and $R < R_E$.

The Euler-Lagrange equations which arise from (4.158) are

$$R_E\left(\frac{1+\zeta F\xi}{2}\right)\theta k_i - w_i = \omega_{,i}, \qquad w_{i,i} = 0,$$
$$R_E\left(\frac{1+\zeta F\xi}{2}\right)w - \frac{S\zeta\xi_2}{2}\theta - \zeta\xi_2\theta_{,z} + \zeta\xi\Delta\theta = 0, \tag{4.160}$$

where ω is a Lagrange multiplier. The stability threshold obtained by solving these equations is to be compared with the linear instability one found by solving the corresponding equations for linearized instability theory. For a growth rate like $e^{\sigma t}$ these are

$$\pi_{,i} = R_L k_i \theta - w_i - \frac{\lambda S}{\Omega}(w_i + \delta_{i3}w_3),$$
$$w_{i,i} = 0, \tag{4.161}$$
$$\sigma\theta = F R_L w - S\theta_{,z} + \Delta\theta.$$

Both systems of equations, (4.160) and (4.161) must be solved numerically. We do not do this here. However, we instead observe that if we take $\zeta = 1$, $\xi_1 = 1$, and $0 < \xi_2 = S \ll 1$ (i.e. for S small) then equations (4.160) become

$$R_E\left[\frac{1+(1-Sz)F(z)}{2}\right]\theta k_i - w_i = \omega_{,i}, \qquad w_{i,i} = 0,$$
$$R_E\left[\frac{1+(1-Sz)F(z)}{2}\right]w - \frac{S^2}{2}\theta - S\theta_{,z} + (1-Sz)\Delta\theta = 0. \tag{4.162}$$

For small S,

$$F(z) \sim 1 - \frac{S}{2}\left(1 - 2z\right) + \frac{S^2}{2}\left(z^2 - z - \frac{1}{3}\right) + O(S^3),$$

and we observe that as $S \to 0$, the nonlinear energy equations (4.162) become the same as the linear instability equations (4.161). Thus, in this limit the linear instability and nonlinear energy stability thresholds are the same and subcritical instabilities will not occur. For S small, therefore, we expect the linear instability and global nonlinear stability boundaries to be very close when (4.160) and (4.161) are solved numerically.

There are many other interesting effects we could cite, some clearly connected with throughflow. For example, (Qiao and Kaloni, 1998) study the combined effect of throughflow and an inclined temperature gradient on convection in a porous medium. Convection with an inclined temperature gradient has been studied extensively by (Nield, 1991a; Nield, 1994), (Nield et al., 1993) and by (Manole et al., 1994). The variation of gravity coupled with other effects is analysed by (Alex et al., 2001), (Kaloni and

Qiao, 2001), (Saravanan and Kandaswamy, 2003). Time-periodic effects are important and the influence of a horizontally periodic temperature gradient on convection in a porous medium is studied by (Capone and Rionero, 2000), while (Malashetty and Swamy, 2007a) investigate stability of a rotating porous layer with time periodic temperature modulation. (Bhadauria, 2007) investigates the effect of temperature modulation via a linearized instabiity analysis, employing a Brinkman - Forchheimer model. The effect of confinement on penetrative convection in an anisotropic porous medium is investigated by (Mahidjiba et al., 2003) and (Malashetty and Swamy, 2007b) analyse the effect of rotation and anisotropy on the onset of convection. Convection in a porous medium in a layer which is itself inclined is analysed by (Rees and Bassom, 2000) and incorporating anisotropic effects by (Rees and Postelnicu, 2001). Convection in an open topped enclosure occupied by a porous medium is analysed by (Holzbecher, 2004).

5
Stability of Other Porous Flows

5.1 Convection and flow with micro effects

5.1.1 Biological processes

An increasingly important topic is that of biological processes in porous media. (Khaled and Vafai, 2003) in their review mention many applications of porous media studies in biological situations, including mass diffusion in tissues and in the brain, blood flow, heat transfer, the modelling of stroke imaging, and other issues. (Wood and Ford, 2007) in their editorial to a special issue of the journal Advances in Water Resources draw particular attention to the many ways in which biological processes are interacting with porous media. Among the many topics discussed in the journal issue referred to above we observe that (Wood et al., 2007) develop and analyse a model where biological interactions occur at the interface between the fluid and solid components of a porous medium. (Sun and Wheeler, 2007) develop model equations which describe the transport of viruses in porous media. They are particularly interested in developing a discontinuous Galerkin finite element method for solution of their equations. (Ford and Harvey, 2007) consider the issue of chemotactic movement of a biological species in a porous medium. They specifically discuss bacteria which may swim, such as *E. coli*. Obviously motion of bio-organisms in porous media does have to be such that the pore size is suitable for the organisms to be able to move. If the porous medium model is appropriately chosen, given the size of the swimming micro-organisms, then bio-convection could be another area for study, cf. (Avramenko and Kuznetsov, 2004). In the

B. Straughan, *Stability and Wave Motion in Porous Media*,
DOI: 10.1007/978-0-387-76543-3_5, © Springer Science+Business Media, LLC 2008

next section we consider a specific case of cell movement which occurs in a porous environment.

5.1.2   Glia aggregation in the brain

The brain may be thought of as a spongy porous medium. Cell movement in the brain is known to occur and may even be connected with Alzheimer's disease via the formation of senile plaques. One of the cell types that is key to the pathology is microglia (Potter, 1992). In response to β-amyloid there is a localized concentration of microglia (Itagaki et al., 1989) and an inflammatory response, which involves the local activation of microglia and the release of neurotoxins and factors that chemotactically attract more microglia from peripheral sites to the amyloid lesions (Rogers et al., 2002). In an attempt to model the chemotactic response of microglial cells, (Luca et al., 2003) have developed a model for the density of microglia in a brain. This model also involves concentrations of probable attractants, such as IL-1β, and of a possible repellent, TNF-α, in order to predict amyloid plaque density and size. This is thus a specific model of chemotaxis in a porous medium, the phenomenon mentioned in section 5.1.1. In fact, both chemoattraction and chemorepulsion are involved.

The work of (Luca et al., 2003) presents the chemoattraction - chemorepulsion model

$$
\begin{aligned}
\frac{\partial m}{\partial t} &= \frac{\partial}{\partial x_i}\left(\mu \frac{\partial m}{\partial x_i}\right) - \frac{\partial}{\partial x_i}\left(\chi_1 m \frac{\partial \phi}{\partial x_i}\right) + \frac{\partial}{\partial x_i}\left(\chi_2 m \frac{\partial \psi}{\partial x_i}\right) \\
\frac{\partial \phi}{\partial t} &= \frac{\partial}{\partial x_i}\left(D_1 \frac{\partial \phi}{\partial x_i}\right) + a_1 m - b_1 \phi \\
\frac{\partial \psi}{\partial t} &= \frac{\partial}{\partial x_i}\left(D_2 \frac{\partial \psi}{\partial x_i}\right) + a_2 m - b_2 \psi.
\end{aligned}
\tag{5.1}
$$

In these equations m, ϕ, ψ represent, respectively, the cell density of microglia, the concentration of attractant (interleukin - 1, IL-1β), and concentration of repellent (tumour necrosis factor - α, TNF-α). (Luca et al., 2003) concentrate on the case where the functions $\mu, \chi_1, \chi_2, D_1, D_2, a_1, a_2, b_1, b_2$ are constant.

It is important to note that even though χ_1 and χ_2 are constant the chemoattraction and chemorepulsion terms are still nonlinear and, hence, equation (5.1)$_1$ is a nonlinear partial differential equation. (Luca et al., 2003) developed a linear instability analysis and also investigated numerically simulated solutions. (Quinlan and Straughan, 2005) developed a fully nonlinear stability analysis. This leads to a nonlinear threshold below which aggregation (and, therefore, senile plaque formation) cannot occur.

Equations (5.1) are defined on a bounded domain $\Omega \subset \mathbb{R}^2$ (or $\mathbb{R}^3$) and the boundary of Ω is denoted by Γ. (Luca et al., 2003) non-dimensionalize equations (5.1) to the form

$$\frac{\partial m}{\partial t} = \Delta m - A_1 \nabla \cdot (m\nabla\phi) + A_2 \nabla \cdot (m\nabla\psi)$$

$$\epsilon_1 \frac{\partial \phi}{\partial t} = \Delta\phi + a^2(m - \phi) \tag{5.2}$$

$$\epsilon_2 \frac{\partial \psi}{\partial t} = \Delta\psi + m - \psi,$$

where the constants $A_1, A_2, \epsilon_1, \epsilon_2$ and a^2 are given by

$$A_1 = \frac{\chi_1 a_1 \bar{m}}{\mu b_1}, \quad A_2 = \frac{\chi_2 a_2 \bar{m}}{\mu b_2}, \quad \epsilon_1 = \frac{\mu}{D_1}, \quad \epsilon_2 = \frac{\mu}{D_2}, \quad a = \frac{L_2}{L_1}.$$

Here $\bar{m}$ is the scale for the microglia density, and L_1, L_2 are length scales for attractant and repellent, respectively. The quantity $\bar{m}$ is, in fact, the constant density of microglia in the steady state before any aggregation may occur.

The boundary conditions which hold are zero flux through Γ, so for interleukin-1β and tumour necrosis factor-α one has

$$\nabla\phi \cdot \mathbf{n} = \frac{\partial \phi}{\partial n} = 0, \quad \nabla\psi \cdot \mathbf{n} = \frac{\partial \psi}{\partial n} = 0, \quad \text{on } \Gamma, \tag{5.3}$$

where $\mathbf{n}$ is the unit outward normal to Γ. The condition of conservation of microglia, namely equation $(5.2)_1$ used on the boundary Γ yields,

$$\frac{\partial m}{\partial n} - A_1 m \frac{\partial \phi}{\partial n} + A_2 m \frac{\partial \psi}{\partial n} = 0 \quad \text{on } \Gamma$$

and so because of (5.3) one derives the boundary condition for m, namely

$$\frac{\partial m}{\partial n} = 0 \quad \text{on } \Gamma. \tag{5.4}$$

The non-dimensional constant steady state from which aggregation will ensue is

$$\bar{m} = 1, \quad \bar{\phi} = 1, \quad \bar{\psi} = 1. \tag{5.5}$$

(Quinlan and Straughan, 2005) derive the fully nonlinear perturbation equations from (5.2) by putting $m = m + \bar{m}$, $\phi = \phi + \bar{\phi}$, $\psi = \psi + \bar{\psi}$. The perturbation quantity m has zero mean as a function of time. The nonlinear perturbation equations are then

$$\frac{\partial m}{\partial t} = \Delta m - A_1 \Delta\phi + A_2 \Delta\psi - A_1 \nabla \cdot (m\nabla\phi) + A_2 \nabla \cdot (m\nabla\psi)$$

$$\epsilon_1 \frac{\partial \phi}{\partial t} = \Delta\phi + a^2(m - \phi) \tag{5.6}$$

$$\epsilon_2 \frac{\partial \psi}{\partial t} = \Delta\psi + m - \psi.$$

The functions m, ϕ, ψ must satisfy the boundary conditions (5.3) and (5.4), and the boundary-initial value problem for equations (5.6) is completed by the addition of the initial data

$$m(\mathbf{x},0) = m_0(\mathbf{x}), \quad \phi(\mathbf{x},0) = \phi_0(\mathbf{x}), \quad \psi(\mathbf{x},0) = \psi_0(\mathbf{x}). \qquad (5.7)$$

(Quinlan and Straughan, 2005) develop a fully nonlinear stability analysis which yields a meaningful threshold guaranteeing aggregation of microglia will not occur. In order to derive their goal they employ the solution measure

$$E(t) = \frac{1}{2}\|m\|^2 + \frac{\lambda\epsilon_1}{2}\|\nabla\phi\|^2 + \frac{\lambda_2\epsilon_2}{2}\|\nabla\psi\|^2, \qquad (5.8)$$

where λ and λ_2 are positive constants which have to be selected optimally.

The work of (Luca et al., 2003) presents a very interesting model for microglia aggregation. As yet I am not aware of any model which incorporates the deformation or motion of brain tissue. Whether this is important remains to be seen.

The deformation of a porous matrix may well be important in several porous media problems, such as drying during cheese production. (Simal et al., 2001) analyse an interesting model for the water and salt diffusion during drying of a cheese. To the best of my knowledge the effect of matrix deformation due to drying of the porous cheese structure has not yet been examined. Nevertheless, the work of (Simal et al., 2001) is a very interesting piece of mathematical and computational analysis for a highly mundane problem.

5.1.3 Micropolar thermal convection

The problem of flow in a saturated porous material when there are particles embedded in the fluid is one demanding increasing attention. Such problems are studied, mainly in connection with thermal convection by (Sharma and Gupta, 1995), (Sharma and Kumar, 1997), (Sharma and Kumar, 1998), (El-Hakiem, 1999), (Siddheshwar and Sri Krishna, 2003), (Sunil et al., 2004; Sunil et al., 2005a; Sunil et al., 2005b; Sunil et al., 2005c; Sunil et al., 2006; Sunil et al., 2008). Clearly, the model has to be chosen carefully because the motion of the fluid and particles will depend on the relative size of the particles to the pores. However, with increasing use of nanofluids where certain fluids have copper oxide particles in suspension and greatly increased heat transfer properties, and their potential use in, for example, high porosity metal foams, cf. (Zhao et al., 2004), the area certainly merits attention. If a high porosity model such as a Brinkman one is chosen and the particle size is relatively small then the model should be plausible. (Hoffmann et al., 2007) is a very interesting article which examines in detail the boundary conditions which should be imposed on the particle spin in a micropolar fluid theory.

(Siddheshwar and Sri Krishna, 2003) study thermal convection in a micropolar fluid saturating a porous medium which is modelled via a Darcy - Brinkman theory. We briefly review their work.

(Siddheshwar and Sri Krishna, 2003) employ a linear temperature relation of form $\rho = \rho_0[1 - \alpha(T - T_0)]$ and a Boussinesq approximation and their equations modelling thermal convection of a micropolar fluid in a saturated porous medium may be written

$$\frac{\rho_0}{\Phi}v_{i,t} + \frac{\rho_0}{\Phi^2}v_j v_{i,j} = -p_{,i} + \rho_0 \alpha g T k_i - \left(\frac{\zeta + \mu}{K}\right)v_i$$
$$+ \left(\frac{2\zeta}{\Phi} + \mu'\right)\Delta v_i + \frac{\zeta}{\Phi}\epsilon_{ijk}\omega_{k,j},$$
$$v_{i,i} = 0, \tag{5.9}$$
$$\rho_0 I \omega_{i,t} + \frac{\rho_0 I}{\Phi}v_j \omega_{i,j} = (\lambda' + \eta')\omega_{j,ji} + \eta' \Delta \omega_i + \zeta(\epsilon_{ijk}v_{k,j} - 2\omega_i),$$
$$M T_{,t} + v_i T_{,i} = \frac{\beta}{\rho_0 C_v \Phi}\epsilon_{ijk}\omega_{k,j}T_{,i} + \chi \Delta T.$$

In these equations v_i, p, ω_i and T are the velocity, pressure (modified by the constants in the density law), the particle spin, and the temperature field. The other coefficients are constants and are Φ porosity, μ viscosity of the fluid, μ' effective viscosity of the porous medium, ζ is a coupling term, η' bulk viscosity coefficient, λ' shear spin viscosity coefficient, β micropolar heat conduction coefficient, C_v specific heat at constant volume, K is permeability, and the coefficients M and χ are the heat capacity ratio and effective thermal diffusivity given by

$$M = \frac{(1 - \Phi)(\rho C_p)_s + \Phi(\rho C_p)_f}{(\rho C_p)_f}, \qquad \chi = \frac{(1 - \Phi)\chi_s + \Phi\chi_f}{(\rho C_p)_f},$$

where C_p is the specific heat at constant pressure and subscript s and f refer to solid and fluid, respectively.

The steady state whose stability is investigated is one for which **v** and $\boldsymbol{\omega}$ are zero while T is linear in z.

Perturbations to v_i, ω_i, T and p are introduced and the non-dimensional equations for the perturbations u_i, ν_i, θ and π are derived by (Siddheshwar and Sri Krishna, 2003) as

$$\frac{\lambda}{Pr}u_{i,t} + \frac{1}{Pr}u_j u_{i,j} = -\pi_{,i} - Da\, u_i - R_m \theta k_i$$
$$+ (N_1 + N_1')\Delta u_i + N_1 \epsilon_{ijk}\nu_{k,j},$$
$$u_{i,i} = 0, \tag{5.10}$$
$$\frac{\lambda N_2}{Pr}\nu_{i,t} + \frac{N_2}{Pr}u_j \nu_{i,j} = N_4 \nu_{j,ji} + N_3 \Delta \nu_i + N_1 \epsilon_{ijk}\nu_{k,j} - 2N_1 \nu_i,$$
$$\theta_{,t} + u_i \theta_{,i} = -u_i + N_5 \epsilon_{ijk}\nu_{k,j} + \Delta \theta,$$

where the non-dimensional coefficients (constants) are given by (Siddheshwar and Sri Krishna, 2003).

(Siddheshwar and Sri Krishna, 2003) linearized (5.10) and developed a detailed linear instability theory. Many results are given in their paper, and a detailed comparison with the results of an equivalent Newtonian fluid theory is presented. The analysis is necessarily much different from that for a Newtonian fluid since, for example, oscillatory convection may occur in a micropolar fluid, cf. (Payne and Straughan, 1989). In addition, (Siddheshwar and Sri Krishna, 2003) develop a weakly nonlinear analysis. They reduce system (5.10) to a system of three equations in the stream function ψ, and the terms ν_2 and θ. They then write

$$\begin{pmatrix} \psi \\ \nu_2 \\ \theta \end{pmatrix} = \begin{pmatrix} A(t) \sin \pi \alpha x \\ B(t) \sin \pi \alpha x \\ C(t) \sin \pi \alpha x \end{pmatrix} \sin \pi z + \begin{pmatrix} 0 \\ 0 \\ D(t) \end{pmatrix} \sin 2\pi z$$

and derive a system of nonlinear ordinary differential equations for the amplitudes $A(t), B(t), C(t)$ and $D(t)$. This Lorentz system is solved numerically and many results are presented and discussed in detail by (Siddheshwar and Sri Krishna, 2003).

Magnetic and rotation effects are incorporated in the analyses of (Sunil et al., 2005c; Sunil et al., 2006) and (Sunil and Mahajan, 2008). (Silva, 2006) establishes some interesting convergence results in a Leray problem for a micropolar fluid.

5.2   Porous flows with viscoelastic effects

5.2.1   Viscoelastic porous convection

Flow of a viscoelastic fluid in a porous material is an area with a multitude of applications. Obtaining oil from rocks or soil below the Earth's surface is one area which affects nearly everyone. To model flow of a viscoelastic fluid in a porous medium is a highly non-trivial task. For example, (Lopez de Haro et al., 1996) describes an averaging procedure where they take a Maxwell viscoelastic fluid in a medium with pores. In the linear case they show the velocity field satisfies an equation of form

$$t_m \frac{\partial^2 v_i}{\partial t^2} + \frac{\partial v_i}{\partial t} = -\frac{\partial P}{\partial x_i} + \eta \Delta v_i,$$

for a suitable pressure term. The viscoelastic effect is clearly evident through the t_m term. In fact, structural stability questions for this model are addressed in (Payne and Straughan, 1999b).

Since averaging procedures tend to lead to complicated models the applied literature has employed more ad hoc models. (Kim et al., 2003) studies thermal convection in a porous layer saturated with viscoelastic fluid. The layer is contained between the planes $z = 0$ and $z = d$ with the lower boundary hotter than the upper, i.e. the standard Bénard problem, but for a viscoelastic fluid in a porous layer. They use a Darcy law but one which is modified to include delay effects. Their momentum equation employs a Boussinesq approximation in the buoyancy term and is

$$\frac{\mu}{K}\left(\epsilon\frac{\partial}{\partial t} + 1\right)v_i = \left(\lambda\frac{\partial}{\partial t} + 1\right)\left(-p_{,i} + \rho_0\alpha g T k_i\right) \tag{5.11}$$

where ρ_0, α, g are constants, being density, thermal expansion coefficient, and gravity. The constants μ and K are viscosity and permeability and ϵ, λ are positive constants which represent the viscoelastic effect. The model is completed by adding the continuity and energy balance equations as

$$\begin{aligned} v_{i,i} &= 0, \\ T_{,t} + v_i T_{,i} &= \kappa\Delta T. \end{aligned} \tag{5.12}$$

The steady state is one which has $v_i \equiv 0$ and T linear in z. (Kim et al., 2003) develop linear instability and weakly nonlinear analyses. In terms of a Darcy number, Da, and Rayleigh number, Ra, the non-dimensional linearised perturbation equations are

$$\begin{aligned} \frac{1}{Da}\left(\epsilon\frac{\partial}{\partial t} + 1\right)\Delta w &= Ra\left(\lambda\frac{\partial}{\partial t} + 1\right)\Delta^*\theta, \\ \theta_{,t} &= w + \Delta\theta, \end{aligned} \tag{5.13}$$

where $w = u_3$ is the perturbation to the third component of velocity, θ is the temperature perturbation, and $\Delta^* = \partial^2/\partial x^2 + \partial^2/\partial y^2$. The boundary conditions are

$$w = \theta = 0 \qquad \text{at} \qquad z = 0, 1$$

with the solution satisfying a plane tiling periodicity in the (x, y) plane. Upon utilizing a time dependence like $e^{\sigma t}$ in (5.13) the problem can be reduced to solving the eigenvalue problem

$$(\epsilon\sigma + 1)(D^2 - a^2)(D^2 - a^2 - \sigma)\theta = a^2 Ra(\lambda\sigma + 1)\theta, \tag{5.14}$$

where a is a wave number and $D = d/dz$. Of course, σ occurs nonlinearly in (5.14) and the solution to the eigenvalue problem is correspondingly more difficult than in the porous case with a Newtonian fluid. (Kim et al., 2003) find that unlike the Newtonian case, overstability is possible in the viscoelastic porous convection scenario. For fixed λ, the viscoelastic effect is shown to be destabilizing. In addition to presenting a linearised instability analysis, (Kim et al., 2003) also delivers a weakly nonlinear theory for bifurcation of time independent and periodic solutions.

Further analysis of oscillatory convection in a porous layer saturated with a viscoelastic fluid is due to (Yoon et al., 2004).

Another analysis of convection of a viscoelastic fluid in a porous medium is due to (Rudraiah et al., 1989). These writers employed equations similar to (5.11) and (5.12), but they allowed for an acceleration term on the right of (5.11). They gave specific expressions for the relaxation and retardation times in (5.11) and also encountered oscillatory convection.

5.2.2 Second grade fluids

(Jordan and Puri, 2003) study the problem of a porous material occupying the half space $y > 0$, the porous medium being filled with a viscoelastic fluid of second grade type. The whole domain is fixed and at time $t = 0^+$ the plane $y = 0$ is subject to a constant speed $U_0 \neq 0$ in the x-direction. They seek a solution for the velocity field of form $\mathbf{v} = (u(y,t), 0, 0)$. The model employed by (Jordan and Puri, 2003) effectively adds a Darcy term to the right hand side of the momentum equation, so the velocity field satisfies the system

$$\rho \dot{v}_i = -p_{,i} + \hat{S}_{ji,j} - \frac{\mu \phi}{K} v_i,$$
$$v_{i,i} = 0,$$
(5.15)

where $\hat{S}_{ji}$ is the extra stress for a fluid of second grade. In fact, $\hat{S}_{ji}$ is given by the equation

$$\hat{\mathbf{S}} = \mu \mathbf{A}_1 + \alpha_1 \mathbf{A}_2 + \alpha_2 \mathbf{A}_1^2,$$
(5.16)

where

$$A_{ij}^1 = v_{i,j} + v_{j,i}, \quad A_{ij}^2 = \dot{A}_{ij}^1 + A_{ik}^1 v_{k,j} + v_{k,i} A_{kj}^1,$$
(5.17)

with a superposed dot denoting the material time derivative.

Upon selecting a solution of form $\mathbf{v} = (u(y,t), 0, 0)$ (Jordan and Puri, 2003) show that (5.15), (5.16) yield the boundary initial value problem

$$\rho u_t - u_{yy} - \ell^2 u_{yyt} + \beta^2 u = 0,$$
$$u(0,t) = H(t), \quad u(\infty, t) = 0, \quad t > 0,$$
$$u(y,0) = 0, \quad y > 0,$$
(5.18)

where $H(t)$ is the Heaviside function. System (5.18) is non-dimensionalized in such a way that ℓ^2, β^2 depend on the coefficients U_0, μ, α_1, ϕ and K.

(Jordan and Puri, 2003) solve (5.18) by a Laplace transform method. They interpret their results by employing asymptotic analysis and numerical calculations. (Akyildiz, 2007) develops a Laguerre - Galerkin numerical method for a Stokes' problem like that mentioned above, for a Newtonian fluid in a non-Darcy porous material.

(Puri and Jordan, 2006) develop an interesting analysis for a model for flow of a dipolar fluid in a porous medium. Their model is one which employs a modification of the dipolar fluid to incorporate a Darcy law term to account for the porous material.

(Hayat et al., 2005), (Hayat and Khan, 2005) study flow of a second grade fluid past a porous plate.

5.2.3   Generalized second grade fluids

A generalization of the constitutive equation for a second grade fluid as employed in section 5.2.2 is believed capable of modelling flow in a situation involving a coal - water slurry, see e.g. (Massoudi et al., 2008). Clearly a slurry involving coal and water or coal and oil is some kind of porous or granular material.

The equations of an incompressible second grade fluid given by (Massoudi et al., 2008) are those for balances of mass, momentum, and energy, and are

$$v_{i,i} = 0,$$

$$\rho\left(\frac{\partial v_i}{\partial t} + v_j \frac{\partial v_i}{\partial x_j}\right) = \frac{\partial T_{ji}}{\partial x_j} + \rho b_i,$$

$$\rho\left(\frac{\partial \epsilon}{\partial t} + v_i \frac{\partial \epsilon}{\partial x_i}\right) = T_{ij} L_{ij} - \frac{\partial q_i}{\partial x_i} + \rho r.$$

Here $\rho, v_i, T_{ji}, b_i, \epsilon, q_i$ and r are the density, velocity, stress tensor, body force, internal energy, heat flux vector, and heat supply for the fluid (slurry), and $L_{ij} = v_{i,j}$. (Massoudi et al., 2008) point out that the constitutive equations for the stress tensor for a *generalized second grade fluid* were given by (Man and Sun, 1987) and by (Man, 1992) and are a generalization of equation (5.16), being

$$\mathbf{T} = -p\mathbf{I} + \mu\Pi^{m/2}\mathbf{A}_1 + \alpha_1\mathbf{A}_2 + \alpha_2\mathbf{A}_1^2, \tag{5.19}$$

or

$$\mathbf{T} = -p\mathbf{I} + \mu\Pi^{m/2}(\mathbf{A}_1 + \alpha_1\mathbf{A}_2 + \alpha_2\mathbf{A}_1^2), \tag{5.20}$$

where $\mathbf{A}^1$ and $\mathbf{A}^2$ are given by equations (5.17) and

$$\Pi = \frac{1}{2}\operatorname{tr}\mathbf{A}_1^2.$$

The coefficient μ is the dynamic viscosity, α_1, α_2 are normal stress coefficients, and m is a constant which may be positive or negative. With $m < 0$ the (slurry) fluid is said to be shear - thinning whereas if $m > 0$ it is shear - thickening.

(Massoudi et al., 2008) consider a non-isothermal theory and find numerical solutions for flow between two vertical plates at different temperatures. Stability and nonexistence results for the generalized second grade fluid model with the constitutive equations (5.19) or (5.20) are given by (Franchi and Straughan, 1993b) and further details of these and related results are given in the book by (Straughan, 1998), chapter 4.

5.3   Storage of gases

5.3.1   Carbon dioxide storage

An extremely interesting physical problem which could have important consequences for every one of us is studied by (Ennis-King et al., 2005) and by (Xu et al., 2006). Both sets of writers develop the same model and their analyses are very similar.

The problem of concern to both sets of writers involves how to dispose of carbon dioxide. Due to worldwide concern over global warming and how to reduce emissions of greenhouse gases into the Earth's atmosphere, engineers are contemplating the storage of carbon dioxide in naturally occurring underground geological locations. Such locations may be where coal is very difficult to mine, where oil or gas reservoirs have been partly or fully exhausted by current recovery techniques, or deep underground brine aquifiers. Carbon dioxide in a supercritical state (where the pressure and temperature are above the critical point values of $p = 73.82$ bar, $T = 31.04°C$) could be injected deep underground into the chosen geological sites. However, there is concern over whether CO_2 stored in this way will seep out and eventually leak back into the atmosphere. The CO_2 could simply leak sideways under the bedrock, or since CO_2 reduces the fluid density (as opposed to many other gases which increase it) the decrease in density could lead to convective mixing in the porous layer well beneath the Earth's surface. A strong mixing could actually lead to enhanced containment as pointed out by (Ennis-King et al., 2005).

In order to study the convective motion of CO_2 in a porous medium (Ennis-King et al., 2005) and (Xu et al., 2006) employ a Darcy porous medium model which allows for variable permeability of the porous matrix. In the system of (Xu et al., 2006) the carbon dioxide is supposed stored in a infinitely wide porous layer of depth H, with z measured downward, the layer being $\{(x, y) \in \mathbb{R}^2\} \times \{z \in (0, H)\}$. Gravity acts downward and the fluid density, ρ, is assumed linear in C, the concentration of dissolved CO_2, viz. $\rho = \rho_0(1 + \beta C)$, ρ_0, β, constants. The partial differential equations describing the motion of fluid in the porous layer then consist of momentum and continuity equations coupled with a transport equation for C, namely,

$$\mu K_{ij}^{-1} v_j = -p_{,i} + \rho_0 \beta k_i C, \qquad v_{i,i} = 0,$$
$$\phi \frac{\partial C}{\partial t} + v_i C_{,i} = \phi D \Delta C, \tag{5.21}$$

where v_i is the fluid (seepage) velocity, ϕ is porosity, p is the pressure taking into account the constant terms in the density, k_i is the vector $(0, 0, 1)$ remembering z is measured downward, and D is a diffusion coefficient.

The boundary conditions adopted are zero flux at the lower boundary $z = H$, i.e. $\partial C/\partial z = 0$, and $C =$constant, at the upper one $z = 0$. Since the CO_2 can move, the basic state whose stability is under investigation is time dependent. While the basic velocity $v_i = 0$, the basic concentration of CO_2 is assumed to be a function of z and t, and so satisfies the system

$$\frac{\partial C}{\partial t} = \frac{\partial^2 C}{\partial z^2}, \tag{5.22}$$

$$C(0,t) = 1, \qquad \frac{\partial C}{\partial z}(1,t) = 0, \qquad t \geq 0, \tag{5.23}$$

$$C(z,0) = 1, \qquad z \geq 0. \tag{5.24}$$

In (5.22), (5.23) and (5.24), C has been non-dimensionalized and (5.24) represents the initial conditions C must satisfy.

The solution to (5.22) – (5.24) is given by (Xu et al., 2006) and by (Ennis-King et al., 2005) as

$$C(z,t) = 1 - \frac{4}{\pi} \sum_{n=1}^{\infty} \frac{1}{(2n-1)} \exp\left\{-\left(n - \frac{1}{2}\right)^2 \pi^2 t\right\} \sin\left(n - \frac{1}{2}\right) \pi z. \tag{5.25}$$

Both sets of writers (Ennis-King et al., 2005) and (Xu et al., 2006) perform both a linearised instability analysis and a global nonlinear energy stability analysis of the perturbations to (5.21) employing the base solution (5.25). If u_i, π, ϕ denote non-dimensional perturbations to v_i, p and C then the linearised perturbation equations of (Xu et al., 2006) are

$$u_i = -\pi_{,i} + k_i Ra\,\phi, \qquad u_{i,i} = 0,$$
$$\phi_{,t} + wC_{,z} = \gamma\Delta^*\phi + \phi_{,zz}, \tag{5.26}$$

where $w = u_3$ and γ represents a variation between horizontal and vertical permeability. The boundary conditions are

$$\phi = 0,\; z = 0, \qquad \phi_{,z} = 0,\; z = 1; \qquad w = 0,\; z = 0, 1. \tag{5.27}$$

Due to the time-dependence of C the system (5.26), (5.27), is solved by writing

$$\phi(z,t) = \sum_{k=1}^{N} a_k(t) \sin\left(k - \frac{1}{2}\right)\pi z,$$
$$w(z,t) = \sum_{k=1}^{N} b_k(t) \sin k\pi z, \tag{5.28}$$

and a system of ODEs is derived for the time-dependent coefficients a_k, b_k. This system is solved numerically by a Runge-Kutta scheme. In (Xu et al., 2006) they find an *instability time* t for each Rayleigh number and wave number such that the normalized time average

$$\Phi(t) = \left(\frac{\int_0^1 \phi^2(z,t)dz}{\int_0^1 \phi^2(z,0)dz}\right)^{1/2}$$

has zero derivative at $\hat{t}$, i.e. $d\Phi/dt = 0$ there. The critical time, t_c, for the onset of instability is then defined by minimizing the set of $\{\hat{t}\}$ over the wave numbers. This yields a critical time and wavenumber for a given Rayleigh number.

The global analysis commences with the nonlinear form of (5.26) and employs the functional

$$E(t) = \frac{\lambda}{2} \int_V \phi^2 dV, \tag{5.29}$$

where $\lambda > 0$ is a coupling parameter to be selected judiciously and V is a period cell for the perturbation. By calculating a differential equation from dE/dt and requiring $dE/dt \leq 0$, $\forall t \geq 0$, a critical Rayleigh number is found, $Ra_c(t)$, which depends on time t. The Euler-Lagrange equations which arise are solved numerically by a Galerkin method, minimizing over the wave numbers and maximizing over λ. Since $Ra_c(t)$ depends on time (Xu et al., 2006) argue that for a given time one determines a critical Rayleigh number. This in turn defines a critical time for the global nonlinear stability analysis.

(Ennis-King et al., 2005) develop a very similar analysis and also look at the case of a semi-infinite domain, $z \to \infty$. Both sets of writers present extensive results incorporating anisotropic permeability and compare the linear instability thresholds with the global stability ones. Unlike many convection problems in porous media where the basic state is time - independent, the linear - nonlinear boundaries display a wide variation.

The recent comments by (Nield, 2007) and the reply by (Xu et al., 2007) are interesting and are worth taking into account. (Nield, 2007) argues that one ought not to work with a Rayleigh number based on horizontal or vertical permeability separately, but one should instead define a Rayleigh number based on a permeability which is a combination of both the horizontal and vertical permeabilities. (Xu et al., 2007) respond by arguing that since they are dealing with a time dependent base flow one has to be careful if one tries to present an analogy with a problem involving slowly changing boundary conditions. They present numerical computations varying both the horizontal and vertical permeabilities to observe the dependence of the critical time on the variation of these permeabilities. On this basis they deduce that the critical time is more sensitive to changes in the vertical permeability, although they also suggest that the idea of using a combined permeability as suggested by (Nield, 2007) does deserve further investigation.

5.3.2 Hydrogen storage

Another area of gas storage which involves porous media is that of hydrogen retainment. The need to store hydrogen efficiently is important since it is believed it may become a suitable replacement for fossil fuels such as coal or oil, cf. (Armandi et al., 2007). In fact, (Armandi et al., 2007) is an interesting article which reports experiments involving storing hydrogen in various kinds of porous carbons. It is clear from the results of (Armandi et al., 2007) that the microstructure of the porous carbon material is very important. The surface area at the microscopic level would appear to be important in the ability of the material to store hydrogen. As far as I am aware, the problem of developing and studying a mathematical model for hydrogen storage as described by (Armandi et al., 2007) is open. This could be an interesting future area.

5.4   Energy growth

5.4.1   Soil salinization

The topic of salinization in soil is addressed from various angles by (Bear and Gilman, 1995), (Gilman and Bear, 1996), (Wooding et al., 1997a; Wooding et al., 1997b), and (van Duijn et al., 2002). In fact, these writers each develop a theory to describe the important process of soil salinization. Salinization is caused by the evaporation of moisture through the surface of the soil. In dry regions of the Earth, where the rainfall is small and the water table lies relatively close to the surface, the mean flow of water through the unsaturated soil is in the upward direction. Often, groundwater is saline and upward flow of this saline solution results in salts being transported to the soil surface. Evaporation of water from the soil layer leaves increased density of salts in the soil with the result that their concentration near the surface increases. Salinization can lead to the formation of salt lakes and so the subject is of much interest in geotechnical engineering.

In regions with little rainfall groundwater typically evaporates through the soil surface due to the ambient hot temperature conditions and the presence of a shallow water table may lead to salts accumulating near the soil surface because of effective upward movement. In fact, this phenomenon should really be modelled using a theory of unsaturated porous media. The work of (Gilman and Bear, 1996) models the unsaturated region as a porous layer with a uniform liquid distribution and this essentially transforms the problem to one for a saturated porous medium, although the permeability, diffusivity and dispersivity must be chosen appropriately. These writers note that the subject of convection in unsaturated porous media is a largely untouched area which is increasingly occupying attention. Their model employs a concentration dependent viscosity because the viscosity dependence on concentration is typically 1.5 to 3 times greater than that of pure water, and they argue, therefore, that one should not neglect this effect. As (Gilman and Bear, 1996) observe the density variation is much smaller, of order 15 to 30%, and so a constant density model may be acceptable apart from where the density occurs in the buoyancy term.

The salinization model of (Gilman and Bear, 1996) employs the non-dimensional equations

$$(1 + \gamma\omega)v_i = - p_{,i} - Ra\,\omega k_i,$$
$$v_{i,i} = 0,$$
$$\frac{\partial \omega}{\partial t} + \frac{\partial}{\partial x_i}(v_i\omega) = - \frac{\partial J_i}{\partial x_i},$$
(5.30)

where v_i, p and ω are velocity, pressure and salt concentration, respectively, $\mathbf{k} = (0,0,1)$, $(x,y) \in \mathbb{R}^2$, and $z \in (0,1)$. The Rayleigh number is Ra and J_i is the salt flux vector chosen as

$$J_i = -D_{ij}\omega_{,j}.$$
(5.31)

The diffusion tensor, D_{ij}, in (5.31) is a function of the velocity and is

$$D_{ij} = (1 + \alpha_T|\mathbf{v}|)\delta_{ij} + (\alpha_L - \alpha_T)\frac{v_i v_j}{|\mathbf{v}|},$$
(5.32)

where $|\mathbf{v}| = \sqrt{v_i v_i}$ and α_L, α_T are longitudinal and transverse dispersivities. The form of (5.32) accounts for the fact that the dry conditions induce transport of saline solution upward via an anisotropic velocity dependence. The basic momentum equation $(5.30)_1$ is Darcy's law but the viscosity is a linear function of ω with coefficient γ.

(Gilman and Bear, 1996) have their layer with $z = 0$ as the boundary of the water table below the salinization layer and such that $z = 1$ is the surface of the soil. Their boundary conditions involve the Peclet number, Pe, which is a measure of the evaporation rate at the surface, and they involve the salt concentration at saturation, ω_*. The boundary conditions are:

on the water table/active convection layer boundary, $z = 0$,

$$p = 0, \quad \omega = \omega_0;$$
(5.33)

on the ground surface, $z = 1$,

$$v_3 = Pe,$$
(5.34)

and

$$\text{either} \quad Pe\,\omega + J_3 = 0, \quad \text{if } \omega \leq \omega_*; \quad \text{or} \quad \omega = \omega_*.$$
(5.35)

In the latter case $Pe\,\omega + J_3 \geq 0$. The boundary conditions (5.33) imply that at the water table the salt concentration is constant along with the pressure. Boundary conditions (5.34) and (5.35) account for that fact that a priori we can have at the ground surface either zero flux, or the salt concentration at saturation which may result in salt precipitation there.

(Gilman and Bear, 1996) investigate instability and possible convective motion in the porous layer, by determining a steady solution of form $\tilde{\mathbf{v}} = (0,0,Pe)$, $\omega = \tilde{\omega}(z)$. They find that the form of $\tilde{\omega}(z)$ depends on which boundary condition is in force in (5.35). If we define ω_0 to be the concentration at the lower surface of the layer, $z = 0$, and define Pe^* as the quantity $Pe^* = \log(\omega_*/\omega_0)$, with Pe' given by $Pe' = Pe/(1 + \alpha_L Pe)$, then $\tilde{\omega}(z)$ will have form

$$\tilde{\omega}(z) = \omega_0\, e^{Pe'z}, \quad \text{for } Pe' < Pe^*;$$
$$\tilde{\omega}(z) = \omega_*\, e^{Pe'(z-1)}, \quad \text{for } Pe' \geq Pe^*. \tag{5.36}$$

(Gilman and Bear, 1996) introduce a perturbation u_i, ω, π to the steady solution $\tilde{v}_i, \tilde{\omega}, \tilde{p}$. Their linearized equations for u_i, ω, π are

$$(1 + \gamma\tilde{\omega})u_i = -\pi_{,i} - (Ra + \gamma Pe)k_i\omega,$$
$$u_{i,i} = 0,$$
$$\frac{\partial \omega}{\partial t} = (1 + \alpha_T Pe)\Delta\omega + (\alpha_L - \alpha_T)Pe\frac{\partial^2 \omega}{\partial z^2} - Pe\frac{\partial \omega}{\partial z} \tag{5.37}$$
$$+ \alpha_T\tilde{\omega}'\frac{\partial u_3}{\partial z} - (\tilde{\omega}' - \alpha_L\tilde{\omega}'')u_3.$$

They introduce a modified Rayleigh number Ra' by $Ra' = Ra + \gamma Pe$, and a function $F(z)$ by $F(z) = [1/Pe'(1 + \gamma\tilde{\omega})]\,d\tilde{\omega}/dz$. The function u_i is then eliminated from (5.37) to derive a coupled system of partial differential equations for π and ω. These are

$$\Delta\pi - \gamma Pe'F(z)\frac{\partial \pi}{\partial z} = Ra'\Big(\gamma Pe'F(z)\omega - \frac{\partial \omega}{\partial z}\Big),$$
$$\frac{\partial \omega}{\partial t} = \Delta\omega + \alpha_L Pe\frac{\partial^2 \omega}{\partial z^2} - Pe\frac{\partial \omega}{\partial z} \tag{5.38}$$
$$+ Pe'(1 - \alpha_L Pe')F(z)\Big(\frac{\partial \pi}{\partial z} + Ra'\omega\Big).$$

Since (5.38) are linear they seek a normal mode solution of form $\pi(\mathbf{x}, t) = P(z)\,e^{\sigma t}\,f(x, y)$, $\omega(\mathbf{x}, t) = \Omega(z)\,e^{\sigma t}\,f(x, y)$, where σ is the growth rate and f is a plane tiling planform. Thus, the instability analysis of (Gilman and Bear, 1996) utilizes the eigenvalue equations

$$(D^2 - \gamma Pe'FD - k^2)P = Ra'(\gamma Pe'F - D)\Omega,$$
$$[(1 + \alpha_L Pe)D^2 - PeD - k^2]\Omega \tag{5.39}$$
$$+ Pe'(1 - \alpha_L Pe')F(DP + Ra'\Omega) = \sigma\Omega,$$

where $D = d/dz$ and k is a wavenumber. Note that P and Ω represent perturbations of pressure and salt concentration, respectively. Equations (5.39) are defined on $(0, 1)$ and the boundary conditions for small Peclet numbers or that of a transient state before the onset of salt precipitation, are

$$P = 0, \quad \Omega = 0 \quad \text{at } z = 0,$$
$$DP + Ra'\Omega = 0, \quad D\Omega - Pe'\Omega = 0 \quad \text{at } z = 1. \tag{5.40}$$

(Gilman and Bear, 1996) solve (5.39), (5.40) numerically by setting $\sigma = 0$ and solving for k^2 as a function of Pe', Ra'. They use a finite difference scheme and need necessarily to transform the exponential dependence out of the problem by using the variable $\zeta = \exp\{Pe'(z-1)\}$, to remove large coefficients of Ra' and Pe' in the boundary conditions, otherwise numerical convergence problems are encountered. An alternative numerical scheme to solve (5.39), (5.40) very efficiently and directly with $\sigma \in \mathbb{C}$, is presented by (Payne and Straughan, 2000a).

5.4.2   Other salinization theories

(Wooding et al., 1997a) develop an analysis of salinization based on a Darcy - law model but one where their flow rate u_i is not solenoidal. If their layer is in the (x, y) plane then their non-dimensional equations may be written

$$
\begin{aligned}
u_{i,i} + \gamma\phi S_{,t} &= 0, \\
-p_{,i} - k_i S &= M^{-1} u_i, \\
S_{,t} + \frac{1}{\phi(1+\gamma S)}\, u_i S_{,i} &= \frac{1}{R_s}\, (D_{ij} S_{,j})_{,i}
\end{aligned}
\tag{5.41}
$$

where S and p are the salinity and pressure, respectively. The quantity ϕ is porosity and γ and M are positive constants. The vector $\mathbf{k} = (0, 0, 1)$, D_{ij} is a diffusivity tensor, and R_s is a salt Rayleigh number.

(van Duijn et al., 2002), (Pieters et al., 2004) and (Pieters and van Duijn, 2006) study in depth a model appropriate to salinization. They work on the spatial domain $\{(x, y) \in \mathbb{R}^2\} \times \{z > 0\}$, with z pointing downward, i.e. the half - space below the plane $z = 0$ which is the ground surface, and so gravity acts downward. In terms of a fluid velocity v_i, pressure p, and density ρ, their equations are

$$
\begin{aligned}
v_{i,i} &= 0, \\
p_{,i} - k_i \rho g + \frac{\mu}{K} v_i &= 0, \\
\phi \rho_{,t} + v_i \rho_{,i} &= D\Delta\rho
\end{aligned}
\tag{5.42}
$$

where $\mathbf{k} = (0, 0, 1)$ (recalling z is measured downward), μ, K, ϕ and D are viscosity, permeability, porosity, and a diffusion coefficient. The boundary conditions of (van Duijn et al., 2002) are

$$
v_i = -E k_i, \quad \rho = \rho_m \qquad \text{at } z = 0
\tag{5.43}
$$

where E is the evaporation rate at the surface and ρ_m is the maximum density at the outflow boundary. They also employ the initial condition

$$
\rho(\mathbf{x}, 0) = \rho_r
\tag{5.44}
$$

where ρ_r is the fluid density in the "ambient" state.

The above system is non-dimensionalized and written in terms of a Rayleigh number, Ra, as

$$U_{i,i} = 0,$$
$$P_{,i} - k_i S + U_i = 0, \qquad (5.45)$$
$$S_{,t} + Ra\, U_i S_{,i} = \Delta S,$$

where S is the non-dimensional density (salinity). These equations hold in $\{(x,y) \in \mathbb{R}^2\} \times \{z > 0\} \times \{t > 0\}$ with the boundary and initial conditions being

$$U_i = -\frac{1}{Ra} k_i, \quad S = 1, \qquad \text{when } z = 0,$$
$$S(\mathbf{x}, 0) = 0. \qquad (5.46)$$

The boundary velocity $-Ra^{-1}k_i$ is denoted by U_i^0, i.e. $U_i^0 = -Ra^{-1}k_i$.

The basic solution studied by (van Duijn et al., 2002), (Pieters et al., 2004) and (Pieters and van Duijn, 2006) is one where the base velocity is the uniform throughflow $U_i = U_i^0$. The basic salinity field, $S(z,t)$, must then be determined as a solution to the problem

$$\frac{\partial S}{\partial t} = \frac{\partial^2 S}{\partial z^2} + \frac{\partial S}{\partial z}, \qquad (z,t) \in \mathbb{R}^+ \times \mathbb{R}^+,$$
$$S = 1, \quad z = 0, t > 0, \qquad (5.47)$$
$$S = 0, \quad z > 0, t = 0.$$

The solution to (5.47) is

$$S_0(z,t) = \frac{1}{2} e^{-z} \operatorname{erfc}\left(\frac{z-t}{2\sqrt{t}}\right) + \frac{1}{2} \operatorname{erfc}\left(\frac{z+t}{2\sqrt{t}}\right). \qquad (5.48)$$

However, (van Duijn et al., 2002), (Pieters et al., 2004) and (Pieters and van Duijn, 2006), also investigate the situation when S_0 is replaced by its asymptotic solution as $t \to \infty$, namely,

$$S_0(z) = e^{-z}. \qquad (5.49)$$

(van Duijn et al., 2002) introduce perturbations s, u_i, π to $S_0, \mathbf{U}_0, P_0$ and these satisfy the equations

$$u_{i,i} = 0,$$
$$\pi_{,i} - k_i s + u_i = 0, \qquad (5.50)$$
$$s_{,t} + Ra\, u_i s_{,i} = \Delta s + s_{,z} - Ra\, w \frac{\partial S_0}{\partial z}.$$

The boundary conditions are that u_i, π, s vanish as $z \to \infty$ and

$$s = u_i = 0, \qquad z = 0. \qquad (5.51)$$

(Pieters et al., 2004) mainly investigates the linearised system arising from (5.50), (5.51). Thus, they remove π, u_1 and u_2 and with $w = u_3$ they analyse in depth the system

$$s_{,t} = \Delta s + s_{,z} - Ra\, w\frac{\partial S_0}{\partial z}$$

$$\Delta w = \Delta^* s,$$

(5.52)

where Δ^* is the horizontal Laplacian. In fact, by writing s and w in normal mode form

$$s = s(z,t)e^{mx+ny}, \qquad w = w(z,t)e^{mx+ny},$$

they use a Green's function, a contraction mapping argument, and the Banach fixed point theorem to deduce existence, regularity and uniqueness for a solution to (5.52). In fact, for any $T > 0$, they show the normal mode solution to (5.52) exists, is unique, and $(s, w) \in C^\infty(\mathbb{R}^+ \times (0, T])$.

A careful weakly nonlinear analysis and bifurcation study of the salinization system (5.50) is given by (Pieters and Schuttelaars, 2007).

5.4.3   Time growth of parallel flows

Before progressing directly to the analysis of (van Duijn et al., 2002) it is a good juncture to digress briefly and report on some recent findings where the kinetic energy for a certain class of fluid flows can have relatively very large transient growth, even though the flow is eventually stable according to linear theory.

As pointed out in section 4.3.1, and in (Straughan, 1998), the equations governing hydrodynamic stability are typically of the form

$$Au_t = L_S u + L_A u + N(u),$$

with the notation described there. In general, genuinely unconditional nonlinear energy stability results are few for problems where the nonlinearities are not simply the convective ones, or where the linear operator $L = L_S + L_A$ is far from being symmetric. In fact, for parallel shear flows the situation is very open. While we introduce this as a prelude to the work of (van Duijn et al., 2002), the work we describe for parallel flows has much bearing on the material discussed in sections 5.8.1 and 6.7, where parallel flow is studied in a porous medium context.

Consider a viscous incompressible fluid contained in the infinite three - dimensional spatial layer $I = \{(x, z) \in \mathbf{R}^2\} \times \{y \in (-1, 1)\}$. One of the areas of hydrodynamic stability where the nonlinear theory of energy stability has been least successful is in the study of parallel flows in the domain I, e.g. in the study of the stability of flows with a base solution like $\mathbf{v} = (U(y), 0, 0)$. Typical of such flows are Couette flow where $U(y) = y$ and Poiseuille flow for which $U(y) = 1 - y^2$. Couette flow is that which arises when the top plate $y = 1$ is sheared at a constant velocity relative to the bottom one, whereas Poiseuille flow is achieved by application of a constant pressure gradient in the x-direction, keeping the planes $y = \pm 1$ fixed.

The theory of the above flows is governed by the Navier - Stokes equations for the velocity and pressure fields, v_i, p. These partial differential equations are, in a suitably non-dimensionalised form,

$$\frac{\partial v_i}{\partial t} + Re\, v_j \frac{\partial v_i}{\partial x_j} = -\frac{\partial p}{\partial x_i} + \Delta v_i, \qquad \frac{\partial v_i}{\partial x_i} = 0, \qquad (5.53)$$

where $Re(= V\hat{L}/\nu)$ is the Reynolds number. (The quantities $V, \hat{L}$ and ν are a typical velocity, depth of the layer before non-dimensionalisation, and the kinematic viscosity of the fluid.) To study the stability of Couette or Poiseuille flow one may set $\mathbf{U} = (U(y), 0, 0)$ and then derive equations for the perturbation velocity and pressure fields (u_i, π) defined by $v_i = U_i + u_i$, $p = \bar{p} + \pi$, where $\bar{p}$ is the pressure corresponding to the base velocity U_i. The perturbation velocity and pressure u_i and π then satisfy the partial differential equations

$$\frac{\partial u_i}{\partial t} + Re\left(U \frac{\partial u_i}{\partial x} + \delta_{i1} U' v\right) + Re\, u_j \frac{\partial u_i}{\partial x_j} = -\frac{\partial \pi}{\partial x_i} + \Delta u_i, \qquad \frac{\partial u_i}{\partial x_i} = 0, \ (5.54)$$

where $U' = dU/dy$ and where $\mathbf{u} = (u, v, w)$.

The classical theory of linear instability for the system of partial differential equations (5.54) writes the functions u_i and π in the form

$$u_i = u_i(y) e^{i(ax+bz-act)}, \qquad \pi = \pi(y) e^{i(ax+bz-act)},$$

and discards the quadratic term $Re\, u_j \partial u_i / \partial x_j$. In this manner one derives from (5.54) the system of ordinary differential equations

$$\begin{aligned}
(ReU - c)iau + ReU'v &= -ia\pi + \left(D^2 - [a^2 + b^2]\right)u, \\
(ReU - c)iav &= -D\pi + \left(D^2 - [a^2 + b^2]\right)v, \\
(ReU - c)iaw &= -ib\pi + \left(D^2 - [a^2 + b^2]\right)w, \\
iau + Dv + ibw &= 0,
\end{aligned} \qquad (5.55)$$

where $D = d/dy$ and $\mathbf{u}(y) = (u(y), v(y), w(y))$. The traditional approach at this point has been to invoke Squire's theorem, arguing that the transformations $Re \to \tilde{Re}\, \tilde{a}/a$, $c \to \tilde{c}\tilde{a}/a$, $\tilde{a} = \sqrt{a^2 + b^2}$, reduce (5.55) to a two-dimensional form

$$\begin{aligned}
(ReU - c)iau + ReU'v &= -ia\pi + (D^2 - a^2)u, \\
(ReU - c)iav &= -D\pi + (D^2 - a^2)v, \\
iau + Dv &= 0.
\end{aligned} \qquad (5.56)$$

The no-slip boundary condition at the plates is interpreted mathematically by requiring $u = v = 0$ for $y = \pm 1$. Next introduce a stream function ψ by $u = \partial \psi / \partial y$, $v = -\partial \psi / \partial x$, and then introduce the function $\phi(y)$ by $\psi = \phi(y)\, e^{ia(x-ct)}$. In this manner, one deduces from (5.56) that the linear instability problem reduces to studying the fourth order ordinary differential equation for $\phi(y)$,

$$(D^2 - a^2)^2\phi = iaRe(U - c)(D^2 - a^2)\phi - iaReU''\phi, \qquad y \in (-1, 1). \quad (5.57)$$

This is the celebrated Orr-Sommerfeld equation. The boundary conditions become

$$\phi = D\phi = 0, \qquad y = \pm 1. \quad (5.58)$$

Equation (5.57) subject to the boundary conditions (5.58) constitutes an eigenvalue problem for the growth rate $c = c_r + ic_i$. If $c_i > 0$ the flow is linearly unstable. The solution to (5.57), (5.58) for the spectrum $\{c^{(k)}\}$ is a hard numerical problem, see e.g. the exposition in (Dongarra et al., 1996).

After many years of studying instability via (5.57) a different philosophy was advocated in the early 90's. (Butler and Farrell, 1992), (Gustavsson, 1991), and (Reddy and Henningson, 1993) have given convincing arguments to assert that the transient onset of instability is not governed solely by the leading eigenvalue of system (5.57), (5.58). In other words, the stability of flows such as Couette and Poiseuille are not completely controlled by the dominant eigenfunction $\phi^{(1)}$ for which $c^{(1)}$ has largest imaginary part. Indeed, they argue that one ought to consider the full three-dimensional linear system (5.55).

It is instructive to rederive the three-dimensional system in the manner of (Butler and Farrell, 1992). Written out in full, the linear equations which arise from (5.54) are

$$
\begin{aligned}
\frac{\partial u}{\partial t} + Re\left(U\frac{\partial u}{\partial x} + U'v\right) &= -\frac{\partial \pi}{\partial x} + \Delta u, \\
\frac{\partial v}{\partial t} + ReU\frac{\partial v}{\partial x} &= -\frac{\partial \pi}{\partial y} + \Delta v, \\
\frac{\partial w}{\partial t} + ReU\frac{\partial w}{\partial x} &= -\frac{\partial \pi}{\partial z} + \Delta w, \\
\frac{\partial u}{\partial x} + \frac{\partial v}{\partial y} + \frac{\partial w}{\partial z} &= 0.
\end{aligned}
\qquad (5.59)
$$

(Butler and Farrell, 1992) argue that it is sufficient to consider the component of velocity normal to the plane of the flow, i.e. v, and the component of vorticity in the normal (y) direction. The normal component of vorticity is $\omega_2 = (\operatorname{curl} \mathbf{u})_2$ which may be written as ω, and then

$$\omega = \frac{\partial u}{\partial z} - \frac{\partial w}{\partial x}. \quad (5.60)$$

Equipped with a knowledge of v and ω one may then calculate the variables u, w and π. The physical reason for v and ω being the main variables to influence the instability process is that a streamwise vortex is seen to be influential in experiments. From a mathematical viewpoint, v and ω arise naturally in the kinetic energy.

(Butler and Farrell, 1992) employ a different non-dimensionalisation which writes $(5.59)_{1-3}$ as

$$\frac{\partial u_i}{\partial t} + U\frac{\partial u_i}{\partial x} + U'v\delta_{1i} = -\frac{\partial \pi}{\partial x_i} + \frac{1}{Re}\Delta u_i. \tag{5.61}$$

The evolution equation for ω is found directly from (5.61) by differentiating the equation for u with respect to z and subtracting from this the x derivative of the equation for w. A single equation for v is obtained from (5.61) by taking curlcurl of that equation and retaining the second component of the result. In this manner we derive the following equations for v and ω derived and used by (Butler and Farrell, 1992),

$$\frac{\partial \omega}{\partial t} + U\frac{\partial \omega}{\partial x} - \frac{1}{Re}\Delta\omega = -U'\frac{\partial v}{\partial z},$$
$$\frac{\partial}{\partial t}\Delta v + U\frac{\partial}{\partial x}\Delta v - U''\frac{\partial v}{\partial x} - \frac{1}{Re}\Delta^2 v = 0. \tag{5.62}$$

The boundary conditions for v and ω are that

$$v = \frac{\partial v}{\partial y} = \omega = 0, \qquad \text{at } y = \pm 1.$$

(Butler and Farrell, 1992) write v and ω in the form $v = v(y)\,e^{i(ax+bz)+\sigma t}$, $\omega = \omega(y)\,e^{i(ax+bz)+\sigma t}$, and this in (5.62) leads to the following system of ordinary differential equations for v and ω,

$$(D^2 - k^2)^2 v - iaReU(D^2 - k^2)v + iaReU''v = Re\sigma(D^2 - k^2)v,$$
$$(D^2 - k^2)\omega - iaReU\omega - ibReU'v = Re\sigma\omega, \tag{5.63}$$

where $k^2 = a^2 + b^2$. System (5.63) is to be solved on the interval $y \in (-1, 1)$ subject to the boundary conditions

$$v = Dv = 0, \quad \omega = 0, \qquad y = \pm 1. \tag{5.64}$$

We shall refer to (5.63), (5.64) as the Butler-Farrell eigenvalue problem. The eigenvalues are $\sigma^{(i)}$ with eigenfunctions $\{v^{(i)}, \omega^{(i)}\}$.

To understand the relevance of the Butler - Farrell eigenvalue problem we return to a discussion of stability of Couette and Poiseuille flows. For Couette flow the linearised theory of instability based on the Orr-Sommerfeld equation predicts the flow is always stable, i.e. $c_i < 0$ for all eigenvalues. For Poiseuille flow linear theory based on the Orr-Sommerfeld equation yields instability for $Re > 5772.22$, see e.g. (Orszag, 1971). The critical wavenumber for instability is $a_c = 1.02056$, (Orszag, 1971). When one bases a *nonlinear energy stability* theory on the kinetic energy then unconditional (i.e. for all initial data) nonlinear stability is found for Couette flow when $R_E < 20.7$, whereas nonlinear stability follows for Poiseuille flow when $R_E < 49.6$, cf. (Joseph, 1976b). Experimental work, on the other hand, has visualised instabilities for Reynolds numbers of the order of 1000. Thus, the linear instability theory is of little use in predicting accurately the onset of instability since the values of $Re = \infty$ and $Re = 5772.22$ are too large. Energy stability theory, on the other hand, is far too conservative in the stability boundary it yields. This is a case where the nonlinear energy method has not to date proved too useful.

To explain why instability in Couette or Poiseuille flow is seen in practice at much lower Reynolds numbers than those predicted by classical linear instability theory, investigations by (Mack, 1976), (Gustavsson, 1981; Gustavsson, 1986), (Gustavsson and Hultgren, 1980), and (Shanthini, 1989), studied the spectrum of the Orr-Sommerfeld operator to see if resonances between eigenvalues could be responsible. A resonance occurs where an eigenvalue is exactly repeated. In that case one of the eigenfunctions of the repeated eigenvalue contains a linear t growth term. (The situation is analogous to the well known case of a second order ordinary differential equation with constant coefficients. When there are repeated roots of the auxilliary equation one of the eigenfunctions grows linearly in t.) This could conceivably lead to strong transient algebraic growth of the perturbation at short times which is eventually damped out exponentially according to linear theory. While several resonances were found in numerical studies, no substantial growth would appear to have been predicted: the comments of (Butler and Farrell, 1992), p. 1645, on this matter are very pertinent. (A resonance in the numerical sense was interpreted as two eigenvalues being a pre-requested distance apart in the complex plane.) The idea that resonances could be responsible for transient solution growth means that it is not sufficient to investigate only the eigenvalue (and eigenfunction) of the Orr-Sommerfeld problem which has greatest imaginary part. This has lead to the development of numerical methods which can accurately yield *all* the eigenvalues and eigenfunctions of (5.57) and (5.58), or at least we find sufficient eigenvalues at the "top end" of the spectrum. By the "top end" of the spectrum we mean those eigenvalues which have largest imaginary parts. We typically calculate all eigenvalues for which $c_i > -1$.

In figures 5.1 and 5.2 below we show the top end of the spectrum for (5.57) and (5.58), in the case of Poiseuille flow with $Re = 5772.22$, $a = 1.02056$ (figure 5.1), and for Couette flow with $Re = 900$, $a = 1.2$ (figure 5.2). The even eigenfunctions satisfy $\phi(y) = \phi(-y)$ whereas the odd ones are such that $\phi(-y) = -\phi(y)$.

The eigenvalues in figure 5.1 extend downward with $c_r \to 2/3$ and form a countably infinite set. Those in figure 5.2 do likewise with $c_r = 0$. The eigenvalues in figures 5.1 and 5.2 are calculated using the technique advocated by (Dongarra et al., 1996), see chapter 9. It is important to note that, in both the Couette and Poiseuille flow eigenvalue problems, the eigenvalues and eigenfunctions near the branch of the "Y-shape" are difficult to calculate numerically. This is because the eigenfunctions of nearby eigenvalues are close to being linearly dependent. (Farrell, 1988b) has shown that for Poiseuille flow the skew-symmetric eigenfunctions corresponding to those eigenvalues nearest the branch of the "Y" contribute most to the

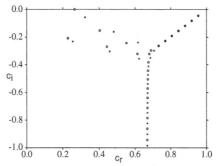

Figure 5.1. Poiseuille flow spectrum, $U = 1 - y^2, Re = 5772.22, a = 1.02056$. $\circ =$ even eigenmodes, $\times =$ odd modes

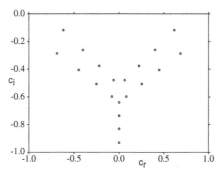

Figure 5.2. The leading eigenvalues in the spectrum for Couette flow, $U = y$, with $Re = 900, a = 1.2$

growth in time of the kinetic energy, and this is thus further evidence for needing an accurate eigenvalue solver yielding many eigenvalues.

The need to calculate many eigenvalues and eigenfunctions accurately is now very necessary, as the work of (Butler and Farrell, 1992), (Reddy and Henningson, 1993), (Schmid and Henningson, 1994), and (Hooper and Grimshaw, 1996) shows. We concentrate on a description of the work of (Butler and Farrell, 1992) who demonstrate that the kinetic energy can have very large growth in a relatively short time, even for flows which are well below the threshold for instability according to linear theory. The variational method employed by (Butler and Farrell, 1992) is based on earlier work of (Farrell, 1988b; Farrell, 1988a; Farrell, 1989) on growth in various two-dimensional geophysical fluid flows.

(Butler and Farrell, 1992) consider the kinetic energy of a perturbation to Couette or Poiseuille flow, i.e. they study the function

$$E(t) = \frac{1}{2V} \int_{-1}^{1} \int_{0}^{A} \int_{0}^{B} (u^2 + v^2 + w^2) \, dz \, dx \, dy, \qquad (5.65)$$

where A and B are the x and z wavelengths and V is the volume of the energy cell. Due to the fact that the system is linearised they can work with

v and ω as $v(x, y, z, t) = \hat{v}(y, t) e^{i(ax+bz)}$, $\omega(x, y, z, t) = \hat{\omega}(y, t) e^{i(ax+bz)}$, where $\hat{v}, \hat{\omega}$ can be complex, although only the real parts are used. (Butler and Farrell, 1992) show that the kinetic energy (5.65) may be written in terms of v and ω as

$$E(t) = \frac{1}{8} \int_{-1}^{1} \left[\hat{v}^* \hat{v} + \frac{1}{k^2} \left(\frac{\partial \hat{v}^*}{\partial y} \frac{\partial \hat{v}}{\partial y} + \hat{\omega}^* \hat{\omega} \right) \right] dy, \qquad (5.66)$$

where a $*$ denotes complex conjugate and $k^2 = a^2 + b^2$. To compute numerical calculations (Butler and Farrell, 1992) discretize v and ω by writing

$$v = \sum_{j=1}^{2N} \gamma_j \left[\tilde{v}_j \exp(\sigma_j t) \right] \exp[i(ax+bz)],$$

$$\omega = \sum_{j=1}^{2N} \gamma_j \left[\tilde{\omega}_j \exp(\sigma_j t) \right] \exp[i(ax+bz)], \qquad (5.67)$$

where γ_j is the spectral projection on the jth mode of the Butler - Farrell eigenvalue problem (5.63), (5.64). The technique of (Butler and Farrell, 1992) to find the eigenfunction and eigenvalue of the jth mode is to discretize using finite differences and then employ the QR algorithm on the generalised matrix eigenvalue problem. The techniques of (Reddy and Henningson, 1993) and (Hooper and Grimshaw, 1996) also find energy growth although the numerical method underpinning the work of (Reddy and Henningson, 1993) is a Chebyshev collocation one, whereas (Hooper and Grimshaw, 1996) employ a Chebyshev tau technique. Hence, (Butler and Farrell, 1992) adopt the notation $v \equiv V_{mj}\gamma_j e^{i(ax+bz)}$, $\omega \equiv \Omega_{mj}\gamma_j e^{i(ax+bz)}$, where $V_{mj} = \tilde{v}_{mj}e^{\sigma_j t}$, $\Omega_{mj} = \tilde{\omega}_{mj}e^{\sigma_j t}$, with m denoting a value between 1 and N and referring to the finite difference point $y_{m+1} = m\Delta y$ in the interval $y \in (-1, 1)$. This allows (Butler and Farrell, 1992) to approximate the energy (5.66) by a finite dimensional form

$$E(t) = \frac{\Delta y}{8} \left[\gamma_p^* V_{mp}^* V_{mj}\gamma_j + \frac{1}{k^2} \left\{ \gamma_p^* \frac{\partial V_{mp}^*}{\partial y} \frac{\partial V_{mj}}{\partial y} \gamma_j + \gamma_p^* \Omega_{mp}^* \Omega_{mj}\gamma_j \right\} \right], \quad (5.68)$$

where summation over the various subscripts is understood. The form of $E(t)$ is conveniently rewritten, (Butler and Farrell, 1992), as

$$E(t) = \gamma_j^* E_{ji}(t)\gamma_i. \qquad (5.69)$$

The form for the Hermitian matrix E_{ji} may be found from (5.68) and the time dependence of E_{ij} has been explicitly pointed out.

The idea of (Butler and Farrell, 1992) is to fix the wavenumbers a and b, fix the Reynolds number Re, and then find the linear perturbation which maximises $E(t)$ at time t subject to the constraint that the initial energy has the numerical value of 1. This yields a maximisation problem for the function F given by $F = \gamma_j^* E_{ji}(t)\gamma_i + \lambda(\gamma_j^* E_{ji}(0)\gamma_i - 1)$, for some Lagrange

multiplier λ. The solution to this maximisation problem is found from the
Euler - Lagrange equations

$$E_{ij}(t)\gamma_j + \lambda E_{ij}(0)\gamma_j = 0. \tag{5.70}$$

The eigenvalues λ represent the ratio $E(t)/E(0)$ for an eigenvector γ_i. (The
technical details of a practical way to do this calculation using the Chebyshev tau method are given in the very readable account of (Hooper and
Grimshaw, 1996).) The calculation of λ involves finding $E_{ij}(t)$ and this
in turn involves calculation of the eigenvalues and eigenfunctions of the
Butler - Farrell eigenvalue problem. (Butler and Farrell, 1992) have completed extensive calculations of (5.70) and have computed the growth rate
for many situations in Couette flow, in Poiseuille flow, and even in Blasius
flow.

For Couette flow, (Butler and Farrell, 1992) find with $Re = 1000$ that
the global optimal for time $\tau = 117$ units is achieved with $a = 0.035$ and
$b = 1.60$. This yields an energy ratio of $E(\tau)/E(0) = 1185$. This is certainly an impressive growth of the kinetic energy. Since only linear theory
is considered, the linear energy eventually decays, as shown schematically
in figure 5.3.

We stress that (Butler and Farrell, 1992) show that a two-dimensional
perturbation yields an energy growth of $O(13)$ and thus the perturbation causing largest energy disturbance is truly three-dimensional. Similar
results pertaining to energy growth are found by (Reddy and Henningson, 1993), (Hooper and Grimshaw, 1996), for Couette and Poiseuille
flow, and by (Schmid and Henningson, 1994) for the problem of Poiseuille
flow in a circular pipe. Other related recent interesting results for shear
flows and channel flows are discussed in (Blyth et al., 2006), (Blyth and

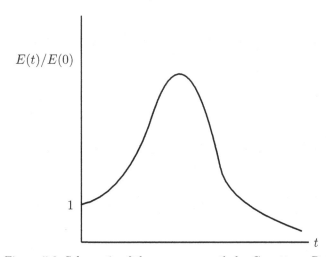

Figure 5.3. Schematic of the energy growth for Couette or Poiseuille flow

Pozrikidis, 2004a; Blyth and Pozrikidis, 2004b), (Crisciani, 2004), (Hifdi et al., 2004a; Hifdi et al., 2004b), (Yecko, 2004), (Luo and Pozrikidis, 2006), and in (Malik and Hooper, 2007), while (Kim and Choi, 2006) determine a critical time at which a fastest growing instability occurs for a spin down to rest hydrodynamic stability problem. (Potherat, 2007) and (Li et al., 2007) are interesting recent studies of shear flows involving a magnetic field and an electric field, respectively.

For the problem of Poiseuille flow, (Butler and Farrell, 1992) obtain similar behaviour to that of figure 5.3 for growth of the kinetic energy. For $Re = 5000$, a value which is stable according to classical linear instability theory, they determine optimal energy growth at time $\tau = 379$ units. The respective wavenumbers are $a = 0$ and $b = 2.044$. The energy ratio is $E(\tau)/E(0) = 4897$.

The work of (Butler and Farrell, 1992), (Reddy and Henningson, 1993) and (Schmid and Henningson, 1994) shows that when the effects of non-linearity are fully understood and added to this it could well describe quantitatively the formation of patches of turbulent fluid flow. In the context of fully nonlinear flow, (Butler and Farrell, 1994) have analysed the nonlinear development of two-dimensional perturbations which employ the optimal configuration of the linear problem, namely that which gains most energy. To do this they employ a vorticity - streamfunction numerical simulation using a Fourier spectral discretization in x with finite differences in y. These results are very interesting and display finite amplitude solutions which persist for a long time.

5.4.4   Stability analysis for salinization

(van Duijn et al., 2002) is an important paper in nonlinear energy stability theory. These writers find linear instability bounds for the linearised system arising from (5.50), (5.51) using the asymptotic base solution $S_0(z)$ given by (5.49). However, they also derive nonlinear energy stability thresholds for the full system (5.50), (5.51) when S_0 is given by (5.48), treating t as a parameter, and they compare the boundaries so found to those of the linear results. While these results are very revealing in themselves, they also treat $(5.50)_2$ as a constraint in the energy theory. This is a beautiful idea and leads to much improved nonlinear energy stability thresholds as compared with the traditional approach where the integrated form of $(5.50)_2$ is employed, cf. section 4.3.4.

The analysis of (van Duijn et al., 2002) is based on an L^2 norm of s. Decay of $\|s\|$ also leads to decay of $\|\mathbf{u}\|$ through $(5.50)_2$. Thus, (van Duijn et al., 2002) work with the equation, arising from (5.52)

$$\frac{d}{dt}\frac{1}{2}\|s\|^2 = -\|\nabla s\|^2 - Ra(S_{,z}^0 s, w), \qquad (5.71)$$

where $(\cdot, \cdot)$ is the inner product on $L^2(V)$, V being the period cell for (u_i, s, π), and $\| \cdot \|$ is the associated norm.

If one defines R_E by

$$R_E^{-1} = \sup_H \frac{-(S^0_{,z}s, w)}{\|\nabla s\|^2}, \tag{5.72}$$

then, since S^0 is a decaying exponential, R_E exists and one sees from (5.71) that

$$\frac{d}{dt}\frac{1}{2}\|s\|^2 \leq -\|\nabla s\|^2\left(1 - \frac{Ra}{R_E}\right). \tag{5.73}$$

Thus, if $Ra \leq R_E$ then $d\|s\|^2/dt < 0$ and there is stability (in a sense). A key finding of (van Duijn et al., 2002) is that the space H one selects is vital to the analysis. If one simply multiplies $(5.50)_2$ by u_i and integrates over V then one finds

$$\|\mathbf{u}\|^2 - (s, w) = 0. \tag{5.74}$$

However, if one eliminates π from (5.50), then one finds

$$\Delta w - \Delta^* s = 0, \tag{5.75}$$

as a pointwise constraint. (van Duijn et al., 2002) analyse in depth the maximum problem (5.72) with H being in turn the spaces

$$H_1 = \{(s, \mathbf{u}) | x, y \text{ periodic with respect to } V,$$
$$s = \mathbf{u} = 0 \text{ at } z = 0, \infty, \quad u_{i,i} = 0,$$
$$\text{and} \quad \|\mathbf{u}\|^2 - (s, w) = 0\}$$

and

$$H_2 = \{(s, w) | x, y \text{ periodic with respect to } V,$$
$$s = w = 0 \text{ at } z = 0, \infty,$$
$$\text{and} \quad \Delta w - \Delta^* s = 0 \text{ in } V\}.$$

The Euler-Lagrange equations for the maximum problem (5.72) using both H_1 and H_2 are solved numerically by (van Duijn et al., 2002). With $S_0 = e^{-z}$ they find the space H_2 leads to much improved stability thresholds. Actual values may be found in (van Duijn et al., 2002) but a schematic of the situation is contained in figure 5.4.

Employing t as a parameter, (van Duijn et al., 2002) also solve the maximum problem (5.72) numerically with $S_0(z, t)$ given by (5.48). This too is very revealing. They find the critical Rayleigh number threshold decreases as t increases in a manner indicated in figure 5.5 . Actual numerical values may be found in (van Duijn et al., 2002).

The paper of (van Duijn et al., 2002) also presents computations of a two-dimensional solution when $Ra > R_L$, the linear instability threshold, and they find growing salt finger-like shapes. They also compute a kinetic

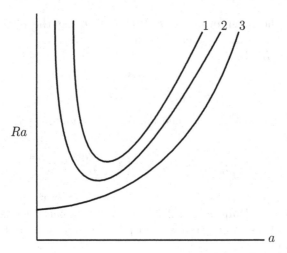

Figure 5.4. Schematic of Rayleigh number against wavenumber. Curve 1 is linear result, curve 2 is energy result with space H_2, curve 3 with H_1, $S_0 = e^{-z}$

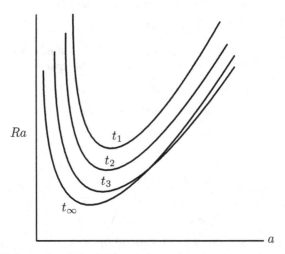

Figure 5.5. Schematic of Rayleigh number against wavenumber. Energy curves using (5.48), $t_1 < t_2 < t_3 < \ldots < t_\infty$

energy, $E = \|\mathbf{u}\|^2/2$, numerically and investigate how $E(t)$ behaves in time for Ra fixed.

5.4.5   Transient growth in salinization

(Pieters and van Duijn, 2006) represents and extension of the work of (van Duijn et al., 2002). However, it is an important one in that, to the best of my knowledge, it is the first study of the transient growth of solutions in porous media even though the Rayleigh number is such that there is *linear* stability. Thus, this is in a sense, the first study in a porous medium of time growing solutions in some sense akin to those in parallel flows found by (Butler and Farrell, 1992), cf. section 5.4.3. In addition, (Pieters and van Duijn, 2006) also present a neat transformation to symmetrize the system and obtain an optimal nonlinear stability result. This amounts to working with a weighted energy, and is of much interest in its own right.

(Pieters and van Duijn, 2006) concentrates on the base function $S_0 = e^{-z}$ and studies in detail the linearised problem one derives from (5.50), (5.51), i.e. (5.52). They introduce normal modes in (5.52) and thus study the spectrum of the linear operator associated to the system

$$
\begin{aligned}
s_{,t} &= D^2 s + Ds - a^2 s + Ra\, e^{-z} B(s), \quad z > 0, t > 0, \\
B^{-1}(w) &\equiv -a^{-2} D^2 w + w = s, \quad z > 0, t > 0, \\
w(0,t) &= s(0,t) = 0, \quad t > 0, \\
s(z,0) &= f(z), \quad z > 0.
\end{aligned}
\tag{5.76}
$$

They introduce the variable $u(z,t) = e^{z/2} s(z,t)$ and show that the linear operator associated to the right hand side of (5.76) becomes a symmetric operator on $L^2(\mathbb{R}^+)$ when (u, w) are employed instead of (s, w). In this manner, they are able to study the spectrum in detail and derive its properties in terms of the wavenumber a.

A very interesting part of the analysis of (Pieters and van Duijn, 2006) is where they study the behaviour of the energy $E(t) = \|s(t)\|^2$ when the Rayleigh number is above the energy threshold R_E but below the linear threshold R_L, i.e. $R_E < Ra < R_L$. In this case there is linear stability, but (Pieters and van Duijn, 2006) are able to demonstrate that $E(t)$ may initially grow, and they obtain bounds for this growth. They show that transient growth may be eliminated by working with the weighted energy

$$
E_w = \int_{\mathbb{R}^+} e^{\alpha z} s^2(z,t) dz
$$

for f in a suitable weighted L^2 space.

Much of the material of section 5.4 is lucidly covered in detail in chapter 3 of (Pieters, 2004). Another interesting geophysical problem which could have important consequences for CO_2 absorption is solute transport in a peat layer. A non-isothermal Darcy porous medium model for convection as a transport mechanism in peat moss layers is developed in detail in chapter 5 of (Pieters, 2004), see also (Rappoldt et al., 2003). This is another problem where the basic "ground state" solution involves a temperature field which depends on both the vertical spatial coordinate, z, and the time, t, and again the analysis is very interesting and carefully carried out. When vibrations are involved there is clearly time dependence to be included and the article by (Govender, 2006b) addresses such a problem for a porous layer in a cylinder.

5.5   Turbulent convection

5.5.1   Turbulence in porous media

Turbulent motion of a fluid in a porous medium may well become a very important topic, especially due to use of heat pipes to transfer heat away from very hot components in e.g. computers. (A silicon chip does not work above a certain temperature and a well used computer needs heat rapidly transferred from vital components.) There are many types of heat pipe, some using a copper tube with a copper wick (porous medium) inside. Also, high porosity metal foams, cf. (Zhao et al., 2004) may be potentially used in heat transfer devices. When heat is transferred the resulting motion may well be turbulent and with high temperature gradients is, in fact, turbulent convection in a porous medium. Thus, in this section we describe very interesting work of (Doering and Constantin, 1998) which derives rigorous bounds for the Nusselt number in turbulent porous convection. (Doering and Constantin, 1998) write that they use Darcy's law and are not actually considering a fully turbulent regime which might need some kind of Forchheimer equation. They write that what they mean by turbulent convection is the absence of spatial and temporal coherence in the solutions. However, the mathematics of (Doering and Constantin, 1998) is quite beautiful and may well be utilized in other more exotic theories than that of Darcy.

While (Doering and Constantin, 1998) also consider inertia in Darcy's law, we restrict attention to the case of zero inertia. They consider a porous medium occupying the infinite layer in the non-dimensional region $\{(x, y) \in \mathbb{R}^2\} \times \{z \in (0, 1)\}$. They restrict attention to periodic solutions with non-dimensional x and y periods, Λ_x, Λ_y, respectively, although they do mention that other horizontal boundary conditions may be employed. Thus, their period cell is the box $V = (0, \Lambda_x) \times (0, \Lambda_y) \times (0, 1)$. We shall adopt their notation and when we write $\int F \, dx \, dy \, dz$ it is understood that this means integration over V.

The non-dimensional equations considered by (Doering and Constantin, 1998) are

$$
\begin{aligned}
T_{,t} + u_i T_{,i} &= \Delta T, \\
u_i + p_{,i} &= Ra \, T k_i, \qquad u_{i,i} = 0,
\end{aligned}
\tag{5.77}
$$

where T, u_i, p are temperature, velocity and pressure, $\mathbf{k} = (0, 0, 1)$ and Ra is the Rayleigh number $Ra = g\alpha(T_L - T_U)Kd/\nu\kappa$. Here d is the dimensional depth of the layer, $g, \alpha, \nu, K, \kappa$ have their usual meaning and T_L and T_U are the temperatures on the lower and upper planes before non-dimensionalization, with $T_L > T_U$. The non-dimensional boundary conditions may be written

$$
T = 1, w = 0, z = 0; \qquad T = 0, w = 0, z = 1;
$$

where $w = u_3$. To complete the boundary-initial value problem one needs the initial condition

$$T(\mathbf{x}, 0) = T_0(\mathbf{x}),$$

where T_0 is a prescribed function. Since we know T_0 then p_0 may be found from $\Delta p = Ra\, T_{,z}$ and then u_i^0 follows from (5.77).

(Doering and Constantin, 1998) observe that the heat flux $\mathbf{J}$ is given by

$$J_i = u_i T - T_{,i}.$$

In turbulent convection a particular average of this is the Nusselt number and this in some sense characterises the problem. (Doering and Constantin, 1998) define their Nusselt number, Nu, by an average of J_3 so that

$$Nu = \sup_{T_0} \limsup_{t \to \infty} \frac{1}{t} \int_0^t ds \, \frac{1}{\Lambda_x \Lambda_y} \int dx\, dy\, dz \left(wT - \frac{\partial T}{\partial z} \right), \qquad (5.78)$$

where T_0 denotes the initial value for $T(\mathbf{x}, 0)$. It is shown in (Doering and Constantin, 1998) that Nu may be expressed in terms of T as

$$Nu = \sup_{T_0} \limsup_{t \to \infty} \frac{1}{t} \int_0^t ds \, \frac{1}{\Lambda_x \Lambda_y} \int dx\, dy\, dz \, |\nabla T|^2. \qquad (5.79)$$

In section 4.2.1 we have seen that for $Ra \leq R_c^2 = 4\pi^2$ the conduction solution prevails and then $Nu = 1$. However, for $Ra > R_c^2$ thermal convection will commence. Experimentally it is found that this occurs first as stationary convection rolls and as Ra is increased these rolls are replaced by oscillatory solutions and eventually (possibly) by chaotically turbulent states. It is to the latter that the analysis of (Doering and Constantin, 1998) is primarily addressed.

5.5.2 The background method

(Doering and Constantin, 1998) base their analysis on a "background" method. The idea is to introduce a field $\tau(\mathbf{x})$ which satisfies the boundary conditions for T, i.e. $\tau = 0$ when $z = 1$, $\tau = 1$ when $z = 0$, and τ satisfies the equations

$$U_i \tau_{,i} = 0,$$
$$U_i + P_{,i} = Ra\, \tau k_i, \qquad U_{i,i} = 0. \qquad (5.80)$$

Here U_i and P are suitable "velocity and pressure" fields with $U_3 = 0$ at $z = 0, 1$. The key is to pick U_i, τ in a clever way. Next, (Doering and Constantin, 1998) decompose the actual temperature, velocity and pressure fields T, u_i and p into the background fields plus a fluctuation part. They set

$$T = \tau + \theta, \quad u_i = U_i + v_i, \quad p = P + q, \qquad (5.81)$$

where θ, v_i and q are the fluctuations. In addition to periodicity the fluctuations satisfy the boundary conditions

$$\theta = 0, \quad v_3 = 0 \text{ at } z = 0, 1. \tag{5.82}$$

Upon using the differential equations (5.77) one finds the equations satisfied by the fluctuations,

$$\theta_{,t} + v_i\theta_{,i} + U_i\theta_{,i} + v_i T_{,i} = \Delta\theta + \Delta\tau,$$
$$v_i + q_{,i} = Ra\,\theta k_i, \qquad v_{i,i} = 0. \tag{5.83}$$

The analysis of (Doering and Constantin, 1998) begins with $(5.83)_1$ to derive

$$\frac{d}{dt}\frac{1}{2}\int \theta^2 dx\,dy\,dz = -\int (\tau_{,i}\theta_{,i} + |\nabla\theta|^2 + \theta v_i\tau_{,i}) dx\,dy\,dz. \tag{5.84}$$

Also, they note that since $T = \tau + \theta$,

$$-\int \tau_{,i}\theta_{,i}\,dx\,dy\,dz = \frac{1}{2}\int (|\nabla\theta|^2 + |\nabla\tau|^2 - |\nabla T|^2) dx\,dy\,dz. \tag{5.85}$$

Upon adding these equations one may derive

$$\frac{d}{dt}\frac{1}{2}\int \theta^2 dx\,dy\,dz + \int |\nabla T|^2\,dx\,dy\,dz =$$
$$\int |\nabla\tau|^2 dx\,dy\,dz - \int (|\nabla\theta|^2 + 2\theta v_i\tau_{,i}) dx\,dy\,dz. \tag{5.86}$$

Their next step is to define the quadratic function

$$H_\tau^{(\lambda)}\{\theta\} \equiv \frac{1}{\Lambda_x\Lambda_y}\int (\lambda|\nabla\theta|^2 + \theta v_i\tau_{,i}) dx\,dy\,dz \tag{5.87}$$

for $\lambda > 0$. They then divide (5.86) by $\Lambda_x\Lambda_y$, average over time and take the supremum over T_0 to see, using the definition for Nu, (5.79), that

$$Nu = \frac{1}{\Lambda_x\Lambda_y}\int |\nabla\tau|^2 dx\,dy\,dz$$
$$+ \sup_{T_0}\limsup_{t\to\infty}\frac{1}{t}\int_0^t \left[-2H_\tau^{(1/2)}\{\theta(\cdot, s)\}\right] ds. \tag{5.88}$$

Now they argue that one selects the function τ such that $H_\tau^{(1/2)} \geq 0$ and hence one can bound Nu by

$$Nu \leq \frac{1}{\Lambda_x\Lambda_y}\int |\nabla\tau|^2 dx\,dy\,dz = 1 + \frac{1}{\Lambda_x\Lambda_y}\int |\nabla\tau + \mathbf{k}|^2 dx\,dy\,dz.$$

The idea is to minimize over the τ functions to obtain

$$Nu \leq \inf_\tau\left\{\frac{1}{\Lambda_x\Lambda_y}\int |\nabla\tau|^2 dx\,dy\,dz\right|$$
$$H_\tau^{(1/2)} \geq 0, \tau(z = 0) = 1, \tau(z = 1) = 0\right\}. \tag{5.89}$$

They then note that this calculation is made more tractable by choosing $\tau = \tau(z), U_i = 0$. The bound in inequality (5.89) may then be replaced by

$$Nu - 1 \leq \inf_{\tau} \left\{ \int_0^1 |\tau'(z) + 1|^2 dz \,\Big|\, H_\tau^{(1/2)} \geq 0, \ \tau(0) = 1, \ \tau(1) = 0 \right\}.$$

The real goal of (Doering and Constantin, 1998) is to improve on this bound and in this regard they refer to work of (Nicodemus et al., 1997a; Nicodemus et al., 1997b) (see also (Nicodemus et al., 1998; Nicodemus et al., 1999)) and so introduce a parameter $c > 1$ and generalize the argument leading to (5.86). They form $(5.84) \times c + 2 \times (5.85)$ to see that

$$\frac{d}{dt} \frac{c}{2} \int \theta^2 dx \, dy \, dz + \int |\nabla T|^2 \, dx \, dy \, dz = \int |\nabla \tau|^2 dx \, dy \, dz \tag{5.90}$$
$$- \int \left[(c-2)\tau_{,i}\theta_{,i} + (c-1)|\nabla\theta|^2 + c\theta v_i \tau_{,i} \right] dx \, dy \, dz.$$

This expression is averaged and one shows

$$Nu = \frac{1}{\Lambda_x \Lambda_y} \int |\nabla \tau|^2 dx \, dy \, dz$$
$$+ \sup_{T_0} \limsup_{t \to \infty} \frac{1}{t} \int_0^t \left(-\frac{1}{\Lambda_x \Lambda_y} \right. \tag{5.91}$$
$$\left. \int \left[(c-1)|\nabla\theta|^2 + c\theta v_i \tau_{,i} - (c-2)\theta \, \Delta\tau \right] dx \, dy \, dz \right) ds.$$

The minimum is taken over θ and v_i of the second term on the right in (5.91) and after some detailed calculations involving the Euler-Lagrange equations for this term, (Doering and Constantin, 1998) show that this procedure leads to the bound

$$Nu - 1 \leq \inf_{a \in (0,1)} \inf_{\tau} \left\{ \frac{1}{4a(1-a)} \int_0^1 \left[\tau'(z) + 1\right]^2 dz \,\Big|\, \right.$$
$$\left. H_\tau^{(a)} \geq 0, \ \tau(0) = 1, \ \tau(1) = 0 \right\} \tag{5.92}$$

where a, $0 < a < 1$, is related to c by $a = (c-1)/c$.

The next stage of (Doering and Constantin, 1998) is to actually select functions $\tau(z)$ to turn (5.92) into a useful bound on the Nusselt number.

5.5.3 Selecting τ.

(Doering and Constantin, 1998) consider the class of functions

$$\tau_\delta(z) = \begin{cases} 1 - z/2\delta, & 0 \leq z \leq \delta, \\ 1/2, & \delta \leq z \leq 1 - \delta, \\ (1-z)/2\delta, & 1 - \delta \leq z \leq 1, \end{cases}$$

arguing that the gradients will be strongest in the boundary layers. They then perform a bounding analysis and Ra is restricted by their choice of τ since they need $H_\tau^{(a)} \geq 0$. After some calculations they show that

$$
\begin{aligned}
Nu &\leq 1 + \frac{1}{4a(1-a)} \int_0^1 \left[\tau_\delta'(z) + 1 \right]^2 dz \\
&= \frac{Ra}{128a^2(1-a)} - \frac{(1-2a)^2}{4a(1-a)},
\end{aligned}
$$

where $Ra \geq 32a$. The Ra term is minimized and this leads to the asymptotic estimate

$$
Nu \leq \frac{27}{512} Ra \left(1 + O\left(\frac{1}{Ra}\right) \right), \quad Ra \to \infty.
$$

In section 7 of (Doering and Constantin, 1998) they modify the variational problem and find an alternative way to bound Nu. This analysis is involved but particularly appealing, involving a reduced class of constraints. We do not go into details, but in terms of the elliptic integrals

$$
K(m) = \int_0^1 \frac{dt}{[(1-t^2)(1-mt^2)]^{1/2}}, \qquad E(m) = \int_0^1 \left(\frac{1-mt^2}{1-t^2}\right)^{1/2} dt
$$

they define the variables $\eta(m) = 8K(m)[K(m) - E(m)]$, $\sigma(m) = 2mK(m)^3/[K(m) - E(m)]$, and then obtain the following rigorous bound on the Nusselt number,

$$
Nu \leq 1 + \frac{4\eta(m)[a\eta(m) + Ra - 4a\sigma(m)]}{3Ra^2(1-a)}. \tag{5.93}
$$

By optimizing over a they produce the asymptotic bound

$$
Nu \leq \frac{9}{256} Ra \left(1 + O\left(\frac{1}{Ra}\right) \right), \quad Ra \to \infty. \tag{5.94}
$$

The paper of (Doering and Constantin, 1998) evaluates (5.93) numerically and graphs of several bounding curves are presented there. In particular, they relate their bounds to available experimental results and the agreement is very good indeed.

The paper of (Otero et al., 2004) contains extensive computations for the turbulence problem of (Doering and Constantin, 1998). In addition, they extend the variational upper bound on the heat transport. The (Nu, Ra) bound of (Doering and Constantin, 1998) is better in the range Ra less than approximately 2000, whereas the bound of (Otero et al., 2004) is an improvement on the bound of (Doering and Constantin, 1998) for Ra above this value.

The effects of rotation are incorporated into an analysis which has some resemblance to that of (Doering and Constantin, 1998) by (Wei, 2004).

This writer derives the bound

$$Nu \leq \frac{Ra}{16\sqrt{Ta+1}},$$

where Ta is a Taylor number (measuring the rotation of the layer).

An interesting article which deals with the way thermal convection switches from an ordered state to one which may be turbulent, and involves non - Darcy effects is that of (Vadasz et al., 2005b). These writers study the bifurcation problem and obtain very interesting results.

Another related analysis of bounding heat transport may be found in the work of (Doering et al., 2006b). These writers tackle the problem of convection when the Prandtl number is infinite. This is a very interesting paper which employs a logarithmic background temperature profile in the analysis. In fact, the τ function has linear in z dependence in the boundary layers close to 0 and 1, but behaves like $\log[z/(1-z)]$ in the remainder of the layer. Bounds on the energy dissipation are derived by (Doering and Foias, 2002).

5.6 Multiphase flow

5.6.1 Water-steam motion

In this section we consider the motion of a liquid water - steam mixture in a porous medium. Such a problem has many applications to e.g. heat pipes, and in geothermal systems, cf. (Pestov, 1998), (Amili and Yortsos, 2004) and the references therein. In fact, (Pestov, 1998) starts by considering geysers in California, Italy, and New Zealand, and explains that really such underground situations should be modelled by a two layer system where liquid or steam may be above or below a porous layer containing both phases coexisting. We revisit the two layer problem specifically in a heat pipe context in section 6.6 when the model of (Amili and Yortsos, 2004) is reviewed.

In fact, (Pestov, 1998) studies a model for water/steam motion in porous media and shows that a two layer situation may develop naturally. Her model is very interesting mathematically because it reduces to a parabolic equation for the pressure disturbance and a forced hyperbolic equation for the relative permeability disturbance.

The equations governing coexistent vapour and liquid in a porous medium are complicated. (Amili and Yortsos, 2004) write the mass and energy balances, together with Darcy law momentum balances as

$$\frac{\partial}{\partial t}(\phi\rho_v s_v + \phi\rho_\ell s_\ell) + (v_i^v \rho^v + v_i^\ell \rho^\ell)_{,i} = 0,$$

$$\frac{\partial}{\partial t}\left\{\phi(\rho_v h_v s_v + \rho_\ell h_\ell s_\ell) + (1-\phi)\rho_r h_r\right\} + (\rho_\ell v_i^\ell h_\ell + \rho_v v_i^v h^v)_{,i} = k_e \Delta T,$$

$$v_i^v = -\frac{kk_{rv}}{\mu_v}(p_{,i} - \rho_v g k_i),$$

$$v_i^\ell = -\frac{kk_{r\ell}}{\mu_\ell}(p_{,i} - \rho_\ell g k_i),$$

where v and ℓ (sub or superscript) denote vapour and liquid phases, ϕ is porosity, v_i velocity, p pressure, g gravity, $\mathbf{k} = (0,0,1)$, ρ density, k permeability, μ viscosity, k_r are the relative permeabilities, and s are the saturations (i.e. the volume fractions of the pore volume occupied by a particular phase). (Pestov, 1998) assumes capillary pressure is unimportant, conduction is negligible, vapour enthalpy and latent heat variations may be neglected and the liquid phase density is constant. She presents non-dimensional forms of the continuity and energy equations as (she also gives another form of the energy equation)

$$\frac{\partial}{\partial t}(\tilde{m}\varphi s_v + s_\ell) + (J_i^v + J_i^\ell)_{,i} = 0,$$

$$\frac{\sigma\tilde{\delta}\epsilon}{\phi}\varphi^{\epsilon-1}\bar{p}\frac{\partial\hat{p}}{\partial t} - \hat{\ell}\left(\frac{\partial s_\ell}{\partial t} + J_{i,i}^\ell\right) = 0 \tag{5.95}$$

where s are the saturations, $\hat{p}$ is the pressure, $\varphi = \bar{p}\hat{p}+1$, $\bar{p}$ being a pressure jump, J_i are fluxes and the other terms are constants defined in (Pestov, 1998). The fluxes are defined via Darcy's law as

$$J_i^v = -\frac{k_{rv}\varphi}{\mu\tilde{Q}}\left(\gamma\hat{p}_{,i} - \tilde{m}\varphi k_i\right), \qquad J_i^\ell = -\frac{k_{r\ell}}{\tilde{m}\tilde{Q}}\left(\gamma\hat{p}_{,i} - k_i\right), \tag{5.96}$$

where $k_{rv}, k_{r\ell}$ are relative permeabilities. Her equations hold in the non-dimensional layer $z \in (0,1)$, $(x,y) \in \mathbb{R}^2$, with $t > 0$. Boundary conditions are assumed at the top and bottom of the layer. (Pestov, 1998) writes $k_{r\ell} = \psi(s_\ell)$, $k_{rv} = 1-\psi(s_\ell)$ for a function ψ and then reduces her equations to a system in two unknowns, the saturation pressure p and relative permeability $k_{r\ell}$. She writes these as a steady solution, $p^0, k_{r\ell}^0$, and a perturbation of form

$$p = p^0(z) + p'(\mathbf{x},t), \qquad k_{r\ell} = k_{r\ell}^0(z) + k'(\mathbf{x},t) = \psi(s_\ell^0 + s').$$

The perturbations p', k', s' are assumed small. Deriving equations for p', k' she finds

$$(1 - \tilde{m}\varphi^0)\frac{\partial s'}{\partial t} + c_\ell\left(1 + \frac{c_v}{c_\ell}\varphi^0\right)\frac{\partial k'}{\partial z} = \frac{\bar{p}}{\bar{p}_{\ell s}\mu\tilde{Q}}\Phi_v + \frac{\bar{p}}{\bar{p}_{\ell s}\tilde{m}\tilde{Q}}\Phi_\ell, \tag{5.97}$$

$$\frac{\partial s'}{\partial t} + c_\ell\frac{\partial k'}{\partial z} = \frac{\bar{p}}{\bar{p}_{\ell s}\tilde{m}\tilde{Q}}\Phi_\ell, \tag{5.98}$$

where

$$
\Phi_v = \frac{\partial}{\partial z}\left[k_{rv}^0 \varphi^0 \left(\frac{\partial p'}{\partial z} + \frac{1}{\varphi^0}\frac{d\varphi^0}{dz}p' - 2\tilde{m}\bar{p}_{\ell s}p' \right) \right]
$$
$$
+ k_{rv}^0 \varphi^0 \Delta p' - \bar{p}_{\ell s}\mu\tilde{Q}\left[\tilde{m}s_v^0 + \nu(\varphi^0)^{\epsilon-1}\right]\frac{\partial p'}{\partial t},
$$
$$
\Phi_\ell = \frac{\partial}{\partial z}\left(k_{r\ell}^0 \frac{\partial p'}{\partial z} \right) + k_{r\ell}^0 \Delta p' + \bar{p}_{\ell s}\tilde{m}\tilde{Q}\nu(\varphi^0)^{\epsilon-1}\frac{\partial p'}{\partial t}.
$$

She argues that $\tilde{m}$ and c_v/c_ℓ are much smaller than 1 and (5.97) may, therefore, by replaced by

$$
\frac{\partial s'}{\partial t} + c_\ell \frac{\partial k'}{\partial z} = \frac{\bar{p}}{\bar{p}_{\ell s}\mu\tilde{Q}}\Phi_v + \frac{\bar{p}}{\bar{p}_{\ell s}\tilde{m}\tilde{Q}}\Phi_\ell. \tag{5.99}
$$

Hence, in this linear approximation, the system (5.98), (5.99) decouples into the equation $\Phi_v = 0$, which is a parabolic equation for p, and equation (5.98) which becomes a forced hyperbolic equation for k'. The equation $\Phi_v = 0$ is

$$
k_1 \zeta^\epsilon \frac{\partial p'}{\partial t} - \zeta^2 \Delta p' - \frac{\partial}{\partial \zeta}\left(\zeta^2 \frac{\partial p'}{\partial \zeta} \right) = 0 \tag{5.100}
$$

for a suitable constant k_1 and a rescaled z−variable, ζ. (Pestov, 1998) presents boundary conditions for (5.100). She solves this equation by writing $p' = e^{-\lambda t}f(x, y)P(\zeta)$, for $f_{xx} + f_{yy} = -a^2 f$, and then P satisfies a Sturm-Liouville problem. (Pestov, 1998) solves this Sturm-Liouville problem in two cases. Firstly for $a = 0$, infinite horizontal wavelength, whence she can write the solution in terms of Bessel functions. Secondly for $a \neq 0$. This requires a detailed asymptotic analysis and is presented in (Pestov, 1998). Once the pressure perturbation is known she returns to equation (5.98) and writes it in the form

$$
\frac{\partial k'}{\partial t} + C\frac{\partial k'}{\partial \zeta} = k_2 \frac{\partial p'}{\partial t} + k_3 \frac{\partial p'}{\partial \zeta}, \tag{5.101}
$$

where C, k_2, k_3 are known. She solves equation (5.101) analytically (in terms of the pressure function) by integrating along the characteristics $\zeta = Ct + r$. Numerical results are presented in some detail and the saturation distance is calculated. The work of (Pestov, 1998) is a beautiful analysis which shows how a two layer situation may evolve from a vapour-liquid coexistent layer in a porous medium. Mathematically, the analysis is really interesting. However, she also relates this carefully to multi-layer structures observed in geothermal reservoirs.

The influence of the choice of the Reynolds tensor on the derivation of multiphasic incompressible fluid models is analysed carefully by (Bresch et al., 2007).

5.6.2 Foodstuffs, emulsions

There are many applications of multiphase flows in porous media. We briefly mention some others in addition to those of section 5.6.1. (Zorrilla and Rubiolo, 2005a) develop a model for immersion chilling and freezing of foods. This model involves deriving global equations for continuity, linear momentum, angular momentum, and energy, and continuity and energy equations for each of the components involved in the freezing process of the food product. These involve the solid food component, water and solute (both in liquid phase), and ice. This is a complicated model and numerical methods and solutions are reported. Further numerical solutions are given by (Zorrilla and Rubiolo, 2005b).

(Dincov et al., 2004) considers a mathematical model for microwave heating of food. They present the relevant form of Maxwell's equations which govern the microwaves in the heating process. Relevant energy equations are presented for a phase consisting of a solid and a liquid and for a gas phase. They interestingly incorporate momentum transfer in the food by employing a form of Darcy's law for the liquid and one for the gas. The model is again non-trivial and numerical solutions are presented from use of a finite volume method.

(Nakanishi and Tanaka, 2007) describe some very interesting methods for producing oxides. These are so called sol-gel processes. The changing of the porous structure due to phase changes is clearly important in the process. I am unaware of any mathematical model for this process, but it is an area where such mathematical analysis may well help.

Transport of micro-emulsions in porous media is important in many mundane areas. (Cortis and Ghezzehei, 2007) describe a model for emulsion transport in a porous medium. Their model consists of an equation for the colloid concentration in the bulk, $c(x,t)$, and a coupled equation for the colloid concentration, s, which is adsorbed or given up at a solid surface. The equations of (Cortis and Ghezzehei, 2007) are

$$\frac{\partial c}{\partial t} + \frac{\rho_b}{\theta}\frac{\partial s}{\partial t} + v\left[\lambda c + \frac{\partial c}{\partial x} - \alpha\frac{\partial^2 c}{\partial x^2}\right] = 0,$$

$$\frac{\partial s}{\partial t} = \frac{\theta}{\rho_b}\,k_f\,c - k_r s.$$

Here θ is the porosity, v is the pore velocity, and the coefficients $\lambda, \alpha, \rho_b, k_f$ and k_r are given by (Cortis and Ghezzehei, 2007). A solution method is considered by (Cortis and Ghezzehei, 2007) and an example involving a sand and a man-made emulsion is presented.

5.7   Unsaturated porous medium

5.7.1   Model equations

Flows in unsaturated porous materials are harder to study than those in saturated materials and consequently the literature is less. However, there are some models for flow in unsaturated porous media, and due to many applications this will undoubtedly be an area of much future research.

(Kapoor, 1996) used a conservation law model and developed a linearized instability analysis for flow in an unsaturated porous material. We here concentrate on reviewing work of (van Duijn et al., 2004) who investigate stability in a linearized setting, but by Lyapunov function, or energy stability, methods. Other pertinent references may be found in these papers.

The spatial domain of (van Duijn et al., 2004) is a domain in $\mathbb{R}^3$ where $\Omega = \Omega_\perp \times (0, H)$ with $\Omega_\perp$ being a bounded region in $\mathbb{R}^2$ with smooth boundary $\partial\Omega_\perp$. The $z-$axis is measured downward, in the direction of gravity. The basic model of (van Duijn et al., 2004) begins with a conservation law for the volume fraction of water, θ, in the unsaturated porous medium. They write,

$$\frac{\partial \theta}{\partial t} = -F_{i,i}, \tag{5.102}$$

where $F_i = \theta v_i$ is the volumetric flux of the fluid (water) and v_i is the velocity of the water. They employ alternative formulations in addition to θ and so introduce the pressure head $\Psi = (p_w - p_g)/\gamma g$, where γ, g, p_w and p_g are density of water, gravity, pressure of water, pressure of the gaseous phase in the unsaturated porous medium. They assume p_g is constant. The function Ψ is assumed a monotonically increasing function of θ. They also introduce the potential Φ by

$$\Phi = \int_{-\infty}^{\Psi} k \, d\Psi = \int_0^\theta D \, d\theta$$

and give constitutive equations for F_i as (they describe these as Darcy's law)

$$F_i = -k(\Psi)\nabla\Psi + k\delta_{i3}, \quad F_i = -D\nabla\theta + k\delta_{i3}, \quad F_i = -\nabla\Phi + k\delta_{i3}, \tag{5.103}$$

where k denotes the hydraulic conductivity, and $D = k \, d\Psi/d\theta$.

(van Duijn et al., 2004) non-dimensionalize in terms of a saturation $S = (\theta - \theta_r)/(\theta_0 - \theta_r)$, where θ_0 is the value of θ at saturation and θ_r is the irreducible volumetric water content. Note that $S(\theta_r) = 0 \leq S \leq S(\theta_0) = 1$. In terms of a Rayleigh number $R = Hk_0/D_0(\theta_0 - \theta_r)$, k_0 and D_0 being the values of k and D at saturation, they show that (5.102) together with (5.103) lead to one of the following non-dimensional, *equivalent* forms of (5.102),

$$\frac{\partial S(\Psi)}{\partial t} = c(\Psi)\frac{\partial \Psi}{\partial t} = \left[k(\Psi)\Psi_{,i} - Rk(\Psi)\delta_{i3}\right]_{,i}, \tag{5.104}$$

or

$$\frac{\partial S}{\partial t} = \left[D(S)S_{,i} - Rk(S)\delta_{i3}\right]_{,i}, \tag{5.105}$$

or

$$\frac{\partial S(\Phi)}{\partial t} = \frac{1}{D(\Phi)}\frac{\partial \Phi}{\partial t} = \left[\Phi_{,i} - Rk(\Phi)\delta_{i3}\right]_{,i}. \tag{5.106}$$

(van Duijn et al., 2004) give functional relationships for k, D, Φ and Ψ for three classes of soils. These are the Broadbridge and White class, the Gardner class, and that for a Burgers class of soils. That for a Burgers class of soils has

$$k(S) = S^2, \quad D(S) = 1, \quad \Psi(S) = 1 - \frac{1}{S}, \quad \Phi(S) = S. \tag{5.107}$$

The boundary conditions considered by (van Duijn et al., 2004) are that

$$\frac{\partial S}{\partial n_i} = 0 \text{ on } \partial\Omega_\perp \times (0,1), \quad S = S_T \text{ at } z = 0, \quad S = S_B \text{ at } z = 1,$$

recalling $z = 0$ is the top of the domain, $z = 1$ the bottom. They show that if $k(\Phi)$ is a Lipshitz continuous function of Φ then a steady solution $\Phi_0 = \Phi_0(z)$ exists for each value of $R > 0$. This solution corresponds to downward flow if $\Phi_T > \Phi_B$ and upward flow if $\Phi_T < \Phi_B$.

5.7.2 Stability of flow

To study stability (van Duijn et al., 2004) let $\phi = \phi(\mathbf{x}, t)$ be a perturbation to Φ_0 so that $\Phi = \Phi_0(z) + \phi$. The function ϕ is zero on the boundaries where $z = 0, 1$ and $\partial\phi/\partial n_i = 0$ on $\partial\Omega_\perp \times (0,1)$. (van Duijn et al., 2004) then linearize (5.106) to obtain the following perturbation equation satisfied by ϕ,

$$S'(\Phi_0)\frac{\partial \phi}{\partial t} = \Delta\phi - R\frac{\partial}{\partial z}\left[k'(\Phi_0)\phi\right]. \tag{5.108}$$

From this they are able to use an energy stability argument to show that

$$\frac{1}{2}\frac{d}{dt}\|\phi(t)\|^2 \leq -\frac{(1-\lambda)}{K_2}\|\phi\|^2 \leq 0,$$

where $\|\cdot\|$ is the weighted L^2 norm given by

$$\|\phi(t)\|^2 = \int_\Omega S'(\Phi_0)\phi^2 dV,$$

K_2 is a constant and $\lambda = -R(1-z)k'(\Phi_0)/2 < 1$. They then show

$$\|\phi(t)\|^2 \leq \exp\left(-\frac{2(1-\lambda)t}{K_2}\right)\|\phi(0)\|^2, \tag{5.109}$$

thereby showing that the steady solution Φ_0 is linearly exponentially stable. This is an important result, because it is for the relatively untouched area of unsaturated porous flows. (van Duijn et al., 2004) interpret (5.109) in terms of all three soil classes mentioned earlier.

Another interesting aspect of the work of (van Duijn et al., 2004) is that by changing norms they are able to investigate classes of solution for which the "energy" function eventually decays, but may exhibit growth before eventually decaying. To do this they work with a perturbation saturation s given by $S = S_0 + s$ and derive bounds for $\|s(t)\|^2$ where $\| \cdot \|$ is now the usual norm on $L^2(\Omega)$. Again, they derive sharp estimates and interpret them in the light of the specific soil classes. In particular, they note that for the Burgers class of soils there is *no need* to linearize. In that case, the *nonlinear* perturbation equation for s is

$$\frac{\partial s}{\partial t} = \Delta s - 2R\frac{\partial}{\partial z}(S_0 s) - R\frac{\partial}{\partial z}s^2 \,.$$

When one multiplies this by s and integrates over Ω the term $(s, s_z^2) = 0$ and the linearized decay result they establish in general holds exactly in the *nonlinear* case, for a Burgers soil.

(van Duijn et al., 2004) also establish a linear exponential decay result by means of an energy method for a theory of unsaturated porous media which incorporates *memory effects in time*. This theory is due to (Hassanizadeh and Gray, 1990; Hassanizadeh and Gray, 1993). In terms of the saturation S this equation is (cf. (5.105))

$$\frac{\partial S}{\partial t} = [D(S)S_{,i}]_{,i} + \tau\left(k(S)\frac{\partial S_{,i}}{\partial t}\right) + R\frac{\partial}{\partial z}k(S) = 0.$$

5.7.3 Transient growth

In the context of Burgers soils (van Duijn et al., 2004) establish a very interesting result. This concerns the situation where the steady solution is unconditionally nonlinearly stable and the norms involving ϕ and s decay monotonically in time. Instead they work with a perturbation ψ to the pressure head, so that $\Psi = \Psi(z) + \psi(\mathbf{x}, t)$. They derive an exponential bound for $\|\psi\|$, the $L^2(\Omega)$ norm. However, they derive the equation for ψ and then write $\psi = \psi(z, t)\exp[i(mx + ny)]$ and find $\psi(z, t)$ satisfies an equation of form

$$\frac{\partial \psi}{\partial t} = \frac{\partial^2 \psi}{\partial z^2} - a^2\psi + A_1(\Psi_0)\frac{\partial \psi}{\partial z} + A_2(\Psi_0)\psi \equiv A(\Psi_0)\psi, \qquad (5.110)$$

where a is the wavenumber, A_1, A_2 are functions of R and Ψ_0, and the linear operator A is defined as shown.

(van Duijn et al., 2004) derive an energy balance of form

$$\frac{1}{2}\frac{d}{dt}\int_0^1 \psi^2 dz = \int_0^1 \left(A(\Psi_0)\psi(t)\right)\psi(t)\, dz \,. \qquad (5.111)$$

They consider the maximum problem

$$\sigma_{\max} = \sup_{\psi(0) \neq 0} \frac{\int_0^1 \big(A(\Psi_0)\psi(0)\big)\psi(0)\,dz}{\int_0^1 \psi^2(0)dz}.$$

They study the solution to this maximization problem and then from (5.111) are able to deduce a region where $E(t) = (1/2)\int_0^1 \psi^2(t)dz$ initially grows, even though there is eventual decay. This is thus, another example of transient growth, not dissimilar to that reported in section 5.4.3.

5.8 Parallel flows

5.8.1 Poiseuille flow

The problem of Poiseuille flow in a porous medium was addressed by (Nield, 2003). He correctly observes that this class of flows is likely to be very important in high porosity materials. Indeed, in view of this, he advocates using a Brinkman model to study Poiseuille flow in a porous material.

We re-investigate the (Nield, 2003) problem here. The basic equations are

$$\rho\Big(\frac{\partial v_i}{\partial t} + v_j \frac{\partial v_i}{\partial x_j}\Big) = -\frac{\partial p}{\partial x_i} + \mu \Delta v_i - \frac{\phi\mu}{K} v_i, \qquad \frac{\partial v_i}{\partial x_i} = 0,$$

where v_i, P are velocity and pressure, ρ is the constant density, μ, ϕ, K are dynamic viscosity, porosity and permeability. These equations are non-dimensionalized with a length scale L, velocity scale V, time scale L/V, and a Reynolds number $R = \rho V L/\mu$. They may then be written

$$R\Big(\frac{\partial v_i}{\partial t} + v_j \frac{\partial v_i}{\partial x_j}\Big) = -\frac{\partial p}{\partial x_i} + \Delta v_i - M^2 v_i, \qquad \frac{\partial v_i}{\partial x_i} = 0, \qquad (5.112)$$

where v_i is now a dimensionless velocity and $M^2 = \phi L^2/K$. The porous medium is saturated with fluid and contained in the infinite layer $\{(x,y) \in \mathbb{R}^2\} \times \{z \in (-1,1)\}$. The no-slip boundary conditions apply so that

$$v_i = 0 \text{ at } z = \pm 1, \qquad (5.113)$$

and a constant pressure gradient $G = -\partial p/\partial x > 0$ is applied in the $x-$direction. This gives rise to the basic solution

$$\bar{\mathbf{v}} = (U(z), 0, 0) \quad \text{where} \quad U = \frac{G}{M^2}\Big(1 - \frac{\cosh Mz}{\cosh M}\Big). \qquad (5.114)$$

When the Darcy term disappears this should reduce to the classical Poiseuille solution for Navier-Stokes theory, namely $U = (1 - z^2)/2$, which one can recover from (5.114) in the limit $M \to 0$. In fact for small M, U

has the asymptotic form

$$U \sim \frac{1}{2}(1 - z^2) + M^2\left(-\frac{5}{24} + \frac{z^2}{4} - \frac{z^4}{24}\right)$$
$$+ M^4\left(-\frac{29}{720} + \frac{z^2}{48} + \frac{z^4}{48} - \frac{z^6}{720}\right) + O(M^6). \tag{5.115}$$

To investigate linearized instability of the basic solution (5.114) we follow (Nield, 2003) and introduce perturbations $u_i(\mathbf{x}, t), \pi(\mathbf{x}, t)$ so that $v_i = \bar{v}_i + u_i$, $p = \bar{p} + \pi$ and then u_i, π is found to satisfy

$$R(u_{i,t} + u_j U_{i,j} + U_j u_{i,j}) = -\pi_{,i} + \Delta u_i - M^2 u_i,$$
$$u_{i,i} = 0. \tag{5.116}$$

Again, we follow (Nield, 2003), study the two-dimensional instability problem and write $u_i = u_i(z) \exp[ia(x-ct)]$, $\pi = \pi(z) \exp[ia(x-ct)]$. Equation $(5.116)_1$ becomes

$$R[-iacu_i + \delta_{i1}U'w + Uiau_i] = -\pi_{,i} + \Delta u_i - M^2 u_i,$$

and then writing $\mathbf{u} = (u, w)$ we find the full system of equations (5.116) reduces to

$$[\mathcal{L} - iaR(U - c)]u = RU'w + ia\pi,$$
$$[\mathcal{L} - iaR(U - c)]w = D\pi, \tag{5.117}$$
$$iau + Dw = 0,$$

where $D = d/dz$, and the operator $\mathcal{L}$ is

$$\mathcal{L} = D^2 - a^2 - M^2. \tag{5.118}$$

In fact, our equation $(5.117)_2$ differs from equation (17b) of (Nield, 2003) in that he has $D^2 - a^2$ in his equivalent equation rather than $\mathcal{L}$.

One now eliminates u and π from (5.117) to derive the fourth order equation

$$(D^2 - a^2)^2 w - M^2(D^2 - a^2)w = iaR(U - c)(D^2 - a^2)w - iaRU''w, \tag{5.119}$$

where $z \in (-1, 1)$. This is our Orr-Sommerfeld equation for Poiseuille flow in a Brinkman porous medium. It differs from the classical Orr-Sommerfeld equation, cf. (Dongarra et al., 1996), p. 404, or see equation (5.57), only by the term involving M^2. To obtain results on instability of Poiseuille flow in a porous medium one must solve equation (5.119) numerically subject to the boundary conditions

$$w = Dw = 0, \qquad z = \pm 1. \tag{5.120}$$

The basic flow is given by (5.114), but in the case of small M one can employ (5.115).

As (Nield, 2003) correctly observes, instability results for Poiseuille flow in porous media, here given by solving (5.119), (5.120), will undoubtedly become important, especially due to the use of high porosity metallic foams in industrial devices such as heat pipes. Eigenvalue bounds may be obtained as (Joseph, 1976b), section 44, does for the classical Orr-Sommerfeld equation. Such bounds are still important, as (Puri, 2005) shows when deriving estimates for Poiseuille flow of a dipolar fluid.

The area of Poiseuille, or more generally parallel, flows in a porous medium is an area I believe will become increasingly important in future. There are many ramifications to arise from variants of such flows in Navier-Stokes theory, such as involving heating, (Choi et al., 2004), temperature-dependent viscosity, (Akyildiz and Bellout, 2005), (Massoudi and Phuoc, 2004), (Vaidya and Wulandana, 2006), (Webber, 2007), slip boundary conditions, (Webber, 2006; Webber, 2007; Webber, 2008), (Webber and Straughan, 2006), cylindrical geometry, (Kim et al., 2006), ramp heating, (Kim et al., 2005), swirl and decelerating flows, (Kim and Choi, 2004), rotating shear flow, (Yecko, 2004), surfactants, (Blyth et al., 2006), (Blyth and Pozrikidis, 2004b), two layer flows, (Blyth and Pozrikidis, 2004a), channel entrance flow, (Hifdi et al., 2004a; Hifdi et al., 2004b), and granular materials (Massoudi and Phuoc, 2007), to mention some. Of course, given the interest of numerical methods to solve the classical Orr-Sommerfeld equation it will be interesting to see their application to system (5.119), (5.120). Many of the accurate schemes are studied in the works of (Orszag, 1971), (Dongarra et al., 1996), (Straughan and Walker, 1996b), (Ivansson, 2003), (Theofilis, 2003), (Mehta, 2004), (Theofilis et al., 2004), (Hirata et al., 2006), (Elbarbary, 2007), and (Valerio et al., 2007), where many other references may be found, see also chapter 9 of this book.

5.8.2   Flow in a permeable conduit

The problem of Poiseuille flow is an important one in porous media, but it is also important in underground flow. An aquifer is typically a layer of water bearing permeable rock, sand or gravel capable of providing significant amounts of water. An aquitard, on the other hand, is a bed of very low permeability, possibly a water saturated sediment or rock whose permeability is so low that it cannot transmit any useful amount of water. When an aquifer occurs naturally in an underground aquitard it may represent a means of transport for any contaminant present. The process of diffusion of the contaminant out of the aquifer and into the surrounding aquitard (or the reverse) is, therefore, important. The conduit(s) of the aquifer may have the shape of a plane layer, but may also be approximated by a cylinder with a circular or other shaped cross section. A recent paper of (Harrington et al., 2007) highlights flow in an underground conduit and explains how such flow is important in connection with contaminant transport. They develop a model for flow in a conduit which consists of a cylinder with cross sectional geometry Ω, the perimeter Γ carrying a solute. The axis of the cylinder is assumed to be the $x-$axis. Inside the conduit is a solute of concentration $C_c(x,t)$ where the solute is assumed well mixed so that y and z variations may be ignored. The concentration of solute on the boundary, $C_m(x,y,z,t)$, arises from contaminant in the matrix outside the conduit. The y and z dependences are recast into r, θ coordinates, or other coordinates, depending on the shape of Ω. The equations of (Harrington et al., 2007) are

$$\frac{R_c}{D_c} \frac{\partial C_c}{\partial t} = \frac{\partial^2 C_c}{\partial x^2} - \frac{v}{D_c} \frac{\partial C_c}{\partial x} - \frac{\lambda R_c}{D_c} C_c + \frac{\phi_m D_m}{\phi_c D_c \Omega_A} \int_\Gamma \frac{\partial C_m}{\partial n} \, dS$$

in the conduit, and

$$\frac{R_m}{D_m} \frac{\partial C_m}{\partial t} = \Delta C_m - \frac{\lambda R_m}{D_m} C_m$$

in the matrix, where Δ is the Laplacian in terms of y, z coordinates. The conduit is of (semi) infinite length and the initial and boundary conditions are

$$C_c(x,0) = 0, \qquad C_m(x,y,z,0) = 0,$$
$$C_c(0,t) = C_0, \qquad \lim_{x \to \infty} C_c(x,t) = 0, \qquad \lim_{r \to \infty} C_m(x,r,\theta,t) = 0,$$

with

$$C_m = C_c \qquad \text{on} \quad \Gamma.$$

The coefficients $R_c, D_c, \lambda, \phi_m, D_m, \phi_c, R_m$ are defined by (Harrington et al., 2007), Ω_A is the area of the domain Ω, and the coefficient v is the flow speed which is assumed known. (Harrington et al., 2007) show how to solve the system above when Ω is a circle, an ellipse, and an infinite channel given by parallel lines $y = \pm b, z \in \mathbb{R}$. They compare their simulations with a practical study site in Saskatchewan in Canada. This analysis is very interesting and investigates in detail the chloride distribution.

It would be an interesting mathematical analysis to study the spatial behaviour (in x) of a solution to this model for an arbitrary geometry Ω.

Contaminant transport, especially that associated with radioactive waste, is a very important subject. (Giacobbo and Patelli, 2007) present an interesting approach to this subject which involves a stochastic model coupled to a Darcy - Richards formulation for an unsaturated porous medium.

6
Fluid - Porous Interface Problems

6.1 Models for thermal convection

The object of this chapter is to study flow in a fluid which is in contact with a porous medium. We suppose the fluid also saturates the porous medium. The topic in question is of immense importance due to many mundane applications. For example, flow in underground channels or streambeds where contaminant or solute may be transported in stream water, see e.g. (Ewing et al., 1994), (El-Habel et al., 2002), (Boano et al., 2007). Another mundane example concerns production of composite materials where fibrous layers are infused with resin and the composite is produced by heat and pressure in an autoclave, see e.g. (Blest et al., 1999). The increasing use of composite materials in automobile and aeroplane production certainly justifies further investigation of flow and convection of a fluid sandwiched between porous layers. A further example which has consequences for everyone is melt water formation above and below ice sheets and ice shelves in the Arctic and Antarctic, and the possible increased melting due to thermal convection, see e.g. (Bogorodskii and Nagurnyi, 2000), (Carr, 2003a; Carr, 2003b). The last topic is discussed further in section 6.4.

While there are numerous applications for a theory of convection / flow in a fluid next to a fluid saturated porous material, there are also many theories to attempt to describe this scenario and this is a very active area of current research, see e.g. (Chandesris and Jamet, 2006), (Chang, 2004; Chang, 2005; Chang, 2006), (Chang et al., 2006), (Das et al., 2002),

B. Straughan, *Stability and Wave Motion in Porous Media*,
DOI: 10.1007/978-0-387-76543-3_6, © Springer Science+Business Media, LLC 2008

(Discacciati et al., 2002), (Goharzadeh et al., 2005), (Govender, 2006a), (Hill and Straughan, 2008), (Hirata et al., 2006), (Hirata et al., 2007), (Hoppe et al., 2007), (Layton et al., 2003), (Le Bars and Worster, 2006), (Miglio et al., 2003), (Riviere, 2005), and the references therein. It is not the goal of this book to review all of the models, nor is it the goal to attempt to assess which model may be preferable for a particular task. We review some of the key models and present some new numerical findings for models which we have found particularly tractable and which may be widely applicable to engineering type problems. Several new results are presented throughout this chapter, for example, section 6.2 presents new numerical results for surface tension driven convection in a fluid overlying a porous layer, while section 6.3 is new and investigates the convection problem by modelling the various coefficients which arise as functions of the porosity.

6.1.1 Extended Navier-Stokes model

This is a model which employs the Navier-Stokes equations in the fluid and adds in a Darcy term to model flow in the porous medium, cf. (Ewing et al., 1994). These writers observe that one way to couple liquid flow and such flow in a porous medium is to employ appropriate boundary conditions at the fluid - porous medium interface. However, another approach, favoured by (Ewing et al., 1994) is to extend the Navier-Stokes equations and introduce a Darcy term. As (Ewing et al., 1994) write, this approach has been mainly employed in the area of numerical simulation of convection/diffusion of alloys involving melting and solidification. (Ewing et al., 1994) in particular, study the problem of flow over a step where the region after the step consists of a fluid overlying a porous medium, as shown in figure 6.1.

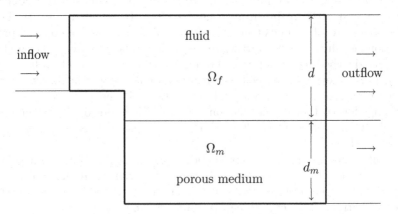

Figure 6.1. L - shaped flow domain

(Ewing et al., 1994) treat an incompressible fluid containing a contaminant flowing in from the left, passing over a step and then flowing in a channel where the fluid overlies a porous medium of depth d_m, as shown in figure 6.1. Their model employs the computational domain $\Omega_f \cup \Omega_m$, the L-shaped region in figure 6.1 shown in darker outline. Their equations are the steady, two-dimensional Navier-Stokes equations coupled with a Darcy term, coupled to equations for contaminant transport, namely,

$$u\frac{\partial u}{\partial x} + u\frac{\partial u}{\partial y} = \nu\Delta u - \frac{1}{\rho}\frac{\partial p}{\partial x} - \frac{\nu}{k}u,$$

$$u\frac{\partial v}{\partial x} + v\frac{\partial v}{\partial y} = \nu\Delta v - \frac{1}{\rho}\frac{\partial p}{\partial y} - \frac{\nu}{k}v,$$

$$\frac{\partial u}{\partial x} + \frac{\partial v}{\partial y} = 0,$$
$$\tag{6.1}$$

$$\frac{\partial c}{\partial t} + f\left(u\frac{\partial c}{\partial x} + v\frac{\partial c}{\partial y}\right) = \frac{\partial}{\partial x_\alpha}\left(D(\mathbf{x})\frac{\partial c}{\partial x_\alpha}\right),$$

where α sums from 1 to 2. The spatial domain for equations (6.1) is $\Omega = \Omega_f \cup \Omega_m$, and the functions k and f are such that

$$k = \begin{cases} \infty, & \mathbf{x} \in \Omega_f, \\ k_p, & \mathbf{x} \in \Omega_m, \end{cases} \qquad f = \begin{cases} 1, & \mathbf{x} \in \Omega_f, \\ \phi, & \mathbf{x} \in \Omega_m, \end{cases}$$

where k_p represents a Darcy flow term and ϕ is the porosity in the porous region Ω_m. Many numerical solutions of this problem for appropriate boundary conditions are given by (Ewing et al., 1994), the computations being performed by a "fictitious regions" method.

6.1.2 Nield (Darcy) model

In this section we examine the problem where a fluid overlies a layer of porous material saturated by the same fluid. The layer is such that the temperature of its upper surface is fixed at T_U, say, while the temperature of the lower surface is likewise fixed at a higher temperature T_L. If the temperature gradient is large enough thermal convection will arise. This is an important problem which has led to much understanding of flow of a fluid adjacent to a porous medium. Two fundamental papers which have both had a major impact on many subsequent workers in this area are those of (Beavers and Joseph, 1967) and (Nield, 1977). That of (Beavers and Joseph, 1967) presents a boundary condition which is applicable at the fluid - porous medium boundary. The Beavers - Joseph condition is analysed in section 2.10 and spatial decay is discussed in section 3.5. We shall employ the notation of (Straughan, 2001c; Straughan, 2002a), which is a generalization of that of (Chen and Chen, 1988).

Suppose the fluid occupies the layer $z \in (0, d)$ while the porous medium fills the layer $z \in (-d_m, 0)$, x and y occupying the whole of $\mathbb{R}^2$. The fundamental boundary condition proposed by (Beavers and Joseph, 1967) in this context is

$$\frac{\partial u^\beta}{\partial z} = \frac{\alpha}{\sqrt{K}} (u^\beta - u_m^\beta), \quad \beta = 1, 2, \ z = 0. \tag{6.2}$$

Here u^β are the x and y components of fluid velocity, u_m^β are the equivalent velocity components in the porous medium, α is a constant depending on the porous medium, and K is the permeability. This is a boundary condition which has been employed with a lot of success. (Straughan, 2002a) analyses the above condition and the (properly invariant) extension due to (Jones, 1973) in some detail. His numerical findings indicate that for a good many problems little difference is observed whether one uses the Beavers - Joseph or the Jones condition. This finding is also noted numerically by (McKay and Straughan, 1993) and by (Chang et al., 2006).

The fundamental model for thermal convection in a fluid overlying a porous medium was developed by (Nield, 1977). He employs the Beavers - Joseph boundary condition (6.2). The Nield model is investigated in some detail by (Chen and Chen, 1988). The work of these writers is also predominant in this field in that they discovered the linear instability curves for the onset of thermal convection may be bi-modal. This means that they possess two local minima. Introduce the parameter $\hat{d}$ by

$$\hat{d} = \frac{d}{d_m} = \frac{\text{depth of fluid layer}}{\text{depth of porous layer}}. \tag{6.3}$$

(Chen and Chen, 1988) discovered that for a porous medium comprised of small glass beads, when $\hat{d} \leq 0.13$ the instability is initiated in the porous medium, whereas for $\hat{d}$ larger than this the instability will commence in the fluid. The bi-modal character of the two layer problem has since been verified by many writers.

With the notation of this section the Nield model employs the Navier-Stokes equations in $\Omega_f = \mathbb{R}^2 \times \{z \in (0, d)\} \times \{t > 0\}$, and the Darcy equations for thermal convection in a porous medium $\Omega_m = \mathbb{R}^2 \times \{z \in (-d_m, 0)\} \times \{t > 0\}$. Thus, in Ω_f

$$\frac{\partial u_i}{\partial t} + u_j \frac{\partial u_i}{\partial x_j} = -\frac{1}{\rho_0} \frac{\partial p}{\partial x_i} + \nu \Delta u_i + \bar{\alpha} g T k_i,$$

$$\frac{\partial u_i}{\partial x_i} = 0, \tag{6.4}$$

$$\frac{\partial T}{\partial t} + u_i \frac{\partial T}{\partial x_i} = \frac{k_f}{(\rho_0 c_p)_f} \Delta T,$$

whereas in Ω_m

$$\frac{1}{\phi}\frac{\partial u_i^m}{\partial t} = -\frac{1}{\rho_0}\frac{\partial p^m}{\partial x_i} - \frac{\nu}{K}u_i^m + \bar{\alpha}gT_mk_i,$$

$$\frac{\partial u_i^m}{\partial x_i} = 0, \qquad\qquad (6.5)$$

$$(\rho_0 c_p)^*\frac{\partial T_m}{\partial t} + (\rho_0 c_p)_f u_i^m\frac{\partial T_m}{\partial x_i} = k^*\Delta T_m.$$

In these equations $u_i, p, T, u_i^m, p^m, T_m$ denote velocity, pressure and temperature in the fluid, and velocity, pressure and temperature in the porous medium, respectively. The coefficients $\rho_0, \nu, \bar{\alpha}, g, k_f, c_p$ are density, kinematic viscosity, thermal expansion coefficient, gravity, thermal conductivity and specific heat at constant pressure. Throughout, we employ subscript or superscript f or m to denote a fluid or porous quantity, ϕ is the porosity, and a $*$ denotes a weighted porous medium value. For example, if X denotes k or $\rho_0 c_p$, then

$$X^* = \phi X_f + (1 - \phi)X_m. \qquad\qquad (6.6)$$

The (Nield, 1977) model has enjoyed huge success, especially in connection with linearized instability theory, and we return to specific details of such calculations in section 6.2.

The term $\phi^{-1}\partial u_i^m/\partial t$ in equation $(6.5)_1$ has mostly been neglected throughout this book. This is an inertia term (acceleration) which for many flows in porous media is believed to be negligible. The effect of the inertia term is explicitly investigated by (Khadrawi and Al-Nimr, 2005).

6.1.3   Forchheimer model

If the flow velocity is not very small then it may be argued that one will need to modify the Darcy model and replace it with a one of Forchheimer, Brinkman, or Brinkman-Forchheimer type. (Chen, 1990) does exactly this in his linearized instability analysis of the equivalent problem of section 6.1.2, although he also allows for a vertical throughflow throughout the layer. Thus, the (Chen, 1990) model still utilizes equations (6.4), and equations $(6.5)_2$ and $(6.5)_3$. However, he replaces equation $(6.5)_1$ by the equation

$$\frac{1}{\phi}\frac{\partial u_i^m}{\partial t} + \frac{B}{K}|\mathbf{u}^m|u_i^m = -\frac{1}{\rho_0}\frac{\partial p^m}{\partial x_i} - \frac{\nu}{K}u_i^m + \bar{\alpha}gT_mk_i, \qquad (6.7)$$

where B is a Forchheimer coefficient.

Since (Chen, 1990) studies the effect of throughflow on convection his basic state is one in which $\mathbf{u} = (0, 0, W)$, W constant, and this creates a steady temperature profile which is not linear in z. The effect of throughflow is felt in the linearized perturbation equations and, in particular, the Forchheimer term does not disappear. Additionally, (Chen, 1990) uses a

Beavers - Joseph boundary condition. However, one has to be careful, because his boundary conditions also contain the Forchheimer effect. The paper of (Chen, 1990) contains many numerical results.

6.1.4 Brinkman model

The possibility of considering a Brinkman equation to model the flow in a porous medium, rather than the Darcy equation $(6.5)_1$ was considered by (Nield, 1983). He explains that care must be taken with such an approach. In (Nield, 1991b) he also analyses the limitations which may occur if one replaces $(6.5)_1$ by a Brinkman-Forchheimer equation. A detailed linear instability analysis of the thermal convection problem for the two-layer situation of section 6.1.2 is given by (Hirata et al., 2007). These writers essentially employ equations (6.4) and (6.5), however, they replace $(6.5)_1$ by a Brinkman equation of form

$$\frac{1}{\phi}\frac{\partial u_i^m}{\partial t} = -\frac{1}{\rho_0}\frac{\partial p^m}{\partial x_i} - \frac{\nu}{K}u_i^m + \frac{\mu_{eff}}{\rho_0}\Delta u_i^m + \bar{\alpha}gT_m k_i, \qquad (6.8)$$

in which μ_{eff} is an effective viscosity. Because of the presence of the higher derivative term $(\mu_{eff}/\rho_0)\Delta u_i^m$ they are able to dispense with the Beavers-Joseph boundary condition and instead employ continuity conditions at the interface. The paper of (Hirata et al., 2007) provides many numerical calculations.

It is worth pointing out that (Chen and Chen, 1992) performed a numerical simulation of thermal convection in a two layer fluid / porous system like that of section 6.1.2. They employed equations (6.4) and (6.5), but replaced $(6.5)_1$ by a Brinkman-Forchheimer equation of form

$$\frac{1}{\phi}\frac{\partial u_i^m}{\partial t} + \frac{B}{K}|\mathbf{u}^m|u_i^m = -\frac{1}{\rho_0}\frac{\partial p^m}{\partial x_i} - \frac{\nu}{K}u_i^m + \frac{\nu}{\phi}\Delta u_i^m + \bar{\alpha}gT_m k_i. \qquad (6.9)$$

Again, the presence of the Δu_i^m terms allows then to employ continuity conditions at the porous medium - fluid interface. (Chen and Chen, 1992) expand the horizontal components of their solution in a Fourier series and use an implicit finite difference scheme to compute the solution, for the nonlinear problem. Many numerical results are presented by (Chen and Chen, 1992) and generally the agreement with the linearized instability results of (Chen and Chen, 1988) is very good.

6.1.5 Nonlinear equation of state

(Carr and Straughan, 2003) analyse the problem of thermal convection in a fluid overlying a porous layer. They adopt the model of section 6.1.2 but, they change the linear equation of state adopted there, namely $\rho = \rho_0[1 - \bar{\alpha}(T - T_0)]$ to one appropriate to a layer of water in the temperature range $0°C - 14°C$, i.e. where the water exhibits a maximum density effect.

Their equation of state in both the fluid and porous medium layers has form

$$\rho = \rho_0 \left[1 - \bar{\alpha}(T - 4)^2 \right]. \tag{6.10}$$

The model of (Carr and Straughan, 2003) is addressed to study convection where the porous medium represents a thawed layer of ground which overlies permafrost, the thawed porous layer being overlain by a layer of water. This thus models the scenario of patterned ground formation as found near the edges of shallow alpine lakes, cf. (McKay and Straughan, 1993) and the references therein.

The equations of (Carr and Straughan, 2003) are essentially (6.4) and (6.5) but with the quadratic density - temperature relationship (6.10). The equations of (Carr and Straughan, 2003) are, in the fluid,

$$\frac{\partial u_i}{\partial t} + u_j \frac{\partial u_i}{\partial x_j} = -\frac{1}{\rho_0} \frac{\partial p}{\partial x_i} + \nu \Delta u_i - 8\bar{\alpha}gTk_i + \bar{\alpha}gT^2 k_i,$$

$$\frac{\partial u_i}{\partial x_i} = 0, \tag{6.11}$$

$$\frac{\partial T}{\partial t} + u_i \frac{\partial T}{\partial x_i} = \frac{k_f}{(\rho_0 c_p)_f} \Delta T,$$

and in the porous layer

$$0 = -\frac{1}{\rho_0} \frac{\partial p^m}{\partial x_i} - \frac{\nu}{K} u_i^m - 8\bar{\alpha}gT_m k_i + \bar{\alpha}gT_m^2 k_i,$$

$$\frac{\partial u_i^m}{\partial x_i} = 0, \tag{6.12}$$

$$(\rho_0 c_p)^* \frac{\partial T_m}{\partial t} + (\rho_0 c_p)_f u_i^m \frac{\partial T_m}{\partial x_i} = k^* \Delta T_m.$$

The seemingly innocuous change from the model (6.4), (6.5) to the model when (6.11), (6.12) are employed leads to a major difference in instability results. (Carr and Straughan, 2003) show that the presence of the porous medium leads to penetrative convection for much lower surface temperatures (at $z = d$) than that found for penetrative convection in either a single fluid layer, or a single layer of porous material saturated with water. Indeed, as the surface temperature at $z = d$ is increased a multi-cellular structure develops at the onset of convection. The fluid - porous medium two layer problem leads to a complicated cell structure. For example, with the lower boundary, $z = -d_m$ kept at 0°C, upper surface ($z = d$) temperatures of 13°C and 13.2°C lead to very different cellular structures. The former has 21 thin cells in the vertical direction in the fluid whereas the latter has one wider cell within the porous medium with 15 cells of equivalent width in the fluid. Thus, the bi-modal nature of the (Chen and Chen, 1988) analysis is still found, but the porous medium influences the penetrative convection markedly.

(Carr, 2003a; Carr, 2004) also analyses penetrative convection in a fluid overlying a porous layer by employing an internal heat source model rather than a nonlinear equation of state. The model consists of equations (6.4), (6.5) but the energy equations $(6.4)_3$, $(6.5)_3$ are replaced by

$$\frac{\partial T}{\partial t} + u_i \frac{\partial T}{\partial x_i} = \frac{k_f}{(\rho_0 c_p)_f} \Delta T + 2Q,$$

$$\frac{(\rho_0 c_p)^*}{(\rho_0 c_p)_f} \frac{\partial T_m}{\partial t} + u_i^m \frac{\partial T_m}{\partial x_i} = \frac{k^*}{(\rho_0 c_p)_f} \Delta T_m + 2Q^m,$$

where Q, Q^m are appropriate heat sources or sinks. This is a very interesting study which can have multiple stably / unstably stratified layers not dissimilar to those of section 4.5.

6.1.6 Reacting layers

(McKay, 1998) considers the instability problem of section 6.1.2 but he allows heat generation due to chemical reactions in the fluid or porous layers. His equations are (6.4) and (6.5) but with heat source terms which depend on temperature. In fact, he replaces equations $(6.4)_3$ and $(6.5)_3$ by

$$\frac{\partial T}{\partial t} + u_i \frac{\partial T}{\partial x_i} = \frac{k_f}{(\rho_0 c_p)_f} \Delta T + Q \exp\left(\frac{-E}{RT}\right),$$

$$\frac{(\rho_0 c_p)^*}{(\rho_0 c_p)_f} \frac{\partial T_m}{\partial t} + u_i^m \frac{\partial T_m}{\partial x_i} = \frac{k^*}{(\rho_0 c_p)_f} \Delta T_m + \phi Q \exp\left(\frac{-E}{RT_m}\right).$$

The coefficients E, R are positive constants, E being the activation energy and R the universal gas constant. (McKay, 1998) relates such porous / fluid reaction situations to practical problems involving removing heat from a nuclear reactor by flooding the core with coolant, or removing heat from radioactive waste products, or delaying the thermal explosion of coal piles or waste dumps. (McKay, 1998) performs a linearized instability analysis of his model and employs a Chebyshev collocation technique in the numerical analysis of the resulting eigenvalue problem.

6.2 Surface tension

6.2.1 Basic solution

As stated in section 6.1.2, the basic model for thermal convection in a two layer fluid - porous system was given by (Nield, 1977). In fact, (Nield, 1977) presented an asymptotic solution for small wavenumber for prescribed heat flux boundary conditions. We here study the (Nield, 1977) model but when the upper surface is free and surface tension effects are taken into account. This model was analysed by (Straughan, 2001c).

The basic equations are (6.4) and (6.5). To make the problem determinate we need boundary conditions. These consist of conditions on the velocity and temperature on the upper (fluid) surface, together with a suitable condition there describing the stress state at that boundary. We also assume there is no flow through the lower boundary and the temperature is assigned there. At the fluid - porous medium interface we assume continuity of normal velocity, continuity of temperature, continuity of heat flux, the Beavers-Joseph condition (6.2) and continuity of normal stress. These conditions are those of (Nield, 1977) and of (Chen and Chen, 1988).

We suppose the boundaries $z = d, z = -d_m$ are held at fixed constant temperatures, T_U, T_L, with $T_L > T_U$. The basic steady state solution is then one for which $u_i \equiv 0, u_i^m \equiv 0$ and $T = T(z), T_m = T_m(z)$. We find this as in (Nield, 1977) and (Chen and Chen, 1988), to be

$$\bar{u}_i = 0, \qquad \bar{u}_i^m = 0,$$

$$\bar{T} = T_0 - (T_0 - T_U)\frac{z}{d}, \quad 0 \leq z \leq d, \tag{6.13}$$

$$\bar{T}_m = T_0 - (T_L - T_0)\frac{z}{d_m}, \quad -d_m \leq z \leq 0.$$

In these expressions T_0 is the temperature at the interface. This is found as in (Nield, 1977), (Chen and Chen, 1988) by requiring continuity of temperature and heat flux at the interface,

$$k_f \frac{d\bar{T}}{dz} = k^* \frac{d\bar{T}_m}{dz} \qquad \text{at} \quad z = 0,$$

and then

$$T_0 = \frac{k^* d T_L + k_f d_m T_U}{k^* d + k_f d_m}.$$

The steady pressures $\bar{p}$ and $\bar{p}^m$ may be found from (6.4) and (6.5).

At the fluid surface $z = d$ (Straughan, 2001c) adopts a radiation type boundary condition in the steady state,

$$\delta_1 \frac{d\bar{T}}{dz} + \delta_2 \bar{T} = c, \qquad \text{at} \quad z = d. \tag{6.14}$$

The coefficients δ_1 and δ_2 depend on the ambient conditions and in bright sunshine δ_1 will be large because heating is mainly by radiation, but in cloudy or foggy conditions δ_2 is likely to be dominant. The variable c is known. (Straughan, 2001c) shows that if one writes δ_1, δ_2 in terms of a constant L, $\delta_1 = 1/(1 + L)$, $\delta_2 = L/(1 + L)$, then (6.13) is consistent with (6.14) provided $T_U = [cd(1+L)+T_0]/(1+Ld)$. He then shows that in terms of a perturbation (u_i, θ, π) to the basic solution $(\bar{u}_i, \bar{T}, \bar{p})$ and (u_i^m, θ_m, π_m) to $(\bar{u}_i^m, \bar{T}_m, \bar{p}_m)$ the boundary condition (6.14) leads to a condition on the perturbation temperature field at the fluid surface, of form

$$\frac{\partial \theta}{\partial z} + L\theta = 0, \qquad \text{on } z = d.$$

The boundary conditions on the velocity in the steady state are zero flow at the lower boundary, so that $w_m = 0$ at $z = -d_m$, where $w_m = u_3^m$. At the interface $z = 0$, $\mathbf{u}.\mathbf{n}$ is continuous, where $\mathbf{n} = (0, 0, 1)$. The surface tension condition involving u_i at $z = d$ is given below.

6.2.2 Perturbation equations

Let now (u_i, θ, π), (u_i^m, θ^m, π^m) be perturbations to the steady solution (6.13). Hence, we put

$$u_i = \bar{u}_i + u_i, \qquad T = \bar{T} + \theta, \qquad p = \bar{p} + \pi,$$
$$u_i^m = \bar{u}_i^m + u_i^m, \qquad T_m = \bar{T}_m + \theta_m, \qquad p_m = \bar{p}_m + \pi_m,$$

in equations (6.4), (6.5) and derive *linearized* equations for u_i, θ, π, u_i^m, θ_m, and π_m. (Straughan, 2001c) observes that this procedure is formally the same as that of (Chen and Chen, 1988). However, (Chen and Chen, 1988) ignore time derivative terms when deriving the boundary conditions and we argue that *a priori* one cannot do this. Because, it is known that surface tension driven convection in a fluid with no porous medium below may lead to convective motion commencing by oscillatory convection.

One introduces a time dependence of form

$$u_i = u_i(\mathbf{x})\, e^{\sigma t}, \quad \theta = \theta(\mathbf{x})\, e^{\sigma t}, \quad \pi = \pi(\mathbf{x})\, e^{\sigma t},$$
$$u_i^m = u_i^m(\mathbf{x})\, e^{\sigma_m t}, \quad \theta_m = \theta_m(\mathbf{x})\, e^{\sigma_m t}, \quad \pi_m = \pi_m(\mathbf{x})\, e^{\sigma t},$$

and then the linearized perturbation equations which one obtains from (6.4) are

$$\rho_0 \sigma u_i = -\frac{\partial \pi}{\partial x_i} + \mu \Delta u_i + \rho_0 \bar{\alpha} g k_i \theta,$$

$$\frac{\partial u_i}{\partial x_i} = 0, \tag{6.15}$$

$$\sigma \theta = \left(\frac{T_0 - T_U}{d}\right) w + \frac{k_f}{(\rho_0 c_p)_f} \Delta \theta.$$

Similarly, from (6.5) one shows that

$$\frac{\rho_0}{\phi} \sigma_m u_i^m = -\frac{\partial \pi^m}{\partial x_i} - \frac{\mu}{k} u_i^m + \rho_0 \bar{\alpha} g k_i \theta_m,$$

$$\frac{\partial u_i^m}{\partial x_i} = 0, \tag{6.16}$$

$$\sigma_m \theta_m = \left(\frac{T_L - T_0}{d_m}\right) \frac{(\rho_0 c_p)_f}{(\rho_0 c_p)^*} w_m + \frac{k^*}{(\rho_0 c_p)^*} \Delta \theta_m.$$

In the above equations $w = u_3$, $w_m = u_3^m$, and $\mu = \nu \rho_0$ is the dynamic viscosity.

(Straughan, 2001c) employs the non-dimensionalization of (Chen and Chen, 1988) and so we now put

$$w = W(z) f(x, y), \quad \theta = \Theta(z) f(x, y),$$
$$w_m = W_m(z) f(x, y), \quad \theta_m = \Theta_m(z) f(x, y),$$

where f is the horizontal planform, such that $\Delta^* f = -a^2 f$, in the fluid, $\Delta^* f_m = -a_m^2 f_m$ in the porous medium, $\Delta^* = \partial^2/\partial x^2 + \partial^2/\partial y^2$ being the horizontal Laplacian. From equations (6.15) we show W and Θ satisfy

$$(D^2 - a^2)^2 W - a^2 Ra\Theta = \frac{\sigma}{Pr} (D^2 - a^2) W,$$
$$(D^2 - a^2)\Theta - W = \sigma\Theta, \tag{6.17}$$

where $z \in (0, 1)$ and $D = d/dz$, while a similar reduction from (6.16) leads to

$$(D^2 - a_m^2) W_m + a_m^2 Ra_m \Theta_m = -\sigma_m \frac{\delta^2}{\phi Pr_m} (D^2 - a_m^2) W_m,$$
$$(D^2 - a_m^2)\Theta_m - W_m = \sigma_m G_m \Theta_m, \tag{6.18}$$

where $z_m \in (-1, 0)$ and $D = d/dz_m$. The Rayleigh number and porous Rayleigh numbers, Ra and Ra_m are defined by

$$Ra = \frac{g\bar{\alpha}\rho_0(T_U - T_0)d^3(\rho_0 c_p)_f}{\mu k_f}, \quad Ra_m = Ra \frac{(\delta \epsilon_T)^2}{\hat{d}^4}, \tag{6.19}$$

Pr, Pr_m are the Prandtl and porous Prandtl numbers, δ is the Darcy number, $\delta = \sqrt{k}/d_m$, and $G_m = (\rho_0 c_p)^*/(\rho_0 c_p)_f$, $\epsilon_T = \lambda_f/\lambda_m$, where the fluid and porous medium thermal diffusivities are defined in terms of the thermal conductivities by $\lambda_f = k_f/(\rho_0 c_p)_f$, $\lambda_m = k^*/(\rho_0 c_p)^*$.

We observe that σ and σ_m are not independent, in fact

$$\sigma_m = \frac{\hat{d}^2}{\epsilon_T} \sigma. \tag{6.20}$$

In this non-dimensionalization the Rayleigh numbers Ra and Ra_m are *negative* and equations (6.17), (6.18) combine to yield a 10th order eigenvalue problem for the eigenvalue σ.

6.2.3 Perturbation boundary conditions

There are two boundary conditions on the bottom of the non-dimensional porous layer, corresponding to no outflow and fixed temperature, and so

$$W_m = \Theta_m = 0, \quad z = -1. \tag{6.21}$$

Zero flow out of the upper surface and implementation of (6.14) lead to

$$W = 0, \quad D\Theta + L\Theta = 0, \quad z = 1. \tag{6.22}$$

Continuity of normal velocity, temperature, and heat flux at the interface lead to

$$W = \hat{d}W_m, \quad \hat{d}\Theta = \epsilon_T^2 \Theta_m, \quad D\Theta = \epsilon_T D_p \Theta_m, \qquad z = 0, \qquad (6.23)$$

where $D_p = d/dz_m$.

(Straughan, 2001c) shows that if the surface tension is a linear function of temperature, $\sigma = \sigma_0 [1 - \gamma(T - T_0)]$, σ_0, γ constants, then the appropriate boundary condition in terms of W and θ is

$$D^2 W = Ma \, \Delta^* \theta, \quad \text{on } z = 1, \qquad (6.24)$$

where Ma is the Marangoni number defined by

$$Ma = \frac{\gamma \sigma_0 (T_U - T_0)d}{\lambda_f \mu}. \qquad (6.25)$$

In the two layer porous - fluid problem studied here, $Ma < 0$.

There are two further conditions to determine on the interface. The first arises by differentiating the Beavers-Joseph conditions (6.2) with respect to x and then with respect to y. One may then find, (Straughan, 2001c)

$$D^2 W - \frac{\alpha \hat{d}}{\delta} DW + \frac{\alpha \hat{d}^3}{\delta} D_p W_m = 0. \qquad (6.26)$$

The last boundary condition we need arises from continuity of normal stress at the interface. Thus, if t_m^i and t_f^i are the stress vectors in the porous and fluid media, we need

$$n_i t_m^i = n_i t_f^i, \qquad \text{on } z = 0.$$

For a Darcy porous medium the stress is effectively a pressure, so $n_i t_m^i = -\pi^m \delta_{i3} n_i$ on $z = 0$, whereas for a Navier-Stokes fluid $n_i t_f^i = -(\pi^f \delta_{i3} - 2\mu d_{i3})n_i$, at $z = 0$. Thus continuity of normal stress yields

$$\pi^m = \pi - 2\mu \frac{\partial w}{\partial z}, \qquad \text{on } z = 0, \qquad (6.27)$$

cf. (Nield, 1977), equation (31). (There is confusion in the literature over this boundary condition. It appears at first sight as though the pressure is discontinuous. However, we interpret the pressure π^m in the porous medium as a pressure averaged over the whole of a representative volume $\tilde{\Omega}$ not just over the pore part occupied by the fluid, Ω_f. The notation of $\tilde{\Omega}$ and Ω_f is as in section 1.6.1. With this interpretation equation (6.27) makes sense.) By differentiation with respect to x^α, $\alpha = 1, 2$ one then shows from this, (Straughan, 2001c),

$$\frac{\hat{d}^4}{\phi Pr_m} \sigma_m D_p W^m + \frac{\hat{d}^4}{\delta^2} D_p W^m$$
$$= \frac{1}{Pr} \sigma \, DW - D^3 W - 3\Delta^* DW, \qquad \text{on } z = 0. \qquad (6.28)$$

An instability analysis for the (Nield, 1977) model taking into account surface tension then reduces to solving the eigenvalue problem comprising of equations (6.17), (6.18) together with the ten boundary conditions, (6.21), (6.22), (6.23), (6.24), (6.26) and (6.28).

Further work on the surface tension driven convection problem in the superposed porous - fluid case may be found in (Shivakumara et al., 2006), while (Rudraiah et al., 2007) also consider the effect of an electric field on the onset of surface tension driven convection in a Brinkman porous - fluid case, see also (Chamkha et al., 2006).

6.2.4   Numerical results

The numerical technique for solving the eigenvalue problem of the last section is discussed in section 9.2.3. (Straughan, 2001c) presents several numerical results. However, in this section we present new numerical findings not given anywhere else.

In the numerical calculations we fix $Pr = 6, G_m = 10, \epsilon_T = 0.7, \phi = 0.3, \alpha = 0.1, \delta = 0.002$ and $L = 10$. Figure 6.2 demonstrates how the critical Rayleigh number varies as the depth ratio $\hat{d}$ is varied. The Marangoni number is fixed as $Ma = -100$. Note that the curve for $\hat{d} = d/d_m = 0.06$ has the minimum of $-Ra_m$ on the left hand branch. This indicates that instability is initiated and dominated by the porous part of the layer. The $-Ra_m$ local minimum on the $\hat{d} = 0.06$ curve occurs much higher than the left hand one. The right hand minimum is associated with the fluid part of the layer. Table 6.1 verifies these findings and the value of $-Ra_m = 21.55$ is clearly much lower than that of $-Ra_m = 95.84$. When $\hat{d} = 0.08$ the porous layer still dominates with $-Ra_m = 19.94$. The right hand branch has a value of $-Ra_m = 32.51$, much higher than the $\hat{d} = 0.06$ curve. However, the porous layer is still dominant in the convection process. As $\hat{d}$ increases from 0.08 to 0.10 the minimum of $-Ra_m$ switches from the left hand part of the curve to a value $-Ra_m = 13.95$ on the right. This indicates that the

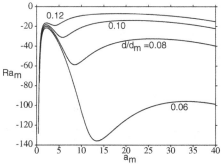

Figure 6.2. Critical porous Rayleigh number versus wavenumber. $Pr = 6$, $G_m = 10$, $\epsilon_T = 0.7$, $\phi = 0.3$, $Ma = -100$, $\alpha = 0.1$, $\delta = 0.002$, $L = 10$, the depth ratio $\hat{d} = d/d_m$ values are shown on the figure

Table 6.1. Minimum $-Ra_m$ values for varying $\hat{d}$. The absolute minimum $-Ra_m$ value for a given $\hat{d}$ is shown in bold

$\hat{d}$	a_m	Ra_m	a_m	Ra_m
0.06	2.2	**−21.55**	33.4	−95.84
0.08	2.2	**−19.94**	26.0	−32.51
0.10	2.2	−18.37	21.2	**−13.95**
0.12	2.2	−16.57	18.0	**−6.97**

deeper fluid layer is now controlling the physics of the onset of convection. When $\hat{d}$ increases to 0.12 this effect is amplified.

In figure 6.3 the chosen values are again $Pr = 6, G_m = 10, \epsilon_T = 0.7, \phi = 0.3, \alpha = 0.1, \delta = 0.002, L = 10$. Now, the depth ratio $\hat{d}$ is fixed at 0.08. The Marangoni number is now varied from 0 to -400. For values of $Ma = 0, -100, -200$, the minimum of $-Ra_m$ is on the left hand branch of the curve, as verified by the values in table 6.2. However, as the surface tension effect is increased and $-Ma$ increases to 300 the minimum of $-Ra_m$ switches to the right hand branch, with a critical value of $-Ra_m = 16.82$. This shows that the surface tension and fluid layer now dominate the instability process. When $Ma = -400$ this effect is increased.

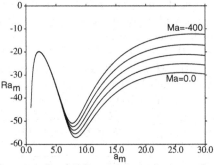

Figure 6.3. Critical porous Rayleigh number vs. wavenumber. $Pr = 6, G_m = 10$, $\epsilon_T = 0.7, \phi = 0.3, \hat{d} = 0.08, \alpha = 0.1, \delta = 0.002, L = 10$, curves for Marangoni numbers $0, -100, -200, -300, -400$

Table 6.2. Minimum $-Ra_m$ values for varying Ma. The absolute minimum $-Ra_m$ value for a given Marangoni number is shown in bold

Ma	a_m	Ra_m	a_m	Ra_m
0.0	2.2	**−19.91**	26.2	−28.92
−100	2.2	**−19.88**	26.4	−25.13
−200	2.2	**−19.85**	26.8	−21.10
−300	2.2	−19.81	27.2	**−16.82**
−400	2.2	−19.78	28.0	**−12.24**

6.3 Porosity effects

6.3.1 Porosity variation

(Straughan, 2002a) studied the way the solution to the fluid - porous layer problem governed by equations (6.4), (6.5), varies as one changes the parameters of the system. In particular, using representative values, the variation of the Beavers-Joseph parameter α was investigated as were the variations in δ, ϵ_T, G_m and values of the porosity of 0.3 and 0.5 were analysed. He also studied how the solution changes when the Beavers-Joseph boundary condition is replaced by the (Jones, 1973) one. The instability problem is like that of the last section, with the upper surface being fixed or free, but in the latter case surface tension effects are not present. In this section we shall analyse the problem studied by (Straughan, 2002a) further and concentrate on how the solution changes as the porosity, ϕ, is varied. The procedure is, however, different from that of (Straughan, 2002a). The point is that we treat δ, ϵ_T and G_m as functions of ϕ. We vary ϕ and then δ, ϵ_T, G_m likewise vary. The critical Rayleigh number is calculated for various values of ϕ.

For clarity, we collect the appropriate equations together now. The equations governing the instability are (6.17) and (6.18), so that

$$(D^2 - a^2)^2 W - a^2 Ra\Theta = \frac{\sigma}{Pr}(D^2 - a^2)W,$$
$$(D^2 - a^2)\Theta - W = \sigma\Theta, \tag{6.29}$$

$$(D^2 - a_m^2)W_m + a_m^2 Ra_m \Theta_m = -\sigma_m \frac{\delta^2}{\phi Pr_m}(D^2 - a_m^2)W_m,$$
$$(D^2 - a_m^2)\Theta_m - W_m = \sigma_m G_m \Theta_m, \tag{6.30}$$

where $z \in (0,1)$ for (6.29) while $z \in (-1,0)$ for (6.30). The boundary conditions are

$$W_m = 0, \quad \Theta_m = 0, \qquad z = -1, \tag{6.31}$$

$$W = 0, \quad \Theta = 0, \quad \text{and either } DW = 0 \text{ or } D^2W = 0, \qquad z = 1, \tag{6.32}$$

$$W = \hat{d}W_m, \quad \hat{d}\Theta = \epsilon_T^2 \Theta_m, \quad D\Theta = \epsilon_T D_p\Theta_m, \qquad z = 0, \tag{6.33}$$

$$D^2W - \alpha\frac{\hat{d}}{\delta}DW + \alpha\frac{\hat{d}^3}{\delta}D_pW_m = 0, \qquad z = 0, \tag{6.34}$$

$$\frac{\hat{d}^4}{\phi Pr_m}\sigma_m D_pW_m + \frac{\hat{d}^4}{\delta^2}D_pW_m$$
$$= \frac{\sigma}{Pr}DW - D^3W - 3\Delta^* DW, \qquad z = 0. \tag{6.35}$$

Conditions (6.32) correspond to prescribed temperature on the upper surface which may be fixed or free, but we are *not* considering the surface tension effect. The following relations are also needed,

$$Ra_m = Ra\frac{(\delta\epsilon_T)^2}{\hat{d}^4}, \quad \sigma_m = \frac{\hat{d}^2}{\epsilon_T}\sigma, \quad a = \hat{d}a_m, \quad Pr_m = \epsilon_T Pr. \quad (6.36)$$

We also need values for ϵ_T, G_m, δ and α. In this section we treat ϵ_T, G_m and δ as functions of the porosity ϕ, a procedure entirely different to that of (Straughan, 2002a). The Beavers - Joseph parameter α is varied in the range found by (Beavers and Joseph, 1967), namely α varies from 0.1 to 4. (Straughan, 2002a) concludes that this parameter has a wide variation and further experiments calculating values for α for different porous media and saturating fluids are certainly needed. We use the facts that $G_m = (\rho_0 c_p)^*/(\rho_0 c_p)_f$, $\epsilon_T = \lambda_f/\lambda_m$, $\lambda_m = k^*/(\rho_0 c_p)^*$, and $\delta = \sqrt{K}/d_m$ to write these as functions of ϕ. The G_m, ϵ_T relations follow from the definition of the starred quantities as in (6.6). To determine δ we note that (Chen, 1990), equation (8), writes that when the porous medium is composed of glass spheres the permeability may be expressed as

$$K = \frac{d_g^2}{172.8}\frac{\phi^3}{(1-\phi)^2}, \quad (6.37)$$

where d_g is the diameter of the spheres forming the porous medium. Thus, we employ a porous layer 3cm thick as in (Chen and Chen, 1988; Chen and Chen, 1992), (Chen, 1990) and take 3mm diameter spheres, again consistent with (Chen and Chen, 1988). (Of course, this fixes the porosity in an experiment, but we here allow it to vary.) This leads to

$$K = 5.21 \times 10^{-4}\frac{\phi^3}{(1-\phi)^2}, \quad (6.38)$$

and to the relation for δ,

$$\delta = \frac{\sqrt{K}}{d_m} = 0.76073 \times 10^{-2}\frac{\phi^{3/2}}{(1-\phi)}. \quad (6.39)$$

Observe that if $\phi = 0.3$, then $\delta = 0.00178$ which is consistent with (Chen and Chen, 1988), (Chen, 1990), (Straughan, 2001c; Straughan, 2002a).

The ϵ_T, G_m equations may be written,

$$G_m = \frac{(\rho_0 c_p)_m}{(\rho_0 c_p)_f} + \left[1 - \frac{(\rho_0 c_p)_m}{(\rho_0 c_p)_f}\right]\phi, \quad (6.40)$$

$$\epsilon_T = \frac{G_m}{k_m/k_f + (1 - k_m/k_f)\phi}. \quad (6.41)$$

In our computations we use values from (Lide, 1991) for ρ_0, c_p, k appropriate to the working fluid being water with the porous medium composed of glass beads. Thus, we use

$$\rho_f = 0.99970 \text{ g cm}^{-3}, \quad c_p^f = 4.1921 \text{ J/g}^\circ\text{K}, \quad k_f = 0.58 \text{ W/m}^\circ\text{K},$$

$$\rho_m = 2.6 \text{ g cm}^{-3}, \quad c_p^m = 1.026 \text{ J/g}^\circ\text{K}, \quad k_m = 1.0886 \text{ W/m}^\circ\text{K}.$$

Employing these values we are led to the relations

$$G_m = 0.6365 - 0.3635\phi,$$
$$\epsilon_T = \frac{0.6365 - 0.3635\phi}{1.8769 + 0.8769\phi}. \tag{6.42}$$

In the computations we prescribe $\hat{d}$ and assume $Pr = 6$ which is a value appropriate for water. The Pr_m value then follows from (6.36). Thus, we vary ϕ and $G_m, \epsilon_T, \delta, Pr_m$ change accordingly.

6.3.2 Numerical results

Throughout this section we assume the upper surface, $z = d$, is fixed. Thus, we employ the boundary condition $DW = 0$ there. Figures 6.4 and 6.5 are computed with $Pr = 6$, the porosity $\phi = 0.3$, $\alpha = 0.1$, and equations (6.39) and (6.42) then yield $G_m = 0.52745$, $\epsilon_T = 0.246475$, $\delta = 1.78572 \times 10^{-3}$. The fluid / porous medium depth ratio $\hat{d}$ is allowed to vary between $\hat{d} = 0.03$ to 0.07. We find a bimodal neutral curve behaviour, as we expect to from the

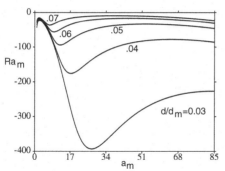

Figure 6.4. Critical porous Rayleigh number versus wavenumber. $Pr = 6$, $\phi = 0.3$, $\alpha = 0.1$, $G_m = 0.52745$, $\epsilon_T = 0.246475$, $\delta = 1.78572 \times 10^{-3}$. The $\hat{d}$ values are shown on the figure

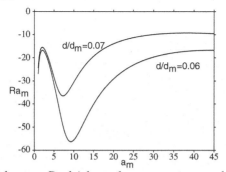

Figure 6.5. Critical porous Rayleigh number versus wavenumber. $Pr = 6$, $\phi = 0.3$, $\alpha = 0.1$, $G_m = 0.52745$, $\epsilon_T = 0.246475$, $\delta = 1.78572 \times 10^{-3}$. (Greater detail, two of the $\hat{d}$ curves)

Table 6.3. Minimum $-Ra_m$ values for varying $\hat{d}$. The absolute minimum $-Ra_m$ value for a given value of $\hat{d}$ is shown in bold. The local maximum value of $-Ra_m$ is shown in slanted type

$\hat{d}$	a_m	Ra_m	a_m	Ra_m	a_m	Ra_m
0.03	2.4	**−22.91**	27.2	*−393.11*	84.0	−227.10
0.04	2.2	**−20.11**	17.4	*−175.70*	64.8	−77.80
0.05	2.2	**−18.22**	12.2	*−93.95*	52.6	−33.47
0.06	2.0	−16.75	9.2	*−56.30*	44.4	**−16.71**
0.07	2.0	−15.49	7.2	*−36.53*	38.4	**−9.27**

original work of (Chen and Chen, 1988). We see from figure 6.4 and table 6.3 that when $\hat{d}$ is in the range $0.03 - 0.05$ the minimum value of $-Ra_m$ occurs with $a_m = 2.4$ or 2.2 and this indicates that convection is initiated in the porous medium. As $\hat{d}$ increases to 0.06 and 0.07 the instability mechanism switches and the absolute minimum of $-Ra_m$ occurs on the right hand part of the neutral curve with $a_m = 44.4$ and 38.4. This indicates that the deeper fluid layer is influencing the instability process more and instability is initiated in the fluid layer. Figure 6.5 shows greater detail for the neutral curves with $\hat{d} = 0.06$ and $\hat{d} = 0.07$.

In figure 6.6 and table 6.4 we show the neutral curves as ϕ is varied with $\hat{d}$ fixed. The parameters are $Pr = 6$, $\alpha = 0.1$, $\hat{d} = 0.06$ and ϕ takes values 0.3 to 0.5. The values of δ, G_m and ϵ_T are calculated from equations (6.39) and (6.42). In fact, for $\phi = 0.3$, $\delta = 1.78572 \times 10^{-3}$, $G_m = 0.52745$, $\epsilon_T = 0.246475$, for $\phi = 0.4$, $\delta = 3.2075 \times 10^{-3}$, $G_m = 0.4911$, $\epsilon_T = 0.220456$, while for $\phi = 0.5$, $\delta = 5.37917 \times 10^{-3}$, $G_m = 0.45475$, $\epsilon_T = 0.196407$. These values are very different from the values displayed in (Straughan, 2002a) and give us yet more insight into the instability process. We observe that as ϕ increases from 0.3 to 0.4 the instability mechanism switches from being driven by the fluid when $\phi = 0.3$ to being initiated by the porous medium when $\phi = 0.4$. When $\phi = 0.5$ the instability is still governed by the porous

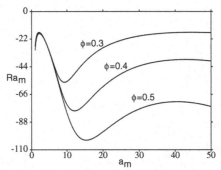

Figure 6.6. Critical porous Rayleigh number versus wavenumber. $Pr = 6$, $\alpha = 0.1$, $\hat{d} = 0.06$. G_m, ϵ_T, δ calculated from (6.39), (6.42). The ϕ values are shown on the figure

Table 6.4. Minimum $-Ra_m$ values for varying ϕ. The absolute minimum $-Ra_m$ value for a given value of ϕ is shown in bold. The local maximum value of $-Ra_m$ is shown in slanted type

ϕ	a_m	Ra_m	a_m	Ra_m	a_m	Ra_m
0.3	2.0	**−16.75**	9.2	*−56.30*	44.4	**−16.71**
0.4	2.2	**−16.92**	12.0	−79.12	42.8	−38.30
0.5	2.2	**−17.64**	15.4	−102.39	40.6	−71.89

medium. Increasing ϕ means greater fluid content in the porous part of the layer. It is worthwhile to observe that as ϕ increases the minimum critical value of $-Ra_m$ increases from 16.71 to 17.64. Hence, even though the instability is switching to being dominated by the porous medium as ϕ increases the total system is more stable. The presence of a greater volume of fluid means the instability occurs at greater Rayleigh numbers.

Figure 6.7 shows the neutral curve behaviour for the Beavers-Joseph parameter, α, varying. The parameters are $Pr = 6$, $\phi = 0.3$, $\hat{d} = 0.06$, $G_m = 0.52745$, $\epsilon_T = 0.246475$, and $\delta = 1.78572 \times 10^{-3}$. In figure 6.7 we show neutral curves for $\alpha = 0.1$ and $\alpha = 0.8$. We additionally computed these curves for $\alpha = 1.5, 2.2, 2.9$ and 3.6. For $\alpha \geq 0.8$ the variation in the neutral curves is little, as may be inferred from table 6.5. For $\alpha = 0.1$ the instability is governed by the fluid layer with $a_m = 44.4$. For $\alpha = 0.8$ and greater the instability is dominated by the porous layer with $a_m = 2.2$. Figure 6.7 and table 6.5 indicate that increasing the Beavers-Joseph parameter α has the effect of making the whole two layer system more stable. However, the effect is small. It would appear that for this range of parameters, changing the Beavers-Joseph coefficient has little effect on the instability process.

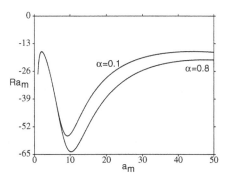

Figure 6.7. Critical porous Rayleigh number versus wavenumber. $Pr = 6$, $\phi = 0.3$, $\hat{d} = 0.06$, $G_m = 0.52745$, $\epsilon_T = 0.246475$, $\delta = 1.78572 \times 10^{-3}$. The α values are shown on the figure

Table 6.5. Minimum $-Ra_m$ values for varying α. The absolute minimum $-Ra_m$ value for a given value of α is shown in bold. The local maximum value of $-Ra_m$ is shown in slanted type

α	a_m	Ra_m	a_m	Ra_m	a_m	Ra_m
0.1	2.0	**−16.75**	9.2	*−56.30*	44.4	**−16.71**
0.8	2.2	**−16.85**	10.2	*−63.70*	47.6	−20.57
1.5	2.2	**−16.86**	10.4	*−65.04*	48.2	−21.36
2.2	2.2	**−16.86**	10.6	*−65.60*	48.4	−21.70
2.9	2.2	**−16.86**	10.6	*−65.91*	48.4	−21.89
3.6	2.2	**−16.87**	10.6	*−66.11*	48.6	−22.01

6.4   Melting ice, global warming

6.4.1   Three layer model

Melting of sea ice in the Arctic or Antarctic is a topic of concern to everyone. (Martin and Kauffman, 1974) studied the formation of under ice melt ponds and their model is based on a convection mechanism. Since thermal convection can result in enhanced heat transfer and such heat transfer can in turn result in enhanced melting of the ice shelves, the subject has been widely studied. Recent analyses are those of (Bogorodskii and Nagurnyi, 2000), (Schmittner et al., 2002), (Carr, 2003a; Carr, 2003b), and many other references may be found there.

(Bogorodskii and Nagurnyi, 2000) indicate that the ice melts on the surface due to radiation heating by the Sun, and below the ice shelf. The siutation is as shown in figure 6.8. Observe that there are meltwater puddles on the ice surface due to direct melting by the heating caused by the Sun's radiation. In addition, it is believed melting also occurs below the ice shelf which results in a layer of water below and this gives rise to a fluid - porous

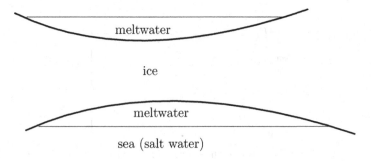

meltwater

ice

meltwater

sea (salt water)

Figure 6.8. Ice shelf melting

medium - fluid convection "sandwich" as studied theoretically by (Nield, 1983).

(Bogorodskii and Nagurnyi, 2000) note that surface meltwater puddles absorb short wave radiation much faster than the ice or snow itself. Such absorption may, therefore, lead to accelerated melting which can cause the ice shelf to break and form drifting ice.

To model the meltwater / ice / meltwater convection process (Bogorodskii and Nagurnyi, 2000) treat the ice as a Darcy porous medium. They assume a symmetric geometry and, therefore, they effectively use the liquid - porous medium - liquid convection model first developed by (Nield, 1983). This model has a three layer structure composed of a layer of sea water of depth d_m overlying a plane layer of ice of depth $2d$, which in turn overlies another layer of sea water which has depth d_m. The ice is regarded as a porous medium of Darcy type. The geometric configuration is as shown in figure 6.9.

Due to the symmetry of the problem (Nield, 1983) and (Bogorodskii and Nagurnyi, 2000) show that it is sufficient to study convection only in the layer $z \in (0, d+d_m)$. Explicit time dependence in the equations is neglected by (Nield, 1983) and by (Bogorodskii and Nagurnyi, 2000) (effectively they are assuming exchange of stabilities so the growth rates σ, σ_m are real). The governing equations are, therefore, in the fluid

$$
u_j \frac{\partial u_i}{\partial x_j} = -\frac{1}{\rho_0} \frac{\partial p}{\partial x_i} + \nu \Delta u_i + \bar{\alpha} g T k_i,
$$
$$
\frac{\partial u_i}{\partial x_i} = 0, \qquad u_j \frac{\partial T}{\partial x_j} = \kappa \Delta T,
$$

(6.43)

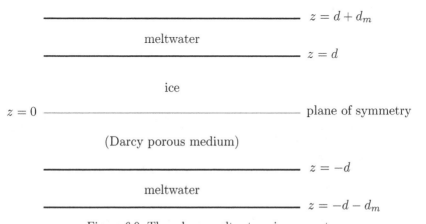

Figure 6.9. Three layer meltwater - ice geometry

whereas in the ice,

$$0 = -\frac{1}{\rho_0}\frac{\partial p_m}{\partial x_i} - \frac{\nu}{K}u_i^m + \bar{\alpha}gT_m k_i,$$

$$\frac{\partial u_i^m}{\partial x_i} = 0, \qquad u_j^m\frac{\partial T_m}{\partial x_j} = \kappa_m\Delta T_m. \tag{6.44}$$

Both (Nield, 1983) and (Bogorodskii and Nagurnyi, 2000) study linear instability of the steady solution to (6.43), (6.44). This solution has $u_i \equiv 0, u_i^m \equiv 0$, with $T(z), T_m(z)$ being linear functions in their respective layers.

The linearized perturbation equations of (Bogorodskii and Nagurnyi, 2000) are transformed to the layers $z \in (h, 1)$ for the fluid, $z \in (0, h)$ for the ice, with $0 < h < 1$. Their perturbation equations in terms of the z-dependent parts $W(z), \Theta(z), W_m(z), \Theta_m(z)$ become

$$D^2W - a^2 Ra\Theta = 0, \qquad D\Theta + W = 0,$$

$$DW_m + a_m^2 Ra\Gamma_1\Theta_m = 0, \qquad D\Theta_m + \Gamma_2^2 W_m = 0,$$

where a, a_m are wavenumbers, Ra is a Rayleigh number, and Γ_1, Γ_2 are positive constants. The boundary conditions are

on $z = 1$, $W = W'' = 0$, $\Theta' = 0$ (or $W' = 0$ instead of $W'' = 0$),

on $z = h$, $W' = 0$, $W = W_m$, $KW''' + W_m' = 0$,

$\Theta = \Theta_m$, $\Gamma_2\Theta_m' = \Theta'$,

on $z = 0$, $W_m' = 0$, $\Theta_m' = 0$.

The last two boundary conditions arise due to geometrical symmetry of the three layer problem.

(Bogorodskii and Nagurnyi, 2000) like (Nield, 1983) develop a long wave asymptotic solution $(a \to 0)$ which is possible with heat flux boundary conditions. Their conclusions are certainly interesting. They study various parameter ranges and conclude that their model favours melting at the lower boundary of sea ice. They also show that sea ice melts from the top, bottom and inside, and importantly the melting is at least twice as intensive as was predicted before their analysis.

6.4.2   Under ice melt ponds

(Martin and Kauffman, 1974) developed and studied a model for the evolution of the meltwater which occurs under the ice, see figure 6.8. (Carr, 2003a; Carr, 2003b) develops a detailed analysis for a similar model. Her model concentrates on the layer of meltwater below the ice and she studies the geometry of figure 6.10.

She studies an idealized situation where the boundary $z = 0$ is fixed, although a thin ice layer may form there and so her geometry may be

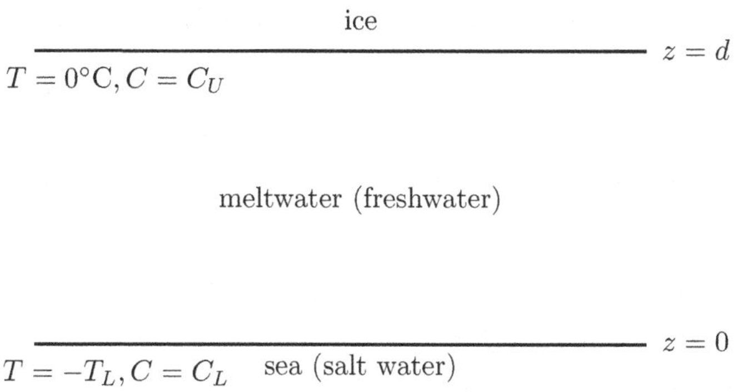

ice

$$T = 0°C, C = C_U$$ ————————————————————— $z = d$

meltwater (freshwater)

$$T = -T_L, C = C_L$$ ————————————————————— $z = 0$

sea (salt water)

Figure 6.10. Under ice meltwater geometry

realistic. The meltwater is relatively less dense than the sea water since it is melted from ice. However, the temperature at the lower boundary $z = 0$ is negative, $-T_L$, and due to the maximum density characteristic of water the warmer water at $z = d$ will have a tendency to fall under gravity and create a convective motion. This is offset by the relatively denser water below. Thus, there is competition between temperature and salt as to whether convective motion (and enhanced ice melting?) ensues. The model adopted by (Carr, 2003a; Carr, 2003b) for the layer $z \in (0, d)$ commences with a density quadratic in temperature T to reflect the maximum density of water, but linear in salt concentration, C, so

$$\rho = \rho_0 \left[1 - \alpha(T - 4)^2 + A(C - \hat{C}) \right].$$

She uses this in the body force in the Navier-Stokes equations (buoyancy term) and her system of equations is

$$\frac{\partial u_i}{\partial t} + u_j \frac{\partial u_i}{\partial x_j} = -\frac{1}{\rho_0} \frac{\partial p}{\partial x_i} + \nu \Delta u_i + g k_i \left[\alpha(T^2 - 8T) - AC \right],$$

$$\frac{\partial u_i}{\partial x_i} = 0,$$

$$\frac{\partial T}{\partial t} + u_i \frac{\partial T}{\partial x_i} = \kappa \Delta T,$$ (6.45)

$$\frac{\partial C}{\partial t} + u_i \frac{\partial C}{\partial x_i} = \kappa_C \Delta C.$$

The boundary conditions adopted are as shown in figure 6.10 with $C_L > C_U$ and these give rise to the steady state

$$\bar{u}_i = 0, \qquad \bar{T} = \frac{T_L}{d} z - T_L, \qquad \bar{C} = C_L - \left(\frac{C_L - C_U}{d} \right) z.$$

(Carr, 2003a; Carr, 2003b) performs a linearized instability analysis and a global nonlinear stability analysis for this basic solution.

Her non-dimensionalized perturbation equations are, in terms of the velocity, temperature, salt and pressure perturbations, u_i, θ, s, π,

$$\frac{\partial u_i}{\partial t} + u_j \frac{\partial u_i}{\partial x_j} = -\frac{\partial \pi}{\partial x_i} + \Delta u_i - 2RM(z)\theta k_i + Pr\theta^2 k_i - R_C s k_i,$$

$$\frac{\partial u_i}{\partial x_i} = 0,$$

$$Pr\left(\frac{\partial \theta}{\partial t} + u_i \frac{\partial \theta}{\partial x_i}\right) = \Delta \theta - Rw,$$

$$Sc\left(\frac{\partial s}{\partial t} + u_i \frac{\partial s}{\partial x_i}\right) = \Delta s + R_C w.$$

In these equations $M(z) = (T_L + 4)/T_L - z, w = u_3, Pr, Sc$ are the Prandtl and Schmidt numbers, and R^2, R_C^2 are the Rayleigh number and salt Rayleigh number, respectively. (Carr, 2003a; Carr, 2003b) provides many numerical results for the linear instability boundary and for global nonlinear energy decay.

The work of (Martin and Kauffman, 1974), (Bogorodskii and Nagurnyi, 2000), (Carr, 2003a; Carr, 2003b) provides a good account of possible mechanisms for convection and enhanced melting of ice shelves. Of course, a full model perhaps combines the (Bogorodskii and Nagurnyi, 2000) and (Carr, 2003a; Carr, 2003b) ones in that one has a three layer system, of convection in the upper meltlayer, with convection in porous ice, with penetrative convection in the meltlayer below the ice. The melting effect at the sea ice - water boundaries may also need to be incorporated.

6.5   Crystal growth

A physical problem which has some resemblance to the 2 layer thermal convection problem of section 6.1.2 is that involving solidification of a binary solution, although the solidification problem is in some ways much more complicated. (Lu and Chen, 1997) is a good article which explains several models appropriate to this problem. As (Lu and Chen, 1997) , (Chung and Chen, 2000b), (Worster, 1992) point out the problem of directional solidification in a binary solution, such as ammonium chloride solution, frequently leads to convection occurring in the "mushy" layer between the solid and liquid regions. This convection typically has "chimneys" of fluid moving in the mushy layer and these can lead to "freckles" forming in the solid. If the solid is a metal used for example in a gas turbine blade then such freckles may be weak points which can lead to fatigue and even fracture. Therefore, an understanding of the convection process during solidification is very important.

(Lu and Chen, 1997) is a lucid article which reviews the history of many models developed to describe mushy layers, from involved models using mixture theories to less complicated ones. In particular, they investigate three ramifications of the two layer model of (Worster, 1992). Thus, (Lu and Chen, 1997) consider a mushy layer which lies above a eutectic solid region and the mushy layer is below the fluid region. The fluid region is infinite in vertical extent, the whole system being infinite in the horizontal directions x and y. The fluid is a binary solution of concentration C_∞, temperature T_∞, and a unidirectional solidification is taking place from below. The mushy layer between the fluid and solid regions is assumed to occupy the region $z \in (0, h)$ with $(x, y) \in \mathbb{R}^2$ and this region moves upward with the interfaces $z = 0, h$ moving with a uniform speed V. In fact, the liquid-mush interface $z = h$ is allowed to deform but in a linearized instability analysis one linearizes about $z = h$, (Worster, 1992).

(Lu and Chen, 1997) transform the equations with respect to the interface speed V. They introduce the solutal and thermal expansion coefficients β^* and α^*, a parameter $\beta = \beta^* - \Gamma \alpha^*$, where Γ is the slope of the liquidus curve, and the eutectic concentration C_E. In this manner the model for convection in the fluid/mushy layer region consists of the Navier-Stokes equations in the fluid layer $\{z \in (h, \infty)\} \times \mathbb{R}^2$ with appropriate porous media equations in the mushy region $\{z \in (0, h)\} \times \mathbb{R}^2$. In the fluid domain $\{z \in (h, \infty)\} \times \mathbb{R}^2$, the equations are

$$\frac{1}{Pr}(\mathcal{D}u_i + u_j u_{i,j}) = \Delta u_i + R_T \theta k_i - R_C \Phi k_i - \frac{\beta}{\beta^*} R_C P_{,i},$$

$$\frac{\partial u_i}{\partial x_i} = 0, \tag{6.46}$$

$$\mathcal{D}\theta + u_i \theta_{,i} = \Delta \theta,$$

$$\mathcal{D}\Phi + u_i \Phi_{,i} = \epsilon \Delta \Phi.$$

Here, $\mathcal{D} = \partial/\partial t - \partial/\partial z$, a derivative introduced by transforming to the moving domain, Pr, ϵ are the Prandtl and Lewis numbers, R_T and R_C are the Rayleigh and solute Rayleigh numbers, $\theta = (T - T_L(C_\infty))/(T_L(C_\infty - T_E))$ is a dimensionless temperature, and $\Phi = (C - C_\infty)/(C_\infty - C_E)$ is the dimensionless concentration. In the mushy layer $\{z \in (0, h)\} \times \mathbb{R}^2$ they adopt Darcy's law, conservation of thermal energy, and conservation of solute, together with the fact that the velocity is solenoidal, to have

$$\frac{u_i}{K(\phi)} = R_m(p_{,i} + \theta k_i),$$

$$\frac{\partial u_i}{\partial x_i} = 0, \tag{6.47}$$

$$\mathcal{D}\theta + u_i \theta_{,i} = \Delta \theta - \mathcal{F}\mathcal{D}\phi,$$

$$\phi \mathcal{D}\Phi + u_i \Phi_{,i} = -(\Phi - \mathcal{C})\mathcal{D}\phi,$$

where ϕ, K are the porosity and permeability, R_m is the porous Rayleigh number, $\mathcal{F}$ is a variable connected to latent heat, $\mathcal{C} = (C_S - C_\infty)/(C_\infty - C_E)$, and u_i represents the (averaged) fluid velocity in the layer. The porosity varies throughout the layer.

Equations (6.46) comprise effectively an eighth order system while (6.47) are essentially fifth order and so solution of this system requires thirteen boundary conditions. These are assumed to be, (Lu and Chen, 1997), far from the interface in the fluid region,

$$\text{as } z \to \infty, \quad \theta \to \theta_\infty, \quad \Phi \to 0, \quad u_i \to 0, \tag{6.48}$$

at the mushy layer - solid interface,

$$\text{on } z = 0, \quad \theta = -1, \quad w = 0, \tag{6.49}$$

w being u_3, and at the interface between the mushy layer and the fluid,

$$\text{on } z = h, \quad \theta = \Phi, \quad n_i\theta_{,i} = n_i\Phi_{,i}, \quad [n_iu_{,i}] = 0, \quad [\theta] = 0,$$
$$[n_i\theta_{,i}] = 0, \quad \phi = 1, \quad [p] = 0, \quad u_\alpha a^{\alpha\beta} x^i_{;\beta} = 0, \tag{6.50}$$

where $[\cdot]$ denotes the jump in a quantity across $z = h$, and the velocity is written $u_i = u_n n_i + u_\alpha a^{\alpha\beta} x^i_{;\beta}$, i.e. resolved into normal and tangential components at the interface.

As (Lu and Chen, 1997) note, the vanishing of the tangential components of velocity at the interface ($z = h$) is used by (Worster, 1992). They also advocate replacing this by an appropriate version of the Beavers-Joseph boundary condition, namely

$$\left.\frac{\partial u^\alpha}{\partial z}\right|_{h+} = \alpha^* \sqrt{\frac{\mathcal{H}}{K(1)}} \left(u^\alpha|_{h+} - u^\alpha|_{h-}\right), \quad \alpha = 1, 2, \tag{6.51}$$

where α^* is a Beavers-Joseph number and $\mathcal{H}$ is a non-dimensional parameter. (Lu and Chen, 1997) refer to equations (6.46), (6.47) together with the boundary conditions (6.48) – (6.50) as model 1, and when the condition $u_\alpha a^{\alpha\beta} x^i_{;\beta} = 0$ of (6.50) is replaced by (6.51) they call this model 2. (Lu and Chen, 1997) also consider a third model in which case the Darcy equation of (6.47) is replaced by a nonlinear Brinkmam law of form

$$\frac{1}{Pr} Du_i + \frac{1}{\phi Pr} u_j u_{i,j} = \Delta u_i - \mathcal{H}\phi \left[\frac{u_i}{K(\phi)} + R_m(p_{,i} + \theta k_i)\right].$$

This equation requires other boundary conditions adapted from continuity across the boundaries and this they refer to as model 3.

The steady state of (Lu and Chen, 1997) is found to be one in which

$$\bar{\theta} = \theta_\infty + (\theta_i - \theta_\infty) e^{-(z-h)}, \quad \bar{\Phi} = \theta_i e^{-(z-h)/\epsilon},$$

where θ_i is the interfacial temperature.

(Lu and Chen, 1997) develop a detailed linear instability analysis of this solution for all three models. The neutral curves are complicated but a bimodal character is still found and convection cells may be in the fluid and in the mushy region.

(Chung and Chen, 2000b) develop a weakly nonlinear analysis of the mushy convection problem and (Chung and Chen, 2000a) study the convection problem when the layer is inclined and rotating. Many other references to work dealing with convection in a solidification system may be found in (Worster, 1992), (Lu and Chen, 1997), (Chung and Chen, 2000b), (Chung and Chen, 2000a). The effects of inertia on instability in an inclined rotating porous layer which may be perceived as a dendrite or mushy layer are investigated by (Riahi, 2007), while (Govender, 2006a) studies a similar problem in a near eutectic approximation with a large Stefan number.

6.6 Heat pipes

In section 5.6 we reviewed work of (Pestov, 1998) where she analysed the stability of a multiphase layer. In this section we briefly analyse the interesting work of (Amili and Yortsos, 2004) on the stability of a two - layer porous region pertaining to a heat pipe. There are many types of heat pipe, some are naturally occurring, some are man made to assist in transferring heat rapidly away from very hot components in machinery. (Amili and Yortsos, 2004) analyse the situation where a two phase (vapour-liquid) region overlies a vapour only region, see figure 6.11.

They treat the two dimensional situation with y vertical, x horizontal, the vapour in the steady state is at rest in the layer $0 < y < H$ with the temperature T a linear function of y, while the temperature in the steady state in the two phase region is constant, namely the saturation temperature T_{sat}. The whole layer depth is y_t, gravity acts downward, and (Amili and Yortsos, 2004) treat the mathematical problem where $y_t \to \infty$,

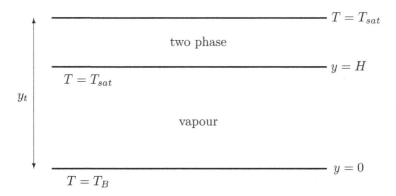

Figure 6.11. Two phase heat pipe

(H fixed). These writers derive the steady state solution for which the temperature, T_0, pressure, p_0, and saturation, s_0, satisfy

$$\frac{dp_0}{dy} = RaT_0, \qquad T_0 = 1 - y, \qquad \text{in } 0 < y < 1,$$

$$\frac{dp_0}{dy} = \frac{1}{\lambda k_{rv}}, \qquad T_0 = 0, \qquad s = s_0, \qquad 1 < y < \infty,$$

(6.52)

where $k_{rv} = 1 - s_\ell$, s_ℓ being the liquid saturation, λ is a constant, Ra is the Rayleigh number, and the layer $(0, H)$ has been non-dimensionalized to $(0, 1)$.

The governing equations presented by (Amili and Yortsos, 2004) are non-dimensionalized and they present them in the non-dimensional regions $0 < y < 1$ (vapour), $1 < y < \infty$ (liquid / vapour), treating the whole region as a porous medium. They derive the following equations in the vapour $\{x \in \mathbb{R}\} \times \{0 < y < 1\}$,

$$v_i^v = -p_{,i} + RaT\delta_{i2},$$

$$v_{i,i}^v = 0,$$

$$\beta_1 \frac{\partial T}{\partial t} + v_i^v T_{,i} = \Delta T,$$

(6.53)

whereas for the two-phase region $\{x \in \mathbb{R}\} \times \{1 < y < \infty\}$, they have

$$\beta_2 \frac{\partial s}{\partial t} + (\bar{\rho}_v v_i^v + v_i^\ell)_{,i} = 0,$$

$$v_i^\ell = -\frac{k_{r\ell}}{\bar{\mu}_v}\left(p_{,i} + \frac{Ra_2}{\bar{\rho}_v}\delta_{i2}\right),$$

$$v_i^v = -k_{rv} p_{,i},$$

$$-\phi \frac{\partial s}{\partial t} + v_{i,i}^v = 0.$$

(6.54)

The coefficients β_1 and β_2 are given by

$$\beta_1 = \frac{\phi \rho_v c_{pv} + (1 - \phi)\rho_r c_{pr}}{\rho_v c_{pv}}, \qquad \beta_2 = \phi(1 - \bar{\rho}_v),$$

v denoting liquid and r rock, the heat pipe being in a rock porous medium. The equations in (6.54) correspond to conservation of mass, vapour and liquid momentum, and conservation of energy. The functions $k_{r\ell}, k_{rv}$ are relative permeabilities and $k_{r\ell} = s_\ell, k_{rv} = 1 - s_\ell$, while v_i^v, v_i^ℓ denote vapour and liquid velocities.

(Amili and Yortsos, 2004) analyse the linearized instability of the base solution (6.52) and introduce the perturbation quantities θ, π, Σ and Δ such that

$$T = T_0 + \epsilon\theta(y)e^{(ikx+\sigma t)}, \qquad p = p_0 + \epsilon\pi(y)e^{(ikx+\sigma t)},$$

$$s = s_0 + \epsilon\Sigma(y)e^{(ikx+\sigma t)}, \qquad \delta = 1 + \epsilon\Delta e^{(ikx+\sigma t)}.$$

Here ϵ is a small parameter, θ, π, Σ are perturbations to the temperature, pressure and saturation, the x-variation is accounted for by the e^{ikx} terms, and δ is a perturbation to the interface $y = 1$. Their perturbation equations are

$$\frac{d^2\pi}{dy^2} - k^2\pi - Ra\frac{d\theta}{dy} = 0,$$

$$\frac{d^2\theta}{dy^2} - \frac{d\pi}{dy} - (k^2 - Ra + \beta_1\sigma)\theta = 0,$$

(6.55)

in the vapour region $0 < y < 1$, and

$$\frac{d^2\pi}{dy^2} - k^2\pi + \frac{1}{k_{r\ell 0}}\left(\frac{dp_0}{dy} + \frac{Ra_2}{\bar{\rho}_v}\right)\frac{d\Sigma}{dy} - \frac{\phi\bar{\mu}_v}{k_{r\ell 0}}\sigma\Sigma = 0,$$

$$\frac{d^2\pi}{dy^2} - k^2\pi - \frac{1}{k_{rv0}}\frac{dp_0}{dy}\frac{d\Sigma}{dy} + \frac{\phi}{k_{rv0}}\sigma\Sigma = 0,$$

(6.56)

in the two phase region $1 < y < \infty$.

This appears to be a fourth order system in $0 < y < 1$ and a third order system in $1 < y < \infty$. However, due to the fact that the interface is allowed to deform Δ occurs in the boundary conditions at the interface. Then, (Amili and Yortsos, 2004) have two boundary conditions at $y = 0$, namely $\theta = 0, \pi_y = 0$, two at $y = \infty$, involving π and Σ, and four at the (linearized) interface $y = 1$. They solve equations (6.55) and (6.56) as a fourth order system in θ, π, θ_y and π_y in $(0, 1)$ and a third order system in π, π_y, Σ in $(1, \infty)$ by an orthonormal shooting method.

(Amili and Yortsos, 2004) present many numerical results. The neutral curves σ vs. k are interesting and show that the basic solution is linearly stable for k small and also for k larger. However, there is an intermediate region, in the wavenumber k, in which instability may arise. This corresponds to convective motion in the layered system in which heat is transferred. The work of (Amili and Yortsos, 2004) represents a very useful contribution to an active area of research with mundane applications.

6.7 Poiseuille flow

6.7.1 Darcy model

Until now we have concentrated on thermal convection problems for a fluid overlying a porous solid. However, there are many mundane situations where one is interested in fluid flow over a saturated porous medium in an isothermal situation. For example, (Allen and Khosravani, 1992), (El-Habel et al., 2002), (Ewing and Weekes, 1998) study flow in an underground channel where the fluid also saturates part of the soil. Understanding such flows is of importance in obtaining drinking water supplies and for avoiding contamination. Such flows are also important in the design of fuel cells

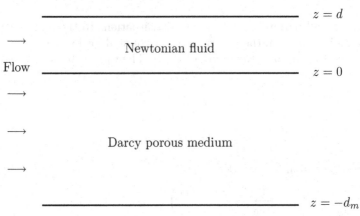

Figure 6.12. Two layer configuration for Poiseuille flow

(Chen, 2003). An emerging area for such flows is also where the porous solid below may freeze, (Basu et al., 2007), cf. section 6.5.

(Chang et al., 2006) commenced a study of hydrodynamic instability problems by considering the Poiseuille flow problem for a Newtonian fluid overlying a porous medium saturated with the same fluid. The configuration for this problem is that of a Newtonian fluid occupying the domain $\mathbb{R}^2 \times \{z \in (0,d)\}$ with the saturated porous medium occupying the spatial domain $\mathbb{R}^2 \times \{z \in (-d_m, 0)\}$, see figure 6.12. A pressure gradient is applied in the $x-$ direction which gives rise to a basic solution corresponding to Poiseuille flow in this scenario. (Chang et al., 2006) studied the linearized instability of this flow.

The equations governing the flow are the Navier-Stokes equations in the domain $\mathbb{R}^2 \times (0,d) \times \{t > 0\}$ with the Darcy equations in $\mathbb{R}^2 \times (-d_m, 0) \times \{t > 0\}$. Thus, we have

$$\frac{\partial u_i}{\partial t} + u_j \frac{\partial u_i}{\partial x_j} = -\frac{1}{\rho}\frac{\partial p}{\partial x_i} + \nu \Delta u_i, \qquad \frac{\partial u_i}{\partial x_i} = 0, \qquad (6.57)$$

in $\mathbb{R}^2 \times \{z \in (0,d)\} \times \{t > 0\}$, and

$$\frac{1}{\Phi}\frac{\partial u_i^m}{\partial t} = -\frac{1}{\rho}\frac{\partial p^m}{\partial x_i} - \frac{\nu}{K}u_i^m, \qquad \frac{\partial u_i^m}{\partial x_i} = 0, \qquad (6.58)$$

in $\mathbb{R}^2 \times \{z \in (-d_m, 0)\} \times \{t > 0\}$. Observe that throughout section 6.7 we use Φ to denote the porosity to avoid confusion with ϕ which is later introduced in connection with the velocity field u_i.

The boundary conditions adopted by (Chang et al., 2006) are no slip at the fixed upper surface and no flow out of the bottom of the porous layer, so that

$$u_i = 0, \quad z = d, \qquad w^m = 0, \quad z = -d_m, \qquad (6.59)$$

together with continuity of normal velocity and pressure at the interface,

$$w = w^m, \quad p^m = p, \qquad z = 0, \tag{6.60}$$

and

$$\frac{\partial u_\gamma}{\partial z} = \frac{\alpha}{\sqrt{K}} (u_\gamma - u_\gamma^m), \quad \gamma = 1, 2, \ z = 0. \tag{6.61}$$

The condition (6.61) is the Beavers-Joseph condition. The other conditions correspond to those of section 6.1, except that in section 6.1 we advocate use of continuity of normal stress instead of continuity of pressure. I believe continuity of normal stress is the correct condition, recollecting that p^m is a pressure averaged over a representative volume $\tilde{\Omega}$ not just over the (pore) fluid volume Ω_f. The notation of $\tilde{\Omega}$ and Ω_f is as in section 1.6.1. However, numerical calculations have revealed that the difference observed in computations between employing continuity of pressure in (6.60) and continuity of normal stress is very small.

If we let dp/dx be a constant pressure gradient then we obtain a basic profile for Poiseuille flow in the two - layer system in which the velocity is not continuous. The steady solution given by (Chang et al., 2006) is

$$\bar{u}(z) = \frac{1}{2} A_1 z^2 + A_2 z + A_3, \quad \bar{v} = \bar{w} = 0, \quad 0 \le z \le d, \tag{6.62}$$

$$\bar{u}^m(z) = -A_1 K, \quad \bar{v}^m = \bar{w}^m = 0, \quad -d_m \le z \le 0, \tag{6.63}$$

where the (constant) coefficients A_1, A_2, A_3 are given by

$$A_1 = \frac{1}{\mu} \frac{dp}{dx},$$

$$A_2 = \alpha A_1 \sqrt{K} - \frac{\alpha A_1 d^2 + 2\alpha^2 A_1 d\sqrt{K}}{2(\alpha d + \sqrt{K})},$$

$$A_3 = -\frac{A_1 d^2 \sqrt{K} + 2\alpha A_1 K d}{2(\alpha d + \sqrt{K})}.$$

6.7.2 Linearized perturbation equations

The linearised non-dimensional perturbation equations given by (Chang et al., 2006) are

$$Re\left(\frac{\partial u_i}{\partial t} + u_j \frac{\partial \bar{u}_i}{\partial x_j} + \bar{u}_j \frac{\partial u_i}{\partial x_j}\right) = -\frac{\partial p}{\partial x_i} + \Delta u_i,$$

$$\frac{\partial u_i}{\partial x_i} = 0, \tag{6.64}$$

for $(x, y) \in \mathbb{R}^2$, $z \in (0, 1)$, $t > 0$, together with

$$\frac{Re^m}{\Phi} \frac{\partial u_i^m}{\partial t^m} = -\frac{1}{\delta^2} u_i^m - \frac{\partial p^m}{\partial x_i^m},$$

$$\frac{\partial u_i^m}{\partial x_i^m} = 0,$$

(6.65)

where $(x^m, y^m) \in \mathbb{R}^2$, $z^m \in (-1, 0)$, $t > 0$. In these equations Re and Re^m are the Reynolds number and the porous Reynolds number, respectively. These numbers are linearly related as in equation (2.13) of (Chang et al., 2006).

The system (6.64), (6.65) is reduced by introducing a normal mode form

$$u_i = u_i(z) \, e^{i(ax+by-act)}, \quad p = \pi(z) \, e^{i(ax+by-act)},$$

$$u_i^m = u_i^m(z_m) \, e^{i(a_m x_m + b_m y_m - a_m c_m t_m)},$$

$$p^m = \pi^m(z_m) \, e^{i(a_m x_m + b_m y_m - a_m c_m t_m)}.$$

A Squire's theorem is invoked to remove the dependence in the y−direction and a two dimensional system is obtained. Then, stream functions and eigenfunctions $\psi, \psi_m, \phi, \phi_m$ are introduced by

$$u = \frac{\partial \psi}{\partial z}, \quad w = -\frac{\partial \psi}{\partial x}, \quad u^m = \frac{\partial \psi^m}{\partial z^m}, \quad w^m = -\frac{\partial \psi^m}{\partial x^m},$$

where ψ and ψ^m may be represented by

$$\psi = \phi(z) \, e^{ia(x-ct)}, \quad \psi^m = \phi^m(z_m) \, e^{ia_m(x_m - c_m t_m)}.$$

In this way, equations (6.64), (6.65) are reduced to investigating the eigenvalue equations

$$(D^2 - a^2)^2 \phi = ia \, Re(U - c)(D^2 - a^2)\phi - ia \, ReU''\phi, \quad z \in (0, 1),$$

$$\left(\frac{1}{\delta^2} - Re^m \frac{ia_m c_m}{\Phi}\right)(D_p^2 - a_m^2)^2 \phi_m = 0, \quad z_m \in (-1, 0),$$

(6.66)

where $D = d/dz$, $D_p = d/dz_m$ and $U(z)$ is $\bar{u}(z)$ rewritten with respect to a velocity scaling, $U(z)$ being given explicitly by equation (2.8) of (Chang et al., 2006).

The coefficients a, a_m and Re, Re^m are connected and system (6.66) is to be solved subject to the boundary conditions,

$$\phi = D\phi = 0, \quad \text{on } z = 1, \quad \phi^m = 0, \quad \text{on } z_m = -1,$$

together with the interface conditions (derivable from equations (6.59) – (6.61))

$$Re\phi = Re^m \phi^m,$$

$$D^2\phi - \frac{\alpha \hat{d}}{\delta} D\phi + \frac{\alpha \hat{d}^2 Re^m}{\delta Re} D_p \phi^m = 0,$$

$$\left(Re^m \frac{ia_m c_m}{\Phi} - \frac{1}{\delta^2} \right) D_p \phi^m$$

$$= \frac{Re}{\hat{d}^3 Re^m} \left[(D^2 - a^2)D\phi - iaRe(U - c)D\phi + iaReU'\phi \right],$$

on the interface $z = z_m = 0$.

To solve equations (6.66) numerically subject to the above boundary conditions it is convenient to multiply through by δ^2 where δ^2 occurs in the denominator (δ^2 is small) and write equation (6.66), as two second order equations of form

$$\begin{aligned}
(D^2 - a^2)\phi - \xi &= 0, \\
(D^2 - a^2)\xi &= iaRe(U - c)\xi - iaReU''\phi.
\end{aligned} \tag{6.67}$$

Equations (6.67) are rewritten in the Chebyshev domain $(-1, 1)$, as are equations (6.66)$_2$ and they are solved using a D^2 Chebyshev tau numerical method (cf. (Dongarra et al., 1996), see also chapter 9) subject to the above boundary condtions.

6.7.3 (Chang et al., 2006) results

(Chang et al., 2006) performed a variety of numerical calculations. In particular, for $\delta = 10^{-3}, \Phi = 0.3, \alpha = 0, 1$ they varied $\hat{d}$ and found an interesting new effect. For $\hat{d} = 0.11$ they found instability governed by the porous layer whereas when $\hat{d} = 0.12$ the fluid layer was dominant. Flow reversal was evident in the top of the porous layer when $\hat{d} = 0.11$ and this is attributed to shear effects near the interface. However, as $\hat{d}$ was increased beyond $\hat{d} = 0.12$ a new instability mode was seen. Thus, they found tri-modal instability. For $\hat{d} = 0.121$ the third mode is seen but the fluid mode still dominates. However, when $\hat{d} = 0.13$ the new (fluid) mode dominates. For $\hat{d} = 0.2, 0.3$ the left hand mode disappears and the new mode continues to dominate the instability. (Chang et al., 2006) interpret the new mode as a even-shear-mode of the Poiseuille flow, based on the eigenfunction behaviour. The dominant eigenfunction in the fluid mode case behaves like an even mode confined to the fluid layer.

(Chang et al., 2006) also investigated the effects of varying the Beavers-Joseph constant α and the Darcy number δ. The third mode was again encountered during variation of these parameters.

6.7.4   Brinkman - Darcy model

To overcome the problem of the discontinuity in the velocity field encountered in the fluid-Darcy layer study just described, (Hill and Straughan, 2008) argue that there is a transition layer between the fluid and the Darcy porous medium. In fact, this idea was proposed much earlier by (Nield, 1983), p. 45. Nield suggests using a Brinkman - type equation in the boundary layer region between the fluid and the Darcy porous medium. (Goharzadeh et al., 2005) also advocate this approach and analyse the problem experimentally. Their findings are very important and they deduce that the thickness of the transition zone is of the same order as the grain size of the material forming the porous medium.

(Hill and Straughan, 2008) develop an instability analysis for the Poiseuille flow problem for a fluid overlying a porous medium, but they adopt a three layer configuration, as shown in figure 6.13. The Newtonian fluid saturates the porous layer below.

The equations of (Hill and Straughan, 2008) are (6.57), (6.58) in the layers contained between $z \in (0, d)$ and $z \in (-d_m, -\beta d_m)$, respectively. However, for completeness we collect the full set of equations below. In the layer contained in $z \in (0, d)$, we have

$$\frac{\partial u_i}{\partial t} + u_j \frac{\partial u_i}{\partial x_j} = -\frac{1}{\rho} \frac{\partial p}{\partial x_i} + \nu \Delta u_i, \qquad \frac{\partial u_i}{\partial x_i} = 0, \qquad (6.68)$$

in the layer $(-\beta d_m, 0)$ there hold

$$\frac{1}{\Phi_b} \frac{\partial u_i^b}{\partial t} = -\frac{1}{\rho} \frac{\partial p^b}{\partial x_i} + \nu_e \Delta u_i^b - \frac{\nu}{K} u_i^b, \qquad \frac{\partial u_i^b}{\partial x_i} = 0, \qquad (6.69)$$

while in the layer $(-d_m, -\beta d_m)$

$$\frac{1}{\Phi} \frac{\partial u_i^m}{\partial t} = -\frac{1}{\rho} \frac{\partial p^m}{\partial x_i} - \frac{\nu}{K} u_i^m, \qquad \frac{\partial u_i^m}{\partial x_i} = 0. \qquad (6.70)$$

In these equations u_i^b, Φ_b, p^b denote the velocity, porosity and pressure in the Brinkman layer, while an m denotes the Darcy layer. The coefficient $\nu_e = \mu_e/\rho$, where μ_e is an effective dynamic viscosity, and $\nu = \mu/\rho$. The porosity Φ_b in the Brinkman layer should drop from 1 in the fluid to Φ in the Brinkman layer. (Hill and Straughan, 2008) adopt the value $\Phi_b = (\Phi+1)/2$. We believe the Φ value in the Brinkman layer will not seriously affect the instability results. If we imagine flow past a rigid porous material like a breeze block then we should adopt $\Phi_b = \Phi$. There are situations where one may wish to have Φ vary continuously throughout the transition layer from the value of 1 in the fluid to Φ in the Darcy layer. Such a situation may arise in flow over a loose porous material such as sand. However, one may then have to account for resuspension effects where one has flow of a superposed fluid over a suspension of solid particles which in turn overlies a porous medium. We do not examine this situation here. The instability

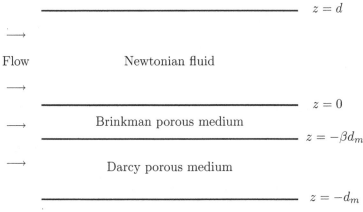

Figure 6.13. Three layer configuration for Poiseuille flow

of the resuspension problem when there is no porous layer present below is considered by (Schaflinger, 1994) and by (Schaflinger et al., 1995), with other references to this particular problem given there.

We need appropriate boundary and interface conditions. Those chosen by (Hill and Straughan, 2008) are

$$u_i = 0 \text{ on } z = d, \qquad w \equiv u_3 = 0 \text{ on } z = -d_m. \tag{6.71}$$

On the interface $z = 0$, they assume continuity of normal and tangential stress, so that

$$-p + 2\mu \frac{\partial w}{\partial z} = -p^b + 2\mu_e \frac{\partial w^b}{\partial z}, \quad z = 0,$$

$$\mu\left(\frac{\partial u_\zeta}{\partial z} + \frac{\partial w}{\partial x_\zeta}\right) = \mu_e\left(\frac{\partial u_\zeta^b}{\partial z} + \frac{\partial w^b}{\partial x_\zeta}\right), \quad z = 0, \quad \zeta = 1, 2. \tag{6.72}$$

On the Brinkman-Darcy interface $z = -\beta d_m$ we have what is essentially the condition for continuity of normal stress

$$-p^m = -p^b + 2\mu_e \frac{\partial w^b}{\partial z}, \quad z = -\beta d_m, \tag{6.73}$$

and a (Jones, 1973) condition

$$\frac{\partial u_\zeta^b}{\partial z} + \frac{\partial w^b}{\partial x_\zeta} = \frac{\alpha}{\sqrt{K}}(u_\zeta^b - u_\zeta^m), \quad z = -\beta d_m, \quad \zeta = 1, 2. \tag{6.74}$$

One could equally well employ a Beavers-Joseph condition instead of equation (6.74).

6.7.5   Steady solution

For a given constant pressure gradient dp/dx, the steady solution of (Hill and Straughan, 2008) is of form $\mathbf{u} = (\bar{u}, 0, 0)$, $\mathbf{u}^b = (\bar{u}^b, 0, 0)$,

$\mathbf{u}^m = (\bar{u}^m, 0, 0)$, where

$$\bar{u}(z) = \frac{c_1}{2}z^2 + c_2 z + c_3,$$

$$\bar{u}^b(z) = c_4 \exp\left(\frac{fz}{\sqrt{K}}\right) + c_5 \exp\left(-\frac{fz}{\sqrt{K}}\right) - Kc_1,$$

$$\bar{u}^m(z) = -Kc_1.$$

In these expressions $f = \sqrt{\mu/\mu_e}$, $c_1 = (1/\mu)dp/dx$,

$$c_2 = A\left\{c_1\left(K - \frac{d^2}{2}\right)\left[(f+\alpha)\exp(\frac{2\beta f d_m}{\sqrt{K}}) - (f-\alpha)\right]\right\},$$

$$c_3 = -\frac{A}{2}\sqrt{K}\,c_1 d\left[(f+\alpha)\exp(\frac{2\beta f d_m}{\sqrt{K}})(fd + 2\sqrt{K}) + (f-\alpha)(fd - 2\sqrt{K})\right],$$

$$c_4 = Afc_1\sqrt{K}\left(K - \frac{d^2}{2}\right)(f+\alpha)\exp(\frac{2\beta f d_m}{\sqrt{K}}),$$

$$c_5 = Afc_1\sqrt{K}\left(K - \frac{d^2}{2}\right)(f-\alpha),$$

$$A = \frac{1}{(f+\alpha)(f\sqrt{K}+d)\,\exp[(2\beta f d_m)/\sqrt{K}] + (f\sqrt{K}-d)(f-\alpha)}.$$

It is evident that u_i is continuous across the interface $z = 0$. However, u_i is still discontinuous across the (fictitious) layer $z = -\beta d_m$. Nevertheless, the basic flow profiles appear much smoother than those of the fluid/Darcy problem, as is seen in figure 2 of (Hill and Straughan, 2008). The discontinuity at the interface $z = -\beta d_m$ is of a lower order and has less effect on the solution than the discontinuity in the fluid/Darcy layer problem. Nonetheless, the error introduced by the discontinuity at $z = -\beta d_m$ is investigated in detail by (Hill and Straughan, 2008).

6.7.6 Linearized perturbation equations

(Hill and Straughan, 2008) non-dimensionalize $\bar{u}, \bar{u}^b, \bar{u}^m$ to have non-dimensional functions $U(z), U_b(z), U_m$. The linearized perturbation equations derived by (Hill and Straughan, 2008) are

$$Re\left(\frac{\partial u_i}{\partial t} + U\frac{\partial u_i}{\partial x} + \delta_{i3}U'w\right) = -\frac{\partial p}{\partial x_i} + \Delta u_i, \qquad \frac{\partial u_i}{\partial x_i} = 0,$$

in $\mathbb{R}^2 \times (0,1) \times \{t > 0\}$,

$$\frac{Re^b}{\Phi_b}\frac{\partial u_i^b}{\partial t} = -\frac{\partial p^b}{\partial x_i} + \frac{1}{f^2}\Delta u_i^b - \frac{1}{\delta^2}u_i^b, \qquad \frac{\partial u_i^b}{\partial x_i} = 0,$$

in $\mathbb{R}^2 \times (-\beta, 0) \times \{t > 0\}$,

$$\frac{Re^m}{\Phi}\frac{\partial u_i^m}{\partial t} = -\frac{\partial p^m}{\partial x_i} - \frac{1}{\delta^2}u_i^m, \qquad \frac{\partial u_i^m}{\partial x_i} = 0,$$

in $\mathbb{R}^2 \times (-1, -\beta) \times \{t > 0\}$, where Re, Re^b, Re^m are Reynolds numbers appropriate to the fluid, Brinkman and Darcy layers. Normal modes are used on the above equations and a version of Squire's theorem is employed to reduce to a two-dimensional system in the x, z directions. In terms of streamfunctions ψ, ψ^b, ψ^m and their associated eigenfunctions ϕ, ϕ^b, ϕ^m the following equations are derived

$$(D^2 - a^2)^2 \phi = Re(U - c)ia(D^2 - a^2)\phi - iaReU''\phi, \quad 0 < z < 1,$$

$$\left(1 - \frac{ia_b c_b Re^b \delta^2}{\Phi_b} - \frac{\delta^2}{f^2}(D_b^2 - a_b^2)\right)(D_b^2 - a_b^2)\phi^b = 0, \quad -\beta < z < 0, \quad (6.75)$$

$$\left(1 - \frac{ia_m c_m Re^m \delta^2}{\Phi}\right)(D_m^2 - a_m^2)\phi^m = 0, \quad -1 < z < -\beta.$$

System (6.75) corresponds to an Orr-Sommerfeld system for the fluid-Brinkman-Darcy problem. This system is 10th order and is solved numerically by (Hill and Straughan, 2008) by using a Chebyshev tau D^2 method, writing $(6.75)_1$ and $(6.75)_2$ as two second order equations. The boundary conditions employed are deduced from equations (6.71) and are

$$\phi = D\phi = 0 \text{ on } z = 1, \qquad \phi^m = 0 \text{ on } z = -1.$$

The interface conditions arise from equations (6.72) – (6.74) and are

$$Re\,\phi = Re^b\,\phi^b,$$

$$ReD\phi = \hat{d}Re^b D_b\phi^b,$$

$$f^2(D^2 + a^2)\phi = \hat{d}^2 \frac{Re^b}{Re}(D_b^2 + a_b^2)\phi^b,$$

$$Re\left[-iaRe(U - c)D\phi + (D^2 - 3a^2)D\phi + U'iaRe\,\phi\right]$$
$$= Re^b \hat{d}^3 \left(\frac{1}{f^2}(D_b^2 - 3a_b^2) + \frac{ia_b c_b Re^b}{\Phi_b} - \frac{1}{\delta^2}\right)\phi^b,$$

on the interface $z = 0$, and

$$Re^b \phi^b = Re^m \phi^m,$$

$$(D_b^2 + a_b^2)\phi^b = \frac{\alpha}{\delta}D_b\phi^b - \frac{\alpha Re^m}{\delta Re^b}D_m\phi^m,$$

$$Re^b \left\{\frac{ia_b c_b Re^b}{\Phi_b} - \frac{1}{\delta^2} + \frac{1}{f^2}(D_b^2 - 3a_b^2)\right\}D_b\phi^b$$
$$= Re^m \left(\frac{ia_m c_m Re^m}{\Phi} - \frac{1}{\delta^2}\right)D_m\phi^m$$

on the interface $z = -\beta$.

6.7.7   Numerical results

The numerical findings of (Hill and Straughan, 2008) are interesting. In particular, they do not observe the third mode which was seen by (Chang et al., 2006), see section 6.7.3. Values employed by (Hill and Straughan, 2008) are $\delta = 5 \times 10^{-3}$, $\alpha = 0.1$, $\beta = 0.1$, $f = 0.8$, $\Phi = 0.3$, and these are consistent with the experimental configuration of (Chen and Chen, 1988) who used 3mm glass beads and a 3cm deep layer. The value of $\beta = 0.1$ is consistent with the findings of (Goharzadeh et al., 2005), i.e. in accordance with a transition layer of depth of order of the grain size of the material comprising the porous layer. Unlike (Chang et al., 2006), (Hill and Straughan, 2008) find the transition layer has a large effect on where the instability switches from being dominated by the porous layer to being dominated by the fluid layer. Indeed, (Hill and Straughan, 2008) find the changeover is for $\hat{d}$ in the range (0.0319, 0.0328). The neutral curves presented by (Hill and Straughan, 2008) show that the mechanism of instability is also very different from that of (Chang et al., 2006). When $\hat{d}$ is changed from 0.031 to 0.0314 there is an isolated "instability island" which appears to the right of the infinite neutral curve. As $\hat{d}$ is increased the "instability island" grows in size and its minimum decreases below that on the unbounded (porous dominance) curve. For $\hat{d} = 0.032$ the instability is already governed by the fluid layer. As $\hat{d}$ is increased further the instability island rejoins the infinite curve but still remains such that its minimum is an absolute minimum. Eigenfunction profiles which confirm this are presented in (Hill and Straughan, 2008). Other parameter variations are studied by (Hill and Straughan, 2008) and, in particular, they present a variation in the Brinkman/Darcy interface parameter β. For very small $\beta(< 10^{-3})$ the porous mode dominates (with $\hat{d} = 0.13, \Phi = 0.3, \delta = 5 \times 10^{-3}$, $\alpha = 0.1, f = 0.8$). Thus, the depth of the transition layer is crucial to the instability analysis. We conclude by remarking that further experimental results on the Poiseuille flow problem for fluid flowing over a porous layer would be helpful. In this way we may be able to determine what is an appropriate depth of the transition layer, this depth may well depend on the applied pressure gradient, or equivalently, on the maximum velocity in the fluid layer.

6.7.8   Forchheimer - Darcy model

Another approach to viewing the Poiseuille flow problem for a Newtonian fluid overlying a Darcy porous medium may again be based on a three layer configuration with a transition layer, but where the intermediate layer is now composed of a porous medium governed by the equations for a Forchheimer flow. In this section we present such a theory. Thus, the situation is as in section 6.7.4 but the Brinkman layer there is replaced by one of Forchheimer type. Since the Forchheimer law is believed to be more suitable than the Darcy one when the velocity is high, such a three layer situation may be appropriate when we are dealing with an instability problem, i.e. the transition between the relatively slow flow in the Darcy porous medium to the potentially unstable flow in the fluid might conceivably be well modelled by Forchheimer flow in a transition layer.

Thus, we consider Poiseuille flow with a Newtonian fluid occupying the layer $\mathbb{R}^2 \times \{z \in (0, d)\}$ while the porous medium occupies the layer $\mathbb{R}^2 \times \{z \in (-d_m, 0)\}$. The porous medium is divided into two types (from a mathematical point of view), a Darcy layer occupying the region $\mathbb{R}^2 \times \{z \in (-d_m, -\beta d_m)\}$ with the domain $\mathbb{R}^2 \times \{z \in (-\beta d_m, 0)\}$ containing a porous medium of Forchheimer type. The geometry is as shown in figure 6.14.

The equations governing the motion of the fluid in the domain $\mathbb{R}^2 \times (-d_m, d)$ are the Navier-Stokes equations in $\{z \in (0, d)\}$,

$$\frac{\partial u_i}{\partial t} + u_j \frac{\partial u_i}{\partial x_j} = -\frac{1}{\rho} \frac{\partial p}{\partial x_i} + \nu \Delta u_i, \qquad \frac{\partial u_i}{\partial x_i} = 0, \qquad (6.76)$$

with equations of Forchheimer type in the layer $(-\beta d_m, 0)$, i.e.

$$\frac{1}{\Phi_F} \frac{\partial u_i^F}{\partial t} + \frac{B}{K} |\mathbf{u}^F| u_i^F = -\frac{1}{\rho} \frac{\partial p^F}{\partial x_i} - \frac{\nu}{K} u_i^F, \qquad \frac{\partial u_i^F}{\partial x_i} = 0, \qquad (6.77)$$

and finally, the Darcy equations occupy the layer $z \in (-d_m, -\beta d_m)$, i.e.

$$\frac{1}{\Phi} \frac{\partial u_i^m}{\partial t} = -\frac{1}{\rho} \frac{\partial p^m}{\partial x_i} - \frac{\nu}{K} u_i^m, \qquad \frac{\partial u_i^m}{\partial x_i} = 0. \qquad (6.78)$$

The quantities u_i, p denote the velocity and pressure, with F and m subscripts or superscripts denoting the analogous quantities in the Forchheimer and Darcy regions, respectively. The variable K is permeability with Φ being the porosity. We here assume $\Phi = \Phi_F$, i.e. we are dealing with such as flow past a breeze block where the porous body is rigid and the porosity

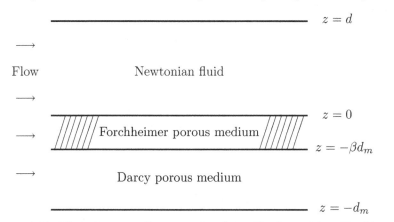

Figure 6.14. Three layer configuration for Poiseuille flow with a Forchheimer transition layer

jumps from Φ in the porous medium to 1 in the fluid. The variable B is a Forchheimer coefficient with the dimension of length.

The boundary conditions we adopt are no slip on the (fixed) fluid surface and no flow out of the bottom of the porous layer, so that

$$u_i = 0 \quad \text{on} \quad z = d, \qquad w \equiv u_3 = 0 \quad \text{on} \quad z = -d_m. \qquad (6.79)$$

The interface conditions adopted are those which correspond to continuity of normal velocity over the fluid - Forchheimer interface, $z = 0$, together with continuity of normal stress and a Beavers-Joseph condition there. Hence, on $z = 0$, we assume

$$
\begin{aligned}
w &= w^F, \qquad p - 2\mu\frac{\partial w}{\partial z} = p^F, \\
\frac{\partial u_\zeta}{\partial z} &= \frac{\alpha}{\sqrt{K}}\,(u_\zeta - u_\zeta^F), \quad \zeta = 1, 2.
\end{aligned}
\qquad (6.80)
$$

The parameter α is a Beavers-Joseph parameter. The continuity of normal stress arises since from equations (6.76) and (6.77) we see the stresses may be given by

$$t_{ij} = -p\delta_{ij} + \mu\left(\frac{\partial u_i}{\partial x_j} + \frac{\partial u_j}{\partial x_i}\right) \qquad \text{and} \qquad t_{ij}^F = -p^F \delta_{ij}.$$

The normal stress is $t_{ij}n_jn_i$ (i.e. the normal component of the stress vector $t_i = n_j t_{ij}$) and we require continuity of this over the interface $z = 0$, i.e.

$$-p\delta_{ij}n_jn_i + \mu\left(\frac{\partial u_i}{\partial x_j} + \frac{\partial u_j}{\partial x_i}\right)n_jn_i = -p^F\delta_{ij}n_jn_i.$$

Since $\mathbf{n} = (0, 0, 1)$, we derive $(6.80)_2$, recalling $u_3 = w$. Next, on the Darcy-Forchheimer interface, $z = -\beta d_m$, we assume continuity of the normal component of velocity and continuity of pressure, so

$$w^m = w^F \qquad \text{and} \qquad p^m = p^F, \quad \text{on } z = -\beta d_m. \qquad (6.81)$$

We seek a steady solution of form

$$
\begin{aligned}
\bar{\mathbf{u}}(z) &= (\bar{u}(z), 0, 0), & z &\in (0, d); \\
\bar{\mathbf{u}}^F(z) &= (\bar{u}^F(z), 0, 0), & z &\in (-\beta d_m, 0); \\
\bar{\mathbf{u}}^m(z) &= (\bar{u}^m(z), 0, 0), & z &\in (-d_m, -\beta d_m).
\end{aligned}
$$

By substituting these expressions into equations (6.76) – (6.78) and using the boundary conditions (6.79) – (6.81) we find

$$
\begin{aligned}
\bar{u}^m &= \frac{KA}{\nu}, \qquad \bar{u}^F = -\frac{\nu}{2B} + \frac{\sqrt{\nu^2 + 4ABK}}{2B}, \\
\bar{u}(z) &= -\gamma_1(z^2 - d^2) + \alpha_1(z - d),
\end{aligned}
\qquad (6.82)
$$

where

$$A = -\frac{1}{\rho}\frac{\partial p}{\partial x}\ \text{(constant)} > 0, \qquad \gamma_1 = \frac{A}{2\nu},$$

$$\alpha_1 = \alpha\left[\frac{Ad^2}{2\nu(\sqrt{K} + d\alpha)} + \frac{\nu}{2B(\sqrt{K} + d\alpha)} - \frac{\sqrt{\nu^2 + 4ABK}}{2B(\sqrt{K} + d\alpha)}\right].$$

(For small B,

$$\bar{u}^F \approx \frac{AK}{\nu} - \frac{A^2B^2K^2}{\nu^3}.)$$

One finds the maximum of $\bar{u}(z)$ occurs where $z = \alpha_1\nu/A$ and then denoting the maximum value of $\bar{u}(z)$ by V we find

$$V = -\gamma_1\left(\frac{\alpha_1^2\nu^2}{A^2} - d^2\right) + \alpha_1\left(\frac{\alpha_1\nu}{A} - d\right).$$

Non-dimensional forms of $\bar{u}, \bar{u}^F$ and $\bar{u}^m$ are defined by scaling $(0, d)$ to $(0, 1)$, i.e. $z = z^*d$, by scaling $(-d_m, 0)$ to $(-1, 0)$ by $z_m = z_m^*d_m$, and by setting

$$U(z^*) = \frac{\bar{u}(z^*)}{V}, \qquad U^F = \frac{\bar{u}^F}{V}, \qquad U^m = \frac{\bar{u}^m}{V}.$$

The steady state profiles for $\hat{d} = 0.1$ are sketched in figure 6.15. To sketch these curves we rewrite U, U^F and U^m in terms of the non-dimensional variables

$$\hat{d} = \frac{d}{d_m}, \qquad \delta = \frac{\sqrt{K}}{d_m}, \qquad \alpha, \qquad \text{and} \qquad B_N = \frac{AB}{\nu^2}d_m^2. \qquad (6.83)$$

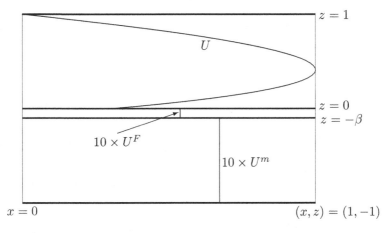

Figure 6.15. Steady solution for Poiseuille flow with a Forchheimer transition layer, $\hat{d} = 0.1$, U^m, U^F are magnified by a factor of 10

In this way we need only prescribe $\hat{d}, \delta, \alpha$ and B_N and then U, U^F, U^m follow. This allows us to compare the steady solution profile with the two-layer one of (Chang et al., 2006), their figure 1, p. 291. The number B_N is a non-dimensional measure of the size of the Forchheimer effect. While we expect this to be very small we have incorporated its effect. In fact, one may show

$$U(z) = \frac{\hat{d}^2(1 - z^2) - 2\hat{\alpha}_1\hat{d}(1 - z)}{(\hat{d} - \hat{\alpha}_1)^2}, \qquad z \in (0, 1),$$

$$U^F = \frac{-B_N^{-1} + B_N^{-1}\sqrt{1 + 4\delta^2 B_N}}{(\hat{d} - \hat{\alpha}_1)^2}, \qquad z \in (-\beta, 0), \qquad (6.84)$$

$$U^m = \frac{2\delta^2}{(\hat{d} - \hat{\alpha}_1)^2}, \qquad z \in (-1, -\beta),$$

where

$$\hat{\alpha}_1 = \frac{\alpha}{2(\delta + \hat{d}\alpha)}\left[\hat{d}^2 + \frac{1}{B_N} - \frac{1}{B_N}\sqrt{1 + 4B_N\delta^2}\right].$$

(Note the asymptotic forms, for $B_N << 1$,

$$\hat{\alpha}_1 \approx \frac{\alpha}{2(\delta + \hat{d}\alpha)}(\hat{d}^2 - 2\delta^2 + 4B_N\delta^2),$$

$$U^F \approx \frac{2\delta^2 - 4B_N\delta^2}{(\hat{d} - \hat{\alpha}_1)^2}.)$$

In figure 6.15 we select $\delta = 10^{-3}, \alpha = 0.1$, and $\hat{d} = 0.1$ which are values of (Chang et al., 2006). We take $\beta = 0.1$ and sketch the steady solution for $B_N = 0.1$. (In fact, the variation in the Forchheimer effect is very small for $B_N < 0.5$.)

To study the instability of the basic solution $\bar{u}, \bar{u}^F$ and $\bar{u}^m$ we introduce perturbations u_i, u_i^F and u_i^m by

$$u_i \rightarrow \bar{u}_i + u_i, \qquad u_i^F \rightarrow \bar{u}_i^F + u_i^F, \qquad u_i^m \rightarrow \bar{u}_i^m + u_i^m.$$

The dimensional linearized perturbation equations then have form

$$u_{i,t} + \bar{u}u_{i,x} + w\bar{u}'\delta_{i1} = -\frac{1}{\rho}\pi_{,i} + \nu\Delta u_i, \qquad (6.85)$$

$$\frac{1}{\Phi}u_{i,t}^F + \frac{B}{K}(\bar{u}^F u_i^F + \delta_{i1}\bar{u}^F u^F) = -\frac{1}{\rho}\pi_{,i}^F - \frac{\nu}{K}u_i^F, \qquad (6.86)$$

$$\frac{1}{\Phi}u_{i,t}^m = -\frac{1}{\rho}\pi_{,i}^m - \frac{\nu}{K}u_i^m, \qquad (6.87)$$

where u_i, u_i^F, u_i^m are solenoidal, and where π, π^F, π^m denote the pressure perturbations and equations (6.85) – (6.87) hold in the region $(x, y) \in \mathbb{R}^2$, $t > 0$, with $z \in (0, d), z_m \in (-\beta d_m, 0), z_m \in (-d_m, -\beta d_m)$, respectively.

These equations are non-dimensionalized with the length, time, velocity and pressure in the fluid having scalings of $L = d, T = d/V, V, P = \mu V/d$ with those in the porous regions being scaled with $L_m = d_m, T_m = d_m/V_m, V_m = \bar{u}_m, P_m = \mu V_m/d_m$, i.e. the same scalings as those of (Chang et al., 2006). We introduce a fluid Reynolds number, Re, and a porous Reynolds number, Re_m, by

$$Re = \frac{Vd}{\nu}, \qquad Re^m = \frac{V_m d_m}{\nu}, \qquad (6.88)$$

and we introduce a non-dimensional Forchheimer number, B_m, and the Darcy number, δ, by

$$B_m = \frac{BV}{\nu}, \qquad \delta = \frac{\sqrt{K}}{d_m}. \qquad (6.89)$$

The non-dimensional form of the linearized perturbation equations (6.85) – (6.87) may then be shown to be

$$Re(u_{i,t} + U u_{i,x} + U' \delta_{i1} w) = -\pi_{,i} + \Delta u_i, \qquad u_{i,i} = 0, \qquad (6.90)$$

in $\mathbb{R}^2 \times \{z \in (0,1)\} \times \{t > 0\}$,

$$\frac{Re^m}{\Phi} u_{i,t}^m = -\pi_{,i}^m - \frac{1}{\delta^2} u_i^m, \qquad u_{i,i}^m = 0, \qquad (6.91)$$

in $\mathbb{R}^2 \times \{z_m \in (-1,-\beta)\} \times \{t > 0\}$, and

$$\frac{Re^m}{\Phi} u_{i,t}^F + \frac{B_m}{\delta^2} U^F (u_i^F + \delta_{i1} u^F) = -\pi_{,i}^F - \frac{1}{\delta^2} u_i^F, \qquad u_{i,i}^F = 0, \qquad (6.92)$$

in $\mathbb{R}^2 \times \{z_m \in (-\beta, 0)\} \times \{t > 0\}$.

The boundary conditions on $z = 1$ and on $z_m = -1$ become

$$u_i = 0, \quad z = 1; \qquad w_m = 0, \quad z = -1. \qquad (6.93)$$

On the Forchheimer-Darcy interface $z = -\beta d_m$ we have

$$w^F = w^m \qquad \text{and} \qquad \pi^F = \pi^m. \qquad (6.94)$$

We henceforth assume we are dealing with a two-dimensional perturbation in the (x, z) plane, i.e. $\mathbf{u} = (u, w)$, etc. Then, we may differentiate (6.94) with respect to x so that

$$\frac{\partial \pi^F}{\partial x} = \frac{\partial \pi^m}{\partial x} \qquad \text{on} \quad z = -\beta d_m.$$

Next, use the differential equations to substitute for these derivatives to see that

$$\frac{1}{\Phi} u_{,t}^F + \frac{2B}{K} \bar{u}^F u^F + \frac{\nu}{K} u^F = \frac{1}{\Phi} u_{,t}^m + \frac{\nu}{K} u^m, \qquad z = -\beta d_m.$$

This boundary condition is now non-dimensionalized and we find the non-dimensional interface conditions on $z = -\beta$ are

$$w^F = w^m, \qquad z = -\beta,$$

$$\frac{Re^m}{\Phi}\frac{\partial u^F}{\partial t} + \frac{1}{\delta^2}(2B_m + 1)u^F = \frac{Re^m}{\Phi}\frac{\partial u^m}{\partial t} + \frac{1}{\delta^2}u^m. \qquad (6.95)$$

The next step is to derive the three interface conditions on $z = 0$. Firstly $w = w^F$ there, so in non-dimensional terms

$$w = w^F \mathcal{V}, \qquad (6.96)$$

where $\mathcal{V}$ is a non-dimensional velocity scaling given by

$$\mathcal{V} = \frac{V^m}{V}. \qquad (6.97)$$

The Beavers-Joseph condition is

$$\frac{\partial u_\varsigma}{\partial z} = \frac{\alpha}{\sqrt{K}}(u_\varsigma - u_\varsigma^F), \quad \varsigma = 1, 2, \qquad z = 0,$$

(this derivation works in three-dimensions and so we leave it as it is). Differentiate this equation in the x and y directions and use the fact that $u_{\varsigma,\varsigma} = -\partial w/\partial z$. Then non-dimensionalize to obtain

$$\frac{\partial^2 w}{\partial z^2} = \frac{\alpha \hat{d}}{\delta}\left(\frac{\partial w}{\partial z} - \hat{d}\mathcal{V}\frac{\partial w^F}{\partial z}\right). \qquad (6.98)$$

Note, the interface conditions (6.96) and (6.98) are dimensionless conditions on $z = 0$.

We next take the normal stress continuity condition

$$\pi - 2\mu\frac{\partial w}{\partial z} = \pi^F \qquad \text{on} \quad z = 0,$$

and differentiate this with respect to x and y to find

$$\frac{\partial \pi^F}{\partial x} = \frac{\partial \pi}{\partial x} - 2\mu\frac{\partial^2 w}{\partial x \partial z},$$

$$\frac{\partial \pi^F}{\partial y} = \frac{\partial \pi}{\partial y} - 2\mu\frac{\partial^2 w}{\partial y \partial z},$$

and we then substitute from the dimensional, linearized perturbation equations for these quantities. The equation which results is then operated on by $\partial/\partial x^\varsigma, \varsigma = 1, 2$, and the equation $u_{\varsigma,\varsigma} = -\partial w/\partial z$ is employed. One may then show that

$$\frac{1}{\Phi}w_{,zt}^F + \left(\frac{\nu}{K} + \frac{2B}{K}\bar{u}^F\right)w_{,z}^F = -\frac{\mu}{\rho}w_{,zzz} - \frac{3\mu}{\rho}w_{,z\alpha\alpha}$$

$$+ w_{,zt} + \bar{u}w_{,zx} + \bar{u}'w_{,x}. \qquad (6.99)$$

This equation holds on $z = 0$ and is in dimensional form. The three interface conditions on $z = 0$ are (6.96), (6.98) and (6.99).

We next write the perturbation equations in terms of a single velocity function. Thus, for a two-dimensional perturbation, equations (6.90) are

$$Re(u_t + Uu_x + U'w) = -\pi_x + \Delta u,$$
$$Re(w_t + Uw_x) = -\pi_z + \Delta w, \tag{6.100}$$
$$u_x + w_z = 0.$$

Due to the last equation we may introduce a stream function $\psi(x, z, t)$ such that $u = \psi_z, w = -\psi_x$. We then put

$$\psi = \phi(z)e^{ia(x-ct)}.$$

Then,

$$u = D\phi\, e^{ia(x-ct)} \quad \text{and} \quad w = -\phi\, iae^{ia(x-ct)}.$$

Then the pressure is eliminated from (6.100) and we find ϕ satisfies the equation

$$(D^2 - a^2)^2\phi = Re\{ia(U - c)(D^2 - a^2)\phi - iaU''\phi\}, \qquad z \in (0, 1). \tag{6.101}$$

For the Darcy perturbation equations (6.91) we have

$$\frac{Re^m}{\Phi}u_t^m = -\pi_x^m - \frac{1}{\delta^2}u^m,$$
$$\frac{Re^m}{\Phi}w_t^m = -\pi_z^m - \frac{1}{\delta^2}w^m, \tag{6.102}$$
$$u_x^m + w_z^m = 0.$$

We again introduce a stream function $\psi^m(x^m, z^m, t^m)$ such that $u^m = \psi_z^m, w^m = -\psi_x^m$, with

$$\psi_m = \phi^m(z^m)e^{ia_m(x_m - c_m t_m)}.$$

By eliminating π_{xz}^m from (6.102) one then shows ϕ^m satisfies the equation

$$\left(\frac{1}{\delta^2} - \frac{ia_m c_m Re^m}{\Phi}\right)(D^2 - a_m^2)\phi_m = 0, \qquad z_m \in (-1, -\beta). \tag{6.103}$$

The two - dimensional Forchheimer equations (6.92) yield

$$\frac{Re^m}{\Phi}u_t^F + \frac{2B_m U^F}{\delta^2}u^F = -\pi_x^F - \frac{1}{\delta^2}u^F,$$
$$\frac{Re^m}{\Phi}w_t^F + \frac{B_m U^F}{\delta^2}w^F = -\pi_z^F - \frac{1}{\delta^2}w^F, \tag{6.104}$$
$$u_x^F + w_z^F = 0.$$

We introduce the stream function ψ^F such that $u^F = \psi_z^F, w^F = -\psi_x^F$, where

$$\psi^F = \phi^F(z^m)e^{ia_m(x_m - c_m t_m)}.$$

Then, eliminating the pressures in $(6.104)_{1,2}$ we show ϕ^F satisfies the differential equation

$$
\left(\frac{B_m U^F + 1}{\delta^2} \right) (D^2 - a_m^2)\phi_F + \frac{B_m U^F}{\delta^2} D^2 \phi_F
$$

$$
- i a_m c_m \frac{Re^m}{\Phi} (D^2 - a_m^2)\phi_F = 0, \qquad z_m \in (-\beta, 0). \tag{6.105}
$$

The system of differential equations to be solved comprises (6.101), (6.103) and (6.105). Note that this system yields an eighth order eigenvalue problem for the eigenvalue c (or c_m). This must be solved subject to eight boundary conditions. These are found from equations (6.93), (6.95), (6.96), (6.98) and (6.99). These boundary conditions are written in terms of ϕ, ϕ^F and ϕ^m and non-dimensionalized. We omit details, but one shows they become,

$$
\text{on } z = 1, \qquad \phi = 0, \quad D\phi = 0, \tag{6.106}
$$

$$
\text{on } z = -1, \qquad \phi_m = 0, \tag{6.107}
$$

on $z = -\beta$,

$$
\phi^F = \phi^m,
$$

$$
\left(\frac{2B_m + 1}{\delta^2} - \frac{i a_m c_m Re^m}{\Phi} \right) D\phi^F = \left(\frac{1}{\delta^2} - \frac{i a_m c_m Re^m}{\Phi} \right) D\phi^m, \tag{6.108}
$$

and on $z = 0$,

$$
\frac{a}{a_m}\phi = \mathcal{V}\phi^F,
$$

$$
D^2\phi = \frac{\alpha \hat{d}}{\delta} \left(D\phi - \frac{\hat{d}\mathcal{V}a_m}{a} D\phi^F \right),
$$

$$
- \left[\frac{a_m^2 c_m}{\Phi} + \frac{i a_m}{Re^m \delta^2}(1 + 2B_m \mathcal{V}U^F) \right] D\phi^F \tag{6.109}
$$

$$
= \left(\frac{ia}{Re^m \hat{d}^3 \mathcal{V}} \right)(D^3\phi - 3a^2 D\phi) + \frac{a^2(U - c)}{\hat{d}^2 \mathcal{V}} D\phi - \frac{iaU'}{\hat{d}^2 \mathcal{V}}\phi.
$$

Numerical results for system (6.101), (6.103), (6.105) together with the boundary conditions (6.106) – (6.109) will be presented in future work.

6.7.9 Brinkman - Forchheimer / Darcy model

We have thus far considered two models for flow over a porous medium incorporating a transition layer. Namely, when the transition layer is of Brinkman type, or of Forchheimer type. Each model has its own virtue, the Forchheimer one incorporating the faster flow as the fluid approaches the pure fluid layer, while the Brinkman one adapts to flow near a wall or when the porosity is close to one. In this section we again consider

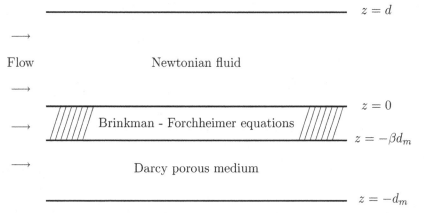

Figure 6.16. Three layer configuration for Poiseuille flow with a Brinkman - Forchheimer transition layer

the Poiseuille flow problem in a three layer configuration incorporating a transition layer, but we now account for both features of the Brinkman and Forchheimer theories by modelling the transition layer with a Brinkman - Forchheimer equation, cf. equation (6.9). Thus, the flow geometry is as shown in figure 6.16.

We observe that the merits, and instances where care should be taken, of using a Brinkman - Forchheimer theory next to a clean fluid are discussed in some detail by (Nield, 1991b). However, he does not specifically discuss Poiseuille flow and does not treat the instability problem.

Thus, in this section we consider Poiseuille flow with a Newtonian fluid occupying the layer $\mathbb{R}^2 \times \{z \in (0, d)\}$ while the porous medium occupies the layer $\mathbb{R}^2 \times \{z \in (-d_m, 0)\}$. The porous medium is now divided into two types (from a mathematical point of view), a Darcy layer occupying the region $\mathbb{R}^2 \times \{z \in (-d_m, -\beta d_m)\}$ with the domain $\mathbb{R}^2 \times \{z \in (-\beta d_m, 0)\}$ containing a porous medium of Brinkman - Forchheimer type, cf. figure 6.16.

The equations governing the motion of the fluid in the domain $\mathbb{R}^2 \times (-d_m, d)$ are the Navier-Stokes equations in $\{z \in (0, d)\}$,

$$\frac{\partial u_i}{\partial t} + u_j \frac{\partial u_i}{\partial x_j} = -\frac{1}{\rho}\frac{\partial p}{\partial x_i} + \nu \Delta u_i, \qquad \frac{\partial u_i}{\partial x_i} = 0, \qquad (6.110)$$

with equations of Brinkman - Forchheimer type in the layer $(-\beta d_m, 0)$, i.e.

$$\frac{1}{\Phi_F}\frac{\partial u_i^{BF}}{\partial t} + \frac{B}{K}|\mathbf{u}^{BF}|u_i^{BF} = -\frac{1}{\rho}\frac{\partial p^{BF}}{\partial x_i} - \frac{\nu}{K}u_i^{BF} + \frac{\nu}{\Phi}\Delta u_i^{BF},$$

$$\frac{\partial u_i^{BF}}{\partial x_i} = 0, \qquad (6.111)$$

and finally, the Darcy equations occupy the layer $z \in (-d_m, -\beta d_m)$, i.e.

$$\frac{1}{\Phi}\frac{\partial u_i^m}{\partial t} = -\frac{1}{\rho}\frac{\partial p^m}{\partial x_i} - \frac{\nu}{K}u_i^m, \qquad \frac{\partial u_i^m}{\partial x_i} = 0. \tag{6.112}$$

Note that u_i, u_i^{BF}, u_i^m denote the velocity in the fluid, Brinkman - Forchheimer, and Darcy layers, respectively, with p, p^{BF} and p^m being similarly the pressures. We also suppose the porosity is the same in the Darcy and Brinkman - Forchheimer layers, namely Φ. The number B is again a Forchheimer coefficient with the dimension of length.

We discuss the steady solution $\mathbf{u} = (u(z), 0, 0)$, $\mathbf{u}^{BF} = (u^{BF}(z), 0, 0)$, $\mathbf{u}^m = (u^m, 0, 0)$, below and its scaled non-dimensional forms U, U^{BF}, U^m. We observe now that one may derive the following non-dimensional linearized perturbation equations

$$Re(u_{i,t} + Uu_{i,x} + U'\delta_{i1}w) = -\pi_{,i} + \Delta u_i, \qquad u_{i,i} = 0, \tag{6.113}$$

in $\mathbb{R}^2 \times \{z \in (0,1)\} \times \{t > 0\}$,

$$\frac{Re^m}{\Phi}u_{i,t}^m = -\pi_{,i}^m - \frac{1}{\delta^2}u_i^m, \qquad u_{i,i}^m = 0, \tag{6.114}$$

in $\mathbb{R}^2 \times \{z_m \in (-1, -\beta)\} \times \{t > 0\}$, and

$$\frac{Re^m}{\Phi}u_{i,t}^{BF} + \frac{B_m}{\delta^2}U^{BF}(u_i^{BF} + \delta_{i1}u^{BF}) = -\pi_{,i}^{BF} - \frac{1}{\delta^2}u_i^{BF} + \frac{1}{\Phi}\Delta u_i^{BF},$$
$$u_{i,i}^{BF} = 0, \tag{6.115}$$

in $\mathbb{R}^2 \times \{z_m \in (-\beta, 0)\} \times \{t > 0\}$.

We again study instability via a two-dimensional perturbation and introduce stream functions ψ, ψ^{BF}, ψ^m and their associated velocity functions ϕ, ϕ^{BF}, ϕ^m as in section 6.7.8. In this way one derives the differential equations for ϕ, ϕ^{BF} and ϕ^m as

$$(D^2 - a^2)^2\phi = Re\{ia(U - c)(D^2 - a^2)\phi - iaU''\phi\}, \tag{6.116}$$

with $z \in (0, 1)$,

$$\left(\frac{1}{\delta^2} - \frac{ia_m c_m Re^m}{\Phi}\right)(D^2 - a_m^2)\phi_m = 0, \tag{6.117}$$

in the lower layer $z_m \in (-1, -\beta)$, and

$$\left(\frac{B_m U^{BF} + 1}{\delta^2}\right)(D^2 - a_m^2)\phi^{BF} + \frac{B_m U^{BF}}{\delta^2}D^2\phi^{BF}$$
$$- ia_m c_m \frac{Re^m}{\Phi}(D^2 - a_m^2)\phi^{BF} \tag{6.118}$$
$$+ \frac{2B_m}{\delta^2}U'_{BF}D\phi^{BF} - \frac{1}{\Phi}(D^2 - a_m^2)^2\phi^{BF} = 0,$$

where z_m is now in the non-dimensional transition layer, namely, $z_m \in (-\beta, 0)$.

Equations (6.116) – (6.118) are the differential equations to be solved. They represent a tenth order system for the eigenvalue c (or c_m) and may be efficiently solved numerically by the D^2 Chebyshev tau method described in section 9.2.1. However, we must first show how one determines the steady solution and the boundary and interface conditions.

We assume the upper boundary is fixed. Thus, in dimensional form

$$u_i = 0 \quad \text{on} \quad z = d. \tag{6.119}$$

On the bottom of the porous layer we suppose no flow out, so

$$u_3 = w^m = 0 \quad \text{on} \quad z = -d_m. \tag{6.120}$$

On the fluid / Brinkman-Forchheimer interface $z = 0$ we assume u_i is continuous and also the stress vector is continuous there. The stress vector $\mathbf{t}$ is given by

$$t_i = n_j t_{ij} \qquad \text{or} \qquad t_i^{BF} = n_j t_{ij}^{BF}$$

where t_{ij}, t_{ij}^{BF} denote the stress tensors. The stress tensors have form

$$t_{ij} = -\pi \delta_{ij} + \mu(u_{i,j} + u_{j,i}),$$
$$t_{ij}^{BF} = -\pi^{BF} \delta_{ij} + \frac{\mu}{\Phi}(u_{i,j}^{BF} + u_{j,i}^{BF}),$$

where we recall we are dealing with perturbations to the basic solution $\bar{\mathbf{u}}, \bar{\mathbf{u}}^{BF}, \bar{\mathbf{u}}^m$. Thus, we require (for a two-dimensional perturbation) the stress vector components t_1 and t_3 to be continuous across $z = 0$. This leads to the four interface conditions

$$u = u^{BF}, \qquad w = w^{BF},$$

$$-\pi^{BF} + \frac{2\mu}{\Phi} \frac{\partial w^{BF}}{\partial z^m} = -\pi + 2\mu \frac{\partial w}{\partial z},$$

$$\frac{\mu}{\Phi}\left(\frac{\partial u^{BF}}{\partial z^m} + \frac{\partial w^{BF}}{\partial x^m}\right) = \mu\left(\frac{\partial u}{\partial z} + \frac{\partial w}{\partial x}\right), \tag{6.121}$$

on $z = 0$.

On the Darcy / Brinkman-Forchheimer interface we assume w is continuous, continuity of normal stress (i.e. of the normal stress vector), and the Beavers-Joseph condition, thus

$$w^m = w^{BF},$$

$$-\pi^m = -\pi^{BF} + \frac{2\mu}{\Phi} \frac{\partial w^{BF}}{\partial z^m}, \tag{6.122}$$

$$\frac{\partial u_\zeta^{BF}}{\partial z^m} = \frac{\alpha}{\sqrt{K}}(u_\zeta^{BF} - u_\zeta^m), \quad \zeta = 1, 2,$$

on $z = -\beta d_m$. (Note we have left the Beavers-Joseph condition in its three-dimensional form, although one may directly handle the two-dimensional equivalent.)

The steady solution to equations (6.110) – (6.112), $\mathbf{u} = (u(z), 0, 0)$, $\mathbf{u}^{BF} = (u^{BF}(z), 0, 0)$, $\mathbf{u}^m = (u^m(z), 0, 0)$, must be found which satisfies the boundary and interface conditions (6.119), (6.120), (6.121) and (6.122). One shows that

$$\bar{u}^m = \frac{AK}{\nu}, \tag{6.123}$$

and using (6.119),

$$\bar{u}(z) = \frac{A}{2\nu}(d^2 - z^2) + \alpha_1(z - d), \qquad z \in (0, d), \tag{6.124}$$

where α_1 is to be determined from the interface conditions (6.121) in conjunction with calculating $\bar{u}^{BF}$. The function $\bar{u}^{BF}(z)$ is not constant and satisfies the nonlinear ordinary differential equation

$$A - \frac{\nu}{K}\bar{u}^{BF} - \frac{B}{K}(\bar{u}^{BF})^2 + \frac{\nu}{\Phi}\bar{u}^{BF}_{zz} = 0, \qquad z \in (-\beta d_m, 0). \tag{6.125}$$

The function $\bar{u}^{BF}(z)$ must be deterimined numerically, either by solving the nonlinear ordinary differential equation (6.125), or by multiplying the equation by $\bar{u}^{BF}_z$ and integrating, and then converting the equation to an implicit integral equation for $\bar{u}^{BF}$. The interface conditions must then be employed to completely determine $\bar{u}^{BF}$ and $\bar{u}$.

The boundary and interface conditions (6.119) – (6.122) may be written in non-dimensional form as

$$u = 0, w = 0, \quad z = 1; \qquad w^m = 0, \quad z = -1;$$

$$w^m = w^{BF}, \quad \frac{\partial^2 w^{BF}}{\partial z_m^2} = \frac{\alpha}{\sqrt{K}}\left(\frac{\partial w^{BF}}{\partial z_m} - \frac{\partial w^m}{\partial z_m}\right), \qquad z = -\beta;$$

$$\frac{Re^m}{\Phi}u_t^m + \frac{1}{\delta^2}u^m = \frac{Re^m}{\Phi}u_t^{BF} + \frac{2B_m}{\delta^2}U^{BF}u^{BF}$$

$$+ \frac{1}{\delta^2}u^{BF} - \frac{1}{\Phi}\Delta u^{BF} + \frac{2}{\Phi}w_{zx}^{BF}, \qquad z = -\beta;$$

$$Vw = V_m w^{BF}, Vu = V_m u^{BF}, \quad z = 0;$$

$$Vd^2\left(\frac{Re^m}{\Phi}u_t^{BF} + \frac{2B_m}{\delta^2}U^{BF}u^{BF} + \frac{1}{\delta^2}u^{BF} - \frac{1}{\Phi}\Delta u^{BF} + \frac{2}{\Phi}w_{zx}^{BF}\right)$$

$$= Re(u_t + Uu_x + U'w) + 2w_{xz}, \quad z = 0;$$

$$\frac{V\hat{d}}{\Phi}(u_z^{BF} + w_x^{BF}) = u_z + w_x, \quad z = 0.$$

The above ten boundary and interface conditions are written in terms of the velocity functions ϕ, ϕ^{BF} and ϕ^m as (in non-dimensional form),

$$\phi = 0, D\phi = 0, \quad z = 1; \qquad \phi^m = 0, \quad z_m = -1; \tag{6.126}$$

$$\phi^m = \phi^{BF}, D^2\phi^{BF} = \frac{\alpha}{\sqrt{K}}(D\phi^{BF} - D\phi^m), \quad z = -\beta; \tag{6.127}$$

$$-ia_m c_m \frac{Re^m}{\Phi} D\phi^{BF} - \frac{1}{\Phi}(D^2 - a_m^2)D\phi^{BF}$$
$$+ \left(\frac{2B_m}{\delta^2}U^{BF} + \frac{1}{\delta^2} + \frac{2a_m^2}{\Phi}\right)D\phi^{BF}$$
$$= -ia_m c_m \frac{Re^m}{\Phi} D\phi^m + \frac{1}{\delta^2}D\phi^m, \quad z = -\beta; \tag{6.128}$$

$$\phi = \frac{a_m}{a}\mathcal{V}\phi^{BF}, D\phi = \hat{d}\mathcal{V}D\phi^{BF}, \quad z = 0; \tag{6.129}$$

$$\mathcal{V}\hat{d}^2\left\{-\frac{1}{\Phi}(D^2 - a_m^2)D\phi^{BF} - ia_m c_m \frac{Re^m}{\Phi}D\phi^{BF}\right.$$
$$\left. + \left(\frac{2B_m U^{BF}}{\delta^2} + \frac{1}{\delta^2} + 2a_m^2\right)D\phi^{BF}\right\}$$
$$= -\left[Re\,ia(U - c) + 2a^2\right]D\phi - Re\,iaU'\phi, \quad z = 0; \tag{6.130}$$

$$\frac{\mathcal{V}\hat{d}}{\Phi}(D^2 + a_m^2)\phi^{BF} = (D^2 + a^2)\phi, \quad z = 0. \tag{6.131}$$

Thus, the eigenvalue problem to be solved consists of equations (6.116) – (6.118) together with the boundary conditions (6.126) – (6.131). Numerical results for this system will appear in future work.

6.8   Acoustic waves, ocean bed

(Caviglia et al., 1992a) treat the problem of propagation of a wave in a region composed of a layer of fluid overlying a layer of porous material. The application which motivates their work is to understand how a sound wave travels through the sea and is then relected / transmitted through the porous sea bed. They model this scenario by assuming the half-space above $z = 0$ is filled with an inviscid fluid while the half space below $z = 0$ is filled with a saturated porous medium, see figure 6.17. To model the porous medium they adopt the linearized version of the theory of (Bowen, 1982), which is a mixture theory which allows for variable porosity as discussed in chapter 1.

In the following sub-section we describe wave propagation in a porous medium allowing the solid structure to also move. The material of section 6.8.1 does not use an equivalent fluid type of theory as is used in chapter 8. Thus, section 6.8.1 is of interest in its own right for sound wave propagation as considered in chapter 8.

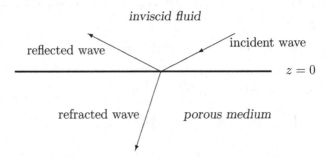

Figure 6.17. Reflection - refraction at the sea bed

6.8.1 Basic equations

A heuristic approach to seeing the connection between porosity and the partial densities is now given. Let V be a volume occupied by a mixture of a fluid and a solid. Suppose the fluid occupies the fraction v_f while the solid occupies v_s, with $v_f + v_s = V$. Then, if $\bar{\rho}_f$ and $\bar{\rho}_s$ denote the actual fluid and solid densities and m_f and m_s are the respective masses, $\bar{\rho}_f = m_f/v_f$ and $\bar{\rho}_s = m_s/v_s$. We define the partial densities ρ_f and ρ_s by $\rho_f = m_f/V$ and $\rho_s = m_s/V$. Then, ρ_f and ρ_s satisfy the volume additivity relation

$$\frac{\rho_f}{\bar{\rho}_f} + \frac{\rho_s}{\bar{\rho}_s} = 1. \tag{6.132}$$

The porosity of the porous medium, ϕ, is defined by

$$\begin{aligned}
\phi &= \frac{\text{fluid volume}}{\text{total volume}} \\
&= \frac{v_f}{V} \\
&= \frac{\rho_f}{\bar{\rho}_f} = 1 - \frac{\rho_s}{\bar{\rho}_s}.
\end{aligned} \tag{6.133}$$

(Caviglia et al., 1992a) consider a mixture of one fluid and one solid in the (Bowen, 1982) theory and we briefly report their equations and findings. The governing equations begin with the balance of mass for the fluid and the solid components,

$$\frac{\partial \rho_f}{\partial t} + (\rho_f v_i^f)_{,i} = 0, \tag{6.134}$$

and

$$\frac{\partial \rho_s}{\partial t} + (\rho_s v_i^s)_{,i} = 0, \tag{6.135}$$

The balances of momentum for the fluid and solid constituents are

$$\frac{\partial(\rho_f v_i^f)}{\partial t} + \frac{\partial(\rho_f v_i^f v_j^f - T_{ij}^f)}{\partial x_j} = p_i^f, \tag{6.136}$$

and

$$\frac{\partial(\rho_s v_i^s)}{\partial t} + \frac{\partial(\rho_s v_i^s v_j^s - T_{ij}^s)}{\partial x_j} = p_i^s. \tag{6.137}$$

In these equations v_i^f, v_i^s are the fluid and solid velocities in the mixture, T_{ij}^f, T_{ij}^s are the fluid and solid partial stress tensors and p_i^f, p_i^s are interaction forces. The stresses are given by the constitutive equations

$$T_{ij}^f = -\lambda_f \frac{(\rho_f - \rho_f^0)}{\rho_f^0} \delta_{ij} + \Gamma_f(\phi - \phi^0)\delta_{ij} + \lambda_{sf} u_{k,k}^s \delta_{ij}, \tag{6.138}$$

and

$$T_{ij}^s = \lambda_s u_{k,k}^s \delta_{ij} + \mu_s(u_{i,j}^s + u_{j,i}^s) + \Gamma_s(\phi - \phi^0)\delta_{ij} - \lambda_{sf} \frac{(\rho_f - \rho_f^0)}{\rho_f^0} \delta_{ij}, \tag{6.139}$$

where $\lambda_f, \Gamma_f, \lambda_{sf}, \lambda_s, \mu_s, \Gamma_s$ are constants, ρ_f^0, ρ_s^0 are constant equilibrium values of ρ_f, ρ_s, and u_i^s is the solid displacement.

The constitutive equations for the interaction forces are taken to be

$$p_i^s = \xi(v_i^f - v_i^s) = -p_i^f, \tag{6.140}$$

where $\xi > 0$ is a constant.

We have variables $u_i^s, v_i^f, \rho_s, \rho_f$ and ϕ and hence need a further equation to complete the theory. In general we can adopt the equation of (Bowen, 1982) for the porosity ϕ, of the form

$$\dot{\phi} = F(\phi, \rho_f, u_{i,j}),$$

for some constitutive function F. We report the linearized equation adopted by (Caviglia et al., 1992a), i.e.

$$\frac{\partial \phi}{\partial t} = -\Lambda_f(\phi - \phi^0) - \frac{\Gamma_s}{\Phi_f} u_{i,i}^s + \frac{\Gamma_f}{\Phi_f}\left(\frac{\rho_f - \rho_f^0}{\rho_f^0}\right), \tag{6.141}$$

where $\Lambda_f > 0$ is constant and Φ_f is a constant. Thermodynamics requires that $\mu_s > 0$ and the matrix A be positive-definite, where

$$A = \begin{pmatrix} \lambda_s + 2\mu_s/3 & \lambda_{sf} & \Gamma_s \\ \lambda_{sf} & \lambda_f & \Gamma_f \\ \Gamma_s & \Gamma_f & \Phi_f \end{pmatrix}.$$

6.8.2 Linear waves in the Bowen theory

(Caviglia et al., 1992a) consider a small amplitude wave moving in the porous medium and then their equations linearized about the equilibrium state $\rho_f^0 = \text{constant}$, $\rho_s^0 = \text{constant}$, $v_i^f \equiv 0$, $v_i^s \equiv 0$, are

$$\frac{\partial \rho_f}{\partial t} + \rho_f^0 v_{i,i}^f = 0,$$

$$\frac{\partial \rho_s}{\partial t} + \rho_s^0 v_{i,i}^s = 0,$$

$$\rho_f^0 \frac{\partial v_i^f}{\partial t} + \frac{\lambda_f}{\rho_f^0} \rho_{f,i} - \lambda_{sf} u_{j,ji}^s - \Gamma_f \beta_{,i} + \xi(v_i^f - v_i^s) = 0,$$

$$\rho_s^0 \frac{\partial v_i^s}{\partial t} - \lambda_s u_{j,ji}^s + \frac{\lambda_{sf}}{\rho_f^0} \rho_{f,i} - \mu_s [\Delta u_i^s + u_{j,ji}^s]$$

$$- \Gamma_s \beta_{,i} + \xi(v_i^s - v_i^f) = 0,$$

$$\frac{\partial \beta}{\partial t} + \Lambda_f(\beta - \beta^0) + \frac{\Gamma_s}{\Phi_f} u_{i,i}^s + \frac{\Gamma_f}{\Phi_f} \frac{\rho_f - \rho_f^0}{\rho_f^0} = 0.$$

(6.142)

Solutions are sought of form

$$\rho_f - \rho_f^0 = \hat{\rho}_f \rho_f^0 \exp[i(\mathbf{k} \cdot \mathbf{x} - \omega t)],$$

$$\rho_s - \rho_s^0 = \hat{\rho}_s \rho_s^0 \exp[i(\mathbf{k} \cdot \mathbf{x} - \omega t)],$$

$$u_i^f = \hat{u}_i^f \exp[i(\mathbf{k} \cdot \mathbf{x} - \omega t)],$$

$$u_i^s = \hat{u}_i^s \exp[i(\mathbf{k} \cdot \mathbf{x} - \omega t)],$$

$$\beta - \beta^0 = \hat{\beta} \exp[i(\mathbf{k} \cdot \mathbf{x} - \omega t)],$$

(6.143)

where $\mathbf{k}$ is a wave number and $\hat{\beta}, \hat{\rho}_s, \hat{\rho}_f, \hat{u}_i^f, \hat{u}_i^s$, are the wave amplitudes.

The representation (6.143) is substituted in (6.142) and equations for $\mathbf{k}$ and ω are obtained. (Caviglia et al., 1992a) consider specifically longitudinal waves which satisfy

$$\mathbf{k} \times \hat{\mathbf{u}}^f = \mathbf{0}, \qquad \mathbf{k} \times \hat{\mathbf{u}}^s = \mathbf{0},$$

and transverse waves which satisfy

$$\mathbf{k} \cdot \hat{\mathbf{u}}^f = 0, \qquad \mathbf{k} \cdot \hat{\mathbf{u}}^s = 0.$$

The transverse wave is found to satisfy the propagation condition

$$\mathbf{k} \cdot \mathbf{k} = \frac{\omega^2}{\mu_s} \frac{\omega \rho_f^0 \rho_s^0 + i\xi(\rho_f^0 + \rho_s^0)}{\omega \rho_f^0 + i\xi}.$$

For the longitudinal waves (Caviglia et al., 1992a) find the propagation condition

$$
\left\{ -\omega^2 \rho_s^0 - i\omega\xi + (\mathbf{k} \cdot \mathbf{k}) \left[2\mu_s + \lambda_s - \frac{\Gamma_s^2}{\Phi_f(\Lambda_f - i\omega)} \right] \right\}
$$
$$
\times \left\{ -\omega^2 \rho_f^0 - i\omega\xi + (\mathbf{k} \cdot \mathbf{k}) \left[\lambda_f + \frac{\Gamma_f^2}{\Phi_f(\Lambda_f - i\omega)} \right] \right\}
$$
$$
- \left\{ i\omega\xi + (\mathbf{k} \cdot \mathbf{k}) \left[\lambda_{sf} + \frac{\Gamma_s \Gamma_f}{\Phi_f(\Lambda_f - i\omega)} \right] \right\}
$$
$$
\times \left\{ i\omega\xi + (\mathbf{k} \cdot \mathbf{k}) \left[\lambda_{sf} - \frac{\Gamma_s \Gamma_f}{\Phi_f(\Lambda_f - i\omega)} \right] \right\} = 0.
$$

(6.144)

This equation is solved for k^2. It is worth observing that when the interaction terms between the fluid and solid are set equal to zero, i.e. $\xi, \lambda_{sf}, \Gamma_f, \Gamma_s = 0$, equation (6.144) reduces to

$$
\left[\omega^2 \rho_s^0 - (2\mu_s + \lambda_s)k^2 \right] \left[\omega^2 \rho_f^0 - \lambda_f k^2 \right] = 0.
$$

This yields the classical relationship between ω^2 and k^2 for an elastic wave (first bracket) and a fluid (second bracket) wave.

In fact, (Caviglia et al., 1992a) show that the longitudinal and transverse waves are the only ones which can occur. They show that there will be two longitudinal waves, a fast and a slow one, and a transverse wave.

6.8.3 Boundary conditions

To study propagation of an incident wave passing through the fluid and into the porous medium we need appropriate conditions at the interface. In fact, by treating the porous medium as a mixture we can derive exact boundary conditions. There is no need to include any experimentally motivated interface condition like the Beavers-Joseph one (equation (6.2)).

The interface between the fluid and porous medium is the plane $z = 0$. (Caviglia et al., 1992a) derive balance laws for the total density $\rho = \rho_f + \rho_s$ and total velocity $\mathbf{v}$ given by $\rho \mathbf{v} = \rho_f \mathbf{v}_f + \rho_s \mathbf{v}_s$. The latter involves the total stress $\mathbf{T} = \mathbf{T}^f + \mathbf{T}^s$. By applying a pill-box argument to the governing equations at the interface (Caviglia et al., 1992a) show that

$$
\left[\rho(\mathbf{v} - \mathbf{V}) \cdot \mathbf{n} \right] = 0, \qquad [\mathbf{T}]\mathbf{n} - [\mathbf{v}]\rho(\mathbf{v} - \mathbf{V}) \cdot \mathbf{n} = 0,
$$

where $\mathbf{V}$ is the interface velocity, $\mathbf{n}$ is the normal at the interface, and $[\cdot]$ denotes the jump in a quantity. From these relations, noting $\mathbf{V} = \mathbf{v}^s$ at the interface they deduce

$$
\rho_F(v_i^F - v_i^s)n_i = \rho_f(v_i^f - v_i^s)n_i,
$$

where F denotes the fluid in the region $z > 0$. From the above relation they deduce that

$$
v_i^F n_i = v_i^f n_i = v_i^s n_i \quad \text{at } z = 0.
$$

The stress continuity condition at the interface, derived by (Caviglia et al., 1992a), in the *linearized case*, is

$$T_{ji}^F n_j = (T_{ji}^f + T_{ji}^s) n_i.$$

6.8.4 Amplitude behaviour

(Caviglia et al., 1992a) consider a wave moving through a perfect fluid in the domain $z > 0$. They study wave motion in the (x, z) plane and have solid and fluid displacements in the region $z < 0$ of form $\mathbf{u}^f = (u_f, 0, w_f)$, $\mathbf{u}^s = (u_s, 0, w_s)$ which are written in terms of potentials as

$$u_s = \frac{\partial \phi_1}{\partial x} + \frac{\partial \phi_2}{\partial x} + \frac{\partial \psi}{\partial z},$$

$$w_s = \frac{\partial \phi_1}{\partial z} + \frac{\partial \phi_2}{\partial z} - \frac{\partial \psi}{\partial x},$$

$$u_f = \mu_1 \frac{\partial \phi_1}{\partial x} + \mu_2 \frac{\partial \phi_2}{\partial x} - a \frac{\partial \psi}{\partial z},$$

$$w_f = \mu_1 \frac{\partial \phi_1}{\partial z} + \mu_2 \frac{\partial \phi_2}{\partial z} + a \frac{\partial \psi}{\partial x}.$$

The equivalent potential in the region $z > 0$ is written as ϕ_F, and then ϕ_F, the longitudinal potentials ϕ_1, ϕ_2 and the transverse potential ψ are written in the form

$$\phi_F = \{ \Phi_i \exp(-ik_z z) + \Phi_r \exp(ik_z z) \} \exp\{ i(k_x x - \omega t) \},$$

$$\phi_1 = \Phi_1 \exp(-ik_{1z} z) \exp\{ i(k_x x - \omega t) \},$$

$$\phi_2 = \Phi_2 \exp(-ik_{2z} z) \exp\{ i(k_x x - \omega t) \},$$

$$\psi = \Psi \exp(-ik_{3z} z) \exp\{ i(k_x x - \omega t) \}.$$

Solutions are found and the interface conditions employed to yield the appropriate transmitted and reflected wave amplitudes. Graphical output of the amplitude behaviour for the densities and the porosity are given in (Caviglia et al., 1992a).

A specific example is considered by (Caviglia et al., 1992a) of a wave moving through water ($z > 0$) into a kerosene-sandstone mixture ($z < 0$). In the kerosene-sandstone porous material the wave phase velocities are calculated as

$$V_1 = 2.2023 \times 10^5 \text{ cm s}^{-1},$$

$$V_2 = 0.3256 \times 10^5 \text{ cm s}^{-1},$$

$$V_3 = 1.1393 \times 10^5 \text{ cm s}^{-1},$$

where V_3 corresponds to the transverse wave, while V_1 and V_2 correspond to the fast and slow longitudinal waves, respectively. Details of the incident, reflected and transmitted wave amplitudes are given in (Caviglia et al., 1992a).

In section 6.8.1 we have considered wave propagation in a saturated porous medium using a mixture theory which includes porosity as an internal variable. In a sense, this leads naturally into the next two chapters where sound wave propagation in other types of porous media is considered in detail.

7

Elastic Materials with Voids

7.1 Acceleration waves in elastic materials

7.1.1 Bodies and their configurations

We consider a body B deformed from a *reference configuration* at time $t = 0$ to a *current configuration* at time t.

Points in the reference configuration are labelled by boldface notation $\mathbf{X}$ or indicial notation X_A. In the current configuration $\mathbf{X} \to \mathbf{x}$. The mapping is thus

$$\mathbf{x} = \mathbf{x}(\mathbf{X}, t) \tag{7.1}$$

or

$$x_i = x_i(X_A, t). \tag{7.2}$$

The coordinates X_A are *material* (or *Lagrangian*) coordinates whereas x_i are *spatial coordinates* (*Eulerian coordinates*).

In elasticity we need the displacement vector $\mathbf{u}$ of a typical particle from $\mathbf{X}$ in the reference configuration to $\mathbf{x}$ at time t, so

$$u_i(X_A, t) = x_i(X_A, t) - X_i. \tag{7.3}$$

The velocity of a particle v_i is

$$v_i(X_A, t) = \frac{\partial x_i}{\partial t}\bigg|_{\mathbf{X} \text{ constant}}.$$

B. Straughan, *Stability and Wave Motion in Porous Media*,
DOI: 10.1007/978-0-387-76543-3_7, © Springer Science+Business Media, LLC 2008

In fluid mechanics we usually use the inverse of (7.1) to write $v_i = v_i(x_j, t)$ - this is the spatial description, i.e. that following the particle.

7.1.2   The deformation gradient tensor

The deformation gradient tensor F_{iA} is defined by

$$F_{iA} = \frac{\partial x_i}{\partial X_A}.$$

From (7.3) we find the displacement gradient as

$$u_{i,A} = \frac{\partial u_i}{\partial X_A} = \frac{\partial x_i}{\partial X_A} - \delta_{iA} = F_{iA} - \delta_{iA}.$$

7.1.3   Conservation of mass

The relation

$$\rho_0 = \rho \det \mathbf{F}$$

is the conservation of mass in Lagrangian form. (Recall that in Eulerian form this is

$$\frac{\partial \rho}{\partial t} + (\rho v_i)_{,i} = 0.)$$

N.B. If the material is incompressible then $\rho = $ constant so $\rho_0/\rho = 1$. Therefore, in an incompressible material the deformation must satisfy

$$\det \mathbf{F} = 1.$$

(See (Spencer, 1980), pp. 91–95.)

7.1.4   The equations of nonlinear elasticity

The tensor π_{iA} is called the *Piola-Kirchoff stress tensor* (useful in elasticity because it refers back to the reference configuration) and is defined by

$$\pi_{iA} = \frac{\partial W}{\partial F_{iA}},$$

where W is the (internal) strain energy function.

The equations of nonlinear elastodynamics referred to the body in its reference configuration have form

$$\rho_0 \ddot{x}_i = \frac{\partial \pi_{iA}}{\partial X_A} + \rho_0 f_i$$

or

$$\rho_0 \frac{\partial^2 x_i}{\partial t^2}\bigg|_{\mathbf{x}} = \frac{\partial}{\partial X_A}\left(\frac{\partial W}{\partial x_{i,A}}\right) + \rho_0 f_i,$$

see (Spencer, 1980), eq. (9.38).

Now, $u_i = x_i(\mathbf{X}, t) - X_i$ and so $\ddot{x}_i = \ddot{u}_i$ and

$$\frac{\partial W}{\partial x_{i,A}} = \frac{\partial W}{\partial u_{r,B}} \frac{\partial u_{r,B}}{\partial x_{i,A}}.$$

But $u_{r,B} = x_{r,B} - \delta_{rB}$ so

$$\frac{\partial u_{r,B}}{\partial x_{i,A}} = \frac{\partial x_{r,B}}{\partial x_{i,A}} = \delta_{ri} \delta_{BA}$$

whence

$$\frac{\partial W}{\partial x_{i,A}} = \frac{\partial W}{\partial u_{r,B}} \delta_{ri} \delta_{AB} = \frac{\partial W}{\partial u_{i,A}}.$$

Thus, the equations of nonlinear elastodynamics may also be written as

$$\rho_0 \ddot{u}_i = \frac{\partial}{\partial X_A} \left(\frac{\partial W}{\partial u_{i,A}} \right) + \rho_0 f_i. \tag{7.4}$$

To proceed, we need to know the functional form of $W = W(u_{i,A})$.

If the deformation is only in one direction, say the x-direction then $u = x - X$, where $u = u_1$, and

$$\epsilon = u_X = \frac{\partial u}{\partial X} = \frac{\partial x}{\partial X} - 1$$

is the one-dimensional strain. The one-dimensional Piola-Kirchoff stress is $\pi = \partial W / \partial u_X$. In particular, for one-dimensional motions, equations (7.4) become

$$\rho_0 u_{tt} = \frac{\partial}{\partial X} \left(\frac{\partial W}{\partial u_X} \right) + \rho_0 f$$

or

$$\rho_0 u_{tt} = \frac{\partial}{\partial X} \left(\frac{\partial W}{\partial \epsilon} \right) + \rho_0 f. \tag{7.5}$$

If W is quadratic in ϵ, then

$$W = \frac{\alpha}{2} u_X^2 = \frac{\alpha}{2} \epsilon^2.$$

Put $f = 0$, then equation (7.5) is linear and reduces to

$$\rho_0 u_{tt} = \alpha u_{XX}.$$

This is just the usual wave equation. Thus, a quadratic strain energy function leads to a linear theory of elasticity.

We shall now focus on a general form for W, namely, $W = W(\epsilon)$.

The equations of nonlinear elasticity are derived in many forms in (Spencer, 1980), (Ogden, 1997), and (Holzapfel, 2000).

7.1.5   Acceleration waves in one-dimension

The general theory of acceleration waves in nonlinear elastodynamics is covered in detail in (Chen, 1973). (Truesdell and Toupin, 1960) and (Truesdell and Noll, 1992) cover many aspects of acceleration waves and singular surfaces, in general.

Although the ideas of acceleration waves have been under constant development for over forty years, they are still being employed with much effect in the current literature. In fact, the use of acceleration waves and related analyses have proved extremely useful in recent investigations of wave motion in various dispersive and random media, in a variety of thermodynamic states, see e.g. (Christov and Jordan, 2005), (Christov et al., 2006; Christov et al., 2007), (Ciarletta and Straughan, 2006), (Eremeyev, 2005), (Fu and Scott, 1988; Fu and Scott, 1991), (Gultop, 2006), (Jordan, 2005a), (Jordan and Christov, 2005), (Jordan and Puri, 2005), (Kameyama and Sugiyama, 1996), (Ostoja-Starzewski and Trebicki, 1999), (Puri and Jordan, 2004), (Quintanilla and Straughan, 2004), (Rai, 2003), (Rajagopal and Truesdell, 1999), (Ruggeri and Sugiyama, 2005), (Su et al., 2005), (Sugiyama, 1994), (Valenti et al., 2004).

Suppose we have an elastic body occupying $\mathbb{R}^3$ and the equations of motion are (7.4). Recall that $_{,A}$ denotes differentiation with respect to X_A, e.g. $u_{i,A} = \partial u_i / \partial X_A$. An *acceleration* wave is a surface S across which $u_{i,tt}, u_{i,tA}, u_{i,AB}, u_{i,ttt}, u_{i,ttA}, u_{i,tAB}, u_{i,ABC}$ suffer at most finite discontinuities, with the functions and first derivatives $u_i, u_{i,t}, u_{i,A}$ continuous everywhere. The body force is at least $C^1(\mathbb{R}^3)$.

To illustrate the basic concepts of acceleration wave analysis, we shall for now restrict attention to a plane acceleration wave moving in the direction of the $x-$axis, with one-dimensional motion.

N.B. Since $u_{i,tt}$ has a finite jump across S we call it an acceleration wave.

For a function $h(x,t)$ we define

$$h^+(x,t) = \lim_{x \to S} h(x,t) \text{ from the right,}$$

$$h^-(x,t) = \lim_{x \to S} h(x,t) \text{ from the left.}$$

In particular, h^+ is the value of h at S approaching from the region which S is about to enter. The jump of h at S, written as $[h]$, is,

$$[h] = h^- - h^+ . \tag{7.6}$$

We commence with the equations of motion for a one-dimensional elastodynamic body,

$$\rho_0 u_{tt} = \frac{\partial}{\partial X}\left(\frac{\partial W}{\partial \epsilon}\right) + \rho_0 f. \tag{7.7}$$

By the chain rule,

$$\frac{\partial}{\partial X}\left(\frac{\partial W}{\partial \epsilon}\right) = \frac{\partial^2 W}{\partial \epsilon^2}\,\epsilon_X = \frac{\partial^2 W}{\partial \epsilon^2}\,u_{XX} = W_{\epsilon\epsilon}u_{XX}.$$

We use this in (7.7) and take the jump of the resulting equation, to find

$$\rho_0[u_{tt}] = W_{\epsilon\epsilon}[u_{XX}] \qquad \text{(note } [\rho_0 f] = 0\text{)}.$$

Next, employ the kinematic condition of compatibility, sometimes known as the Hadamard relation,

$$\frac{\delta}{\delta t}\,[f] = \left[\frac{\partial f}{\partial t}\right] + V\left[\frac{\partial f}{\partial X}\right] \tag{7.8}$$

where $\delta/\delta t$ denotes the time derivative at the wave. (The Hadamard relation is discussed in detail in (Chen, 1973), appendix 1, and also in (Truesdell and Toupin, 1960), section 180.)

Note, since $u \in C^1(\mathbb{R})$, $[u_t] = 0, [u_X] = 0$, so by using the Hadamard relation,

$$0 = \frac{\delta}{\delta t}\,[u_t] = [u_{tt}] + V[u_{tX}],$$

and

$$0 = \frac{\delta}{\delta t}\,[u_X] = [u_{Xt}] + V[u_{XX}].$$

Thus,

$$[u_{tt}] = -V[u_{tX}] = V^2[u_{XX}]. \tag{7.9}$$

From this and the equation of motion we thus find

$$(\rho_0 V^2 - W_{\epsilon\epsilon})\,[u_{XX}] = 0. \tag{7.10}$$

For a non-zero amplitude

$$a(t) = [u_{tt}]$$

we see from (7.9), (7.10), that

$$V^2 = \frac{W_{\epsilon\epsilon}}{\rho_0}.$$

Thus, the speed of the wave is

$$V = \frac{1}{\sqrt{\rho_0}}\sqrt{\frac{\partial^2 W}{\partial \epsilon^2}(u_X^+)}.$$

Note that V depends on the value of u_X^+ before the wave (although $u_X^+ = u_X^-$).

To find the equation governing the amplitude $a(t)$ we differentiate (7.7),

$$\rho_0 u_{ttt} = \frac{\partial}{\partial t}\,(W_{\epsilon\epsilon}u_{XX}) + \rho_0 f_t, \tag{7.11}$$

or, since

$$\frac{\partial}{\partial t}(W_{\epsilon\epsilon}u_{XX}) = W_{\epsilon\epsilon\epsilon}u_{Xt}u_{XX} + W_{\epsilon\epsilon}u_{XXt},$$

then (7.11) becomes

$$\rho_0 u_{ttt} = W_{\epsilon\epsilon\epsilon}u_{Xt}u_{XX} + W_{\epsilon\epsilon}u_{XXt} + \rho_0 f_t.$$

Take the jump of this recalling $f \in C^1$ so $[f_t] = 0$,

$$\rho_0[u_{ttt}] = W_{\epsilon\epsilon\epsilon}[u_{Xt}u_{XX}] + W_{\epsilon\epsilon}[u_{XXt}]. \tag{7.12}$$

From the definition of $[h]$ we may prove the relation for the jump of a product of functions g, h,

$$[gh] = g^+[h] + h^+[g] + [g][h]. \tag{7.13}$$

Also,

$$\frac{\delta a}{\delta t} = [u_{ttt}] + V[u_{ttX}], \tag{7.14}$$

$$\frac{\delta}{\delta t}[u_{tX}] = \frac{\delta}{\delta t}\left(-\frac{a}{V}\right) = [u_{ttX}] + V[u_{tXX}],$$

or

$$-\frac{1}{V}\frac{\delta a}{\delta t} + \frac{a}{V^2}\frac{\delta V}{\delta t} = [u_{ttX}] + V[u_{tXX}]. \tag{7.15}$$

Use (7.13), (7.14) in (7.12),

$$\rho_0\left(\frac{\delta a}{\delta t} - V[u_{ttX}]\right) = W_{\epsilon\epsilon}[u_{XXt}]$$
$$+ W_{\epsilon\epsilon\epsilon}\left(u_{tX}^+[u_{XX}] + u_{XX}^+[u_{tX}] + [u_{XX}][u_{tX}]\right). \tag{7.16}$$

Now substitute from (7.15) for the term $-\rho_0 V[u_{ttX}]$ on the left of (7.16), to obtain

$$2\rho_0\frac{\delta a}{\delta t} - \frac{\rho_0}{V}\frac{\delta V}{\delta t}a + \rho_0 V^2[u_{tXX}] = W_{\epsilon\epsilon}[u_{XXt}]$$
$$+ W_{\epsilon\epsilon\epsilon}\left(u_{tX}^+[u_{XX}] + u_{XX}^+[u_{tX}] + [u_{XX}][u_{tX}]\right). \tag{7.17}$$

Observe now that the terms in $[u_{tXX}]$ on the right and left cancel out since $V^2 = W_{\epsilon\epsilon}/\rho_0$.

The next step is to use (7.9), i.e.

$$[u_{tX}] = -a/V, \qquad [u_{XX}] = a/V^2,$$

in (7.17), to derive

$$2\rho_0\frac{\delta a}{\delta t} - \frac{\rho_0}{V}\frac{\delta V}{\delta t}a = W_{\epsilon\epsilon}\left(\frac{u_{tX}^+}{V^2}a - \frac{u_{XX}^+}{V}a\right) - \frac{W_{\epsilon\epsilon\epsilon}}{V^3}a^2$$

or,

$$2\rho_0 \frac{\delta a}{\delta t} - \left(\frac{\rho_0}{V} \frac{\delta V}{\delta t} + \frac{u_{tX}^+}{V^2} W_{\epsilon\epsilon\epsilon} - \frac{u_{XX}^+}{V} W_{\epsilon\epsilon\epsilon} \right) a + \frac{W_{\epsilon\epsilon\epsilon}}{V^3} a^2 = 0. \qquad (7.18)$$

This is the equation governing the evolutionary behaviour of the amplitude
$a(t)$ - the amplitude equation. It is a Bernoulli equation, which may be
written in the form

$$\frac{\delta a}{\delta t} + \alpha(t)a + \beta(t)a^2 = 0.$$

It may be solved by the substitution $\gamma = 1/a$ to yield the general solution

$$a(t) = \frac{a(0)}{\exp\{\int_0^t \alpha(s)ds\} + \int_0^t \beta(s)\exp\{\int_s^t \alpha(\eta)d\eta\}ds}. \qquad (7.19)$$

7.1.6 Given strain energy and deformation

To solve (7.18) and gain any useful information we need to know
  (a) the form $W = W(\epsilon)$ (constitutive theory)
  (b) the deformation $x(X)$ ahead of the wave, or more generally, $x(X,t)$.
We do this in an illustrative case. Suppose

$$W = \frac{\alpha}{2} u_X^2 + \frac{\beta}{3} u_X^3 + \frac{\gamma}{4} u_X^4,$$

for constants α, β, γ. Then,

$$W_\epsilon = \alpha\epsilon + \beta\epsilon^2 + \gamma\epsilon^3,$$
$$W_{\epsilon\epsilon} = \alpha + 2\beta\epsilon + 3\gamma\epsilon^2,$$
$$W_{\epsilon\epsilon\epsilon} = 2\beta + 6\gamma\epsilon.$$

Suppose the deformation before the wave passes is

$$x = \lambda X,$$

for λ a constant. Then,

$$\epsilon = u_X = \frac{\partial x}{\partial X} - 1 = \lambda - 1.$$

Observe that

$$u_{tX}^+ = \frac{\partial}{\partial t}(\lambda - 1) = 0, \qquad \text{and} \qquad u_{XX}^+ = \frac{\partial}{\partial X}(\lambda - 1) = 0.$$

Since $V^2 = W_{\epsilon\epsilon}/\rho_0$ we find

$$V^2 = \frac{1}{\rho_0} \left\{ \alpha + 2\beta(\lambda - 1) + 3\gamma((\lambda - 1)^2) \right\}. \qquad (7.20)$$

N.B. Equation (7.20) imposes restrictions on α, β, γ, since $V^2 > 0$, so

$$\alpha + 2\beta\epsilon + 3\gamma\epsilon^2 > 0.$$

From equation (7.20), V is constant and so

$$\frac{\delta V}{\delta t} = 0.$$

Thus, the coefficient of a in (7.18) is zero. Hence, (7.18) reduces to

$$\frac{\delta a}{\delta t} + \frac{W_{\epsilon\epsilon\epsilon}}{2\rho_0 V^3} a^2 = 0. \qquad (7.21)$$

But,

$$\frac{W_{\epsilon\epsilon\epsilon}}{2\rho_0 V^3} = \frac{2\beta + 6\gamma(\lambda - 1)}{2\rho_0\{\alpha + 2\beta(\lambda - 1) + 3\gamma(\lambda - 1)^2\}^{3/2}\rho_0^{-3/2}}.$$

Define this (constant) coefficient to be k, i.e.

$$k = \frac{\{\beta + 3\gamma(\lambda - 1)\}\sqrt{\rho_0}}{\{\alpha + 2\beta(\lambda - 1) + 3\gamma(\lambda - 1)^2\}^{3/2}}. \qquad (7.22)$$

Thus, equation (7.21) may be rewritten as

$$\frac{\delta a}{\delta t} + ka^2 = 0.$$

Therefore, at the wave,

$$\int_{a(0)}^{a(t)} \frac{da}{a^2} = -k\int_0^t ds.$$

This gives

$$-\frac{1}{a(t)} + \frac{1}{a(0)} = -kt$$

whence,

$$a(t) = \frac{a(0)}{1 + a(0)kt}.$$

We see that $a(t)$ blows up in a finite time if

$$a(0)k < 0.$$

The blow-up time is

$$T = -\frac{1}{a(0)k}.$$

When the amplitude blows up, $a(t) = u_{tt}^- - u_{tt}^+ \to \infty$ and it is believed a shock wave forms, i.e. u_t develops a discontinuity across $\mathcal{S}$.

For blow-up, since the denominator in (7.22) is positive, we need

$$a(0)\big(\beta + 3\gamma(\lambda - 1)\big) < 0.$$

Thus,
if $a(0) < 0$, we need $\beta + 3\gamma(\lambda - 1) > 0$,
if $a(0) > 0$, we need $\beta + 3\gamma(\lambda - 1) < 0$.

The structure of an acceleration wave as it evolves, and in particular, as the amplitude blows up, is a difficult numerical problem. It has been the subject of numerical calculation by (Jordan and Christov, 2005), (Christov et al., 2006). In general, numerical solution of blow-up in pdes is a subject of much recent interest, see e.g. (Straughan, 1998), (Ushijima, 2000), (Kirane et al., 2005), (Hirota and Ozawa, 2006), (Galakov, 2007), and the references therein.

7.1.7 Acceleration waves in three dimensions

An acceleration wave in an elastic body in three-dimensions is defined as in section 7.1.5. Namely, u_i is C^1 everywhere and the second and higher derivatives of u_i are allowed to have finite discontinuities across a surface $\mathcal{S}$. For simplicity we take the body force $f_i = 0$.

The basic governing equations are then (7.4) with $f_i = 0$, so

$$\rho_0 \ddot{u}_i = \frac{\partial}{\partial X_A}\left(\frac{\partial W}{\partial u_{i,A}}\right)$$

or since $W = W(u_{i,A})$ we find using the chain rule that

$$\rho_0 \ddot{u}_i = \frac{\partial^2 W}{\partial u_{r,B}\partial u_{i,A}} u_{r,BA}. \tag{7.23}$$

We take the jump of equation (7.23) to find

$$\rho_0 [\ddot{u}_i] = \frac{\partial^2 W}{\partial u_{r,B}\partial u_{i,A}} [u_{r,BA}]. \tag{7.24}$$

General compatibility relations for a function $\psi(\mathbf{X}, t)$ are needed across $\mathcal{S}$. These are given in detail in (Truesdell and Toupin, 1960) or in (Chen, 1973). We simply quote those we need. If ψ is continuous in $\mathbb{R}^3$ but its derivative is discontinuous across $\mathcal{S}$ then

$$[\psi_{,A}] = N_A B, \qquad \text{where } B = [N^R \psi_{,R}]. \tag{7.25}$$

When $\psi \in C^1(\mathbb{R}^3)$ then

$$[\psi_{,AB}] = N_A N_B C, \qquad \text{where } C = [N^R N^S \psi_{,RS}]. \tag{7.26}$$

In (7.25) and (7.26), N_A refers to the unit normal to $\mathcal{S}$, but referred back to the reference configuration. Relations (7.25) and (7.26) are derived from (Chen, 1973), equations (4.13), (4.14). The relation corresponding to the Hadamard formula (7.8) in three dimensions is, cf. (Chen, 1973) (4.15),

$$\frac{\delta}{\delta t}[\psi] = [\dot{\psi}] + U_N B \tag{7.27}$$

where $\dot{\psi} = \partial\psi/\partial t|_{\mathbf{X}}$, U_N is the speed at the point on $\mathcal{S}$ with unit normal N_A and B is defined in (7.25).

Since $u_i \in C^1(\mathbb{R}^3)$ we find using (7.27)

$$0 = \frac{\delta}{\delta t}[\dot{u}_i] = [\ddot{u}_i] + U_N[N^A \dot{u}_{i,A}] \tag{7.28}$$

$$0 = \frac{\delta}{\delta t}[u_{i,A}] = [\dot{u}_{i,A}] + U_N[N^B u_{i,AB}]. \tag{7.29}$$

Whence from (7.28) and (7.29) we derive

$$[\ddot{u}_i] = -U_N[N^A \dot{u}_{i,A}] \tag{7.30}$$

$$[N^A \dot{u}_{i,A}] = -U_N[N^A N^B u_{i,AB}]. \tag{7.31}$$

Hence, combining (7.30) and (7.31) one finds

$$[\ddot{u}_i] = U_N^2[N^A N^B u_{i,AB}]. \tag{7.32}$$

By repeated use of (7.25) we find

$$\begin{aligned}
[u_{r,AB}] &= N_A[N^R u_{r,BR}] \\
&= N_A N_B[N^R N^S u_{r,RS}] \\
&= \frac{N_A N_B}{U_N^2}[\ddot{u}_r] \tag{7.33}
\end{aligned}$$

where in the last line we have employed (7.32).

We now define the amplitude of the acceleration wave, $a_i(t)$, by

$$a_i(t) = [\ddot{u}_i].$$

Note that we assume a_i does not vary with surface coordinates over the surface $\mathcal{S}$. Upon employing (7.33) in (7.24) we find

$$\rho_0 a_i = \frac{\partial^2 W}{\partial u_{r,B} \partial u_{i,A}} \frac{N_A N_B}{U_N^2} a_r. \tag{7.34}$$

Now, define the *acoustic tensor* $\mathbf{Q}$ by

$$Q_{ir} = N_A N_B \frac{\partial^2 W}{\partial u_{r,B} \partial u_{i,A}}. \tag{7.35}$$

Then, (7.34) may be rewritten as a wavespeed relation as follows

$$(\rho_0 U_N^2 \delta_{ir} - Q_{ir})a_r = 0. \tag{7.36}$$

This is an eigenvalue/eigenvector equation. The wavespeeds U_N^2 are effectively the eigenvalues of Q_{ij} and a_i are the eigenvectors. In fact, one may investigate conditions for propagation of a plane wave in three-dimensions from equation (7.36). It is necessary to investigate the amplitude. For example, if we can write $a_i = a n_i$, where n_i is the unit normal to $\mathcal{S}$ in the current configuration then we have a longitudinal wave. One can also consider propagation of transverse waves where $a_i = a s_i$, s_i being a tangential vector to

$\mathcal{S}$. The amplitudes of longitudinal and transverse waves may be calculated effectively as in section 7.1.5 although the calculation is more involved.

A very useful comparison of the propagation conditions for an acceleration wave in a nonlinear elastic material and for a plane wave in a linearly elastic body is given by (Ogden, 1997), pp. 473 – 478.

7.2 Acceleration waves, inclusion of voids

7.2.1 *Porous media, voids, applications*

As discussed in chapter 1 there are many types of porous media, e.g. sand, stone, high porosity metallic foams, animal fur, to name a few. Additionally, there are many kinds of theories for describing the evolutionary or static behaviour of porous media, e.g. using homogenization, mixture theories, classical theories of Darcy, Forchheimer, Brinkman. In this section we wish to examine another class of theory which is believed capable of describing certain motions in porous media. This is the theory of elastic materials containing voids developed by (Nunziato and Cowin, 1979). This theory is particularly useful to describe nonlinear wave motion and accounts well for the elastic behaviour of the matrix, being a generalisation of nonlinear elasticity theory. Interestingly, while there are many studies involving the linearised theory of elastic materials with voids, see e.g. (Ciarletta and Iesan, 1993) or (Iesan, 2004), analysis of the fully nonlinear equations is only beginning, see e.g. (Iesan, 2005; Iesan, 2006).

The basic idea of including voids in a continuous body is due to (Goodman and Cowin, 1972), although they developed constitutive theory appropriate to a fluid. This they claim is more appropriate to flow of a granular medium. Acceleration waves in the Goodman-Cowin theory of granular media were studied by (Nunziato and Walsh, 1977; Nunziato and Walsh, 1978). For a reader interested in the theory of voids I would suggest first reading the article of (Goodman and Cowin, 1972), and then progressing to the theory of elastic materials with voids as given by (Nunziato and Cowin, 1979). General descriptions of the theory of elastic materials with voids and various applications are given in the books of (Ciarletta and Iesan, 1993) and (Iesan, 2004). Continuous dependence on the coupling coefficients of the voids theory (a structural stability problem) is studied by (Chirita et al., 2006).

The potential application area for the theory of elastic materials with voids is huge. In particular, wave motion in elastic materials with voids has many applications. (Ciarletta et al., 2007) mention four application areas of immediate interest. To appreciate the potential uses we briefly describe these areas. (Ouellette, 2004) is a beautiful and inspiring article which deals with many applications of acoustic microscopy. We are all aware of optical microscopy, but the potential uses of acoustic microscopy are enormous.

(Ouellette, 2004) points out that the presence of voids presents a serious problem for acoustic microscopy, and a study of wave motion in an elastic material with voids is likely to be very helpful here. She observes that, "acoustic microscopy remains a niche technology and is especially sensitive to variations in the elastic properties of semiconductor materials, such as air gaps, known as delaminations or voids ..." In particular, (Ouellette, 2004) draws attention to several novel applications of acoustic microscopy in diagnostic medicine. She notes that one may, "apply a special ultrasound scanner to deliver pathological assessments of skin tumours or lesions, non-invasively," and especially there is, "no need to kill the specimen as is usually needed in optical microscopy." (Diebold, 2005) further emphasizes these and other applications.

Wave motion is important in the production of ceramics, or certainly in ceramic behaviour. (Saggio-Woyansky et al., 1992) observe that porous ceramics are either reticulate or foam and are made up of a porous network which has relatively low mass, low thermal conductivity, and low density, and (Raiser et al., 1994) report experimental results where microcracking along grain boundaries in ceramics is caused by compressive waves. Since reticulate porous ceramics are used for molten metal filters, diesel engine exhaust filters, as catalyst supports, and industrial hot-gas filters, and both reticulate and foam porous ceramics are used as light-structure plates, in gas combustion burners, and in fire - protection and thermal insulation materials, a study of wave motion in such materials is clearly useful.

A further important application area for elastic materials with voids is in the production of building materials such as bricks. Modern buildings are usually made with lighter, thinner bricks, often with many voids in the building materials. In seismic areas lighter materials are necessary and much applied research activity is taking place. However, the use of lighter materials, especially those with voids is creating an environmental problem because noise transmission through such objects is considerably greater. Consequently, there is much applied research ongoing in the area of acoustic materials with voids, cf.(Garai and Pompoli, 2005), (Maysenhölder et al., 2004), (Wilson, 1997), and any theoretical model for acoustic wave propagation in an elastic material with voids which yields useful results is desirable.

7.2.2   Basic theory of elastic materials with voids

The balance equations for a continuous body containing voids are given by (Goodman and Cowin, 1972). We use the equations as given by (Nunziato and Cowin, 1979) since these are appropriate for an elastic body.

The key thing is to assume that there is a distribution of voids throughout the body B. If $\gamma(\mathbf{X}, t)$ denotes the density of the elastic matrix, then the mass density $\rho(\mathbf{X}, t)$ of B has form

$$\rho = \nu\gamma \tag{7.37}$$

where $0 < \nu \leq 1$ is a volume distribution function with $\nu = \nu(\mathbf{X}, t)$. Since the density or void distribution in the reference configuration can be different we also have

$$\rho_0 = \nu_0\gamma_0$$

where ρ_0, γ_0, ν_0 are the equivalent functions to ρ, γ, ν, but in the reference configuration.

The first balance law is the balance of mass

$$\rho|\det \mathbf{F}| = \rho_0.$$

With π_{Ai} being the Piola-Kirchoff stress tensor and $F_{iA} = x_{i,A}$ as before, the balance of angular momentum states

$$\pi\mathbf{F}^T = \mathbf{F}\pi^T.$$

The balance of linear momentum has form

$$\rho_0\ddot{x}_i = \pi_{Ai,A} + \rho_0 f_i, \tag{7.38}$$

f_i being an external body force. The balance law for the voids distribution is

$$\rho_0 k\ddot{\nu} = h_{A,A} + g + \rho_0\ell, \tag{7.39}$$

where k is an inertia coefficient, h_A is a stress vector, g is an intrinsic body force (giving rise to void creation/extinction inside the body), and ℓ is an external void body force. Actually, (Nunziato and Cowin, 1979) allow the inertia coefficient k to depend on $\mathbf{X}$ and/or t, but, for simplicity, we follow (Goodman and Cowin, 1972) and assume it to be constant.

The energy balance in the body may be expressed as

$$\rho_0\dot{\epsilon} = \pi_{Ai}\dot{F}_{iA} + h_A\dot{\nu}_{,A} - g\dot{\nu} - q_{A,A} + \rho_0 r, \tag{7.40}$$

where ϵ, q_A and r are, respectively, the internal energy function, the heat flux vector, and the externally supplied heat supply function. To understand equation (7.40) we may integrate it over a fixed body B, integrate by parts, and use the divergence theorem to see that

$$\frac{d}{dt}\int_B \rho_0\epsilon dV + \int_B (g\dot{\nu} + h_{A,A}\dot{\nu})dV = \int_B \pi_{Ai}\dot{F}_{iA}dV - \oint_{\partial B} q_A N_A dS + \int_B \rho_0 r dV,$$

where ∂B is the boundary of B. Employing (7.39) with $\ell = 0$ we may rewrite the above as

$$\frac{d}{dt}\int_B (\rho_0\epsilon + \frac{\rho_0 k}{2}\dot{\nu}^2)dV = \int_B \pi_{Ai}\dot{F}_{iA}dV - \oint_{\partial B} q_A N_A dS + \int_B \rho_0 r dV.$$

In this form we recognise the equation as an energy balance equation with a term added due to the kinetic energy of the voids. In fact, (Iesan, 2004), pp. 3–5, shows how one may begin with a conservation of energy law for an arbitrary sub-body of a continuous medium with voids, and then derive equations (7.38), (7.39) and (7.40) from the initial energy balance equation.

It is usual in continuum thermodynamics to also introduce an entropy inequality. We use the Clausius-Duhem inequality

$$\rho_0\dot\eta \geq -\left(\frac{q_A}{\theta}\right)_{,A} + \frac{\rho_0 r}{\theta}, \qquad (7.41)$$

where η is the specific entropy function. Observe that the sign of the first term on the right of (7.41) is different from that of (Nunziato and Cowin, 1979). (One could use a more sophisticated entropy inequality where q_A/θ is replaced by a general entropy flux $\mathbf{k}$, as in (Goodman and Cowin, 1972), but the above is sufficient for our purpose.) In chapters 2 - 6 we have employed T to denote the absolute temperature. However, in the solid mechanics literature and in the literature involving acceleration waves it is more usual to employ θ to denote the absolute temperature. This notation fits in well with employing ϵ, ψ, η to denote internal energy, Helmholtz free energy, and entropy. Therefore, in this chapter, chapter 7, and the next, chapter 8, θ will denote the absolute temperature in a body.

7.2.3   Thermodynamic restrictions

We consider an elastic body containing voids to be one which has as constitutive variables the set

$$\Sigma = \{\nu_0, \nu, F_{iA}, \theta, \theta_{,A}, \nu_{,A}\} \qquad (7.42)$$

supplemented with $\dot\nu$. Thus, the constitutive theory assumes

$$\begin{aligned}
\epsilon &= \epsilon(\Sigma, \dot\nu), & \pi_{Ai} &= \pi_{Ai}(\Sigma, \dot\nu), & q_A &= q_A(\Sigma, \dot\nu), \\
\eta &= \eta(\Sigma, \dot\nu), & h_A &= h_A(\Sigma, \dot\nu), & g &= g(\Sigma, \dot\nu).
\end{aligned} \qquad (7.43)$$

This is different from (Nunziato and Cowin, 1979) who regard η as the independent variable rather than θ and they also assume $q_A = 0$.

To proceed we introduce the Helmholtz free energy function ψ in the manner

$$\epsilon = \psi + \eta\theta. \qquad (7.44)$$

Next, (7.40) is employed to remove the terms $-q_{A,A} + \rho_0 r$ from inequality (7.41) and then utilize (7.44) to rewrite (7.41) as

$$-\rho_0(\dot\psi + \eta\dot\theta) - \frac{q_A\theta_{,A}}{\theta} + \pi_{Ai}\dot F_{iA} + h_A\dot\nu_{,A} - g\dot\nu \geq 0. \qquad (7.45)$$

The chain rule is used together with (7.43) to expand $\dot{\psi}$ and then (7.45) may be written as

$$-\left(\rho_0\frac{\partial\psi}{\partial\nu}+g\right)\dot{\nu}-\frac{q_A\theta_{,A}}{\theta}-\left(\rho_0\frac{\partial\psi}{\partial F_{iA}}-\pi_{Ai}\right)\dot{F}_{iA}$$

$$-\left(\rho_0\frac{\partial\psi}{\partial\theta}+\rho_0\eta\right)\dot{\theta}-\left(\rho_0\frac{\partial\psi}{\partial\nu_{,A}}-h_A\right)\dot{\nu}_{,A} \qquad (7.46)$$

$$-\rho_0\frac{\partial\psi}{\partial\theta_{,A}}\dot{\theta}_{,A}-\rho_0\frac{\partial\psi}{\partial\nu}\ddot{\nu}\geq 0.$$

The next step is to observe that $\dot{F}_{iA},\dot{\theta},\dot{\theta}_{,A},\dot{\nu}_{,A}$ and $\ddot{\nu}$ appear linearly in inequality (7.46). We may then follow the procedure of (Coleman and Noll, 1963) and assign an arbitrary value to each of these quantities in turn, balancing equations (7.38), (7.39) and (7.40) by a suitable choice of the externally supplied functions f_i,ℓ and r. We may in this manner violate inequality (7.46) unless the coefficients of $\dot{F}_{iA},\dot{\theta},\dot{\theta}_{,A},\dot{\nu}_{,A}$ and $\ddot{\nu}$ are each identically zero. Hence, we deduce that

$$\psi\neq\psi(\dot{\nu},\theta_{,A}),$$

$$h_A=\rho_0\frac{\partial\psi}{\partial\nu_{,A}}\quad\Rightarrow\quad h_A\neq h_A(\dot{\nu},\theta_{,A}), \qquad (7.47)$$

$$\pi_{Ai}=\rho_0\frac{\partial\psi}{\partial F_{iA}}\quad\Rightarrow\quad \pi_{Ai}\neq\pi_{Ai}(\dot{\nu},\theta_{,A}), \qquad (7.48)$$

$$\eta=-\frac{\partial\psi}{\partial\theta}\quad\Rightarrow\quad \eta\neq\eta(\dot{\nu},\theta_{,A}),$$

and further

$$\epsilon\neq\epsilon(\dot{\nu},\theta_{,A}).$$

The residual entropy inequality, left over from (7.46), which must hold for all motions is

$$-\left(\rho_0\frac{\partial\psi}{\partial\nu}+g\right)\dot{\nu}-\frac{q_A\theta_{,A}}{\theta}\geq 0.$$

Thus, to specify a material for an elastic body containing voids we have to postulate a suitable functional form for $\psi=\psi(\nu_0,\nu,F_{iA},\theta,\nu_{,A})$. Such a form is usually constructed with the aid of experiments. The functions g and q_A still involve $\dot{\nu}$ and this can lead to behaviour almost viscoelastic-like, see (Nunziato and Cowin, 1979). Other writers, e.g. (Iesan, 2004), (Ciarletta and Iesan, 1993), omit $\dot{\nu}$ from the constitutive list at the outset. In this manner one deduces that g may be given as a derivative of the Helmholtz free energy, (Iesan, 2004), p. 7, although some of the possibly desirable features of viscoelasticity are lost. The wavespeeds of acceleration waves in this case are derived in (Iesan, 2004), (Ciarletta and Iesan, 1993).

7.2.4 Acceleration waves in the isothermal case

In this section we suppose the temperature $\theta = $ constant, then

$$\psi = \psi(\nu_0, \nu, F_{iA}, \nu_{,A}). \tag{7.49}$$

We wish to consider the propagation of an acceleration wave in an elastic body with voids and it is sufficient to consider the momentum equations (7.38) and (7.39) with $f_i = 0$ and $\ell = 0$. These equations thus become

$$\rho_0 \ddot{x}_i = \pi_{Ai,A}, \tag{7.50}$$

$$\rho_0 k \ddot{\nu} = h_{A,A} + g. \tag{7.51}$$

In the present context we define an acceleration wave to be a singular surface $\mathcal{S}$ across which $\ddot{x}_i$, $\dot{x}_{i,A}$, $x_{i,AB}$, $\ddot{\nu}$, $\dot{\nu}_{,A}$, and $\nu_{,AB}$ and higher derivatives suffer a finite discontinuity, but $x_i, \nu \in C^1(\mathbb{R}^3 \times [0, T])$, $[0, T]$ being the time interval. Upon expanding $\pi_{Ai,A}$ and $h_{A,A}$ using (7.49) and (7.47), (7.48) we see that

$$\pi_{Ai,A} = \frac{\partial \pi_{Ai}}{\partial \nu_0} \nu_{0,A} + \frac{\partial \pi_{Ai}}{\partial \nu} \nu_{,A} + \frac{\partial \pi_{Ai}}{\partial F_{rK}} x_{r,KA} + \frac{\partial \pi_{Ai}}{\partial \nu_{,K}} \nu_{,KA} \tag{7.52}$$

$$h_{A,A} = \frac{\partial h_A}{\partial \nu_0} \nu_{0,A} + \frac{\partial h_A}{\partial \nu} \nu_{,A} + \frac{\partial h_A}{\partial F_{iK}} x_{i,KA} + \frac{\partial h_A}{\partial \nu_{,K}} \nu_{,KA}. \tag{7.53}$$

Then, recalling the definition of an acceleration wave and noting g is continuous, we use (7.52) and (7.53) in (7.50) and (7.51) and take the jump of the resulting equations to find

$$\rho_0 [\ddot{x}_i] = \frac{\partial \pi_{Ai}}{\partial F_{rK}} [x_{r,KA}] + \frac{\partial \pi_{Ai}}{\partial \nu_{,K}} [\nu_{,KA}], \tag{7.54}$$

$$\rho_0 k [\ddot{\nu}] = \frac{\partial h_A}{\partial F_{iK}} [x_{i,KA}] + \frac{\partial h_A}{\partial \nu_{,K}} [\nu_{,KA}]. \tag{7.55}$$

We next apply (7.33) of section 7.1.7 to $[x_{r,KA}]$ and $[\nu_{,KA}]$ to see that

$$[x_{r,KA}] = \frac{N_K N_A}{U_N^2} [\ddot{x}_r], \qquad [\nu_{,KA}] = \frac{N_K N_A}{U_N^2} [\ddot{\nu}].$$

Upon utilizing the above two expressions in equations (7.54) and (7.55) and defining the amplitudes a_i and b by

$$a_i(t) = [\ddot{x}_i], \qquad b(t) = [\ddot{\nu}],$$

we derive

$$\rho_0 U_N^2 a_i = Q_{ir} a_r + \frac{\partial \pi_{Ai}}{\partial \nu_{,K}} N_K N_A b, \tag{7.56}$$

$$\rho_0 k U_N^2 b = \frac{\partial h_A}{\partial F_{iK}} N_K N_A a_i + Q_c b, \tag{7.57}$$

where Q_{ir} is the (elastic) acoustic tensor given by

$$Q_{ir} = N_K N_A \frac{\partial \pi_{Ai}}{\partial F_{rK}}$$

and Q_c is an "acoustic variable" associated with the voids, given by

$$Q_c = N_K N_A \frac{\partial h_A}{\partial \nu_{,K}}.$$

If we recall expressions (7.47) and (7.48) for h_A and π_{Ai}, then we find

$$\frac{\partial \pi_{Ai}}{\partial \nu_{,K}} N_K N_A = -N_K N_A \rho_0 \frac{\partial^2 \psi}{\partial \nu_{,K} \partial F_{iA}},$$

$$\frac{\partial h_A}{\partial F_{iK}} N_K N_A = -N_K N_A \rho_0 \frac{\partial^2 \psi}{\partial F_{iK} \partial \nu_{,A}}.$$

Since the second derivatives of ψ are continuous the right hand sides of these expressions are the same and we set each equal to J_i. From equations (7.56) and (7.57) we then deduce the propagation conditions

$$(\rho_0 U_N^2 \delta_{ij} - Q_{ij})a_j = J_i b, \tag{7.58}$$

$$(\rho_0 k U_N^2 - Q_c)b = J_i a_i. \tag{7.59}$$

From this juncture there are various avenues to explore. For example, we could consider

    (a) $a_i = a(t)n_i$, a longitudinal wave,

    (b) $a_i = \hat{a}(t)s_i$, s_i is a tangential vector to $\mathcal{S}$, a transverse wave,

    (c) body has a centre of symmetry, then $J_i = 0$.

(The concept of a centre of symmetry involves the symmetry group of the material and whether orthogonal or proper orthogonal transformations are allowed. This topic is explained in (Spencer, 1980), pp. 106 – 110, (Ogden, 1997), pp. 180 – 183, 209 – 213, and (Truesdell and Noll, 1992), pp. 76 – 81, 149 – 151, although for elastic materials not containing voids.)

For example, in case (a) by taking the inner product of (7.58) with n_i we may deduce the wavespeed equation as

$$(\rho_0 U_N^2 - Q_{ij}n_i n_j)(\rho_0 k U_N^2 - Q_c) - (J_i n_i)^2 = 0.$$

This is a fourth order equation for U_N. It shows there are two waves, a fast wave and a slow wave each of which moves in the positive and negative n_i directions.

For case (c), $J_i = 0$, and we have two distinct, separate waves.

One may now proceed as in section 7.1.5 to differentiate (7.50), (7.51), take the jumps of the results and derive an amplitude equation, in the general 3-D case or in 1-D. The calculation is more involved than in section 7.1.5. However, a very interesting case is when the body has a centre of symmetry for which $J_i = 0$ and then from (7.58), (7.59) we have two distinct waves which propagate with speeds V_1, V_2 in the one-dimensional case, where $V_1^2 = \rho_0^{-1}(\partial\pi/\partial F)$ or $V_2^2 = (\rho_0 k)^{-1}(\partial h/\partial\nu_X)$. Such situations of distinct as opposed to coupled waves are considered in general in continuum mechanics in (Mariano and Sabatini, 2000). If we consider a wave moving into an equilibrium region then it depends which wave moves fastest, the elastic wave with speed V_1 or the void wave with speed V_2. We defer consideration of wave amplitude behaviour until the next section. In that section we consider the added effect of thermodynamics in the wave propagation problem. Then, we find the waves do *not* decouple as they do in the isothermal case studied in the current section.

A very interesting article dealing with the propagation of an acceleration wave in a porous material like sand which may incorporate plastic-like behaviour of a granular material is due to (Weingartner et al., 2006). These writers have the body satisfying the usual momentum equation

$$\rho\frac{dv_i}{dt} = f_i + T_{ji,j}\,,$$

but the stress is given by a constitutive law of form

$$\frac{dT_{ij}}{dt} + T_{ik}\omega_{kj} - \omega_{ik}T_{kj} = L_{ijkh}(\mathbf{T})d_{kh} + N_{ij}(\mathbf{T})\sqrt{d_{rs}d_{rs}}\,,$$

where $d_{ij} = (v_{i,j} + v_{j,i})/2$, $\omega_{ij} = (v_{i,j} - v_{j,i})/2$. The presence of the term $N_{ij}\sqrt{d_{rs}d_{rs}}$ leads to a non-standard analysis for the acceleration wave speed. In fact, ill-posedness and shear banding may occur.

7.3   Temperature rate effects

7.3.1   Voids and second sound

In this section we consider a theory of voids as developed by (Nunziato and Cowin, 1979) but we allow for the possibility of propagation of a temperature wave, by generalizing the voids theory in the thermodynamic framework of (Green and Laws, 1972). In addition to allowing us to explicitly examine the important effects of temperature this allows us to study the propagation of a temperature wave in a porous material. Temperature waves, or second sound as this phenomenon is known, is a subject of much recent activity. There are many theories available to incorporate second sound and (Jaisaardsuetrong and Straughan, 2007) quote eight such frameworks which have been proposed and are under intensive investigation. Among the many theories which allow heat to propagate as a thermal wave (Jaisaardsuetrong and Straughan, 2007) cite the theories of (Green and Laws, 1972), the two temperature theory of (Chen and Gurtin, 1968), the thermal gradient history-dependent theory of (Gurtin and Pipkin, 1968), the dual phase lag theory of (Tzou, 1995b), the τ theory of (Cattaneo, 1948), the theory of (Hetnarsky and Ignaczak, 1999), the internal variable theory used by (Caviglia et al., 1992b), and the more recent theory of (Green and Naghdi, 1991) whose thermodynamics employs an *entropy balance equation*, rather than an entropy inequality, and a thermal displacement variable

$$\alpha(\mathbf{x}, t) = \int_{t_0}^{t} \theta(\mathbf{x}, s) ds. \qquad (7.60)$$

In this section we concentrate on the theory of (Green and Laws, 1972) where a generalized temperature $\phi(\theta, \dot{\theta})$, θ being absolute temperature, is introduced. The (Green and Naghdi, 1991) α-theory is considered in the context of voids in the next section.

The current literature increasingly recognises the importance thermal waves have in the theory of porous media. A very clever way to dry a saturated porous material via second sound is due to (Meyer, 2006) and (Johnson et al., 1994) show how second sound may be employed to calculate physical properties of water saturated porous media. Both of these cover highly important and useful topics. (Kaminski, 1990) reports experimental results for materials with non-homogeneous inner structures which indicate relaxation times of order 11 – 54 seconds rather than order picoseconds as was previously thought. In the field of nanofluids (suspensions) (Vadasz et al., 2005a) stresses the importance of second sound and there is evidence that second sound may be a key mechanism for heat transfer in some biological tissues as the experiments of (Mitra et al., 1995) and the work of (Vedavarz et al., 1992) indicate. There is also much recent interest in second sound in liquid helium, HeII, ^{3}He - ^{4}He, filling a porous medium, see e.g. (Singer et al., 1984), (Buishvili et al., 2002), (Brusov et al., 2003), (Kekutia and Chkhaidze, 2005), and the references therein. Thus, we believe a theory of elastic materials with voids coupled to a suitable thermodynamic theory capable of admitting second sound has a place in modern engineering. One has to be careful how the theory of voids is married to the thermodynamics, however. The incorporation of time derivatives does present a serious problem. For example, (Straughan and Franchi, 1984) and (Franchi and Straughan, 1994) show that thermal convection in a fluid with a Maxwell-Cattaneo theory can lead to strange results depending on which objective time derivative one employs. One would also have to be careful using a time-lag theory as to what derivatives were employed and at which order the expansion was truncated. Therefore, in this and in the next section we use, respectively, the thermodynamics of (Green and Laws, 1972) and (Green and Naghdi, 1991) to develop a thermodynamic theory of elastic materials with voids. The thermodynamics of Green and his co-workers were specifically developed to incorporate into other areas of continuum mechanics and thus we believe these are natural approaches to use.

In this section we develop a thermo-poroacoustic theory which allows for nonlinear elastic effects and for the presence of voids, by using the thermodynamics of (Green and Laws, 1972). This thermodynamics utilises a generalized temperature $\phi(\theta, \dot{\theta})$ rather than just the standard absolute temperature θ.

7.3.2   Thermodynamics and voids

The starting point is to commence with the standard balance equations for an elastic material containing voids, cf. (Nunziato and Cowin, 1979), or equations (7.38), (7.39), (7.40), and we follow the approach of (Ciarletta and Straughan, 2007b),

$$\rho \ddot{x}_i = \pi_{Ai,A} + \rho F_i, \tag{7.61}$$

$$\rho k \ddot{\nu} = h_{A,A} + g + \rho \ell, \tag{7.62}$$

$$\rho \dot{\epsilon} = -q_{A,A} + \pi_{Ai} \dot{x}_{i,A} + h_A \dot{\nu}_{,A} - g \dot{\nu} + \rho r. \tag{7.63}$$

Here X_A denote reference coordinates, x_i denote spatial coordinates, a superposed dot denotes material time differentiation and $_{,A}$ signifies $\partial / \partial X_A$. The variable ρ is the reference density, and we use ρ rather than ρ_0 henceforth, for simplicity. Furthermore, ν is the void fraction, ϵ is the specific internal energy, k is the inertia coefficient, F_i, ℓ and r are externally supplied body force, extrinsic equilibrated body force, and externally supplied heat. The tensor π_{Ai} is the stress per unit area of the X_A–plane in the reference configuration acting over corresponding surfaces at time t (the Piola-Kirchoff stress tensor), q_A is the heat flux vector, and h_A and g are a vector and a scalar function arising in the conservation law for void evolution. (Nunziato and Cowin, 1979) refer to h_A as the equilibrated stress and they call g the intrinsic equilibrated body force.

The thermodynamic development commences with the entropy inequality of (Green and Laws, 1972), and this is

$$\rho \dot{\eta} - \frac{\rho r}{\phi} + \left(\frac{q_A}{\phi} \right)_{,A} \geq 0. \tag{7.64}$$

In this inequality η is the specific entropy and $\phi(> 0)$ is a generalised temperature function which reduces to θ in the equilibrium state. Next, introduce the Helmholtz free energy function ψ by $\psi = \epsilon - \eta \phi$ and rewrite inequality (7.64) using the energy equation (7.63) to obtain

$$-\rho \dot{\psi} - \rho \dot{\phi} \eta + \pi_{Ai} \dot{x}_{i,A} - \frac{q_A \phi_{,A}}{\phi} - g \dot{\nu} + h_A \dot{\nu}_{,A} \geq 0. \tag{7.65}$$

Now, we assume that the constitutive functions

$$\psi, \phi, \eta, \pi_{Ai}, q_A, h_A, g \tag{7.66}$$

depend on the variables

$$x_{i,A}, \nu, \nu_{,A}, \theta, \dot{\theta}, \theta_{,A}. \tag{7.67}$$

Note that we do not include $\dot{\nu}$ in the constitutive list and are so effectively following the voids approach of (Iesan, 2004), (Ciarletta and Iesan, 1993).

One then expands $\dot\psi$ and $\dot\phi$ in (7.65) and argues as in section 7.2.3 to reduce the constitutive equations. Inequality (7.65) expanded is

$$
\dot{x}_{i,A}\left(\pi_{Ai} - \rho\frac{\partial\psi}{\partial x_{i,A}} - \rho\eta\frac{\partial\phi}{\partial x_{i,A}}\right) - \dot\nu\left(\rho\frac{\partial\psi}{\partial\nu} + g + \rho\eta\frac{\partial\phi}{\partial\nu}\right)
$$
$$
- \dot\theta\left(\rho\frac{\partial\psi}{\partial\theta} + \rho\eta\frac{\partial\phi}{\partial\theta}\right) - \ddot\theta\left(\rho\frac{\partial\psi}{\partial\dot\theta} + \rho\eta\frac{\partial\phi}{\partial\dot\theta}\right)
$$
$$
- \dot\theta_{,A}\left(\rho\frac{\partial\psi}{\partial\theta_{,A}} + \rho\eta\frac{\partial\phi}{\partial\theta_{,A}} + \frac{q_A}{\phi}\frac{\partial\phi}{\partial\dot\theta}\right) - \dot\nu_{,A}\left(\rho\eta\frac{\partial\phi}{\partial\nu_{,A}} + \rho\frac{\partial\psi}{\partial\nu_{,A}} - h_A\right)
$$
$$
- \frac{q_A}{\phi}\, x_{i,AB}\frac{\partial\phi}{\partial x_{i,AB}} - \frac{q_A}{\phi}\frac{\partial\phi}{\partial\nu_{,J}}\nu_{,JA} - \frac{q_A}{\phi}\frac{\partial\phi}{\partial\theta_{,J}}\theta_{,JA}
$$
$$
- \frac{q_A}{\phi}\left(\frac{\partial\phi}{\partial\nu}\nu_{,A} + \frac{\partial\phi}{\partial\theta}\theta_{,A}\right) \geq 0. \tag{7.68}
$$

The terms in $x_{i,AB}, \nu_{,JA}$ and $\theta_{,JA}$ appear linearly and so using the fact that ℓ, r and F_i may be selected as we like to balance (7.61) – (7.63), we find

$$
\frac{\partial\phi}{\partial x_{i,A}} = 0, \qquad \frac{\partial\phi}{\partial\nu_{,A}} = 0, \qquad \frac{\partial\phi}{\partial\theta_{,A}} = 0. \tag{7.69}
$$

Thus

$$
\phi = \phi(\theta, \dot\theta, \nu). \tag{7.70}
$$

It is important to observe that the generalized temperature depends on ν in addition to θ and $\dot\theta$. Hence, the void fraction ν directly influences ϕ. Furthermore, the linearity of $\dot{x}_{i,A}, \dot\nu, \ddot\theta, \dot\theta_{,A}$ and $\dot\nu_{,A}$ in (7.68) then allows us to deduce that

$$
\pi_{Ai} = \rho\frac{\partial\psi}{\partial x_{i,A}}, \qquad q_A = -\rho\frac{\partial\psi}{\partial\theta_{,A}}\bigg/\frac{1}{\phi}\frac{\partial\phi}{\partial\dot\theta},
$$
$$
h_A = \rho\frac{\partial\psi}{\partial\nu_{,A}}, \qquad g = -\rho\left(\frac{\partial\psi}{\partial\nu} + \eta\frac{\partial\phi}{\partial\nu}\right), \tag{7.71}
$$

and

$$
\eta = -\frac{\partial\psi}{\partial\dot\theta}\bigg/\frac{\partial\phi}{\partial\dot\theta}. \tag{7.72}
$$

The residual entropy inequality which remains from (7.68) after this procedure, has form

$$
-\dot\theta\left(\rho\frac{\partial\psi}{\partial\theta} + \rho\eta\frac{\partial\phi}{\partial\theta}\right) - \frac{q_A}{\phi}\left(\frac{\partial\phi}{\partial\nu}\nu_{,A} + \frac{\partial\phi}{\partial\theta}\theta_{,A}\right) \geq 0. \tag{7.73}
$$

This inequality places a further restriction on all constitutive equations and motions.

7.3.3   Void-temperature acceleration waves

To study acceleration wave propagation (Ciarletta and Straughan, 2007b) let $[\cdot]$ denote the jump of a function across the singular surface $\mathcal{S}$ (acceleration wave), as in (7.6). They define an acceleration wave for equations (7.61) – (7.63) to be a singular surface $\mathcal{S}$ across which x_i, ν and θ together with their first derivatives are continuous, but the second and higher derivatives suffer a finite discontinuity. They denote by a_i, B, C the wave amplitudes, given as

$$a_i = [\ddot{x}_i], \qquad B = [\ddot{\nu}], \qquad C = [\ddot{\theta}]. \tag{7.74}$$

(Ciarletta and Straughan, 2007b) let N_A be the unit normal vector to $\mathcal{S}$ in the reference configuration and they let U_N denote the corresponding speed of $\mathcal{S}$ at point (X_A, t) in the reference configuration. They expand (7.61) – (7.63) with F_i, ℓ and r zero and take the jumps of the equations to obtain

$$
\begin{aligned}
\rho a_i =& \frac{\partial \pi_{Ai}}{\partial F_{jB}} \frac{1}{U_N^2} N_A N_B a_j + \frac{1}{U_N^2} \frac{\partial \pi_{Ai}}{\partial \nu_{,B}} N_A N_B B \\
& - \frac{1}{U_N} \frac{\partial \pi_{Ai}}{\partial \dot{\theta}} N_A C + \frac{1}{U_N^2} N_A N_B \frac{\partial \pi_{Ai}}{\partial \theta_{,B}} C,
\end{aligned} \tag{7.75}
$$

$$
\begin{aligned}
\rho k B =& \frac{\partial h_A}{\partial F_{iB}} \frac{1}{U_N^2} N_A N_B a_i + \frac{1}{U_N^2} \frac{\partial h_A}{\partial \nu_{,B}} N_A N_B B \\
& - \frac{1}{U_N} \frac{\partial h_A}{\partial \dot{\theta}} N_A C + \frac{1}{U_N^2} N_A N_B \frac{\partial h_A}{\partial \theta_{,B}} C,
\end{aligned} \tag{7.76}
$$

$$
\begin{aligned}
& -\rho \frac{\partial \epsilon}{\partial F_{iA}} \frac{1}{U_N} a_i N_A - \rho \frac{\partial \epsilon}{\partial \nu_{,A}} \frac{1}{U_N} N_A B + \rho \frac{\partial \epsilon}{\partial \dot{\theta}} C - \rho \frac{\partial \epsilon}{\partial \theta_{,A}} \frac{1}{U_N} N_A C \\
& = -\frac{\partial q_A}{\partial F_{iB}} \frac{1}{U_N^2} N_A N_B a_i - \frac{\partial q_A}{\partial \nu_{,B}} \frac{1}{U_N^2} N_A N_B B + \frac{\partial q_A}{\partial \dot{\theta}} \frac{1}{U_N} N_A C \\
& \quad - \frac{\partial q_A}{\partial \theta_{,B}} \frac{1}{U_N^2} N_A N_B C - \pi_{Ai} \frac{1}{U_N} N_A a_i - h_A \frac{1}{U_N} N_A B.
\end{aligned} \tag{7.77}
$$

(Ciarletta and Straughan, 2007b) observe that one may proceed to calculate wavespeeds and amplitudes in a general setting. However, the analysis is more transparent, with no major loss of physics, if one assumes the acceleration wave is moving into an equilibrium region for which ν^+, θ^+ and x_i^+ are constants. They also suppose the body has a centre of symmetry. For such a wave moving into an equilibrium region equations (7.75) – (7.77) simplify to

$$(Q_{ij} - \rho U_N^2 \delta_{ij}) a_j = U_N N_A \frac{\partial \pi_{Ai}}{\partial \dot{\theta}} C, \tag{7.78}$$

$$\left(\rho k U_N^2 - N_A N_B \frac{\partial h_A}{\partial \nu_{,B}} \right) B = N_A N_B \frac{\partial h_A}{\partial \theta_{,B}} C, \tag{7.79}$$

$$\left(\rho \frac{\partial \epsilon}{\partial \theta} U_N^2 + N_A N_B \frac{\partial q_A}{\partial \theta,_B} \right) C$$

$$= -\frac{\partial q_A}{\partial \nu,_B} N_A N_B B + \rho U_N N_A \phi \frac{\partial \eta}{\partial F_{iA}} a_i, \tag{7.80}$$

where Q_{ij} is the acoustic tensor, cf. (7.35), given by

$$Q_{ij} = N_A N_B \frac{\partial \pi_{Ai}}{\partial F_{jB}}. \tag{7.81}$$

(Ciarletta and Straughan, 2007b) note that the wavespeed equations (7.78) – (7.80) are very different to those of (Nunziato and Cowin, 1979), and they also lead to a different outcome from equations (7.58), (7.59) of section 7.2.4. In the above theory the elastic wave (associated with a_i) and the voids wave (associated with B) do *not* decouple as they do in (Nunziato and Cowin, 1979), or in equations (7.58), (7.59). We believe this is due to the richness endowed by the (Green and Laws, 1972) thermodynamics. The key point being that the terms $\partial \pi_{Ai}/\partial \dot{\theta}$ and $\partial h_A/\partial \theta,_B$ do not vanish. Equations (7.78) – (7.80) demonstrate that the waves do not decouple, and thermodynamic effects play an important role.

From equation (7.78) (Ciarletta and Straughan, 2007b) deduce the existence of the propagation direction for a plane wave. They note that, cf. (Chen, 1973), equation (4.10), $N_A = F_{iA}(|\nabla_{\mathbf{x}} s|/|\nabla_{\mathbf{X}} \mathcal{S}|) n_i$ where n_i is the equivalent unit normal in the current configuration. They then define $\beta_{ij} = (|\nabla_{\mathbf{x}} s|/|\nabla_{\mathbf{X}} \mathcal{S}|)(\partial \pi_{Ai}/\partial \dot{\theta}) F_{jA}$ and it then follows as in the equivalent analysis for a purely elastodynamic body in (Lindsay and Straughan, 1979) that a plane wave may propagate in a direction $\mathbf{n}^*$ where $\beta_{ij} n_j^*$ is an eigenvector of Q_{ij}. If we let ν_i be the unit vector in the direction $\beta_{ij} n_j^*$ and set $a_i = A \nu_i$, one may deduce propagation of a generalized longitudinal plane wave in the direction $\mathbf{n}^*$ with amplitude in the direction $\beta_{ij} n_j^*$.

To find the wavespeeds and number of waves we take the inner product of (7.78) with the vector ν_i. The resulting system of equations in A, B, C then has a non-zero solution provided

$$(U_N^2 - U_M^2)(U_N^2 - U_P^2)(U_N^2 - U_T^2)$$
$$- (U_N^2 - U_P^2)U_N^2 K_1 - (U_N^2 - U_M^2)K_2 = 0. \tag{7.82}$$

In this equation the coefficients U_M^2, U_P^2, U_T^2, K_1 and K_2 are given by

$$U_M^2 = N_A N_B \nu_i \nu_j \frac{\partial^2 \psi}{\partial F_{iA} \partial F_{jB}}, \tag{7.83}$$

$$U_P^2 = \frac{N_A N_B}{k} \frac{\partial^2 \psi}{\partial \nu,_A \nu,_B}, \tag{7.84}$$

$$U_T^2 = \frac{N_A N_B}{\phi_{\dot{\theta}} \eta_{\dot{\theta}}} \frac{\partial^2 \psi}{\partial \theta,_A \theta,_B}, \tag{7.85}$$

$$K_1 = \frac{N_A N_K \nu_i \nu_j}{\phi_{\dot\theta} \eta_{\dot\theta}} \frac{\partial^2 \psi}{\partial\dot\theta \partial F_{iA}} \frac{\partial^2 \psi}{\partial\dot\theta \partial F_{jK}},$$ (7.86)

$$K_2 = \frac{N_A N_B N_R N_S}{k\phi_{\dot\theta}\eta_{\dot\theta}} \frac{\partial^2\psi}{\partial\nu_{,A}\partial\theta_{,B}} \frac{\partial^2\psi}{\partial\nu_{,S}\partial\theta_{,R}},$$ (7.87)

where in deriving (7.82) from (7.78) – (7.80), we have employed equations (7.71), (7.72). The quantities U_M, U_P and U_T may be interpreted as follows. Firstly, U_M is the wavespeed of an elastic wave in the absence of other effects, cf. (Chen, 1973). Then, U_P is the wavespeed of a wave associated with the void fraction, cf. (Nunziato and Cowin, 1979), or section 7.2.4. Finally, U_T is the wavespeed of a thermal wave, cf. (Lindsay and Straughan, 1979). The quantities K_1 and K_2 represent cross derivative effects which in turn depend on the form of functional relationship for ψ. When K_1, K_2 are zero then equation (7.82) predicts propagation of three waves with different speeds U_M, U_P, U_T (together with three waves moving in the opposite direction). (Ciarletta and Straughan, 2007b) then argue that when K_1 and K_2 are not too large one may use a continuity argument to conclude (7.82) has three distinct real solutions U_N^2 and three distinct waves continue to propagate. Physically this appears reasonable. The wavespeed values found interpreted alongside experimental results may be useful in suggesting the correct form of functional relationship for ψ to employ in theoretical modelling of thermo-acoustic wave propagation in elastic materials with voids.

7.3.4 Amplitude behaviour

Details of the calculation of the evolutionary behaviour of the amplitudes are not given by (Ciarletta and Straughan, 2007b), only the final results are presented. Therefore, we now calculate the amplitudes A, B and C in the one-dimensional case, again when the region ahead of the wave is in equilibrium and the body has a centre of symmetry. The wavespeed equation is again (7.82) and in the one-dimensional case we have

$$U_M^2 = \psi_{FF}, \qquad U_P^2 = \frac{\psi_{\nu_X \nu_X}}{k}, \qquad U_T^2 = \frac{\psi_{\theta_X \theta_X}}{\phi_{\dot\theta}\eta_{\dot\theta}},$$

$$K_1 = \frac{\psi_{\dot\theta F}^2}{\phi_{\dot\theta}\eta_{\dot\theta}}, \qquad K_2 = \frac{\psi_{\nu_X \theta_X}^2}{k\phi_{\dot\theta}\eta_{\dot\theta}}.$$ (7.88)

The amplitude equations (7.78) – (7.80) become

$$(\rho V^2 - \pi_F)A = -V\pi_{\dot\theta}C,$$

$$(\rho k V^2 - h_{\nu_X})B = h_{\theta_X}C,$$ (7.89)

$$(\rho\epsilon_{\dot\theta}V^2 + q_{\theta_X})C = -Q_{\nu_X}B + \rho V\phi\eta_F A,$$

where V is now the wavespeed. We take F_i, ℓ and r equal to zero in (7.61) – (7.63) and the one-dimensional forms of these equations are

$$\rho\ddot{u} = \pi_X,$$
$$\rho k\ddot{\nu} = h_X + g,$$
$$\rho\dot{\epsilon} = -q_X + \pi\dot{F} + h\dot{\nu}_X - g\dot{\nu},$$
(7.90)

where π, h, q are the one-dimensional forms of π_{Ai}, h_A and q_A, and $F = \partial u/\partial X$. We differentiate each of these equations in turn, with respect to t, and then expand each in turn in terms of the constitutive variables. We then take the jumps of the resulting equations to obtain from each of equations $(7.90)_1$ to $(7.90)_3$ in turn,

$$\rho[\ddot{x}] = -\frac{1}{V^3}\pi_{FF}A^2 + \pi_F[\dot{F}_X] - \frac{\pi_\nu}{V}B - \frac{1}{V^3}\pi_{\nu_X\nu_X}B^2 - \frac{1}{V^3}\pi_{\nu_X\theta_X}BC$$
$$- \frac{\pi_\theta}{V}C + \frac{\pi_{\dot{\theta}F}}{V^2}AC + \pi_{\dot{\theta}}[\dot{\theta}_X] - \frac{\pi_{\theta_X\nu_X}}{V^3}BC - \frac{\pi_{\theta_X\theta_X}}{V^3}C^2,$$
(7.91)

$$\rho k[\ddot{\nu}] = -\frac{2h_{F\nu_X}}{V^3}AC - \frac{2h_{F\theta_X}}{V^3}AB + \frac{2h_{\nu_X\dot{\theta}}}{V^2}BC + \frac{2h_{\dot{\theta}\theta_X}}{V^2}C^2$$
$$- \frac{g_F}{V}A - \frac{g_{\dot{\theta}}}{V}C + h_{\nu_X}[\dot{\nu}_{XX}] + h_{\theta_X}[\dot{\theta}_{XX}],$$
(7.92)

$$\rho\epsilon_{FF}\frac{A^2}{V^2} - \frac{2\rho}{V}\epsilon_{\dot{\theta}F}AC + \rho\epsilon_F[\ddot{F}] + \rho\epsilon_\nu B + \frac{\rho}{V^2}\epsilon_{\nu_X\nu_X}B^2$$
$$+ \frac{2\rho}{V^2}\epsilon_{\nu_X\theta_X}BC + \rho\epsilon_\theta C + \rho\epsilon_{\dot{\theta}\dot{\theta}}C^2 + \rho\epsilon_{\dot{\theta}}[\ddot{\theta}] + \frac{\rho}{V^2}\epsilon_{\theta_X\theta_X}C^2$$
$$= \frac{2}{V^3}q_{F\nu_X}AB + \frac{2}{V^3}q_{F\theta_X}AC - \frac{2}{V^2}q_{\dot{\theta}\nu_X}BC - \frac{2}{V^2}q_{\dot{\theta}\theta_X}C^2$$
(7.93)
$$- q_{\nu_X}[\dot{\nu}_{XX}] - q_{\theta_X}[\dot{\theta}_{XX}] + \pi[\ddot{F}] + \frac{\pi_F}{V^2}A^2 - \frac{1}{V}\pi_{\dot{\theta}}AC$$
$$+ \frac{h_{\nu_X}}{V^2}B^2 + \frac{h_{\theta_X}}{V^2}BC - gB.$$

We now use the following expressions which one may derive with the aid of the Hadamard relation (7.8),

$$[\ddot{x}] = 2\frac{\delta A}{\delta t} + V^2[\dot{F}_X], \qquad [\ddot{\nu}] = 2\frac{\delta B}{\delta t} + V^2[\dot{\nu}_{XX}],$$
$$[\ddot{\theta}] = 2\frac{\delta C}{\delta t} + V^2[\dot{\theta}_{XX}], \qquad [\dot{F}_X] = -\frac{1}{V^2}\frac{\delta A}{\delta t} - \frac{1}{V}[\ddot{F}],$$
$$[\dot{\theta}_X] = -\frac{1}{V}\frac{\delta C}{\delta t} - V[\dot{\theta}_{XX}].$$

These relations are used in (7.75) – (7.77) to rewrite these equations as

$$2\rho\frac{\delta A}{\delta t} + (\rho V^2 - \pi_F)\left(-\frac{1}{V^2}\frac{\delta A}{\delta t} - \frac{1}{V}[\ddot{F}]\right)$$
$$- \pi_{\dot{\theta}}\left(-\frac{1}{V}\frac{\delta C}{\delta t} - V[\dot{\theta}_{XX}]\right) = R_1,$$
(7.94)

$$2\rho k \frac{\delta B}{\delta t} + (\rho k V^2 - h_{\nu x})[\dot{\nu}_{XX}] - h_{\theta x}[\dot{\theta}_{XX}] = R_2, \qquad (7.95)$$

$$2\rho \epsilon_{\dot\theta} \frac{\delta C}{\delta t} + (\rho \epsilon_{\dot\theta} V^2 + q_{\theta x})[\dot{\theta}_{XX}] + \rho \phi \eta_F [\ddot{F}] + q_{\nu x}[\dot{\nu}_{XX}] = R_3, \qquad (7.96)$$

where the terms $R_1, \ldots, R_3$, are given by,

$$R_1 = -\frac{1}{V^3}\pi_{FF}A^2 - \frac{\pi_\nu}{V}B - \frac{1}{V^3}\pi_{\nu_x\nu_x}B^2 - \frac{2}{V^3}\pi_{\nu_x\theta_x}BC$$
$$- \frac{1}{V}\pi_\theta C + \frac{1}{V^2}\pi_{\dot\theta F}AC - \frac{1}{V^3}\pi_{\theta_x\theta_x}C^2, \qquad (7.97)$$

$$R_2 = -\frac{2}{V^3}h_{F\nu_x}AC - \frac{2}{V^3}h_{F\theta_x}AB + \frac{2}{V^2}h_{\nu_x\dot\theta}BC$$
$$+ \frac{2}{V^2}h_{\dot\theta\theta_x}C^2 - \frac{g_F}{V}A - \frac{g_{\dot\theta}}{V}C, \qquad (7.98)$$

$$R_3 = -\frac{\rho}{V^2}\epsilon_{FF}A^2 + \frac{2\rho}{V}\epsilon_{\dot\theta F}AC - \rho\epsilon_\nu B - \frac{\rho}{V^2}\epsilon_{\nu_x\nu_x}B^2$$
$$- \frac{2\rho}{V^2}\epsilon_{\nu_x\theta_x}BC - \rho\epsilon_\theta C - \rho\epsilon_{\dot\theta\dot\theta}C^2 - \frac{\rho}{V^2}\epsilon_{\theta_x\theta_x}C^2$$
$$+ \frac{2}{V^3}q_{F\nu_x}AB + \frac{2}{V^3}q_{F\theta_x}AC - \frac{2}{V^2}q_{\dot\theta\nu_x}BC - \frac{2}{V^2}q_{\dot\theta\theta_x}C^2 \qquad (7.99)$$
$$+ \frac{\pi_F}{V^2}A^2 - \frac{\pi_{\dot\theta}}{V}AC + \frac{h_{\nu x}}{V^2}B^2 + \frac{h_{\theta x}}{V^2}BC - gB.$$

After further use of the Hadamard relation (7.8), and use of (7.89) to eliminate $\delta C/\delta t$ from (7.94), we may rewrite (7.94) – (7.96) in the form

$$2\rho \frac{\delta A}{\delta t} - \left(\frac{\rho V^2 - \pi_F}{V}\right)[\ddot{F}] + \pi_{\dot\theta} V[\dot{\theta}_{XX}] = R_1, \qquad (7.100)$$

$$2\rho k \frac{\delta B}{\delta t} + (\rho k V^2 - h_{\nu x})[\dot{\nu}_{XX}] - h_{\theta x}[\dot{\theta}_{XX}] = R_2, \qquad (7.101)$$

$$2\rho \epsilon_{\dot\theta} \frac{\delta C}{\delta t} + (\rho \epsilon_{\dot\theta} V^2 + q_{\theta x})[\dot{\theta}_{XX}] + \rho \phi \eta_F [\ddot{F}] + q_{\nu x}[\dot{\nu}_{XX}] = R_3. \qquad (7.102)$$

The object is now to remove the terms involving three derivatives in (7.100) – (7.102), i.e. the terms involving $[\ddot{F}] = [\ddot{u}_X]$, $[\dot{\theta}_{XX}]$ and $[\dot{\nu}_{XX}]$. To do this we form the sum $V \times (7.100) + \mu_1 \times (7.101) + \mu_2 \times (7.102)$ and choose μ_1, μ_2 appropriately. In fact we make the choices

$$\mu_2 = \frac{(\rho V^2 - \pi_F)}{\rho\phi\eta_F},$$

$$\mu_1 = \frac{1}{h_{\theta x}}\left\{\pi_{\dot\theta}V^2 + \frac{(\rho V^2 - \pi_F)}{\rho\phi\eta_F}(\rho\epsilon_{\dot\theta}V^2 + q_{\theta x})\right\}.$$

Upon doing this and making use of the wavespeed equation (7.82), we remove the terms involving the derivatives. We next eliminate A and B employing relations (7.89)$_1$ and (7.89)$_2$ recalling the coefficients are constant so that

$$\frac{\delta A}{\delta t} = -\frac{V\pi_{\dot\theta}}{(\rho V^2 - \pi_F)}\frac{\delta C}{\delta t}, \qquad \frac{\delta B}{\delta t} = \frac{h_{\theta_X}}{(\rho k V^2 - h_{\nu_X})}\frac{\delta C}{\delta t}. \qquad (7.103)$$

The result is

$$-\frac{2\rho V^2 \pi_{\dot\theta}}{(\rho V^2 - \pi_F)}\frac{\delta C}{\delta t} + \frac{2\rho k \mu_1 h_{\theta_X}}{(\rho k V^2 - h_{\nu_X})}\frac{\delta C}{\delta t} + 2\rho\epsilon_{\dot\theta}\mu_2 \frac{\delta C}{\delta t} \qquad (7.104)$$
$$= V R_1 + \mu_1 R_2 + \mu_2 R_3.$$

After some simplification equation (7.104) may be written as

$$\frac{\delta C}{\delta t}\frac{2\rho\phi_{\dot\theta}\eta_{\dot\theta}}{\psi_{F\dot\theta}(V^2 - U_P^2)}\left\{K_2 + V^2 K_1 - (V^2 - U_P^2)(V^2 - U_T^2)\right.$$
$$\left. - (V^2 - U_M^2)(V^2 - U_T^2) - (V^2 - U_M^2)(V^2 - U_P^2)\right\} \qquad (7.105)$$
$$+ \hat\alpha C + \hat\beta C^2 = 0,$$

where the coefficients $\hat\alpha$ and $\hat\beta$ are given by

$$\hat\alpha = \rho\psi_{F\theta} + \frac{\rho\psi_{F\nu}\psi_{\nu_X\theta_X}}{k(V^2 - U_P^2)}$$
$$+ \left\{\frac{g_{\dot\theta}}{V}(V^2 - U_M^2) - g_F\psi_{F\dot\theta}\right\}\left\{\frac{\eta_{\dot\theta}(V^2 - U_T^2)}{\eta_F\psi_{\nu_X\theta_X}} + \frac{V^2\psi_{F\dot\theta}}{\psi_{\nu_X\theta_X}(V^2 - U_M^2)}\right\}$$
$$+ \frac{\rho\epsilon_\theta(V^2 - U_M^2)}{\phi\eta_F} + \frac{\rho\psi_{\dot\theta\nu}\psi_{\nu_X\theta_X}}{k\psi_{\dot\theta F}}\frac{(V^2 - U_M^2)}{(V^2 - U_P^2)}, \qquad (7.106)$$

$$\hat{\beta} = \frac{1}{V^3}\left\{\pi_{FF} + \frac{\pi_{\nu_X\nu_X}h_{\theta_X}^2}{\rho^2k^2(V^2-U_P^2)^2} + \frac{2\pi_{\nu_X\theta_X}h_{\theta_X}}{\rho k(V^2-U_P^2)} + \frac{\pi_{\dot{\theta}F}\pi_{\dot{\theta}}V^2}{(V^2-U_M^2)} + \pi_{\theta_X\theta_X}\right\}$$

$$- \frac{2}{V^2h_{\theta_X}}\left\{\pi_{\dot{\theta}}V^2 + \frac{\rho\eta_{\dot{\theta}}}{\eta_F}(V^2-U_M^2)(V^2-U_T^2)\right\}$$

$$\times \left\{\frac{h_{F\nu_X}\pi_{\dot{\theta}}}{\rho(V^2-U_M^2)} + h_{\theta\theta_X} + \frac{h_{F\theta_X}h_{\theta_X}\pi_{\dot{\theta}}}{\rho^2k(V^2-U_P^2)(V^2-U_M^2)} + \frac{h_{\nu_X\dot{\theta}}h_{\theta_X}}{\rho k(V^2-U_P^2)}\right\}$$

$$+ \frac{(V^2-U_M^2)}{\phi\eta_F}\left\{\frac{\epsilon_{FF}\pi_{\dot{\theta}}^2}{\rho(V^2-U_M^2)^2} + \frac{2\epsilon_{\dot{\theta}F}\pi_{\dot{\theta}}}{(V^2-U_M^2)} + \frac{\epsilon_{\nu_X\nu_X}h_{\theta_X}^2}{\rho V^2k^2(V^2-U_P^2)^2}\right.$$

$$+ \rho\epsilon_{\dot{\theta}\dot{\theta}} + \frac{2\epsilon_{\nu_X\theta_X}h_{\theta_X}}{kV^2(V^2-U_P^2)} + \frac{\rho\epsilon_{\theta_X\theta_X}}{V^2} + \frac{2q_{\dot{\theta}\theta_X}}{V^2}$$

$$+ \frac{2q_{F\nu_X}h_{\theta_X}\pi_{\dot{\theta}}}{k\rho^2V^2(V^2-U_M^2)(V^2-U_P^2)} + \frac{2q_{F\theta_X}\pi_{\dot{\theta}}}{\rho V^2(V^2-U_M^2)}$$

$$+ \frac{2q_{\dot{\theta}\nu_X}h_{\theta_X}}{k\rho V^2(V^2-U_P^2)} - \frac{\pi_F\pi_{\dot{\theta}}^2}{\rho^2(V^2-U_M^2)^2} - \frac{\pi_{\dot{\theta}}^2}{\rho(V^2-U_M^2)}$$

$$\left.- \frac{h_{\nu_X}h_{\theta_X}^2}{\rho^2k^2V^2(V^2-U_P^2)^2} + \frac{\pi_{\dot{\theta}}h_{\theta_X}^2}{\rho^2kV(V^2-U_M^2)(V^2-U_P^2)}\right\}. \tag{7.107}$$

Equation (7.105) may be solved as in (7.19), although in this case the expression for $C(t)$ is simpler because the coefficients are constants. If we denote the coefficient of $\delta C/\delta t$ on the left of (7.105) by K, and define $\alpha = \hat{\alpha}/K$, $\beta = \hat{\beta}/K$, then one may show

$$C(t) = \frac{\exp(-\alpha t)C(0)}{1 - C(0)(\beta/\alpha)[\exp(-\alpha t) - 1]}. \tag{7.108}$$

Equation (7.108) gives the behaviour of $C(t)$ directly. It is of interest to note that $C(t)$ may blow up in a finite time if $C(0), \alpha, \beta$ are such that

$$1 = C(0)\frac{\beta}{\alpha}\left[\exp(-\alpha t) - 1\right]$$

has a solution. The blow up time is given by

$$t = -\frac{1}{\alpha}\log\left[1 + \frac{\alpha}{\beta C(0)}\right].$$

The influence of the void evolution function g on the blow up time may be seen directly from the form for $\alpha = \hat{\alpha}/K$ as derived using (7.105) and (7.106). The function g plays no role in determining the wavespeeds, but it has a strong effect on amplitude behaviour via its presence in α in (7.108). Of course, once $C(t)$ is known, we also know $A(t)$ and $B(t)$ completely from equations (7.89)$_1$ and (7.89)$_2$.

7.4 Temperature displacement effects

7.4.1 Voids and thermodynamics

In this section we describe work of (Ciarletta et al., 2007) who employ the theory of (De Cicco and Diaco, 2002). These writers generalize the thermodynamic procedure of (Green and Naghdi, 1993) and use a thermal displacement variable

$$\alpha = \int_{t_0}^{t} \theta(\mathbf{X}, s)ds + \alpha_0, \qquad (7.109)$$

where $\mathbf{X}$ is the spatial coordinate in the reference configuration of the body with θ being the absolute temperature. A general procedure for deriving the equations for a continuous body from a single balance of energy equation is developed by (Green and Naghdi, 1995). These writers derive the conservation equations for balance of mass, momentum, and entropy. The work of (De Cicco and Diaco, 2002), like that of (Green and Naghdi, 1993) starts with an entropy balance equation. (De Cicco and Diaco, 2002) extend the (Green and Naghdi, 1993) thermoelasticity theory to include voids in the manner of (Nunziato and Cowin, 1979). The full nonlinear equations are derived by (De Cicco and Diaco, 2002), although they only utilize a linearized version. We follow (Ciarletta et al., 2007) and rederive the (De Cicco and Diaco, 2002) theory referring to a reference configuration and employing a first Piola-Kirchoff stress tensor, as opposed to the symmetric stress tensor formulation of (De Cicco and Diaco, 2002).

It is worth observing that (Green and Naghdi, 1993) write, ... "*This type of theory, ... thermoelasticity type II, since it involves no dissipation of energy is perhaps a more natural candidate for its identification as thermoelasticity than the usual theory.*" Moreover, (Green and Naghdi, 1993) observe that, ... "*This suggests that a full thermoelasticity theory - along with the usual mechanical aspects - should more logically include the present type of heat flow (type II) instead of the heat flow by conduction (classical theory, type I).*" (The words in brackets have been added for clarity.) We would argue that it is beneficial to develop a fully nonlinear acceleration wave analysis for a Green - Naghdi type II thermoelastic theory of voids.

7.4.2 De Cicco - Diaco theory

The starting point is to consider the momentum and balance of voids equations for an elastic material containing voids, see (7.38), (7.39),

$$\rho \ddot{x}_i = \pi_{Ai,A} + \rho F_i, \qquad (7.110)$$

$$\rho k \ddot{\nu} = h_{A,A} + g + \rho \ell. \qquad (7.111)$$

One needs a balance of energy and from (De Cicco and Diaco, 2002) this is

$$\rho\dot{e} = \pi_{Ai}\dot{x}_{i,A} + h_A\dot{\nu}_{,A} - g\dot{\nu} + \rho s\theta + (\theta\Phi_A)_{,A}. \tag{7.112}$$

In these equations X_A denote reference coordinates, x_i denote spatial coordinates, a superposed dot denotes material time differentiation and $_{,A}$ stands for $\partial/\partial X_A$. The variables ρ, ν, ϵ, k, are the reference density, the void fraction, the specific internal energy, and the inertia coefficient. The terms F_i, ℓ and s denote externally supplied body force, extrinsic equilibrated body force, and externally supplied heat. The tensor π_{Ai} is the stress per unit area of the X_A-plane in the reference configuration acting over corresponding surfaces at time t (the Piola-Kirchoff stress tensor), Φ_A is the entropy flux vector, and h_A and g are a vector and a scalar function arising in the conservation law for void evolution. These are referred to by (Nunziato and Cowin, 1979) as the equilibrated stress and the intrinsic equilibrated body force, respectively.

The next step is to use the entropy balance equation, see (Green and Naghdi, 1993), (De Cicco and Diaco, 2002),

$$\rho\theta\dot{\eta} = \rho\theta s + \rho\theta\xi + (\theta\Phi_A)_{,A} - \Phi_A\theta_{,A} \tag{7.113}$$

where ξ is the internal rate of production of entropy per unit mass, and η, θ are the specific entropy and the absolute temperature. Introduce the Helmholtz free energy function $\psi = \epsilon - \eta\theta$ and then equation (7.112) is rewritten with the aid of (7.113) as

$$\rho\dot{\psi} + \rho\eta\dot{\theta} = \pi_{Ai}\dot{x}_{i,A} + h_A\dot{\nu}_{,A} - g\dot{\nu} + \Phi_A\theta_{,A} - \rho\theta\xi. \tag{7.114}$$

The constitutive theory of (De Cicco and Diaco, 2002) writes the functions

$$\psi, \eta, \pi_{Ai}, \Phi_A, h_A, g, \xi, \tag{7.115}$$

as depending on

$$x_{i,A}, \nu, \nu_{,A}, \dot{\alpha}, \alpha_{,A}. \tag{7.116}$$

The function $\dot{\psi}$ is expanded using the chain rule, and rearranging terms, recollecting $\dot{\alpha} = \theta$, equation (7.114) may be written as

$$\dot{x}_{i,A}\left(\rho\frac{\partial\psi}{\partial x_{i,A}} - \pi_{Ai}\right) + \dot{\nu}_{,A}\left(\rho\frac{\partial\psi}{\partial\nu_{,A}} - h_{Ai}\right) + \dot{\alpha}_{,A}\left(\rho\frac{\partial\psi}{\partial\alpha_{,A}} - \Phi_A\right)$$
$$+ \rho\ddot{\alpha}\left(\frac{\partial\psi}{\partial\dot{\alpha}} - \eta\right) + \dot{\nu}\left(\rho\frac{\partial\psi}{\partial\nu} + g\right) + \rho\theta\xi = 0. \tag{7.117}$$

We now use the fact that $\dot{x}_{i,A}, \dot{\nu}_{,A}, \dot{\alpha}_{,A}, \ddot{\alpha}$ and $\dot{\nu}$ appear linearly in (7.117) and so one derives the forms, cf. (De Cicco and Diaco, 2002), equations (19),

$$\pi_{Ai} = \rho\frac{\partial\psi}{\partial x_{i,A}}, \qquad \Phi_A = \rho\frac{\partial\psi}{\partial\alpha_{,A}}, \qquad h_A = \rho\frac{\partial\psi}{\partial\nu_{,A}},$$
$$g = -\rho\frac{\partial\psi}{\partial\nu}, \qquad \eta = -\frac{\partial\psi}{\partial\theta} = -\frac{\partial\psi}{\partial\dot{\alpha}}, \qquad \xi = 0. \tag{7.118}$$

7.4.3 Acceleration waves

(Ciarletta et al., 2007) adopt the notation $[f] = f^- - f^+$ and define an acceleration wave for equations (7.110) – (7.112) to be a singular surface $\mathcal{S}$ across which x_i, ν and α together with their first derivatives are continuous, but the second and higher derivatives suffer a finite discontinuity. The wave amplitudes a_i, B, C are $a_i = [\ddot{x}_i], B = [\ddot{\nu}], C = [\ddot{\alpha}]$. By expanding equations (7.110) – (7.112) in terms of the constitutive variables and taking jumps of the resulting equations one finds, with F_i, ℓ and s zero, and use of the Hadamard relation (7.27)

$$
\rho a_i = \frac{\partial \pi_{Ai}}{\partial F_{jB}} \frac{1}{U_N^2} N_A N_B a_j + \frac{1}{U_N^2} \frac{\partial \pi_{Ai}}{\partial \nu_{,B}} N_A N_B B
$$
$$
- \frac{1}{U_N} \frac{\partial \pi_{Ai}}{\partial \dot{\alpha}} N_A C + \frac{1}{U_N^2} N_A N_B \frac{\partial \pi_{Ai}}{\partial \alpha_{,B}} C, \tag{7.119}
$$

$$
\rho k B = \frac{\partial h_A}{\partial F_{iB}} \frac{1}{U_N^2} N_A N_B a_i + \frac{1}{U_N^2} \frac{\partial h_A}{\partial \nu_{,B}} N_A N_B B
$$
$$
- \frac{1}{U_N} \frac{\partial h_A}{\partial \dot{\alpha}} N_A C + \frac{1}{U_N^2} N_A N_B \frac{\partial h_A}{\partial \alpha_{,B}} C, \tag{7.120}
$$

$$
-\rho \frac{\partial \epsilon}{\partial F_{iA}} \frac{1}{U_N} a_i N_A - \rho \frac{\partial \epsilon}{\partial \nu_{,A}} \frac{1}{U_N} N_A B + \rho \frac{\partial \epsilon}{\partial \dot{\alpha}} C - \rho \frac{\partial \epsilon}{\partial \alpha_{,A}} \frac{1}{U_N} N_A C
$$
$$
= \theta \frac{\partial \Phi_A}{\partial F_{iB}} \frac{1}{U_N^2} N_A N_B a_i + \theta \frac{\partial \Phi_A}{\partial \nu_{,B}} \frac{1}{U_N^2} N_A N_B B - \theta \frac{\partial \Phi_A}{\partial \dot{\alpha}} \frac{1}{U_N} N_A C \tag{7.121}
$$
$$
+ \theta \frac{\partial \Phi_A}{\partial \alpha_{,B}} \frac{1}{U_N^2} N_A N_B C - \pi_{Ai} \frac{1}{U_N} N_A a_i - h_A \frac{1}{U_N} N_A B,
$$

where U_N is the corresponding speed of $\mathcal{S}$ at point (X_A, t) in the reference configuration.

(Ciarletta et al., 2007) examine the novel effects associated with the current theory by supposing the acceleration wave is advancing into an equilibrium region for which ν^+, α^+ and x_i^+ are constants, and they suppose the body has a centre of symmetry.

In this case equations (7.119) – (7.121) become,

$$
(Q_{ij} - \rho U_N^2 \delta_{ij}) a_j = U_N N_A \frac{\partial \pi_{Ai}}{\partial \dot{\alpha}} C, \tag{7.122}
$$

$$
\left(\rho k U_N^2 - N_A N_B \frac{\partial h_A}{\partial \nu_{,B}} \right) B = N_A N_B \frac{\partial h_A}{\partial \alpha_{,B}} C, \tag{7.123}
$$

$$
\left(\rho \frac{\partial \epsilon}{\partial \dot{\alpha}} U_N^2 - \theta\, N_A N_B \frac{\partial \Phi_A}{\partial \alpha_{,B}} \right) C
$$
$$
= \theta \frac{\partial \Phi_A}{\partial \nu_{,B}} N_A N_B B + \rho U_N N_A \theta \frac{\partial \eta}{\partial F_{iA}} a_i, \tag{7.124}
$$

where Q_{ij} is the acoustic tensor defined by

$$Q_{ij} = N_A N_B \frac{\partial \pi_{Ai}}{\partial F_{jB}}. \tag{7.125}$$

As is found in section 7.3.3 the behaviour of an acceleration wave in the current theory is different from that of (Nunziato and Cowin, 1979). In the above theory the elastic wave (associated with a_i) and the voids wave (associated with B) do *not* decouple as they do in (Nunziato and Cowin, 1979). (Ciarletta et al., 2007) argue that this demonstrates the importance of the temperature displacement effect, and they believe the coupled theory (7.122) – (7.124) is to be expected and one should find three interconnected waves.

(Ciarletta et al., 2007) show that a plane wave may propagate in the direction n_j^* where $\beta_{ij} n_j^*$ is an eigenvector of Q_{ij} and $\beta_{ij} = (|\nabla_x s| / |\nabla_X S|)(\partial \pi_{Ai}/\partial \dot\alpha) F_{jA}$. Then we let ν_i be the unit vector in the direction $\beta_{ij} n_j^*$ and put $a_i = A\nu_i$.

One now forms the inner product of (7.122) with ν_i. This procedure yields a system of equations in A, B and C, and this in turn leads to the following sixth order equation for the wavespeed U_N,

$$(U_N^2 - U_M^2)(U_N^2 - U_P^2)(U_N^2 - U_T^2) \tag{7.126}$$

$$+ (U_N^2 - U_P^2)U_N^2 K_1 + (U_N^2 - U_M^2)K_2 = 0. \tag{7.127}$$

In (7.127) the coefficients U_M^2, U_P^2, U_T^2, K_1 and K_2 are given in terms of the Helmholtz free energy by

$$U_M^2 = N_A N_B \nu_i \nu_j \frac{\partial^2 \psi}{\partial F_{iA} \partial F_{jB}}, \tag{7.128}$$

$$U_P^2 = \frac{N_A N_B}{k} \frac{\partial^2 \psi}{\partial \nu_{,A} \nu_{,B}}, \tag{7.129}$$

$$U_T^2 = \frac{-N_A N_B}{\psi_{\dot\alpha\dot\alpha}} \frac{\partial^2 \psi}{\partial \alpha_{,A} \partial \alpha_{,B}}, \tag{7.130}$$

$$K_1 = \frac{N_A N_K \nu_i \nu_j}{\psi_{\dot\alpha\dot\alpha}} \frac{\partial^2 \psi}{\partial \dot\alpha \partial F_{iA}} \frac{\partial^2 \psi}{\partial \dot\alpha \partial F_{jK}}, \tag{7.131}$$

$$K_2 = \frac{N_A N_B N_R N_S}{k \psi_{\dot\alpha\dot\alpha}} \frac{\partial^2 \psi}{\partial \nu_{,A} \partial \alpha_{,B}} \frac{\partial^2 \psi}{\partial \nu_{,S} \partial \alpha_{,R}}. \tag{7.132}$$

The quantities U_M, U_P, U_T have an analogous interpretation to those of section 7.3.3. In fact, U_M is the wavespeed of an elastic wave in the absence of other effects, U_P is the wavespeed of a wave connected to the void fraction, and U_T is the wavespeed of a thermal displacement wave. The terms K_1 and K_2 represent mixed derivative effects arising from the form prescribed for ψ. If K_1 and K_2 are zero then equation (7.127) allows for propagation of three distinct waves with different speeds U_M, U_P, U_T (together with three waves moving in the opposite direction). (Ciarletta et al., 2007) argue that when K_1 and K_2 are not too large one may use a continuity argument to conclude that equation (7.127) has three distinct real solutions U_N^2 and three distinct waves propagate. They believe physically this is realistic and the values found for U_N^2 together with experimental results should suggest precise functional forms for ψ to use in theoretical modelling of thermo-acoustic wave propagation in elastic materials with voids.

One may follow the procedure of section 7.3.4 to calculate the amplitudes A, B, C as functions of time.

7.5   Voids and type III thermoelasticity

7.5.1   Thermodynamic theory

As we have seen in section 7.4, (De Cicco and Diaco, 2002) have developed a theory of thermoelasticity with voids which is a generalization of the dissipationless theory of thermoelasticity of (Green and Naghdi, 1993). The latter writers refer to this as thermoelasticity of type II, type I being the classical theory where the equation governing the temperature field is effectively parabolic as opposed to hyperbolic in type II theory. The theory of a thermoelastic body with voids corresponding to type I thermoelasticity was developed by D. Iesan, see e.g. (Iesan, 2004). However, (Green and Naghdi, 1992) have developed a further theory of thermoelasticity which employs the thermal displacement variable α and the thermodynamics of (Green and Naghdi, 1991; Green and Naghdi, 1995). This theory leads to what is essentially a second order in time equation for the thermal displacement field, but differently from the type II theory of (Green and Naghdi, 1993) the theory of (Green and Naghdi, 1992) does have damping and hence dissipation. (Green and Naghdi, 1991; Green and Naghdi, 1992) refer to this theory as being of type III.

The goal of this section is to develop a type III theory of thermoelasticity, but allowing for the accommodation of a distribution of voids throughout the body. The essential difference between type II and type III thermoelasticity is that the variable $\dot{\alpha}_{,A}$ is added to the constitutive list (7.116), whereas it is absent in section 7.4. We have not seen the work presented here elsewhere.

We commence with the balance laws for a thermoelastic body with voids, equations (7.38), (7.39) and (7.40). With ρ denoting the density in the reference configuration and referring everything to this configuration, we have the equation of momentum balance

$$\rho \ddot{x}_i = \pi_{Ai,A} + \rho f_i. \tag{7.133}$$

The equation of voids distribution is

$$\rho k \ddot{\nu} = h_{A,A} + g + \rho \ell. \tag{7.134}$$

The equation of energy balance is

$$\rho \dot{\epsilon} = \pi_{Ai} \dot{x}_{i,A} + h_A \dot{\nu}_{,A} - g\dot{\nu} + \rho s\theta - (\theta p_A)_{,A}. \tag{7.135}$$

The notation is as in section 7.4 excepting we let s be the heat supply and $p_A = q_A/\theta$ is the entropy flux vector. We choose this representation to keep in line with (Green and Naghdi, 1991; Green and Naghdi, 1992), and observe that $p_A = -\Phi_A$ where Φ_A is the entropy flux vector of (De Cicco and Diaco, 2002). We follow (Green and Naghdi, 1992) and postulate an entropy balance equation

$$\rho \dot{\eta} = \rho s + \rho \xi - p_{A,A}, \tag{7.136}$$

where ξ is the internal rate of production of entropy per unit mass. The variable θ is the absolute temperature and as in equation (7.109), $\alpha(\mathbf{X}, t)$ is the thermal displacement.

We next introduce the Helmholtz free energy function ψ in terms of the internal energy ϵ, entropy η and temperature θ, by $\psi = \epsilon - \eta\theta$. Then, from (7.135) and (7.136) it is a straightforward matter to derive the reduced energy equation, cf. (Green and Naghdi, 1992), equation (2.5),

$$\rho \dot{\psi} + \rho \eta \dot{\theta} = \pi_{Ai} \dot{x}_{i,A} + h_A \dot{\nu}_{,A} - g\dot{\nu} - \rho \xi \theta - \theta_{,A} p_A. \tag{7.137}$$

A thermoelastic body of type III which contains a distribution of voids is defined to be one for which the functions

$$\psi, \eta, \pi_{Ai}, p_A, h_A, g \text{ and } \xi \tag{7.138}$$

depend on the independent variables

$$F_{iA} = x_{i,A}, \nu, \nu_{,A}, \dot{\alpha}, \alpha_{,A}, \dot{\alpha}_{,A}. \tag{7.139}$$

We do not consider the inhomogeneous situation which would also require inclusion of X_A in the list (7.139), cf. (Iesan, 2004). Observe that we do not include $\dot{\nu}$ in the list (7.139). This follows (Iesan, 2004) and allows us to determine g from ψ.

The procedure now is to expand ψ in terms of the variables in the list (7.139), and recalling $\dot{\alpha} = \theta$, we obtain from (7.137),

$$\begin{aligned}
(\rho \psi_{F_{iA}} - \pi_{Ai})\dot{F}_{iA} + \dot{\nu}(\rho \psi_\nu + g) + \dot{\nu}_{,A}(\rho \psi_{\nu,A} - h_A) \\
+ \ddot{\alpha}(\rho \psi_{\dot{\alpha}} + \rho \eta) + \rho \psi_{\dot{\alpha},A} \ddot{\alpha}_{,A} + \dot{\alpha}_{,A}(p_A + \rho \psi_{\alpha,A}) + \rho \xi \dot{\alpha} = 0.
\end{aligned} \tag{7.140}$$

We observe that $\dot{F}_{iA}, \dot{\nu}_{,A}, \ddot{\alpha}, \ddot{\alpha}_{,A}, \dot{\nu}$, appear linearly in (7.140). Thus, we may deduce that the coefficients of these terms in (7.140) must be zero. The process is akin to that described in Appendix A of (Green and Naghdi, 1992). Thus, we find that

$$\begin{aligned}
\pi_{Ai} = \rho \psi_{F_{iA}}, \qquad g = -\rho \psi_\nu, \qquad h_A = \rho \psi_{\nu,A}, \\
\eta = -\psi_{\dot{\alpha}}, \qquad \psi \neq \psi(\dot{\alpha}_{,A}).
\end{aligned} \tag{7.141}$$

Hence, once we prescribe a functional form for the Helmholtz free energy function ψ we also know the stress tensor, entropy, and the voids functions h_A and g. What remains from (7.140) is

$$\rho\xi\dot{\alpha} + \dot{\alpha}_{,A}(\rho\psi_{\alpha,A} + p_A) = 0. \tag{7.142}$$

This leads to further restrictions on constitutive functions. We now also have that

$$
\begin{aligned}
\psi &= \psi(x_{i,A}, \nu, \nu_{,A}, \dot{\alpha}, \alpha_{,A}), \\
p_A &= p_A(x_{i,A}, \nu, \nu_{,A}, \dot{\alpha}, \alpha_{,A}, \dot{\alpha}_{,A}), \\
\xi &= \xi(x_{i,A}, \nu, \nu_{,A}, \dot{\alpha}, \alpha_{,A}, \dot{\alpha}_{,A}).
\end{aligned}
\tag{7.143}
$$

7.5.2 Linear theory

One may study acceleration waves in the nonlinear theory of section 7.5.1. The acceleration waves in this case do not have a separately propagating temperature wave as is found in section 7.4. The reason is that in some sense type III thermoelasticity behaves more like type I thermoelasticity. For acceleration wave motion in thermoelasticity without voids this is explained in detail by (Quintanilla and Straughan, 2004), and a similar explanation holds here. Nevertheless, the extra damping present in the current theory may be useful in practical problems and with this in mind we now develop the equations for a linear theory.

Let the body have a centre of symmetry although we allow it to be anisotropic. We denote the displacement in this section as u_i, cf. (7.3). We then write ψ as a quadratic function of the variables in the list (7.143). Thus,

$$
\begin{aligned}
\rho\psi = &\frac{1}{2}a_{iAjB}u_{i,A}u_{j,B} - \frac{a_1}{2}\theta^2 - \frac{a_2}{2}\nu^2 + A_{iA}\theta u_{i,A} + B_{iA}\nu u_{i,A} \\
&+ \frac{R_{AB}}{2}\nu_{,A}\nu_{,B} + S_{AB}\nu_{,A}\alpha_{,B} + \frac{T_{AB}}{2}\alpha_{,A}\alpha_{,B},
\end{aligned}
\tag{7.144}
$$

where a_{iAjB}, R_{AB}, T_{AB} have the following symmetries,

$$a_{iAjB} = a_{jBiA}, \qquad R_{AB} = R_{BA}, \qquad T_{AB} = T_{BA}.$$

From (7.141) we now see that

$$
\begin{aligned}
\pi_{Ai} &= a_{iAjB}u_{j,B} + A_{iA}\theta + B_{iA}\nu, & h_A &= R_{AB}\nu_{,B} + S_{AB}\alpha_{,B}, \\
\rho\eta &= a_1\theta - A_{iA}u_{i,A}, & g &= a_2\nu - B_{iA}u_{i,A}.
\end{aligned}
\tag{7.145}
$$

We also write

$$
\begin{aligned}
\rho\xi &= \phi_1\nu + \phi_2\dot{\alpha}, \\
p_A &= -K_{AB}\nu_{,B} - L_{AB}\alpha_{,B} - M_{AB}\dot{\alpha}_{,B}.
\end{aligned}
$$

From (7.142) one may use the cyclic thermomechanical process argument of (Green and Naghdi, 1991), section 9, to infer that L_{AB}, M_{AB}, R_{AB} are non-negative tensor forms, $\phi_2 \leq 0$, $\phi_1 = 0$, and $S_{AB} = K_{AB}$, $T_{AB} = L_{AB}$.

In this manner, equations (7.133), (7.134) and (7.136) lead to the linear equations

$$\rho \ddot{u}_i = (a_{iAjB}u_{j,B})_{,A} + (A_{iA}\theta)_{,A} + (B_{iA}\nu)_{,A},$$
$$\rho k \ddot{\nu} = (R_{AB}\nu_{,B})_{,A} + (K_{AB}\alpha_{,B})_{,A} + a_2\nu - B_{iA}u_{i,A}, \tag{7.146}$$
$$a_1\ddot{\alpha} = A_{iA}\dot{u}_{i,A} + \phi_2\dot{\alpha} + (K_{AB}\nu_{,B})_{,A} + (T_{AB}\alpha_{,B})_{,A} + (M_{AB}\dot{\alpha}_{,B})_{,A}.$$

One may study the boundary - initial value problem for (7.146). For example, uniqueness and stability are easily investigated either by using an energy method, or if definiteness of the elastic coefficients a_{iAjB} is not imposed, by a logarithmic convexity argument. For the latter one will be better employing a time integrated version of α as done by (Ames and Straughan, 1992; Ames and Straughan, 1997) and (Quintanilla and Straughan, 2000), these articles following the introduction of this method for the (Green and Laws, 1972), (Green, 1972), version of thermoelasticity in (Straughan, 1974).

One may also study one-dimensional waves as in (Green and Naghdi, 1992) and then (7.146) essentially reduce to

$$\rho u_{tt} = au_{xx} + A\theta_x + B\nu_x,$$
$$\rho k \nu_{tt} = R\nu_{xx} + K\alpha_{xx} + a_2\nu - Bu_x, \tag{7.147}$$
$$a_1\alpha_{tt} = Au_{tx} + \phi_2\alpha_t + K\nu_{xx} + T\alpha_{xx} + M\alpha_{txx}.$$

The damped character of the temperature wave is evident from (7.147) as is observed in the non voids case by (Green and Naghdi, 1992), page 262. If the displacement and voids effects are absent from (7.147)$_3$, then we see that α satisfies the equation

$$a_1\frac{\partial^2 \alpha}{\partial t^2} - M\frac{\partial^3 \alpha}{\partial t \partial x^2} = \phi_2\frac{\partial \alpha}{\partial t} + T\frac{\partial^2 \alpha}{\partial x^2}.$$

This equation clearly does not permit the possibility of undamped thermal waves, unless $M = \phi_2 = 0$. The damping evident in equations (7.147) may be useful for description of some practical situations.

7.6   Acceleration waves, microstretch theory

We now consider sound wave propagation in the microstretch theory of (Eringen, 1990; Eringen, 2004b). As mentioned in chapter 1 this is a more general voids theory than that of (Nunziato and Cowin, 1979), studied in section 7.2. Eringen's microstretch theory allows the porous particles to have a spin associated with each point in space. For wave motion such an effect could be important and we here include such a study from a nonlinear singular surface viewpoint. A general study of singular surface propagation in a continuous body formed of a thermo-microstretch material which has memory is given by (Iesan and Scalia, 2006).

The theory developed by (Eringen, 1990) includes temperature effects while (Eringen, 2004b) also includes electromagnetic effects which could be important in wave motion in ceramics, for example. However, we here restrict attention to waves in the isothermal theory, ignoring electromagnetic effects. The basic variables of the theory of (Eringen, 1990; Eringen, 2004b) are the displacement u_i, microstretch φ, and the microrotation vector ϕ_i. The microstretch theory of (Eringen, 1990; Eringen, 2004b) is based on balance laws for these quantities. These are balance of momentum,

$$\rho_0 \ddot{u}_i = \pi_{Ai,A} + \rho_0 f_i \tag{7.148}$$

and balance of microstretch

$$\rho_0 \frac{j_0}{2} \ddot{\varphi} = m_{A,A} + T + \rho_0 \ell, \tag{7.149}$$

in which we measure quantities in the current configuration but refer back to the reference configuration. Thus, π_{Ai} is a Piola-Kirchoff stress tensor, f_i is a prescribed body force, j_0 is the microinertia, m_A is a microstretch couple, ℓ is a prescribed microstretch source term and T (denoted by $t - s$ in (Eringen, 2004b)) is the microstretch stress. Here , A denotes $\partial/\partial X_A$. Equations (7.148) and (7.149) clearly have the same structure as equations (7.38) and (7.39) of the voids theory discussed in section 7.2.2. In addition to equations (7.148) and (7.149), the Eringen theory has a balance of spins equation of form

$$\rho_0 J \ddot{\phi}_i = m_{Ai,A} + \epsilon_{iAj} \pi_{Aj} + \rho_0 \ell_i, \tag{7.150}$$

where ℓ_i is an applied body couple density, m_{Ai} is the couple stress tensor, and we have taken the microinertia tensor $J_{ik} = J\delta_{ik}$ for simplicity. The constitutive theory assumes that

$$\pi_{Ai}, m_A, T \text{ and } m_{Ai} \tag{7.151}$$

are functions of the variables

$$F_{iA} = u_{i,A}, \ \phi_i, \ \phi_{i,A}, \ \varphi \text{ and } \varphi_{,A}. \tag{7.152}$$

In fact, (Eringen, 2004b) combines $u_{i,A}$ and ϕ_i into a single strain measure $e_{iA} = u_{i,A} + \epsilon_{Ami}\phi_m$.

An acceleration wave for the microstretch theory is defined to be a surface $\mathcal{S}$ such that u_i, ϕ_i and φ are continuous everywhere in $\mathbb{R}^3$ together with their first derivatives in t and X_A. However, the derivatives $\ddot{u}_i, \dot{u}_{i,A}, u_{i,AB}, \ddot{\phi}_i, \dot{\phi}_{i,A}, \phi_{i,AB}, \ddot{\varphi}, \dot{\varphi}_{,A}, \varphi_{,AB}$, together with their third and higher derivatives may suffer finite discontinuities across $\mathcal{S}$.

We now set the force terms f_i, ℓ and $\ell_i = 0$ and expand $\pi_{Ai,A}, m_{A,A}$ and $m_{Ai,A}$ in equations (7.148), (7.149) and (7.150) according to the constitutive theory (7.151), (7.152). We evaluate the result across the surface $\mathcal{S}$, recollecting u_i, ϕ_i and φ are everywhere C^1 to obtain from

(7.148), (7.150), (7.149),

$$\rho_0[\ddot{u}_i] = \frac{\partial \pi_{Ai}}{\partial e_{jB}}[e_{jB,A}] + \frac{\partial \pi_{Ai}}{\partial \phi_{r,B}}[\phi_{r,BA}] + \frac{\partial \pi_{Ai}}{\partial \varphi_{,B}}[\varphi_{,BA}],$$

$$\rho_0 J[\ddot{\phi}_i] = \frac{\partial m_{Ai}}{\partial e_{jB}}[e_{jB,A}] + \frac{\partial m_{Ai}}{\partial \phi_{r,B}}[\phi_{r,BA}] + \frac{\partial m_{Ai}}{\partial \varphi_{,B}}[\varphi_{,BA}], \qquad (7.153)$$

$$\frac{\rho_0 j_0}{2}[\ddot{\varphi}] = \frac{\partial m_A}{\partial e_{jB}}[e_{jB,A}] + \frac{\partial m_A}{\partial \phi_{r,B}}[\phi_{r,BA}] + \frac{\partial m_A}{\partial \varphi_{,B}}[\varphi_{,BA}],$$

where $[\cdot]$ denotes the jump across $\mathcal{S}$.

Equations (7.153) are the general jump equations for an arbitrary body. However, we believe the results one obtains when the body has a centre of symmetry are revealing. When the body possesses a centre of symmetry, the terms $\partial \pi_{Ai}/\partial \varphi_{,B}$, $\partial m_{Ai}/\partial \varphi_{,B}$, $\partial m_A/\partial F_{jB}$ and $\partial m_A/\partial \phi_{r,B}$ are zero. The Clausius-Duhem inequality, cf. (Eringen, 2004b), shows

$$\pi_{Ai} = \rho_0 \frac{\partial \psi}{\partial e_{iA}}, \qquad m_{Ai} = \rho_0 \frac{\partial \psi}{\partial \phi_{i,A}},$$

$$m_A = \rho_0 \frac{\partial \psi}{\partial \varphi_{,A}}, \qquad T = \rho_0 \frac{\partial \psi}{\partial \varphi},$$

where ψ is the Helmholtz free energy. Thus, for the centro-symmetric case equations (7.153) reduce to

$$[\ddot{u}_i] = \frac{\partial^2 \psi}{\partial e_{iA} \partial e_{jB}}[u_{j,AB}] + \frac{\partial^2 \psi}{\partial e_{iA} \partial \phi_{j,B}}[\phi_{j,BA}],$$

$$J[\ddot{\phi}_i] = \frac{\partial^2 \psi}{\partial \phi_{i,A} \partial e_{jB}}[u_{j,AB}] + \frac{\partial^2 \psi}{\partial \phi_{i,A} \partial \phi_{j,B}}[\phi_{j,BA}], \qquad (7.154)$$

$$\frac{j_0}{2}[\ddot{\varphi}] = \frac{\partial^2 \psi}{\partial \varphi_{,A} \partial \varphi_{,B}}[\varphi_{,BA}].$$

Equation $(7.154)_3$ behaves as does equation (7.59) in the voids theory of (Nunziato and Cowin, 1979). It decouples from the system and thus represents a compression or expansion wave associated to the microstretch distribution throughout the body. However, equations $(7.154)_{1,2}$ do not decouple and lead to an elastic wave as equation (7.58). Here, we see the strong coupling effect of the microrotation.

Define now the wave amplitudes

$$A_i(t) = [\ddot{u}_i], \qquad B_i(t) = [\ddot{\phi}_i], \qquad C(t) = [\ddot{\varphi}].$$

Then, employing (7.26) and (7.32) equations (7.154) yield the system

$$(U_N^2 \delta_{ij} - Q_{ij})A_i = R_{ij}B_j,$$

$$(JU_N^2 \delta_{ij} - M_{ij})B_j = R_{ji}A_j, \qquad (7.155)$$

where Q_{ij} and M_{ij} are acoustic tensors of form

$$Q_{ij} = \frac{\partial^2 \psi}{\partial e_{iA} \partial e_{jB}} N_A N_B, \qquad M_{ij} = \frac{\partial^2 \psi}{\partial \phi_{i,A} \partial \phi_{j,B}} N_A N_B, \qquad (7.156)$$

and

$$R_{ij} = \frac{\partial^2 \psi}{\partial e_{iA} \partial \phi_{j,R}} N_A N_B . \qquad (7.157)$$

In addition, we have

$$\left(\frac{j_0}{2} V^2 - \frac{\partial^2 \psi}{\partial \varphi_{,A} \partial \varphi_{,B}} \right) C = 0. \qquad (7.158)$$

For $C \neq 0$, equation (7.158) clearly shows a microstretch wave may propagate with speed

$$V = \sqrt{2 \frac{\partial^2 \psi}{\partial \varphi_{,A} \partial \varphi_{,B}} \bigg/ j_0} \, .$$

System (7.155) is more complicated, but interesting. It immediately raises the question of what conditions must be imposed on the acoustic tensors Q_{ij} and M_{ij} to ensure an acceleration wave may propagate. Such questions have been addressed by (Chadwick and Currie, 1974; Chadwick and Currie, 1975) in classical thermoelasticity, see also (Lindsay and Straughan, 1979) for the Green-Laws $\phi(\theta, \dot{\theta})$ theory of thermoelasticity. However, system (7.155) is a more complicated system than that of classical thermoelasticity. The microspin of the particles is clearly having an effect.

If we assume there is a plane wave and $\mathbf{A} = A\boldsymbol{\nu}$, $\mathbf{B} = B\boldsymbol{\mu}$, then we find such a wave may propagate if $\boldsymbol{\mu}$ is an eigenvector of $R^T R$ and $\boldsymbol{\nu}$ is an eigenvector of RR^T. Of course, to have such a wave we would require $\boldsymbol{\mu}, \boldsymbol{\nu}$ to be related so that there is only one propagation direction. It is clearly an interesting question to analyse further the conditions for propagation of an acceleration wave in the theory of microstretch elasticity of (Eringen, 1990; Eringen, 2004b), and then calculate the associated amplitude equation(s).

8
Poroacoustic Waves

8.1 Poroacoustic acceleration waves

8.1.1 Equivalent fluid theory

The transmission of acoustic waves is an important problem in the construction industry. There are many applied/engineering research articles
devoted to this problem and theory for a satisfactory explanation of sound
propagation in porous materials is in relative infancy, see e.g. the accounts
in (Ayrault et al., 1999), (Fellah et al., 2003), (Fellah and Depollier, 2000),
(Garai and Pompoli, 2005), (Mouraille et al., 2006), (Moussatov et al.,
2001), (Wilson, 1997). However, since modern building materials are being
designed to be lighter there is a great environmental need to have a consistent theory. One wishes to know what effect different gases will have on
the transmission of sound when entrapped in building materials such as
bricks or plasterboard, and experiments in this line are being conducted,
cf. (Ciarletta and Zampoli, 2006).

(Fellah et al., 2003) write that acoustic wave motion in porous media
may be divided into two cases, that where the elastic matrix of the porous
medium moves, and that where it is rigid. For the former case they refer
to the theory of (Biot, 1956a; Biot, 1956b) which is briefly mentioned in
section 6.8. They write that the latter case is conveniently described by
the equivalent fluid model. To describe their model we note that (Fellah
and Depollier, 2000) write that the frame of the porous solid is assumed
not to deform when an acoustic wave passess through. In this situation
acoustic waves propagate only in the fluid although the density and bulk

B. Straughan, *Stability and Wave Motion in Porous Media*,
DOI: 10.1007/978-0-387-76543-3_8, © Springer Science+Business Media, LLC 2008

modulus change to take account of fluid structure interactions. This is an inspiring paper and develops the model of the equivalent fluid on the basis of a momentum equation and a mass conservation equation, which take form

$$\rho_f \alpha(\omega) \frac{\partial v_i}{\partial t} = -p_{,i}, \qquad \frac{\beta(\omega)}{K_a} \frac{\partial p}{\partial t} = -v_{i,i}. \tag{8.1}$$

In these equations v_i and p are the fluid velocity and acoustic pressure, ρ_f and K_a are fluid density and compressibility modulus of the fluid. The coefficients α and β depend on a frequency ω and are given explicitly in (Fellah and Depollier, 2000), equations (2) and (3).

Firstly, (Fellah and Depollier, 2000) relate equations (8.1) to viscoelastic behaviour by modifying these equations to involve fractional derivatives. Of interest to this book is the work where they consider equations (8.1) in low and high frequency domains. For the low frequency domain and a wave in one - dimension they show that β becomes independent of ω and α is replaced by $\alpha_0 + \alpha_1 (\partial/\partial t)^{-1}$ so that (8.1) may be taken as

$$\rho_f \alpha_0 \frac{\partial v}{\partial t} + \frac{\eta \phi}{k_0} v = -\frac{\partial p}{\partial x}, \qquad \frac{\gamma}{K_a} \frac{\partial p}{\partial t} = -\frac{\partial v}{\partial x}. \tag{8.2}$$

Here η is dynamic viscosity, ϕ porosity, γ is the adiabatic constant, and k_0 is the static permeability. Equations (8.2) are noticeably very similar to the equations for a compressible perfect fluid with a Darcy term added. The Darcy term would be the $\eta \phi v / k_0$ term and this issue is further investigated in section 8.1.2. From (8.2) (Fellah and Depollier, 2000) show v satisfies a damped wave equation of form

$$a v_{tt} + d v_t - v_{xx} = 0,$$

for suitable constants a and d. This leads to a sound speed in the porous medium $V = 1/\sqrt{a} = \sqrt{K_a/\rho_f \alpha_0 \gamma}$.

In the high frequency approximation (Fellah and Depollier, 2000) show that α and β may be replaced by operators of form $\alpha_\infty (\delta(t) + k_1 t^{-1/2}) *$ where α_∞ and k_1 are constants, δ is the Dirac delta function, and $*$ is the time convolution operator. In this case (8.2) are replaced by the linear equations involving time delay,

$$\rho_f \alpha_\infty \frac{\partial v}{\partial t} + \frac{2\rho_f \alpha_\infty}{\Lambda} \left(\frac{\eta}{\pi \rho_f} \right)^{1/2} \int_{-\infty}^{t} \frac{\partial v/\partial s}{\sqrt{t-s}} \, ds = -\frac{\partial p}{\partial x},$$

$$\frac{1}{K_a} \frac{\partial p}{\partial t} + \frac{2(\gamma-1)}{K_a \Lambda'} \left(\frac{\eta}{\pi Pr \rho_f} \right)^{1/2} \int_{-\infty}^{t} \frac{\partial p/\partial s}{\sqrt{t-s}} \, ds = -\frac{\partial v}{\partial x}. \tag{8.3}$$

They derive from these equations a time-delay wave equation of form

$$A v_{tt} + B \int_{-\infty}^{t} \frac{v_{ss}}{\sqrt{t-s}} \, ds + C v_t - v_{xx} = 0. \tag{8.4}$$

The evolution of various sound pulses are computed by (Fellah and Depollier, 2000) on the basis of equation (8.4).

8.1.2 Jordan - Darcy theory

(Jordan, 2005a; Jordan, 2006) proposes a very interesting nonlinear model for acoustic wave propagation in a porous medium by adding a Darcy (friction) term into the equations for a perfect fluid. This is an extension of the procedure of (Fellah and Depollier, 2000) who also analyse a compressible fluid model in a rigid porous matrix, as described in section 8.1.1.

In this chapter we take up the Jordan model but describe the analysis of (Ciarletta and Straughan, 2006) as opposed to that of (Jordan, 2005a) since their work makes no approximations and treats the full nonlinear system of equations. (Ciarletta and Straughan, 2006) commence by showing that the (Jordan, 2005a) model is not *ad hoc*, but may be derived from the continuum theory of a mixture of a fluid and an elastic solid. In fact, they derive equations (2.1) – (2.4) of (Jordan, 2005a) from a continuum thermodynamic theory for fluid flow in an elastic solid by employing the mixture theory of (Eringen, 1994), cf. section 1.9.1.

To see this let ρ^f, ρ^s be the partial densities of fluid and solid in a mixture of a fluid and an elastic solid, let v_i^f, v_i^s be the fluid and solid velocities, and let π^f, t_{ji}^s be the fluid pressure and solid partial stress tensor. Suppose the fluid is a gas and the entropy is constant, so we consider isentropic flow and then the theory of (Eringen, 1994) is based on his equations (2.12) and (2.15) for conservation of mass and balance of momentum, which we rewrite as

$$\rho_{,t}^f + (\rho^f v_i^f)_{,i} = 0, \tag{8.5}$$

$$\rho_{,t}^s + (\rho^s v_i^s)_{,i} = 0, \tag{8.6}$$

$$\rho^f (v_{i,t}^f + v_k^f v_{i,k}^f) = -\pi_{,i}^f - \frac{1}{\xi}(v_i^f - v_i^s), \tag{8.7}$$

$$\rho^s (v_{i,t}^s + v_k^s v_{i,k}^s) = t_{ji,j}^s + \frac{1}{\xi}(v_i^f - v_i^s). \tag{8.8}$$

The fluid pressure π^f and solid partial stress tensor t_{ij}^s are given by Eringen (1994) equations (3.9), (3.23), and have forms

$$\pi^f = \rho \rho^f \frac{\partial \psi}{\partial \rho^f}, \qquad t_{kl}^s = \rho \left(\frac{\partial \psi}{\partial C_{KL}} + \frac{\partial \psi}{\partial C_{LK}} \right) \frac{\partial x_k}{\partial X_K^s} \frac{\partial x_l}{\partial X_L^s},$$

where $\rho = \rho^s + \rho^f$ is the total density, and $\psi = \psi(\rho^f, C_{KL})$ is the Helmholtz free energy. The deformation tensor C_{KL} is defined by $C_{KL} = (\partial x_i/\partial X_K^s)(\partial x_i/\partial X_L^s)$. (Ciarletta and Straughan, 2006) argue that if one maintains the elastic solid of the porous medium fixed then the solid velocity $v_i^s \equiv 0$, so that equations (8.5) and (8.7) reduce to the model of (Jordan, 2005a). This may be interpreted as a nonlinear equivalent fluid model along the lines of section 8.1.1. To interpret Eringen's model in the light of (Jordan, 2005a), equation (2.2), we put $\xi = K/\mu\chi$, where K and χ are the permeability and porosity of the medium and μ is the dynamic viscosity of the gas.

The Jordan - Darcy model consists of equtions (8.5) and (8.7) with $v_i^s \equiv 0$. For simplicity, let us now set $\rho^f = \rho$, $\pi^f = p$, and $v_i^f = v_i$. Then, the equations for the Jordan-Darcy model of flow of a compressible fluid in a porous medium are,

$$\rho_{,t} + (\rho v_i)_{,i} = 0, \tag{8.9}$$

$$\rho(v_{i,t} + v_j v_{i,j}) = -p_{,i} - kv_i, \tag{8.10}$$

where the pressure $p = p(\rho) = (\rho + \rho^s)\rho^{\partial\psi/\partial\rho}$, ρ^s constant, and the Darcy coefficient $k = \mu\chi/K$.

8.1.3 Acceleration waves

(Jordan, 2005a) performs an acceleration wave analysis although he neglects terms $O(\epsilon^2)$ in equations (8.9) and (8.10), where ϵ is the Mach number of the flow. We describe the analysis of (Ciarletta and Straughan, 2006) who perform an acceleration wave study for the full system of equations (8.9) and (8.10). In this section an acceleration wave is defined for the system of equations (8.9) and (8.10) to be a singular surface S across which the velocity v_i and the density ρ are continuous in both x_i and t, although their first and higher derivatives in both x_i and t, in general, possess finite discontinuities. For simplicity we follow (Ciarletta and Straughan, 2006) and restrict attention to a one-dimensional acceleration wave. The analysis is then easy to follow, although the realistic extension to three dimensions is considered in section 8.2. By a one dimensional wave we really mean a plane wave in three dimensions moving along the $x-$axis where the problem only depends on functions defined in x and t.

We now work in one space dimension, put $\mathbf{v} = (u, 0, 0)$, and consider an acceleration wave moving along the $x-$axis. Equations (8.9), (8.10) in $1 -$ D are

$$\rho_t + \rho u_x + u\rho_x = 0, \tag{8.11}$$

$$\rho(u_t + uu_x) = -\frac{dp}{d\rho}\rho_x - ku. \tag{8.12}$$

As in section 7.1.5 $+$ denotes the region ahead of S and $-$ denotes the region behind the wave, the wave moving toward the $+$ direction. The amplitudes $A(t)$ and $B(t)$ of the acceleration wave are defined as

$$A(t) = [u_x] = u_x^- - u_x^+, \quad \text{and} \quad B(t) = [\rho_x]. \tag{8.13}$$

Since S is a singular surface in three-space, orthogonal to the $x-$axis, we may refer to u_x^- and u_x^+ as the limits of u_x on S from the left and right, respectively. The amplitudes A and B are functions of t, and no variation along the wave surface is considered, in keeping with the general analysis of (Chen, 1973).

Now, take the jump of equations (8.11), (8.12) recalling ρ and u are continuous, to find

$$[\rho_t] + u[\rho_x] + \rho[u_x] = 0,$$
$$[u_t] + u[u_x] + P[\rho_x] = 0, \qquad (8.14)$$

where we have set

$$P = \frac{1}{\rho}\frac{dp}{d\rho}. \qquad (8.15)$$

Next, with the aid of the Hadamard relation (7.8) we deduce that since ρ, u are continuous

$$[\rho_t] = -V[\rho_x], \qquad [u_t] = -V[u_x], \qquad (8.16)$$

where V is the wavespeed. Upon using (8.16) in (8.14) we may rearrange the resulting system to arrive at the vector equation

$$\begin{pmatrix} \rho & u - V \\ u - V & P \end{pmatrix} \begin{pmatrix} [u_x] \\ [\rho_x] \end{pmatrix} = \begin{pmatrix} 0 \\ 0 \end{pmatrix}. \qquad (8.17)$$

We require non-zero amplitudes $[u_x], [\rho_x]$ and so the determinant of the matrix in equation (8.17) must vanish. This leads to the following equation for the wavespeeds

$$(u - V)^2 = \rho P = \frac{dp}{d\rho}. \qquad (8.18)$$

From this one sees there are right and left moving waves with speeds

$$V = u \pm \sqrt{p_\rho}. \qquad (8.19)$$

Since u, ρ are continuous there is no need to write u^+, p_ρ^+ in (8.19). However, it is important to realise that the right hand side of (8.19) depends only on u^+ and ρ^+ which we assume are known (quantities ahead of the wave) and so we can determine the wavespeed.

We now focus on the right moving wave so $V = u + \sqrt{p_\rho}$. We may use the Hadamard relation to calculate analytically the wave amplitudes A and B, and this we do next. Note that A and B are directly related since via e.g. (8.17),

$$\rho A = (V - u)B. \qquad (8.20)$$

8.1.4 Amplitude equation derivation

To derive the amplitude equation we differentiate (8.11) and (8.12) in turn with respect to x and then take the jumps of the resulting equations. In this way we obtain the equations

$$[\rho_{tx}] + 2[u_x\rho_x] + u[\rho_{xx}] + \rho[u_{xx}] = 0, \qquad (8.21)$$

$$[u_{tx}] + [u_x^2] + u[u_{xx}] = -P_\rho[\rho_x^2] - P[\rho_{xx}] - \frac{k}{\rho}[u_x] + \frac{k}{\rho^2}u[\rho_x]. \qquad (8.22)$$

In this equation it is understood that ρ, u and $P(\rho)$ are evaluated at ρ^+, u^+. We next use the product relation (7.13) on the terms $[u_x\rho_x]$ in (8.21) and $[u_x^2], [\rho_x^2]$ in (8.22) together with (8.20) to derive

$$\frac{\delta A}{\delta t} - V[u_{xx}] + A^2 + u[u_{xx}] + \left(2u_x^+ + \frac{k}{\rho}\right)A$$
$$+ \left(2P_\rho\rho_x^+ - \frac{ku}{\rho^2}\right)B + P[\rho_{xx}] + P_\rho B^2 = 0, \qquad (8.23)$$

and

$$-\left(\frac{\rho}{u-V}\right)\frac{\delta A}{\delta t} - \left(\frac{\rho}{u-V}\right)\frac{\delta}{\delta t}\left(\frac{u-V}{\rho}\right)B + (u-V)[\rho_{xx}]$$
$$+ \rho[u_{xx}] - 2\left(\frac{\rho}{u-V}\right)A^2 + 2\rho_x^+ A + 2u_x^+ B = 0. \qquad (8.24)$$

Next, multiply equation (8.24) by $-(u-V)/\rho$ and eliminate B using (8.20) to see that

$$\frac{\delta A}{\delta t} - \frac{(u-V)^2}{\rho}[\rho_{xx}] - (u-V)[u_{xx}] + 2A^2$$
$$+ \left\{2u_x^+ - 2\rho_x^+\frac{(u-V)}{\rho} - \frac{\delta}{\delta t}\log\left(\frac{u-V}{\rho}\right)\right\}A = 0. \qquad (8.25)$$

We now add equations (8.23) and (8.25). Observe that the $[u_{xx}]$ terms cancel out and also the $[\rho_{xx}]$ terms vanish because of the wavespeed equation (8.18). Hence, when we eliminate B from the resulting equation using (8.20) we obtain a Bernoulli equation in A. The result is

$$\frac{\delta A}{\delta t} + bA + aA^2 = 0. \qquad (8.26)$$

The coefficients a and b are given explicitly as

$$a = \frac{1}{2}\left(3 + \rho\frac{d}{d\rho}(\log P)\right), \qquad (8.27)$$

$$b = 2u_x^+ + \frac{k}{2\rho}\left(1 + \frac{u}{u-V}\right) + \frac{\rho_x^+}{\rho}(V-u)$$
$$+ \frac{\rho\rho_x^+ P_\rho}{V-u} + \frac{1}{2}\frac{\delta}{\delta t}\log\left(\frac{V-u}{\rho}\right), \qquad (8.28)$$

$$= 2u_x^+ + \frac{k}{2\rho}\left(1 - \frac{u}{\sqrt{P_\rho}}\right) + \frac{\rho_x^+\sqrt{P_\rho}}{\rho}$$
$$+ \frac{\rho\rho_x^+ P_\rho}{\sqrt{P_\rho}} + \frac{1}{2}\frac{\delta}{\delta t}\log\left(\frac{\sqrt{P_\rho}}{\rho}\right). \qquad (8.29)$$

The general solution to equation (8.26) is given in equation (7.19).

To understand exactly the effect of the Darcy term in the Jordan-Darcy model (Ciarletta and Straughan, 2006) consider the situation where the medium ahead of the wave is at rest with a constant density, i.e. $u^+ \equiv 0, \rho^+ \equiv$ constant. Observe that this is a solution to equations (8.11) and (8.12). In this case the right moving wave has speed $V = \sqrt{dp/d\rho}$ and (8.26) holds but now a and b are constants with values

$$a = \frac{3}{2} + \frac{\rho}{2} \frac{d}{d\rho} (\log P), \qquad b = \frac{k}{2\rho}. \tag{8.30}$$

The solution to (8.26) in this case is

$$A(t) = \frac{A(0)}{e^{bt} + A(0)ab^{-1}(e^{bt} - 1)}. \tag{8.31}$$

(Ciarletta and Straughan, 2006) compare the amplitude given by (8.31) with the corresponding amplitude in a perfect fluid, i.e. when no porous medium is present, which is the case when $k = 0$ and so, therefore, $b = 0$. In this case, instead of (8.31) the amplitude solution is

$$A_{PF}(t) = \frac{A_{PF}(0)}{1 + A_{PF}(0)at}. \tag{8.32}$$

(Ciarletta and Straughan, 2006) show that $b > 0$ increases the attenuation of sound. To see this, we note that if $A(0) > 0$ the wave is expansive and the amplitude decays in time. However, if $A(0) < 0$ the wave is compressive. The amplitude in (8.32) *always* blows up in a finite time, a phenomenon associated with shock wave formation, cf. (Fu and Scott, 1991). When the porous medium is present, the wave amplitude may still blow-up, but the blow-up is delayed. In fact (Ciarletta and Straughan, 2006) further consider a polytropic gas so that $p = k_0\rho^\gamma$, with $k_0, \gamma > 0$ constants. In this case

$$a = \frac{\gamma + 1}{2}, \qquad b = \frac{\chi\mu}{2\rho K}. \tag{8.33}$$

(Ciarletta and Straughan, 2006) note that when $A_{PF}(0) < 0$ the wave amplitude *always* blows up in a finite time. However, if $A(0) < 0$ blow-up only occurs if

$$|A(0)| > \frac{b}{a} = \frac{\mu\chi}{\rho(\gamma + 1)K}. \tag{8.34}$$

This clearly demonstrates the attenuation the porous medium creates. (Ciarletta and Straughan, 2006) take values of χ and K for brick from (Nield and Bejan, 2006) and values of μ, ρ and γ for air at 15°C from (Batchelor, 1967), p.594, and show that for such a brick containing air the restriction (8.34) shows that the wave amplitude will not blow-up if $|A(0)| < \zeta$, where ζ takes values in the range 1.90298×10^7 s^{-1} to 2.47123×10^9 s^{-1}. Thus, we believe (8.34) should be useful when assessing sound attenuation in a porous medium which contains a gas infused into the pores.

8.2   Temperature effects

8.2.1   Jordan-Darcy temperature model

In this section we consider the Jordan-Darcy model of section 8.1.2, but we include thermodynamics in order that we may assess directly the effects of temperature on acoustic wave propagation in porous media. As far as we are aware, this material is new and does not appear elsewhere.

The equations for a classical perfect fluid are, cf. (Truesdell and Toupin, 1960), the equations of continuity, momentum, and energy, respectively, namely,

$$\rho_t + (\rho v_i)_{,i} = 0, \tag{8.35}$$

$$\rho(v_{i,t} + v_j v_{i,j}) = -p_{,i} + \rho F_i, \tag{8.36}$$

$$\rho \dot{\epsilon} = -p\, v_{i,i} - q_{i,i} + \rho r, \tag{8.37}$$

where ρ, v_i, ϵ are density, velocity, internal energy, $\epsilon = \epsilon(\rho, \theta)$, and $\dot{\epsilon}$ is the material derivative with respect to time of ϵ. The quantities F_i, r, p and q_i are externally supplied body force, externally supplied heat source or sink, pressure $p = p(\rho, \theta)$, and heat flux. The variable θ denotes the absolute temperature. For a perfect fluid one typically chooses the heat flux to be zero, so we select $q_i \equiv 0$. We present our derivation from equations (8.35) – (8.37) but one could equally well begin with the equivalent equations in (Whitham, 1974), p. 150. For the nonlinear wave analysis to be considered here we lose no generality by taking $F_i = 0$ and $r = 0$. We modify equation (8.36) by adding a Jordan-Darcy term of form $-kv_i$. Thus our system of equations to describe flow in a saturated porous medium is

$$\begin{aligned}
&\rho_t + v_i\rho_{,i} + \rho v_{i,i} = 0, \\
&\rho(v_{i,t} + v_j v_{i,j}) = -p_{,i} - kv_i, \\
&\rho \dot{\epsilon} = -p\, v_{i,i}.
\end{aligned} \tag{8.38}$$

In general $\epsilon = \epsilon(\rho, \theta)$. However, there are many classes of fluid for which one may adopt the simpler relation $\epsilon = \epsilon(\theta)$. In this section we calculate the wavespeeds in the general case $\epsilon = \epsilon(\rho, \theta)$, but for the more technical calculation of the amplitudes we restrict attention to the case $\epsilon = \epsilon(\theta)$. We believe this will make the analysis more transparent, although one could repeat the amplitude calculation with $\epsilon = \epsilon(\rho, \theta)$, *mutatis mutandis*.

The class of fluids for which $\epsilon = \epsilon(\theta)$ is quite large, and we show it holds for a perfect gas for which $p = R\rho\theta$. One may also show the same relation holds for Hirn's gas for which

$$p = R\theta\left(\frac{\rho}{\rho - \delta_2}\right) - \delta_1,$$

for suitable constants $\delta_1, \delta_2 > 0$, and for a Joule and Thomson gas for which

$$p = R\rho\theta - \frac{\delta_1\rho}{\theta},$$

for a suitable constant $\delta_1 > 0$. Other equations of state are given by (Otto, 1929). The internal energy ϵ satisfies the relation $\epsilon = \psi + \eta\theta$ where ψ is the Helmholtz free energy function and the entropy $\eta = -\partial\psi/\partial\theta$. The pressure is given by $p = \rho^2\partial\psi/\partial\rho$. Thus, for a perfect gas,

$$\psi_\rho = \frac{R\theta}{\rho} \quad \text{and so} \quad \psi = R\theta\log\rho + f(\theta)$$

for some function f. Thus, we see that

$$\epsilon = f(\theta) - \theta f'(\theta) = \epsilon(\theta).$$

(Actually $p = (\rho + \rho^s)\rho^{\partial\psi/\partial\rho}$ here and throughout the rest of this chapter.)

A similar deduction holds for Hirn's gas, the Joule-Thomson gas, and many other of the equations of state given by (Otto, 1929). Hence, consideration of the amplitude equation when $\epsilon = \epsilon(\theta)$ is not without interest.

8.2.2 Wavespeeds

We now consider equations (8.38) with $\epsilon = \epsilon(\rho, \theta)$ and study acceleration wave propagation. For equations (8.38) we define an acceleration wave to be a singular surface $\mathcal{S}$ such that ρ, v_i, θ are continuous in the spatial domain $\mathbb{R}^3$, but $\rho_t, \rho_{,i}, v_{i,t}, v_{i,j}, \theta_t, \theta_{,i}$, and higher derivatives may suffer a finite discontinuity.

To determine the wavespeed(s) of an acceleration wave we use $(8.38)_1$ to rewrite $(8.38)_3$ after expanding $\dot{\epsilon}$. Thus, we rewrite system (8.38) as

$$\rho_t + v_i\rho_{,i} + \rho v_{i,i} = 0,$$
$$\rho(v_{i,t} + v_j v_{i,j}) = -\big(p(\rho, \theta)\big)_{,i} - kv_i, \tag{8.39}$$
$$\rho\epsilon_\theta(\theta_t + v_i\theta_{,i}) + (p - \rho^2\epsilon_\rho)v_{i,i} = 0.$$

We take the jump of these equations and find

$$[\rho_t] + v_i[\rho_{,i}] + \rho[v_{i,i}] = 0,$$
$$\rho[v_{i,t}] + \rho v_j[v_{i,j}] = -p_\rho[\rho_{,i}] - p_\theta[\theta_{,i}], \tag{8.40}$$
$$\rho\epsilon_\theta[\theta_t] + \rho v_i\epsilon_\theta[\theta_{,i}] + (p - \rho^2\epsilon_\rho)[v_{i,i}] = 0.$$

Define the amplitudes A_i, B and C to be

$$A_i = [v_{i,j}n_j], \quad B = [n_i\rho_{,i}], \quad C = [n_i\theta_{,i}]. \tag{8.41}$$

With the aid of (176.2) and (176.10) of (Truesdell and Toupin, 1960) we may show that

$$[v_{i,j}] = A_i n_j, \quad [\rho_{,i}] = Bn_i, \quad [\theta_{,i}] = Cn_i. \tag{8.42}$$

We also require the three-dimensional Hadamard relation (7.27) which we rewrite as

$$\frac{\delta}{\delta t}[\psi] = [\psi_t] + V[n_i \psi_{,i}].\qquad(8.43)$$

Using these two relations one may show that

$$[v_{i,i}] = A_i n_i, \quad [\rho_t] = -VB, \quad [\theta_t] = -VC,\qquad(8.44)$$

and

$$[v_{i,t}] = -V[n_j v_{i,j}] = -V A_i.\qquad(8.45)$$

Upon utilizing equations (8.42), (8.44) and (8.45) in equation (8.40) one may show that

$$
\begin{aligned}
(v_i n_i - V)B + \rho n_i A_i &= 0,\\
\rho(v_j n_j - V)A_i + p_\rho n_i B + p_\theta n_i C &= 0,\\
\rho \epsilon_\theta (v_i n_i - V)C + (p - \rho^2 \epsilon_\rho) n_i A_i &= 0.
\end{aligned}
\qquad(8.46)
$$

From $(8.46)_2$ we see that the acceleration wave must be longitudinal, i.e. $A_i = n_i A$, where $A = [n_i n_j v_{i,j}]$ so that (8.46) may be reduced to the matrix form

$$
\begin{pmatrix}
v_i n_i - V & \rho & 0\\
p_\rho & \rho(v_i n_i - V) & p_\theta\\
0 & p - \rho^2 \epsilon_\rho & \rho \epsilon_\theta (v_i n_i - V)
\end{pmatrix}
\begin{pmatrix} B\\ A\\ C \end{pmatrix}
=
\begin{pmatrix} 0\\ 0\\ 0 \end{pmatrix}.
\qquad(8.47)
$$

We require non-zero amplitudes, A, B, C, so that the determinant of the matrix in (8.47) must be zero. This leads to the equation

$$(v_i n_i - V)\{\rho^2 \epsilon_\theta (v_i n_i - V)^2 - \rho^2 \epsilon_\theta p_\rho - p_\theta(p - \rho^2 \epsilon_\rho)\} = 0.\qquad(8.48)$$

Thus, we can have a wave moving with the fluid with $V = v_i n_i$, or a right and a left propagating wave with speeds determined by

$$(v_i n_i - V)^2 = p_\rho + \frac{p_\theta(p - \rho^2 \epsilon_\rho)}{\rho^2 \epsilon_\theta}.\qquad(8.49)$$

The latter waves are of interest here. We see that (8.49) compared to the isothermal wavespeed given by (8.18) shows the thermodynamic correction due to temperature effects. Namely, the term $p_\theta(p - \rho^2 \epsilon_\rho)/\rho^2 \epsilon_\theta$ is the correction due to temperature effects.

8.2.3 Amplitude equation

In the interests of clarity we now restrict attention to a one-dimensional acceleration wave moving along the x-axis. One may think of this as meaning $\mathcal{S}$ is a plane parallel to the (y, z) plane propagating in the x-direction. Thus, we now have $\mathbf{v} = (u(x,t), 0, 0)$, $\rho(x,t), \theta(x,t)$. We also restrict attention to the case where $\epsilon = \epsilon(\theta)$. Then, the wavespeed equation (8.49) reduces to

$$(V - u)^2 = p_\rho + \frac{p p_\theta}{\rho^2 \epsilon_\theta}. \tag{8.50}$$

The wave amplitudes A, B, C are now given by

$$A(t) = [u_x], \quad B(t) = [\rho_x], \quad C(t) = [\theta_x]. \tag{8.51}$$

The governing equations (8.39) in the one-dimensional situation become

$$\rho_t + u\rho_x + \rho u_x = 0,$$
$$\rho(u_t + u u_x) = -\big(p(\rho, \theta)\big)_x - ku, \tag{8.52}$$
$$\rho\epsilon_\theta(\theta_t + u\theta_x) + p u_x = 0.$$

These equations are differentiated with respect to x and expanded using the chain rule. We then take the jumps of the results to find, with

$$P = \frac{p_\rho}{\rho}, \qquad Q = \frac{p_\theta}{\rho}, \tag{8.53}$$

$$[\rho_{tx}] + 2[\rho_x u_x] + \rho[u_{xx}] + u[\rho_{xx}] = 0, \tag{8.54}$$

$$[u_{tx}] + [u_x^2] + u[u_{xx}] + P[\rho_{xx}] + Q[\theta_{xx}] + \frac{k}{\rho}[u_x]$$
$$- \frac{ku}{\rho^2}[\rho_x] + P_\rho[\rho_x^2] + P_\theta[\rho_x \theta_x] + Q_\rho[\rho_x \theta_x] + Q_\theta[\theta_x^2] = 0, \tag{8.55}$$

$$\rho\epsilon_{\theta\theta}[\theta_t\theta_x] + \rho\epsilon_{\theta\theta}u[\theta_x^2] + \rho\epsilon_\theta[\theta_{tx}] + \rho\epsilon_\theta[u_x\theta_x]$$
$$+ \rho\epsilon_\theta u[\theta_{xx}] + p_\rho[\rho_x u_x] + p_\theta[\theta_x u_x] + p[u_{xx}] = 0. \tag{8.56}$$

We next use the product formula (7.13) and the relations which follow from the Hadamard relation (7.8),

$$[u_t] = -V[u_x], \qquad [u_{xt}] = \frac{\delta A}{\delta t} - V[u_{xx}],$$

together with equivalent expressions for ρ and θ, to derive

$$\frac{\delta B}{\delta t} + (u - V)[\rho_{xx}] + \rho[u_{xx}] + 2(\rho_x^+[u_x] + u_x^+[\rho_x] + AB) = 0, \tag{8.57}$$

$$\frac{\delta A}{\delta t} + (u - V)[u_{xx}] + \frac{p_\rho}{\rho}[\rho_{xx}] + \frac{p_\theta}{\rho}[\theta_{xx}] + \frac{k}{\rho}A$$
$$- \frac{ku^+}{\rho^2}B + 2u_x^+ A + A^2 + P_\rho(2\rho_x^+ B + B^2)$$
$$+ (P_\theta + Q_\rho)(\rho_x^+ C + \theta_x^+ B + C^2) + Q_\theta(2\theta_x^+ C + C^2) = 0, \tag{8.58}$$

and

$$\rho\epsilon_\theta \frac{\delta C}{\delta t} + \rho\epsilon_\theta(u - V)[\theta_{xx}] + p[u_{xx}]$$
$$+ (\rho\epsilon_{\theta\theta}\theta_t^+ - \rho\epsilon_{\theta\theta}\theta_x^+ V + 2\rho\epsilon_{\theta\theta}u\theta_x^+ + \rho\epsilon_\theta u_x^+ + p_\theta u_x^+)C$$
$$+ (\rho\epsilon_\theta\theta_x^+ + p_\rho\rho_x^+ + p_\theta\theta_x^+)A + p_\rho u_x^+ B$$
$$+ \rho\epsilon_{\theta\theta}(u - V)C^2 + (\rho\epsilon_\theta + p_\theta)AC + p_\rho AB = 0. \qquad (8.59)$$

The next step is to use the equations which follow from (8.46), namely,

$$B = -\frac{\rho}{(u - V)} A, \qquad C = -\frac{p}{\rho\epsilon_\theta(u - V)} A, \qquad (8.60)$$

together with the derivative equations

$$\frac{\delta B}{\delta t} = -\left(\frac{\rho}{u - V}\right)\frac{\delta A}{\delta t} - A\frac{\delta}{\delta t}\left(\frac{\rho}{u - V}\right),$$
$$\rho\epsilon_\theta \frac{\delta C}{\delta t} = -\frac{p}{(u - V)}\frac{\delta A}{\delta t} - A\rho\epsilon_\theta\frac{\delta}{\delta t}\left(\frac{p}{\rho\epsilon_\theta(u - V)}\right), \qquad (8.61)$$

in (8.57) and (8.59) to find

$$-\left(\frac{\rho}{u - V}\right)\frac{\delta A}{\delta t} - A\frac{\delta}{\delta t}\left(\frac{\rho}{u - V}\right) + (u - V)[\rho_{xx}] + \rho[u_{xx}]$$
$$+ 2(\rho_x^+ A + u_x^+ B + AB) = 0, \qquad (8.62)$$

and

$$-\frac{p}{(u - V)}\frac{\delta A}{\delta t} - A\rho\epsilon_\theta\frac{\delta}{\delta t}\left(\frac{p}{\rho\epsilon_\theta(u - V)}\right) + \rho\epsilon_\theta(u - V)[\theta_{xx}]$$
$$+ (\rho\epsilon_{\theta\theta}\theta_t^+ - \rho\epsilon_{\theta\theta}\theta_x^+ V + 2\rho\epsilon_{\theta\theta}u\theta_x^+ + \rho\epsilon_\theta u_x^+ + p_\theta u_x^+)C$$
$$+ (\rho\epsilon_\theta\theta_x^+ + p_\rho\rho_x^+ + p_\theta\theta_x^+)A + p_\rho u_x^+ B + p[u_{xx}]$$
$$+ \rho\epsilon_{\theta\theta}(u - V)C^2 + (\rho\epsilon_\theta + p_\theta)AC + p_\rho AB = 0. \qquad (8.63)$$

We now wish to remove the second derivative terms $[u_{xx}], [\rho_{xx}]$ and $[\theta_{xx}]$ from (8.58), (8.62) and (8.63). To do this we form the sum (8.58)+μ_1(8.62)+μ_2(8.63), for μ_1, μ_2 to be chosen. Upon forming this sum the second derivative terms yield

$$[u_{xx}](u - V + \rho\mu_1 + \mu_2 p)$$
$$+ [\rho_{xx}]\left(\frac{p_\rho}{\rho} + (u - V)\mu_1\right) + [\theta_{xx}]\left(\frac{p_\theta}{\rho} + \mu_2\rho\epsilon_\theta(u - V)\right).$$

We select

$$\mu_1 = -\frac{p_\rho}{\rho(u - V)}, \qquad \mu_2 = -\frac{p_\theta}{\rho^2\epsilon_\theta(u - V)},$$

and this removes the $[\rho_{xx}], [\theta_{xx}]$ terms. However, we easily check that with this choice of μ_1, μ_2 the coefficient of $[u_{xx}]$ is zero, thanks to the wavespeed equation (8.50). What remains is a sum involving $\delta A/\delta t$, A, B, C and products in $A^2, B^2, C^2, AB, AC, AB$. The B and C terms are removed using (8.60) and we are left with a Bernoulli equation for $A(t)$, namely,

$$\frac{\delta A}{\delta t} + bA + a^2 A = 0. \tag{8.64}$$

The coefficients a and b have forms

$$\begin{aligned}
a =1&+ \frac{p_\theta}{2\rho\epsilon_\theta} + \frac{1}{2(u-V)^2}\left(\rho p_{\rho\rho} \right. \\
&\left. + \frac{p^2}{\rho^4\epsilon_\theta^2}(2\rho p_{\rho\theta} - p_\theta + \rho p_{\theta\theta}) - \frac{p^2 p_\theta \epsilon_{\theta\theta}}{\rho^3\epsilon_\theta^3} \right),
\end{aligned} \tag{8.65}$$

$$\begin{aligned}
b = &\frac{k}{2\rho}\left(1 + \frac{u}{u-V}\right) + \frac{p_\rho}{2\rho(u-V)}\frac{\delta}{\delta t}\left(\frac{\rho}{u-V}\right) \\
&+ \frac{p_\theta}{2\rho(u-V)}\frac{\delta}{\delta t}\left(\frac{p}{\rho\epsilon_\theta(u-V)}\right) \\
&+ u_x^+\left\{1 + \frac{p_\rho}{(u-V)^2} + \frac{1}{2\rho\epsilon_\theta(u-V)^2}\left(p_\theta p_\rho + \frac{p p_\theta}{\rho} + \frac{p p_\theta^2}{\rho^2\epsilon_\theta}\right)\right\} \\
&+ \rho_x^+\left\{-\frac{p_{\rho\rho}}{(u-V)} - \frac{1}{2\rho^2\epsilon_\theta(u-V)}\left(2p p_\theta + p_\theta p_\rho - \frac{p p_\theta}{\rho}\right)\right\} \\
&+ \frac{\theta_x^+}{2}\left\{-\frac{2p_{\rho\theta}}{(u-V)} - \frac{p_\theta^2}{\rho^2\epsilon_\theta(u-V)} + \frac{p p_\theta \epsilon_{\theta\theta}(2u-V)}{\rho^2\epsilon_\theta^2(u-V)^2} - \frac{2p p_{\theta\theta}}{\rho^2\epsilon_\theta(u-V)}\right\} \\
&+ \frac{\theta_t^+ p p_\theta \epsilon_{\theta\theta}}{2\rho^2\epsilon_\theta^2(u-V)^2}.
\end{aligned} \tag{8.66}$$

The solution to equation (8.64) is given in (7.19). If we consider the wave moving into an equilibrium region for which $u^+ = 0$, $\rho^+ =$constant, $\theta^+ =$constant, then the solution to (8.64) is (8.31). In that case the coefficient b reduces to $b = k/2\rho$ while the coefficient a still has form (8.65), but $u-V = -V$. One should note that a and b reduce to the forms (8.27), (8.28) when temperature effects are neglected, i.e. the theory does reduce to that of section 8.1.2. One thing we point out is the attenuating effect the porous medium has on the acoustic wave propagation. This is evident through the term involving the Darcy coefficient k in (8.66). Equation (8.66) does show the extra effect the thermodynamics have, but the effect of the porous medium is always evident.

8.3   Heat flux delay

8.3.1   Cattaneo poroacoustic theory

We now wish to consider the equations of section 8.2.1, namely equations (8.38) but with a non-zero heat flux **q**. The reason for this is that we also wish to consider the propagation of temperature waves in addition to sound waves. As mentioned in section 7.3.1 there has been much recent interest in temperature wave propagation in porous media. The history of temperature wave propagation in a perfect fluid is briefly reviewed in (Lindsay and Straughan, 1978).

In this section we wish to consider temperature wave propagation simultaneously with sound wave propagation in a porous medium by adapting the approach of (Cattaneo, 1948). The (Cattaneo, 1948) approach has been the subject of many recent investigations and a convenient recent reference is that of (Jordan, 2005b) who applies the Cattaneo method to traffic flow, see also (Alvarez-Ramirez et al., 2006). (Jordan, 2007) is an interesting related article. (Su and Dai, 2006) compare the solution to a phase lagged equation with an oscillating heat source to that of the damped wave equation with the same source.

We begin with equations (8.38) but keep the heat flux term in, so we have

$$\rho_t + v_i \rho_{,i} + \rho v_{i,i} = 0,$$
$$\rho(v_{i,t} + v_j v_{i,j}) = -p_{,i} - k v_i, \qquad (8.67)$$
$$\rho \dot{e} = -p\, v_{i,i} - q_{i,i},$$

cf. equation (8.37). The classical theory at this point would have a constitutive equation for q_i, and typically this would be one of Fourier law type, so

$$q_i(\mathbf{x}, t) = -\kappa \theta_{,i}(\mathbf{x}, t). \qquad (8.68)$$

In general, the thermal conductivity κ is a function of temperature, θ. However, we here assume κ is constant. The logic behind the (Cattaneo, 1948) theory (see (Jordan, 2005b) and also the comments in section 8.3.5) is that the heat flux is not directly proportional to the temperature gradient at the *same instant of time*. Instead, there is a delay in time in this reaction so that (8.68) is replaced by

$$q_i(\mathbf{x}, t + \tau) = -\kappa \theta_{,i}(\mathbf{x}, t), \qquad (8.69)$$

for some small delay time $\tau > 0$. One may expand this relation in a Taylor series and then

$$q_i(\mathbf{x}, t) + \tau q_{i,t}(\mathbf{x}, t) \approx -\kappa \theta_{,i}(\mathbf{x}, t). \qquad (8.70)$$

Thus, in this section we generalize this relation to one for a moving body and assume we have equations (8.67) coupled with the equation

$$\dot{q}_i + \frac{1}{\tau} q_i = -\frac{\kappa}{\tau} \theta_{,i}, \qquad (8.71)$$

where the dot denotes the material derivative and this equation holds at the time t. This approach is discussed in another fluid dynamical context by (Straughan and Franchi, 1984).

The above approach is somewhat *ad hoc* and so now we present a justification via a thermodynamic argument.

8.3.2 Thermodynamic justification

To produce a thermodynamic procedure which is consistent with (8.71) we adopt the techniques of (Caviglia et al., 1992b). Thus, we have the equations (cf. (8.67)) but we include an externally supplied body force F_i and an externally supplied heat supply r. Thus,

$$\rho_t + v_i \rho_{,i} + \rho v_{i,i} = 0,$$
$$\rho(v_{i,t} + v_j v_{i,j}) = -p_{,i} - kv_i + \rho F_i, \qquad (8.72)$$
$$\rho \dot{\epsilon} = -p\,v_{i,i} - q_{i,i} + \rho r.$$

The Clausius-Duhem (entropy) inequality has form

$$\rho \dot{\eta} \geq -\left(\frac{q_i}{\theta}\right)_{,i} + \frac{\rho r}{\theta}. \qquad (8.73)$$

If we introduce the Helmholtz free energy ψ via $\epsilon = \psi + \eta \theta$ then using $(8.72)_3$ we may rewrite inequality (8.73) as

$$-\rho(\dot{\psi} + \eta \dot{\theta}) - \frac{1}{\theta} q_i \theta_{,i} - p v_{i,i} \geq 0. \qquad (8.74)$$

We introduce an internal variable ξ_i and then following (Caviglia et al., 1992b) postulate that ξ_i satisfies an evolution equation of form

$$\dot{\xi}_i = -m\theta_{,i} - n\xi_i. \qquad (8.75)$$

The coefficients m and n are scalars depending on the variables of the constitutive theory, with $n > 0$.

The constitutive theory we consider is that

$$\psi, \eta, p \text{ and } q_i$$

depend on the independent variables

$$\rho, \theta, \theta_{,i}, \xi_i. \qquad (8.76)$$

Expanding ψ, inequality (8.74) may be rewritten

$$-\rho\psi_\rho \dot{\rho} - pv_{i,i} - \rho\dot{\theta}(\psi_\theta + \eta) - \rho\psi_{\theta_{,i}} \dot{\overline{\theta_{,i}}} - \rho\psi_{\xi_i} \dot{\xi}_i - \frac{q_i \theta_{,i}}{\theta} \geq 0. \qquad (8.77)$$

Then, recalling $\dot{\rho} = -\rho v_{i,i}$, from equation $(8.72)_2$, the above inequality may be replaced by

$$(\rho^2 \psi_\rho - p)v_{i,i} - \rho\dot{\theta}(\psi_\theta + \eta) - \rho\psi_{\theta_{,i}} \dot{\overline{\theta_{,i}}} - \rho\psi_{\xi_i} \dot{\xi}_i - \frac{q_i \theta_{,i}}{\theta} \geq 0. \qquad (8.78)$$

We now argue that $\dot{\overline{\theta_{,i}}}$ may be selected independently of the other variables in (8.78), balancing equation $(8.72)_3$ by a suitable choice of r. Thus, the coefficient of $\dot{\theta}_{,i}$ must be zero. Likewise, choosing r appropriately, $\dot{\theta}$ may be selected arbitrarily so that the coefficient of this term must also be zero. Then, we apply a similar argument to the first term in (8.78) and select $v_{i,i}$ arbitrarily balancing equation $(8.72)_2$ by our choice of F_i. In this way one shows that

$$\psi_{\theta,i} = 0, \quad \eta = -\psi_\theta, \quad \text{and} \quad p = \rho^2 \psi_\rho. \tag{8.79}$$

The inequality which remains from (8.78) is

$$-\rho\psi_{\xi_i}\dot{\xi}_i - \frac{q_i\theta_{,i}}{\theta} \geq 0. \tag{8.80}$$

Upon insertion of equation (8.75) in inequality (8.80) one obtains

$$\left(m\rho\psi_{\xi_i} - \frac{q_i}{\theta} \right)\theta_{,i} + n\rho\psi_{\xi_i}\xi_i \geq 0. \tag{8.81}$$

We may now select $\theta_{,i}$ arbitrarily balancing $(8.72)_3$ by a suitable choice of r. However, care must be taken since ξ_i and $\theta_{,i}$ are connected via (8.75). Nevertheless, v_i may be selected arbitrarily balancing $(8.72)_2$ by F_i and so we may choose ξ_i as we like. Then, from inequality (8.81) we deduce

$$q_i = m\rho\theta\psi_{\xi_i} \quad \text{and} \quad \psi_{\xi_i}\xi_i \geq 0. \tag{8.82}$$

Now, arguing as in (Caviglia et al., 1992b) we consider a stationary solution so that in equilibrium

$$q_i = -k_1\theta_{,i}, \quad \dot{\xi}_i = 0, \tag{8.83}$$

then, *in equilibrium*, (8.75) yields

$$\xi_i = -\frac{m}{n}\theta_{,i}$$

and thus utilizing (8.83), this *suggests* we may take

$$q_i = \frac{nk_1}{m}\xi_i$$

and then upon using $(8.82)_1$ we would have

$$\psi_{\xi_i} = \frac{nk_1}{\rho\theta m^2}\xi_i \,.$$

In the case where n and m do not depend on ξ_i this relation leads to the form for ψ,

$$\psi(\rho, \theta, \xi_i) = \hat{\psi}(\theta, \rho) + \frac{nk_1}{2\rho\theta m^2}\xi^2, \tag{8.84}$$

where $\xi^2 = \xi_i \xi_i$. Whence, from (8.79) we have, for additionally n and m independent of ρ and θ,

$$p = \rho^2 \psi_\rho = \rho^2 \hat{\psi}_\rho - \frac{nk_1}{2\theta m^2 \rho^2} \xi^2,$$

$$\eta = -\hat{\psi}_\theta + \frac{nk_1}{2\rho \theta^2 m^2} \xi^2, \qquad (8.85)$$

$$\epsilon = \hat{\psi} - \theta \hat{\psi}_\theta + \frac{nk_1}{\rho \theta m^2} \xi^2.$$

At first sight results (8.84), (8.85) may appear strange, since they suggest dependence on ξ_i and, therefore, q_i. However, this is in complete agreement with what is shown by (Coleman et al., 1982) in another context.

8.3.3 Acceleration waves

We have given a thermodynamic justification of a system of equations like (8.67), (8.71). We now return to study wave motion in this system. One could deal with the full system developed in section 8.3.2 but we deal with a simplified model for which $\epsilon = \epsilon(\theta)$ and $p = p(\rho, \theta)$. This is for clarity, and we still find acoustic and thermal wave propagation.

Hence, we recollect equations (8.67) and (8.71) at this point in the form

$$\begin{aligned}
&\rho_t + v_i \rho_{,i} + \rho v_{i,i} = 0, \\
&\rho(v_{i,t} + v_j v_{i,j}) = -p_{,i} - k v_i, \\
&\rho \epsilon_\theta \dot{\theta} = -p v_{i,i} - q_{i,i}, \\
&q_i + \tau \dot{q}_i = -\kappa \theta_{,i},
\end{aligned} \qquad (8.86)$$

where we treat τ and κ as constants. We define an acceleration wave for equations (8.86) to be a singular surface, $\mathcal{S}$, in $\mathbb{R}^3$ such that ρ, v_i, θ and q_i are continuous everywhere but $\rho_t, \rho_{,i}, v_{i,t}, v_{i,j}, \theta_t, \theta_{,i}, q_{i,t}, q_{i,j}$ and their higher derivatives suffer a finite discontinuity across $\mathcal{S}$. The amplitudes A_i, B, C and D_i are defined by

$$\begin{aligned}
A_i = [v_{i,j} n_j], \qquad B = [n_i \rho_{,i}], \\
C = [n_i \theta_{,i}], \qquad D_i = [q_{i,j} n_j]
\end{aligned} \qquad (8.87)$$

where $[\cdot]$ denotes the jump defined in (7.6).

Taking the jumps in equations (8.86) we find

$$\begin{aligned}
&[\rho_t] + v_i[\rho_{,i}] + \rho[v_{i,i}] = 0, \\
&\rho[v_{i,t}] + \rho v_j[v_{i,j}] + p_\rho[\rho_{,i}] + p_\theta[\theta_{,i}] = 0, \\
&\rho \epsilon_\theta[\theta_t] + \rho \epsilon_\theta v_i[\theta_{,i}] + p[v_{i,i}] + [q_{i,i}] = 0, \\
&[q_{i,t}] + v_j[q_{i,j}] + \frac{\kappa}{\tau}[\theta_{,i}] = 0.
\end{aligned} \qquad (8.88)$$

From the compatibility relations, (8.42), and the Hadamard relation (7.27) we have

$$[v_{i,j}] = A_i n_j, \quad [\rho_{,i}] = B n_i, \quad [\theta_{,i}] = C n_i, \quad [q_{i,i}] = D_i n_i,$$
$$[v_{i,i}] = A_i n_i, \quad [\rho_t] = -VB, \quad [v_{i,t}] = -VA_i, \tag{8.89}$$

V being the wavespeed of $\mathcal{S}$. Thus, employing (8.89) in (8.88) one may derive

$$(v_i n_i - V)B + \rho A_i n_i = 0,$$
$$\rho(v_j n_j - V)A_i + p_\rho n_i B + p_\theta n_i C = 0,$$
$$\rho \epsilon_\theta (v_j n_j - V)C + p n_i A_i + D_i n_i = 0, \tag{8.90}$$
$$(v_j n_j - V)D_i + \frac{\kappa}{\tau} n_i C = 0.$$

Equations $(8.90)_2$ and $(8.90)_4$ allow us to deduce that $A_i = A n_i$, $D_i = D n_i$ and thus (8.90) reduce to

$$\begin{pmatrix} v_i n_i - V & \rho & 0 & 0 \\ p_\rho & \rho(v_i n_i - V) & p_\theta & 0 \\ 0 & p & \rho \epsilon_\theta (v_i n_i - V) & 1 \\ 0 & 0 & \kappa/\tau & v_i n_i - V \end{pmatrix} \begin{pmatrix} B \\ A \\ C \\ D \end{pmatrix} = \begin{pmatrix} 0 \\ 0 \\ 0 \\ 0 \end{pmatrix}.$$

We require the amplitudes $(B, A, C, D)^T \neq \mathbf{0}$ and thus we must have

$$\begin{vmatrix} v_i n_i - V & \rho & 0 & 0 \\ p_\rho & \rho(v_i n_i - V) & p_\theta & 0 \\ 0 & p & \rho \epsilon_\theta (v_i n_i - V) & 1 \\ 0 & 0 & \kappa/\tau & v_i n_i - V \end{vmatrix} = 0.$$

Upon expanding the determinant we find that

$$\rho^2 \epsilon_\theta (v_i n_i - V)^4 - (v_i n_i - V)^2 \left\{ \frac{\kappa \rho}{\tau} + p p_\theta + \rho^2 \epsilon_\theta p_\rho \right\} + \frac{\kappa \rho p_\rho}{\tau} = 0. \tag{8.91}$$

We now assume the wave is advancing into an equilibrium region so that ahead of the wave

$$\rho \equiv \text{constant}, \quad v_i \equiv 0, \quad \theta \equiv \text{constant}, \quad q_i \equiv 0,$$

and hence

$$\rho^+ = \text{constant}, \quad v_i^+ = 0, \quad \theta^+ = \text{constant}, \quad q_i^+ = 0.$$

In this situation the wavespeed equation (8.91) reduces to

$$V^4 - V^2 \left\{ \frac{\kappa}{\tau \rho \epsilon_\theta} + \frac{p p_\theta}{\rho^2 \epsilon_\theta} + p_\rho \right\} + \frac{\kappa p_\rho}{\tau \rho \epsilon_\theta} = 0. \tag{8.92}$$

We wish to interpret equation (8.92) and hence briefly consider the propagation of a temperature wave. For a rigid heat conductor equations (8.86) would have $\rho \equiv$ constant, $u_i \equiv 0$, so that

$$\rho \epsilon_\theta [\theta_t] = -[q_{i,i}],$$

$$[q_{i,t}] + \frac{\kappa}{\tau}[\theta_{,i}] = 0.$$

Then, following the analysis above one finds

$$- \rho \epsilon_\theta V C + D_i n_i = 0,$$

$$- V D_i + \frac{\kappa}{\tau} n_i C = 0.$$

This leads to a wavespeed equation of form

$$V^2 = \frac{\kappa}{\tau \rho \epsilon_\theta}.$$

Thus, we define the thermal wavespeed, U_T, and the mechanical wavespeed, U_M, as

$$U_T^2 = \frac{\kappa}{\tau \rho \epsilon_\theta}, \qquad U_M^2 = p_\rho,$$

(see section 8.1.3 for U_M^2) and then equation (8.92) is conveniently rewritten as

$$(V^2 - U_M^2)(V^2 - U_T^2) - \kappa V^2 = 0, \tag{8.93}$$

where

$$\kappa = \frac{p p_\theta}{\rho^2 \epsilon_\theta}.$$

We expect $p_\theta > 0$ (cf. e.g. $p = R\rho\theta$) and since $\epsilon_\theta = -\theta\psi_{\theta\theta}$ we expect $\epsilon_\theta > 0$ and so $\kappa > 0$, cf. the choice $\psi = c(\theta - \theta \log \theta)$. In this case equation (8.93) leads to

$$(V^2 - U_M^2)(V^2 - U_T^2) = |\kappa|V^2 > 0.$$

Hence, (8.93) has two solutions V_1^2, V_2^2 with

$$0 < V_2^2 < \min\{U_M^2, U_T^2\} < \max\{U_M^2, U_T^2\} < V_1^2.$$

Thus, the present theory predicts a slow wave with speed V_2, and a fast wave with speed V_1, both of which have right and left moving waves (due to the presence of V^2).

Before moving on to the calculation of the amplitudes, we observe that (Straughan and Franchi, 1984) suggest replacing equation (8.71) with one involving another objective derivative for q_i, namely

$$\tau \left(q_{i,t} + v_j q_{i,j} - \frac{1}{2} q_k \{v_{i,k} - v_{k,i}\} \right) = -q_i - \kappa\theta_{,i}.$$

One may repeat the above analysis with this replacement and nothing changes concerning the wavespeed relation.

8.3.4 Amplitude derivation

We now show how one may calculate the amplitudes A, B, C and D, but restrict attention to a one-dimensional wave $\mathcal{S}$. In one space dimension equations (8.86) become (with $\mathbf{v} = (u, 0, 0), \mathbf{q} = (q, 0, 0)$)

$$\rho_t + u\rho_x + \rho u_x = 0,$$

$$u_t + u u_x + \frac{p_\rho}{\rho}\rho_x + \frac{p_\theta}{\rho}\theta_x + \frac{k}{\rho}u = 0,$$

$$\theta_t + u\theta_x + \frac{p}{\rho\epsilon_\theta}u_x + \frac{q_x}{\rho\epsilon_\theta} = 0,$$

$$q_t + u q_x + \frac{1}{\tau}q + \frac{\kappa}{\tau}\theta_x = 0.$$

(8.94)

To make the exposition as clear as possible we restrict attention to a wave moving into an equilibrium region for which $u^+ = 0, \rho^+ \equiv \text{constant}, \theta^+ \equiv$ constant. The equation for q becomes $q_t + q/\tau = 0$ which may be solved to see that $q(t) = e^{-t/\tau}q(0)$. In what follows we assume $q^+ \equiv 0$.

To determine the amplitudes we differentiate each of (8.94) with respect to x and take the jumps, recollecting the conditions ahead of the wave to find

$$[\rho_{tx}] + 2[u_x][\rho_x] + \rho[u_{xx}] = 0,$$

$$[u_{tx}] + [u_x]^2 + P_\rho[\rho_x]^2 + (Q_\rho + P_\theta)[\theta_x][\rho_x] + P[\rho_{xx}]$$
$$+ Q_\theta[\theta_x]^2 + Q[\theta_{xx}] + \frac{k}{\rho}[u_x] = 0,$$

$$\epsilon_\theta[\rho_x][\theta_t] + \rho\epsilon_{\theta\theta}[\theta_x][\theta_t] + \rho\epsilon_\theta[\theta_{tx}] + \rho\epsilon_\theta[u_x][\theta_x]$$
$$+ p_\rho[\rho_x][u_x] + p_\theta[\theta_x][u_x] + p[u_{xx}] + [q_{xx}] = 0,$$

$$[q_{tx}] + [u_x][q_x] + \frac{1}{\tau}[q_x] + \frac{\kappa}{\tau}[\theta_{xx}] = 0,$$

(8.95)

where we have employed the product relation (7.13) and have defined P and Q by

$$P = \frac{p_\rho}{\rho}, \qquad Q = \frac{p_\theta}{\rho}.$$

The amplitudes in this section are defined by

$$B = [\rho_x], \quad A = [u_x], \quad C = [\theta_x], \quad D = [q_x]$$

and then we use relations like

$$\frac{\delta B}{\delta t} = [\rho_{tx}] + V[\rho_{xx}]$$

obtained from the Hadamard relation (7.8) together with equivalent expressions involving A, C, D. In this manner one derives from (8.95),

$$\frac{\delta B}{\delta t} - V[\rho_{xx}] + \rho[u_{xx}] + 2AB = 0, \tag{8.96}$$

$$\frac{\delta A}{\delta t} - V[u_{xx}] + A^2 + P_\rho B^2 + (Q_\rho + P_\theta)BC$$
$$+ P[\rho_{xx}] + Q_\theta C^2 + Q[\theta_{xx}] + \frac{k}{\rho}A = 0, \tag{8.97}$$

$$- V\epsilon_\theta BC - V\rho\epsilon_{\theta\theta}C^2 + \rho\epsilon_\theta\left(\frac{\delta C}{\delta t} - V[\theta_{xx}]\right)$$
$$+ \rho\epsilon_\theta AC + p_\rho AB + p_\theta AC + p[u_{xx}] + [q_{xx}] = 0, \tag{8.98}$$

$$\frac{\delta D}{\delta t} - V[q_{xx}] + AD + \frac{1}{\tau}D + \frac{\kappa}{\tau}[\theta_{xx}] = 0. \tag{8.99}$$

We need to eliminate the second derivative terms like $[\rho_{xx}]$ and also eliminate three of A, B, C or D. To do this we note that the amplitudes are related (by taking the jumps of (8.94)) as follows,

$$A = \frac{V}{\rho}B, \qquad C = \left(\frac{V^2 - p_\rho}{p_\theta}\right)B, \qquad D = \frac{\kappa}{\tau p_\theta V}(V^2 - p_\rho)B. \tag{8.100}$$

We form the combination $(8.96) + \lambda_1(8.97) + \lambda_2(8.98) + \lambda_3(8.99)$ and select

$$\lambda_1 = \frac{V}{P}, \qquad \lambda_2 = \frac{V^2 - \rho P}{pP}, \qquad \lambda_3 = \frac{V^2 - \rho P}{VPp}.$$

After using the wavespeed equation (8.92) the second derivative terms disappear and one derives the amplitude equation

$$\frac{\delta B}{\delta t} + 2AB + \frac{V}{P}\left[\frac{\delta A}{\delta t} + A^2 + P_\rho B^2 + (Q_\rho + P_\theta)BC + Q_\theta C^2 + \frac{k}{\rho}A\right]$$
$$+ \left(\frac{V^2 - \rho P}{pP}\right)\left\{-V\epsilon_\theta BC - \rho V\epsilon_{\theta\theta}C^2 + \rho\epsilon_\theta\frac{\delta C}{\delta t} + \rho\epsilon_\theta AC\right.$$
$$\left. + p_\rho AB + p_\theta AC\right\}$$
$$+ \left(\frac{V^2 - \rho P}{pPV}\right)\left\{\frac{\delta D}{\delta t} + AD + \frac{1}{\tau}D\right\} = 0. \tag{8.101}$$

Upon using the relations (8.100) the above equation may be written in the form

$$
\begin{aligned}
&\frac{\delta B}{\delta t}\left\{1 + \frac{V^2}{p_\rho} + (V^2 - p_\rho)^2\frac{1}{pPp_\theta}\left(\rho\epsilon_\theta + \frac{\kappa}{\tau V^2}\right)\right\} \\
&+ B\frac{1}{P}\left\{\frac{V^2 k}{\rho^2} + \frac{(V^2 - p_\rho)^2\kappa}{pV^2\tau^2 p_\theta}\right\} \\
&+ B^2\left\{\frac{2V}{\rho} + \frac{V^3}{\rho p_\rho} + \frac{VP_\rho}{P}\right\} \\
&+ (V^2 - p_\rho)\frac{V}{P}\left(\frac{(Q_\rho + P_\theta)}{p_\theta} + \frac{p_\rho}{\rho p}\right) \\
&+ (V^2 - p_\rho)^2\frac{V}{Pp_\theta}\left(\frac{Q}{p_\theta} + \frac{p_\theta}{\rho p} + \frac{\kappa}{\rho p V^2\tau}\right) \\
&- (V^2 - p_\rho)^3\frac{\rho V\epsilon_{\theta\theta}}{pPp_\theta^2}\right\} = 0.
\end{aligned}
\tag{8.102}
$$

Since the coefficients of the terms $\delta B/\delta t$, B and B^2 are constant the solution to this equation is found exactly as in equation (8.31), and derivations regarding amplitude blow-up may be made as in that section.

It is interesting to analyse equation (8.102) in the limit $\kappa \to 0$ and the isothermal case for which $V^2 \to p_\rho$. Then equation (8.102) reduces to

$$
\frac{\delta B}{\delta t} + \frac{k}{2\rho}B + B^2\left\{\frac{3V}{2\rho} + \frac{V}{2}(\log P)_\rho\right\} = 0.
$$

This is in complete agreement with equation (8.26) with the coefficients given by (8.30), since in terms of $A = VB/\rho$, this becomes

$$
\frac{\delta A}{\delta t} + \frac{k}{2\rho}A + A^2\left\{\frac{3}{2} + \frac{\rho}{2}(\log P)_\rho\right\} = 0.
$$

8.3.5 Dual phase lag theory

There has been much recent interest in developing theories of heat propagation which extend the phase lag heat flux law of (8.69) and, in particular, which consider extensions of the Taylor series for the heat flux in equation (8.70). Much of this stems from the work of (Tzou, 1995b; Tzou, 1995a), and we cite in particular, (Han et al., 2006), (Jou and Criado-Sancho, 1998), (Quintanilla, 2002b), (Quintanilla and Racke, 2006; Quintanilla and Racke, 2007), (Serdyukov, 2001), (Serdyukov et al., 2003) and the references therein. The key would appear to be the assertion that (8.69) be replaced by an equation of form

$$
q_i(\mathbf{x}, t + \tau_q) = -\kappa\theta_{,i}(\mathbf{x}, t + \tau),
\tag{8.103}
$$

where τ_q and τ will have (in general) different values. Various truncations of the Taylor series expansion are considered. For example, (8.103) is replaced by

$$q_i(\mathbf{x}, t) + \tau_q q_{i,t}(\mathbf{x}, t) = -\kappa \theta_{,i}(\mathbf{x}, t) - \kappa \tau \theta_{,it}(\mathbf{x}, t), \tag{8.104}$$

(Han et al., 2006), (Jou and Criado-Sancho, 1998), (Serdyukov, 2001), (Serdyukov et al., 2003). While the above representation is attributed to various writers by the above cited papers we should point out that this relation is, in fact, effectively given by (Cattaneo, 1948), see also (Fichera, 1992), even if a sign is different. Combined with the energy equation for a rigid heat conductor,

$$\rho \epsilon_\theta \theta_t = -q_{i,i}, \tag{8.105}$$

equation (8.104) yields (for $\rho \epsilon_\theta = c$, constant)

$$c\theta_t + c\tau_q \theta_{tt} = \kappa \Delta \theta + \kappa \tau \Delta \theta_t. \tag{8.106}$$

(Quintanilla, 2002b) and (Serdyukov et al., 2003) consider adding a further term in the expansion of $q_i(t + \tau_q)$ to the left of (8.104) so that

$$q_i(\mathbf{x}, t) + \tau_q q_{i,t}(\mathbf{x}, t) + \frac{\tau_q^2}{2} q_{i,tt}(\mathbf{x}, t) = -\kappa \theta_{,i}(\mathbf{x}, t) - \kappa \tau \theta_{,it}(\mathbf{x}, t). \tag{8.107}$$

Together with (8.105) this leads to the hyperbolic equation

$$\frac{c\tau_q^2}{2} \theta_{ttt} + c\tau_q \theta_{tt} + c\theta_t = \kappa \Delta \theta + \kappa \tau \Delta \theta_t. \tag{8.108}$$

A very interesting derivation of equation (8.108) for gas flow through a package of heat conducting plates is given by (Serdyukov et al., 2003). These writers use a Cattaneo theory for the plates and a Newton cooling - like law for the gas, of form

$$\rho c(\tau \theta_{tt} + \theta_t) = \kappa \Delta \theta - \beta_1(\theta - \theta_g),$$
$$\rho_g c_g \theta_t^g = \beta_2(\theta - \theta_g),$$

where θ and θ_g are the temperatures of the plates and gas, respectively.

(Quintanilla and Racke, 2006; Quintanilla and Racke, 2007) consider a further extension to (8.107) of form

$$q_i(\mathbf{x}, t) + \tau_q q_{i,t}(\mathbf{x}, t) + \frac{\tau_q^2}{2} q_{i,tt}(\mathbf{x}, t)$$
$$= -\kappa \theta_{,i}(\mathbf{x}, t) - \kappa \tau \theta_{,it}(\mathbf{x}, t) - \frac{\kappa \tau^2}{2} \theta_{,itt}(\mathbf{x}, t). \tag{8.109}$$

Due to the intense interest in these dual phase lag theories we wish to briefly consider a possible extension involving the model for poroacoustic wave propagation. In particular, we consider a relation like (8.103) but with κ dependent on temperature, θ, as it invariably is, so we consider

$$q_i(\mathbf{x}, t + \tau_q) = -\kappa\big(\theta(\mathbf{x}, t + \tau)\big)\theta_{,i}(\mathbf{x}, t + \tau).$$

We consider the equivalent of the expansion (8.103) and so write

$$q_i + \tau_q \dot{q}_i + \frac{\tau_q^2}{2} \ddot{q}_i = -\left(\kappa(\theta) + \tau \dot{\theta} \kappa'(\theta) \right) \left(\theta_{,i} + \tau \dot{\theta}_{,i} \right). \qquad (8.110)$$

One could argue to not consider the term $\tau^2 \dot{\theta} \kappa'(\theta) \dot{\theta}_{,i}$ since it is $O(\tau^2)$. However, in a rigid heat conductor theory the inclusion of this term is necessary if one wishes to derive a Bernoulli equation containing a quadratic term and thus a theory capable of finite time amplitude blow-up.

Thus, a potential theory might involve equation (8.67) with $p = p(\rho, \theta)$, $\epsilon = \epsilon(\theta)$, and equation (8.110). We would, in this case, define an acceleration wave to be a singular surface, $\mathcal{S}$, across which ρ, v_i are continuous, θ, q_i and their *first* derivatives are continuous, but first derivatives (and higher) of ρ, v_i and second derivatives (and higher) of θ, q_i suffer at most finite discontinuities.

For an acceleration wave in one-dimension we find

$$(u - V)[\rho_x] + \rho[u_x] = 0,$$

$$(u - V)[u_x] + \frac{p_\rho}{\rho}[\rho_x] = 0,$$

$$\epsilon_\theta[\dot{\theta}_x] = \left(-\frac{p_\rho}{\rho} + \frac{p}{\rho^2} \right)[\rho_x u_x] - \frac{p_\theta \theta_x^+}{\rho}[u_x] - \frac{[q_{xx}]}{\rho} + \frac{q_x}{\rho^2}[\rho_x], \qquad (8.111)$$

$$\frac{\tau_q^2}{2}[\ddot{q}] = -(\kappa + \tau \dot{\theta}^+ \kappa')\tau[\dot{\theta}_x].$$

We see that equations $(8.111)_{1,2}$ are disconnected and yield a wavespeeed on their own. The other two equations then yield a wavespeed for an unconnected wave. Thus, care must be taken when attempting to develop a Tzou-like theory for poroacoustic - thermal wave propagation.

We have not considered thermodynamics in our model and a more sophisticated approach must due so (cf. the Cattaneo model and the work of (Morro and Ruggeri, 1988), (Coleman et al., 1982)). A more complete model may need the dependence on quantities like q_i by terms like the pressure, cf. section 8.3.2.

8.4   Temperature rate effects

8.4.1   Green-Laws theory

In this section we describe work of (Ciarletta and Straughan, 2007a) who generalize the Jordan-Darcy model of sections 8.1 and 8.2 to include the possibility of propagation of temperature waves. To do this they employ the thermodynamics of (Green and Laws, 1972) which uses a generalized temperature $\phi(\theta, \dot{\theta})$. As noted in section 7.3.1 the current literature recognizes the role that thermal waves play in wave propagation, particularly in porous media. We repeat that (Johnson et al., 1994) employ second sound to calculate physical properties of porous media, (Meyer, 2006) discusses how temperature waves may be employed to dry porous media, and thermal waves may be important in biological tissues, (Mitra et al., 1995).

The basic theory for (Green and Laws, 1972) thermodynamics used to develop a perfect fluid is given in (Lindsay and Straughan, 1978) and (Ciarletta and Straughan, 2007a) add a Darcy term into the equations and study acceleration wave propagation in one space dimension.

With a superposed dot denoting the material time derivative the equations of mass continuity, conservation of momentum, and conservation of energy, are presented by (Ciarletta and Straughan, 2007a). These are continuity of mass

$$\dot{\rho} + \rho \frac{\partial v_i}{\partial x_i} = 0, \tag{8.112}$$

conservation of momentum,

$$\rho \dot{v}_i = \frac{\partial t_{ki}}{\partial x_k} - k v_i, \tag{8.113}$$

and conservation of energy,

$$\rho \dot{\epsilon} = t_{ki} d_{ik} - \frac{\partial q_i}{\partial x_i}, \tag{8.114}$$

if one chooses zero body force and heat supply function. Standard notation is employed so $\rho, v_i, t_{ki}, \epsilon, q_i, d_{ik}$ and k are density, velocity, stress tensor, internal energy, heat flux, symmetric part of the velocity gradient, $d_{ik} = (v_{i,k} + v_{k,i})/2$, and the (constant) Darcy coefficient, respectively. The constitutive theory of (Ciarletta and Straughan, 2007a) is taken to be the same as that of (Lindsay and Straughan, 1978). The Helmholtz free energy is $\psi = \epsilon - \eta\phi$, and then $\epsilon, \psi, t_{ki}, q_i$ and the generalized temperature ϕ are functions of the constitutive variables $\rho, \theta, \dot{\theta}$, and λ where $\lambda = \theta_{,i}\theta_{,i}/2$. One may employ the entropy inequality of (Green and Laws, 1972) to derive the constitutive restrictions presented by (Lindsay and Straughan, 1978), and these are

$$\phi = \phi(\theta, \dot{\theta}), \qquad \psi = \psi(\rho, \theta, \dot{\theta}, \lambda), \tag{8.115}$$

$$\eta = -\frac{\partial \psi}{\partial \dot{\theta}} \bigg/ \frac{\partial \phi}{\partial \dot{\theta}} = \eta(\rho, \theta, \dot{\theta}, \lambda), \tag{8.116}$$

$$q_i = -K\theta_{,i}, \tag{8.117}$$

$$K = \rho\phi \frac{\partial \psi}{\partial \lambda} \bigg/ \frac{\partial \phi}{\partial \dot{\theta}} = K(\rho, \theta, \dot{\theta}, \lambda), \tag{8.118}$$

$$t_{ik} = -p\delta_{ik} - \rho \frac{\partial \psi}{\partial \lambda} \theta_{,i}\theta_{,k}, \tag{8.119}$$

$$p = \rho^2 \frac{\partial \psi}{\partial \rho} = p(\rho, \theta, \dot{\theta}, \lambda). \tag{8.120}$$

The coefficient K is the thermal diffusivity and p is the pressure.

The residual entropy inequality which remains from the (Green and Laws, 1972) one is given by (Lindsay and Straughan, 1978) as

$$-\left(\frac{\partial\psi}{\partial\theta} + \eta\frac{\partial\phi}{\partial\theta}\right)\dot\theta + 2K\frac{\partial\phi}{\partial\theta}\frac{\lambda}{\phi} \geq 0. \qquad (8.121)$$

Since this must hold for all motions one then finds in equilibrium

$$\left(\frac{\partial\psi}{\partial\theta} + \eta\frac{\partial\phi}{\partial\theta}\right)\bigg|_E = 0, \quad \left(\frac{\partial\eta}{\partial\dot\theta}\frac{\partial\phi}{\partial\dot\theta}\right)\bigg|_E - \left(\frac{\partial\eta}{\partial\dot\theta}\right)\bigg|_E \geq 0, \quad K|_E \geq 0, \quad (8.122)$$

where $f|_E = f(\rho, \theta, \dot\theta, \lambda)|_E$ denotes the value of f in thermal equilibrium, i.e. where $\dot\theta = \lambda = 0$, and cf. (Green and Laws, 1972), $\partial\phi/\partial\theta|_E = 1$.

Upon employing (8.115) – (8.120), the energy balance law (8.114) may be written as

$$K\theta_{,ii} + \frac{\partial K}{\partial\rho}\rho_{,i}\theta_{,i} + 2\lambda\frac{\partial K}{\partial\theta} + \frac{\partial K}{\partial\dot\theta}\dot\lambda + \frac{\partial K}{\partial\lambda}\lambda_{,i}\theta_{,i}$$

$$-\rho\left(\frac{\partial\psi}{\partial\theta} + \eta\frac{\partial\phi}{\partial\theta}\right)\dot\theta - \rho\phi\frac{\partial\eta}{\partial\rho}\dot\rho - \rho\phi\frac{\partial\eta}{\partial\theta}\dot\theta - \rho\phi\frac{\partial\eta}{\partial\dot\theta}\ddot\theta \qquad (8.123)$$

$$-\rho\left(\frac{\partial\psi}{\partial\lambda} + \phi\frac{\partial\eta}{\partial\lambda}\right)\dot\lambda + \rho\phi\frac{\partial}{\partial\dot\theta}\left(\frac{\partial\psi}{\partial\lambda}\bigg/\frac{\partial\phi}{\partial\dot\theta}\right)\theta_{,i}\theta_{,j}d_{ij} = 0.$$

(Ciarletta and Straughan, 2007a) assume ψ and ϕ are such that $\partial\eta/\partial\dot\theta > 0$, as suggested by (8.122).

8.4.2   Wavespeeds

(Ciarletta and Straughan, 2007a) developed the wavespeed and amplitude equations in a one dimensional setting, without presenting details. We now derive the wavespeeds in three dimensions.

Our basic governing equations are now (8.112), (8.113) and (8.123). For these equations we define an acceleration wave to be a singular surface S across which the velocity v_i, the density ρ, and the temperature gradient $\theta_{,i}$ are continuous, but their first and higher derivatives, in general, possess finite discontinuities.

We follow (Ciarletta and Straughan, 2007a) and restrict attention to a wave moving into a region at rest and at constant temperature, so that $v_i^+ = 0$, $\rho^+ = $ constant, $\theta^+ = $ constant. It is believed no essential loss of physics occurs via this procedure, unless perhaps, one is analysing curved waves. However, the aim of this book is to describe the implications for wave motion in porous media. Unlike (Ciarletta and Straughan, 2007a) we do, however, develop the wavespeeds from a three-dimensional analysis.

We expand equations (8.112) and (8.113) and take the jumps of these. We also take the jump of (8.123). Recalling that $v_i^+ = 0$ and ρ^+ and θ^+

are constant one obtains

$$[\rho_t] + \rho[v_{i,i}] = 0, \tag{8.124}$$

$$\rho[v_{i,t}] = -\frac{\partial p}{\partial \rho}[\rho_{,i}] - \frac{\partial p}{\partial \theta}[\theta_{,ti}], \tag{8.125}$$

$$K[\theta_{,ii}] - \rho\phi\eta_\rho[\rho_t] - \rho\phi\eta_{\dot\theta}[\theta_{tt}] = 0. \tag{8.126}$$

From use of the Hadamard relation (7.27) we know that

$$\begin{aligned}
&[\rho_t] = -V[n_i\rho_{,i}], \qquad [v_{i,t}] = -V[n_j v_{i,j}], \\
&[\theta_{tt}] = -V[n_i\theta_{,ti}] = V^2[n_i n_j\theta_{,ij}],
\end{aligned} \tag{8.127}$$

and by use of the compatibility conditions (176.2), (176.10) of (Truesdell and Toupin, 1960) we have

$$[\theta_{,ii}] = [n_i n_j\theta_{,ij}]. \tag{8.128}$$

Define the amplitudes A_i, B and C by

$$A_i = [n_j v_{i,j}], \qquad B = [n_i\rho_{,i}], \qquad C = [n_i n_j\theta_{,ij}]. \tag{8.129}$$

Upon utilizing (8.127) and (8.128) in (8.124) – (8.126) one finds

$$\begin{aligned}
&-VB + \rho n_i A_i = 0, \\
&-\rho V A_i = -p_\rho n_i B + p_{\dot\theta} V n_i C, \\
&KC + \rho\phi\eta_\rho V B - \rho\phi\eta_{\dot\theta} V^2 C = 0.
\end{aligned} \tag{8.130}$$

We may always write the vector A_i as a sum of components in the normal and tangential directions to the wave, say $A_i = An_i + A_\perp s_i$, s_i being a tangential vector. The right hand side of $(8.130)_2$ involves only n_i and so we see that $A_i = An_i$, where $A = [n_i n_j v_{i,j}]$. Thus, (8.130) is a linear system of equations for A, B and C. Since we require these to be non-zero we find (8.130) leads to the wavespeed equation

$$(V^2 - p_\rho)(\rho\phi\eta_{\dot\theta} V^2 - K) + \rho\phi\eta_\rho p_{\dot\theta} V^2 = 0. \tag{8.131}$$

We divide this equation by $\rho\phi\eta_{\dot\theta}$ and define the variables U_T, U_M and κ by

$$\begin{aligned}
&U_M^2 = p_\rho = (\rho^2\psi_\rho)_\rho, \qquad U_T^2 = \frac{K}{\rho\phi\eta_{\dot\theta}} = \frac{\psi_\lambda}{\phi_{\dot\theta}\eta_{\dot\theta}}, \\
&\kappa = \frac{\eta_\rho p_{\dot\theta}}{\eta_{\dot\theta}} = \frac{\rho^2\phi_{\dot\theta}\psi_{\rho\dot\theta}^2}{(\phi_{\dot\theta}\psi_{\dot\theta\dot\theta} - \psi_{\dot\theta}\phi_{\dot\theta\dot\theta})}.
\end{aligned} \tag{8.132}$$

Then equation (8.131) may be conveniently written as

$$(V^2 - U_M^2)(V^2 - U_T^2) + \kappa V^2 = 0. \tag{8.133}$$

The quantities U_M and U_T are the speeds of a mechanical wave and a temperature wave, respectively, in the absence of either effect, cf. (Lindsay and Straughan, 1978). We might expect $\kappa < 0$ and then (8.132) gives rise to two waves (both left and right moving) with speeds U_1^2, U_2^2, where

$$U_2^2 < \min\{U_T^2, U_M^2\} < \max\{U_T^2, U_M^2\} < U_1^2.$$

8.4.3 Amplitude behaviour

The final amplitude equation is presented in (Ciarletta and Straughan, 2007a). We now briefly derive this, although we restrict attention to a wave propagating in one-dimension and into an equilibrium region for which $\mathbf{v} = (u, 0, 0), u^+ = 0, \theta^+ = $ constant, and $\rho^+ = $ constant.

For this case, the governing equations (8.112), (8.113), (8.123) become

$$\rho_t + \rho u_x + u\rho_x = 0, \tag{8.134}$$

$$\begin{aligned}
\rho(u_t + uu_x) = &-p_\rho \rho_x - p_\theta \theta_x - p_{\dot\theta}\dot\theta_x - p_\lambda \lambda_x \\
&- (\rho\psi_\lambda \theta_x^2)_x - ku.
\end{aligned} \tag{8.135}$$

$$\begin{aligned}
&K\theta_{xx} + K_\rho \rho_x \theta_x + 2\lambda K_\theta + K_{\dot\theta}\dot\lambda + K_\lambda \lambda_x \theta_x \\
&\quad - \rho(\psi_\theta + \eta\phi_\theta)\dot\theta - \rho\phi\eta_\rho \dot\rho - \rho\phi\eta_\theta(\theta_t + u\theta_x) \\
&\quad - \rho\phi\eta_{\dot\theta}(\theta_{tt} + 2u\theta_{tx} + u_t\theta_x + uu_x\theta_x + u^2\theta_{xx}) \\
&\quad - \rho(\psi_\lambda + \phi\eta_\lambda)(\lambda_t + u\lambda_x) + 2\rho\phi\frac{\partial}{\partial\dot\theta}\left(\frac{\psi_\lambda}{\phi_{\dot\theta}}\right)u_x\lambda = 0. \tag{8.136}
\end{aligned}$$

In terms of the one-dimensional amplitudes $A = [u_x], B = [\rho_x]$ and $C = [\theta_{xx}]$ equations (8.134) – (8.136) yield

$$\begin{aligned}
\rho V A - p_\rho B + p_{\dot\theta} V C &= 0, \\
\rho A - V B &= 0, \\
(K - \rho\phi\eta_{\dot\theta} V^2)C + \rho\phi\eta_\rho V B &= 0,
\end{aligned} \tag{8.137}$$

and a non-zero solution of this system means we need

$$\begin{vmatrix}
\rho V & -p_\rho & V p_{\dot\theta} \\
\rho & -V & 0 \\
0 & \rho\phi\eta_\rho V & K - \rho\phi V^2\eta_{\dot\theta}
\end{vmatrix} = 0$$

which leads to the wavespeed equation (8.133) in the one-dimensional case. We also observe that (8.137) lead to direct relations between A and B and A and C as follows

$$B = \frac{\rho}{V}A, \qquad C = \frac{\rho}{p_{\dot\theta}}\left(\frac{p_\rho}{V^2} - 1\right)A. \tag{8.138}$$

These equations allow us to eliminate B and C and derive one equation for A. Once we solve for $A(t)$, the functions $B(t)$ and $C(t)$ are also known from (8.138).

To derive the amplitude equation we differentiate each of (8.134) – (8.136) in turn with respect to x and take the jump of each equation.

Using the Hadamard relation and the fact that the region ahead of the wave is in equilibrium one may derive the following equations. From the continuity of mass (8.134),

$$\frac{\delta B}{\delta t} - V[\rho_{xx}] + \rho[u_{xx}] + 2AB = 0. \tag{8.139}$$

From the momentum equation (8.113) there follows,

$$- VBA + \rho[u_{tx}] + \rho A^2 = -p_\rho[\rho_{xx}] - p_{\rho\rho}B^2 + 3p_{\rho\dot\theta}VBC - p_\theta C$$
$$- p_{\dot\theta}([\theta_{txx}] + 2AC) - p_{\dot\theta\dot\theta}V^2C^2 - p_\lambda C^2 - kA - 2\rho\psi_\lambda C^2,$$

while the balance of energy equation (8.136) leads to

$$K[\theta_{xxx}] + 2K_\rho BC - 2K_{\dot\theta}VC^2 + \phi\eta_\rho VB^2 - \rho(\phi_{\dot\theta}\eta_\rho + \phi\eta_{\rho\dot\theta})V^2BC$$
$$+ \rho\phi\eta_{\rho\rho}VB^2 - \rho\phi\eta_\rho[\rho_{tx}] - \rho\phi\eta_\rho AB + \rho\phi\eta_\theta VC - \rho\eta_{\dot\theta}V^2BC$$
$$+ \rho(\phi_{\dot\theta}\eta_{\dot\theta} + \phi\eta_{\dot\theta\dot\theta})V^3C^2 - \rho V^2\phi\eta_{\dot\theta\rho}BC - \rho\phi\eta_{\dot\theta}[\theta_{ttx}]$$
$$+ 3\rho\phi\eta_{\dot\theta}VAC + \rho(\psi_\lambda + \phi\eta_\lambda)VC^2 = 0.$$

The next stage involves utilizing (8.138) to eliminate B and C in favour of A. However, one must also employ the wavespeed equation (8.133) to remove the higher derivative terms like $[u_{xx}]$, $[\rho_{xx}]$ and $[\theta_{xxx}]$. To do this we follow a procedure like that of section 8.2.3 where $V(8.58) + \mu_1(8.62) + \mu_2(8.63)$ was formed. We must also use the Hadamard relation to derive formulae like

$$[\theta_{xtt}] = -2V\frac{\delta C}{\delta t} + V^2[\theta_{xxx}].$$

After some calculation one may arrive at an amplitude equation of form (8.64) but now the constants a and b are different and given below. In fact,

$$b = \frac{k}{2\rho}\left\{\frac{p_\rho}{V^2} + \frac{2\phi\eta_{\dot\theta}(p_\rho - V^2)}{(K - \rho\phi\eta_{\dot\theta}V^2)}\right\}^{-1}$$
$$+ \frac{(p_\rho - V^2)\{V^2\rho\phi\eta_\theta + (K - \rho\phi\eta_{\dot\theta}V^2)p_\theta/V^2p_{\dot\theta}\}}{2\{\rho\phi\eta_{\dot\theta}(p_\rho - V^2) + (K - \rho\phi\eta_{\dot\theta}V^2)p_\rho/V^2\}} \tag{8.140}$$

and

$$
a = \frac{1}{2\{\rho\phi\eta_{\dot{\theta}}(p_\rho - V^2) + (p_\rho/V^2)(K - \rho\phi\eta_{\dot{\theta}}V^2)\}} \times
$$

$$
\left\{ p_{\dot{\theta}}\phi\frac{\partial}{\partial\rho}\left(\rho^2\frac{\partial\eta}{\partial\rho}\right) + \frac{(K - \rho\phi\eta_{\dot{\theta}}V^2)}{V^2}\left(2\frac{\partial p}{\partial\rho} + \rho\frac{\partial^2 p}{\partial\rho^2}\right) \right.
$$

$$
+ \left(\frac{p_\rho}{V^2} - 1\right)\left\{\left(2 - \frac{3\rho p_{\dot{\theta}\rho}}{p_{\dot{\theta}}}\right)(K - \rho\phi\eta_{\dot{\theta}}V^2)\right.
$$

$$
\left. + 2\rho\frac{\partial K}{\partial\rho} + V^2\{2\rho\phi\eta_{\dot{\theta}} - \rho^2\eta_\rho(\phi_{\dot{\theta}} + 2\dot{\theta}\phi)\}\right\}
$$

$$
\tag{8.141}
$$

$$
+ \left(\frac{p_\rho}{V^2} - 1\right)^2 \left(\frac{(K - \rho\phi\eta_{\dot{\theta}}V^2)}{p_{\dot{\theta}}^2}\right)(\rho V^2 p_{\dot{\theta}\dot{\theta}} + \rho p_\lambda + 2\rho^2\psi_\lambda)
$$

$$
\left. + \frac{\rho^2 V^2}{p_{\dot{\theta}}}\{V^2(\phi_{\dot{\theta}}\eta_{\dot{\theta}} + \phi\eta_{\dot{\theta}\dot{\theta}} - 2K_{\dot{\theta}}) + \psi_\lambda + \phi\eta_\lambda\}\right)\right\}.
$$

Again, the solution for A is found from (8.31).

(Ciarletta and Straughan, 2007a) observe that the amplitude equation derived above, which has the same form as (8.64), is completely consistent with the isothermal theory of sections 8.1.2 – 8.1.4. For, if we omit θ and λ from coefficients (8.140) and (8.141) we obtain that $V^2 \to \partial p/\partial\rho$, $b \to K/2\rho$, and $a \to 3/2 + \rho^2(\partial/\partial\rho)(\rho^{-1}\partial p/\partial\rho)/2\partial p/\partial\rho$. (Ciarletta and Straughan, 2007a) also note that the solution to the amplitude equation of the current section displays the damping effect of the thermodynamic variables. Because, even when $k \to 0$ and the porous medium disappears, there is attenuation of the wave amplitude due to what remains in the b term. Thus, one may then assess the combined effect of the porous medium, via the k term, and the (Green and Laws, 1972) theory, via the presence of the new coefficients in a and b. Together with suitable experimental results, this should lead to useful information concerning the forms for $\phi(\theta, \dot{\theta})$ and $\psi(\rho, \theta, \dot{\theta}, \lambda)$.

8.5   Temperature displacement effects

8.5.1   Green-Naghdi thermodynamics

In this section we develop another theory for transmission of acoustic waves in a porous medium, allowing for the propagation of a temperature wave. To do this we combine the Jordan method of introducing a Darcy term into a perfect fluid as in section 8.1.2 together with a thermodynamic approach to presenting a dissipationless fluid via (Green and Naghdi, 1991) thermodynamics. The theory of (Green and Naghdi, 1991) has been applied to a fluid by (Quintanilla and Straughan, 2008), although these writers did not consider a porous medium. Another study of acceleration waves in a different class of Green and Naghdi fluids is due to (Jordan and Straughan, 2006).

The starting point is to consider the theory of (Quintanilla and Straughan, 2008). We thus introduce the thermal displacement as in section 7.4.1, so $\alpha = \int_{t_0}^{t} \theta(\mathbf{X}, s)ds$ is the thermal displacement variable of (Green and Naghdi, 1991), (Green and Naghdi, 1996), where θ is the absolute temperature. The variable α is defined as a function of $\mathbf{X}$, but in fluid dynamics it is more natural to use the spatial coordinate $\mathbf{x}$ and throughout this section we do this. (Green and Naghdi, 1996) do present a fluid theory, but that of (Quintanilla and Straughan, 2008) is more general and we present an approach analogous to theirs.

The governing equations are the energy balance law, the entropy balance equation, the equation of balance of mass, and the momentum equation. The reduced energy balance equation is

$$T_{ij}L_{ij} - p_i\gamma_i - \rho(\dot{\psi} + \eta\dot{\theta}) - \rho\theta\xi = 0, \qquad (8.142)$$

where $T_{ij}, p_i, \rho, \psi, \eta$ and ξ are, respectively, the stress tensor, entropy flux vector, density, Helmholtz free energy function, entropy, and the internal rate of production of entropy, defined in the current configuration. Also, $L_{ij} = \partial v_i/\partial x_j$, where v_i is the velocity field, $\gamma_i = (\dot{\alpha})_{,i}$, and for later use we define the symmetric and skew-symmetric parts of L_{ij} as $d_{ij} = (L_{ij} + L_{ji})/2$, $\omega_{ij} = (L_{ij} - L_{ji})/2$. The (Green and Naghdi, 1991; Green and Naghdi, 1996) entropy balance law is

$$\rho\dot{\eta} = \rho s + \rho\xi - \frac{\partial p_i}{\partial x_i}. \qquad (8.143)$$

The function s is the external rate of supply of entropy per unit mass. The balance of mass equation is

$$\dot{\rho} + \rho\frac{\partial v_i}{\partial x_i} = 0, \qquad (8.144)$$

while that for balance of linear momentum is

$$\rho\dot{v}_i = \frac{\partial T_{ji}}{\partial x_j} + \rho b_i, \qquad (8.145)$$

with b_i being an externally supplied body force. In fact, in this section we modify equation (8.145) since we wish to include a Darcy term to reflect the fact that the fluid (gas) is moving in a porous medium. Thus, we present instead of (8.145),

$$\rho\dot{v}_i = \frac{\partial T_{ji}}{\partial x_j} - kv_i, \qquad (8.146)$$

with k being a Darcy coefficient.

The presentation of (Quintanilla and Straughan, 2008) is different from that of (Green and Naghdi, 1996) who assume ψ, η, T_{ij}, p_i and ξ depend on the variables $\rho, L_{ij}, \theta, \alpha_{,i}$ and γ_i. However, they also assume p_i is linear in γ_i, T_{ij} is quadratic in d_{ij}, ξ is quadratic in d_{ij} and γ_i, and the Helmholtz free energy ψ has the form

$$\psi = \frac{1}{2} m \delta_i \delta_i + f(\rho, \theta) \tag{8.147}$$

where $\delta_i = \alpha_{,i}$ and m is a constant. In this manner (Green and Naghdi, 1996) analyse a restricted class of dissipationless flows by assuming the Reynolds, Peclet and m numbers are suitably large. (Quintanilla and Straughan, 2008) is more general in that they define a dissipationless fluid to be one which omits $\gamma_i = \theta_{,i} = (\dot\alpha)_{,i}$ as a variable from the constitutive theory. This is a logical way to proceed in the light of the manner in which (Green and Naghdi, 1993) develop their theory of thermoelasticity without energy dissipation, cf. the analogous voids theory in section 7.4. The approach of (Quintanilla and Straughan, 2008) has a richer structure which allows for a fully nonlinear constitutive theory in which variables like the entropy flux vector are defined naturally in terms of the Helmholtz free energy rather than having a preimposed form.

The constitutive theory supposes that

$$T_{ij}, \psi, \eta, p_i \text{ and } \xi \tag{8.148}$$

depend on the independent variables

$$\rho, L_{ij}, \theta, \alpha_{,i} \,. \tag{8.149}$$

The velocity gradient satisfies $L_{ij} = d_{ij} + \omega_{ij}$, and so the energy balance equation (8.142) is written as

$$\left[T_{ij} + \delta_{ij}\rho^2\psi_\rho + \frac{\rho}{2}(\psi_{\alpha,i}\alpha_{,j} + \psi_{\alpha,j}\alpha_{,i})\right]d_{ij} + T_{ij}\omega_{ij}$$
$$- \gamma_i(p_i + \rho\psi_{\alpha,i}) - \rho\psi_{L_{ij}}\overline{L_{ij}} - \dot\theta\rho(\psi_\theta + \eta) \tag{8.150}$$
$$- \rho\theta\xi + \frac{\rho}{2}\omega_{ij}(\psi_{\alpha,j}\alpha_{,i} - \psi_{\alpha,i}\alpha_{,j}) = 0.$$

Since T_{ij} is symmetric, $T_{ij}\omega_{ij} = 0$. Now, $\overline{L_{ij}}$ and $\gamma_i = (\dot\alpha)_{,i}$ appear linearly in (8.150) and thus we conclude $\partial\psi/\partial L_{ij} = 0$ and $p_i = -\rho\,\partial\psi/\partial\alpha_{,i}$. This is very different from (Green and Naghdi, 1996) who *assume* a form for p_i. In the present case, the form of the entropy flux vector follows once the Helmholtz free energy function is chosen. Further, $\dot\theta = \ddot\alpha$ appears linearly and then it follows from (8.150) that $\eta = -\partial\psi/\partial\theta$. Therefore,

$$p_i = -\frac{\partial\psi}{\partial\alpha_{,i}}, \quad \eta = -\frac{\partial\psi}{\partial\theta} \tag{8.151}$$

and the velocity gradient does not appear in ψ so

$$\psi = \psi(\rho, \theta, \alpha_{,i}). \tag{8.152}$$

The general form for the function ξ from (8.149) requires that ξ depends on $\rho, \theta, \alpha_{,i}, d_{ij}, \omega_{ij}$. However, to produce a dissipationless theory which may be compared to the dissipationless theory of thermoelasticity of (Green and Naghdi, 1993), we discard the dependence on the velocity gradient terms (which represent viscous dissipation), and so drop the d_{ij} and ω_{ij} dependence to have $\xi = \xi(\rho, \theta, \alpha_{,i})$. Given this form for ξ and relations (8.151) and (8.152), what remains of equation (8.150) is

$$\left[T_{ij} + \delta_{ij}\rho^2 \psi_\rho + \frac{\rho}{2}(\psi_{\alpha,i}\alpha_{,j} + \psi_{\alpha,j}\alpha_{,i}) \right] d_{ij}$$
$$- \rho\theta\xi + \frac{\rho}{2}\omega_{ij}(\psi_{\alpha,j}\alpha_{,i} - \psi_{\alpha,i}\alpha_{,j}) = 0. \tag{8.153}$$

Next ω_{ij} and d_{ij} appear linearly in (8.153) and this leads to

$$\psi_{\alpha,i}\alpha_{,j} = \psi_{\alpha,j}\alpha_{,i}, \tag{8.154}$$

and then

$$T_{ij} = -p\delta_{ij} - \frac{\rho}{2}(\psi_{\alpha,i}\alpha_{,j} + \psi_{\alpha,j}\alpha_{,i}), \tag{8.155}$$

where p is a pressure given by $p = \rho^2 \partial\psi/\partial\rho$. Due to the forms in (8.154) and (8.155), equation (8.153) then implies $\xi = 0$, a relation which agrees with what is found by (Green and Naghdi, 1996).

The basic equations of balance of mass, balance of momentum, and balances of energy and entropy may then be written as (for zero body force and heat supply),

$$\rho_t + \rho_{,i}v_i + \rho v_{i,i} = 0,$$
$$\rho(v_{i,t} + v_j v_{i,j}) = -p_{,i} - \frac{1}{2}\left[\rho(\psi_{\alpha,j}\alpha_{,i} + \psi_{\alpha,i}\alpha_{,j})\right]_{,j} - kv_i, \tag{8.156}$$
$$- \rho\{(\psi_\theta)_{,t} + v_i(\psi_\theta)_{,i}\} = (\rho\psi_{\alpha,i})_{,i}.$$

These are the central equations of this section with which we analyse nonlinear wave motion.

8.5.2 Acceleration waves

In this section an acceleration wave is a two-dimensional surface $\mathcal{S}$ in $\mathbb{R}^3$ such that $v_i, \rho, \alpha, \dot{\alpha}$, and $\alpha_{,i}$ are continuous throughout $\mathbb{R}^3$, but their derivatives $\dot{v}_i, v_{i,j}, \dot{\rho}, \rho_{,i}, \ddot{\alpha}, \dot{\alpha}_{,i}$ and $\alpha_{,ij}$, along with higher derivatives, suffer a finite discontinuity (jump) across $\mathcal{S}$. As usual the jump of a function f is denoted by $[f] = f^- - f^+$, with f^+ or f^- referring to the limit of f as the wave is approached from ahead of the wave, or from behind, respectively.

Equations (8.156) are expanded recalling $\psi = \psi(\rho, \theta, \alpha_{,i})$ and the jumps are taken across $\mathcal{S}$. Then we obtain from (8.156)$_1$,

$$[\rho_t] + v_i[\rho_{,i}] + \rho[v_{i,i}] = 0, \tag{8.157}$$

from the momentum equation $(8.156)_2$

$$
\begin{aligned}
\rho[v_{i,t}] + \rho v_j[v_{i,j}] &= -p_\rho[\rho_{,i}] - p_\theta[\theta_{,i}] - p_{\alpha,j}[\alpha_{,ji}] \\
&\quad -\frac{1}{2}[\rho_{,j}](\alpha_{,j}\psi_{\alpha,i} + \alpha_{,i}\psi_{\alpha,j}) - \frac{1}{2}\rho\psi_{\alpha,i}[\Delta\alpha] - \frac{1}{2}\rho\psi_{\alpha,j}[\alpha_{,ij}] \\
&\quad -\frac{1}{2}\rho\alpha_{,j}(\psi_{\alpha,i\rho}[\rho_{,j}] + \psi_{\alpha,i\theta}[\theta_{,j}] + \psi_{\alpha,i\alpha,k}[\alpha_{,kj}]) \\
&\quad -\frac{1}{2}\rho\alpha_{,i}(\psi_{\alpha,j\rho}[\rho_{,j}] + \psi_{\alpha,j\theta}[\theta_{,j}] + \psi_{\alpha,j\alpha,k}[\alpha_{,kj}]),
\end{aligned} \tag{8.158}
$$

and from the energy equation $(8.156)_3$,

$$
\begin{aligned}
-\rho\psi_{\theta\theta}[\dot{\theta}] - \rho\psi_{\theta\rho}[\dot{\rho}] - \rho\psi_{\theta\alpha,i}[\overline{(\alpha_{,i})}] \\
= [\rho_{,i}]\psi_{\alpha,i} + \rho(\psi_{\rho\alpha,i}[\rho_{,i}] + \psi_{\theta\alpha,i}[\theta_{,i}] + \psi_{\alpha,i\alpha,j}[\alpha_{,ji}]).
\end{aligned} \tag{8.159}
$$

It is important to note that since v_i is continuous the term $[v_i] = 0$ in the jump equation arising from $(8.156)_2$, and so the Darcy coefficient effect is not present in (8.158), and will not feature directly in the wavespeed.

One may study the motion of an acceleration wave advancing into a region of known properties. However, we restrict attention to the case where the wave is moving into an equilibrium region for which $v_i^+ \equiv 0$, $\rho^+ \equiv$ constant, $\theta^+ \equiv$ constant, and hence $\alpha_{,i}^+ \equiv 0$, and the body possesses a centre of symmetry.

For a three-dimensional acceleration wave we define the amplitudes A^i, B and C by

$$
A^i = [v^i_{,j}n^j], \qquad B = [n^i\rho_{,i}], \qquad C = [\alpha_{,ij}n^in^j], \tag{8.160}
$$

where n^i is the unit normal to $\mathcal{S}$ in the $+$ direction. Using the compatibility relations (176.2) and (176.10) of (Truesdell and Toupin, 1960) one may write

$$
A^in_j = [v^i_{,j}], \qquad Bn_i = [\rho_{,i}], \qquad Cn_in_j = [\alpha_{,ij}]. \tag{8.161}
$$

Upon using the fact that the body has a centre of symmetry, and the wave is advancing into an equilibrium region, equations (8.157) – (8.159) reduce to

$$
-VB + \rho A^in_i = 0, \tag{8.162}
$$

$$
-\rho V A_i = -p_\rho Bn_i + p_\theta Vn_iC, \tag{8.163}
$$

$$
-\rho\psi_{\theta\theta}V^2C + \rho\psi_{\rho\theta}VB = \rho\psi_{\alpha,i\alpha,j}n^in^jC, \tag{8.164}
$$

where V is the wavespeed of $\mathcal{S}$.

The momentum jump equation (8.163) implies that $A_i = n_iA$, where $A = [v^j_{,i}n^in_j]$, and so the wave must be longitudinal. Thus, (8.162) – (8.164) are a system of equations in A, B, C which may be written as

$$
\begin{pmatrix}
\rho & -V & 0 \\
\rho V & -p_\rho & p_\theta V \\
0 & -\psi_{\rho\theta}V & \psi_{\theta\theta}V^2 + \psi_{\alpha,i\alpha,j}n^in^j
\end{pmatrix}
\begin{pmatrix}
A \\ B \\ C
\end{pmatrix}
=
\begin{pmatrix}
0 \\ 0 \\ 0
\end{pmatrix}.
$$

We require a non-zero solution $(A, B, C)^T \neq \mathbf{0}$ to this system and so

$$\begin{vmatrix} \rho & -V & 0 \\ \rho V & -p_\rho & p_\theta V \\ 0 & -\psi_{\rho\theta} V & \psi_{\theta\theta} V^2 + \psi_{\alpha,i\alpha,j} n^i n^j \end{vmatrix} = 0.$$

This leads to the wavespeed equation

$$(V^2 - U_M^2)(V^2 - U_T^2) + \kappa V^2 = 0, \tag{8.165}$$

where

$$U_M^2 = p_\rho, \qquad U_T^2 = -\frac{\psi_{\alpha,i\alpha,j} n^i n^j}{\psi_{\theta\theta}}, \qquad \kappa = \frac{p_\theta \psi_{\rho\theta}}{\psi_{\theta\theta}} = \frac{\rho^2 (\psi_{\rho\theta})^2}{\psi_{\theta\theta}}. \tag{8.166}$$

The quantity $U_M = \sqrt{p_\rho}$ is the wavespeed of an acoustic wave in the classical theory, cf. section 8.1.3, and U_T is the wavespeeed of a thermal wave in a (Green and Naghdi, 1991) rigid heat conductor, cf. (Jaisaardsuetrong and Straughan, 2007).

8.5.3 Wave amplitudes

We now present results for a one-dimensional wave moving along the x−axis. Hence, let $\mathbf{v} = (u(x,t),0,0)$ with $\rho(x,t)$, $\alpha(x,t)$. The amplitudes become

$$A(t) = [u_x] = u_x^- - u_x^+, \qquad B = [\rho_x], \qquad C = [\alpha_{xx}]. \tag{8.167}$$

We must consider the one-dimensional version of equations (8.156) and these are

$$\rho_t + u\rho_x + \rho u_x = 0, \tag{8.168}$$

$$u_t + uu_x = -\frac{p_x}{\rho} - \frac{\rho_x}{\rho} \alpha_x \psi_{\alpha_x} - (\alpha_x \psi_{\alpha_x})_x - k\frac{u}{\rho}, \tag{8.169}$$

$$- \rho\{(\psi_\theta)_t + u(\psi_\theta)_x\} = (\rho\psi_{\alpha_x})_x. \tag{8.170}$$

The idea is to write $p = \rho^2 \psi_\rho$ and expand these equations recalling $\psi = \psi(\rho, \dot{\alpha}, \alpha_x)$, then differentiate with respect to x and take the jumps of the results. For example, with $P = p_\rho/\rho$, (8.169) is

$$u_t + uu_x = - P\rho_x - \frac{p_\theta}{\rho} \theta_x - \frac{p_{\alpha_x}}{\rho} \alpha_{xx} - \frac{\rho_x}{\rho} \alpha_x \psi_{\alpha_x} - \psi_{\alpha_x} \alpha_{xx}$$

$$- \alpha_x(\psi_{\alpha_x\rho}\rho_x + \psi_{\alpha_x\theta}\theta_x + \psi_{\alpha_x\alpha_x}\alpha_{xx}) - \frac{k}{\rho} u.$$

One differentiates this equation with respect to x and takes the jump recalling the body has a centre of symmetry and the wave is moving into a region of equilibrium.

After some calculation, from (8.168) – (8.169) in turn one may show that

$$[\rho_{tx}] + 2[u_x\rho_x] + \rho[u_{xx}] = 0, \tag{8.171}$$

and

$$[u_{tx}] + [u_x]^2 = -P[\rho_{xx}] - P_\rho[\rho_x^2] - P_\theta[\rho_x \dot\alpha_x] - \frac{p_\theta}{\rho}[\theta_{xx}]$$

$$- \frac{p_{\theta\rho}}{\rho}[\rho_x \theta_x] - \frac{p_{\theta\theta}}{\rho}[\theta_x^2] + \frac{p_\theta}{\rho^2}[\theta_x \rho_x] \qquad (8.172)$$

$$- \frac{p_{\alpha_x \alpha_x}}{\rho}[\alpha_{xx}^2] - 2\psi_{\alpha_x \alpha_x}[\alpha_{xx}^2] - \frac{k}{\rho}[u_x],$$

while from (8.170) we derive

$$- \psi_{\theta\theta\rho}[\rho_x \alpha_{tt}] - \psi_{\theta\theta\theta}[\theta_x \alpha_{tt}] - \psi_{\theta\theta}([\alpha_{ttx}] + [u_t \alpha_{xx}] + 2[u_x \alpha_{xt}])$$

$$- \psi_{\theta\rho}([\rho_{tx}] + [u_x \rho_x]) - \psi_{\rho\theta\rho}[\rho_x \rho_t] - \psi_{\rho\theta\theta}[\theta_x \rho_t] - \psi_{\theta\alpha_x\alpha_x}[\alpha_{xt}\alpha_{xx}]$$

$$= \frac{\psi_{\alpha_x \alpha_x}}{\rho}[\alpha_{xx}\rho_x] + \psi_{\rho\alpha_x\alpha_x}[\alpha_x \rho_x] + \psi_{\theta\alpha_x\alpha_x}[\alpha_{xx}\theta_x]$$

$$+ \alpha_{\alpha_x\alpha_x}[\alpha_{xxx}] + \psi_{\theta\alpha_x\alpha_x}[\theta_x \alpha_{xx}] + \psi_{\rho\alpha_x\alpha_x}[\rho_x \alpha_{xx}]. \qquad (8.173)$$

We now use the Hadamard relation (7.8), the product relation (7.13) and the fact that

$$B = \frac{\rho}{V} A, \qquad C = \frac{\rho(p_\rho - V^2)}{p_\theta V^2} A.$$

After some calculation, one may obtain from (8.171) – (8.173), the equations

$$\frac{\rho}{V}\frac{\delta A}{\delta t} + \frac{2\rho}{V} A^2 + \rho[u_{xx}] - V[\rho_{xx}] = 0, \qquad (8.174)$$

$$\frac{p_\rho}{V^2}\frac{\delta A}{\delta t} - V[u_{xx}] + P[\rho_{xx}] - \frac{p_\theta V}{\rho}[\alpha_{xxx}] = -\frac{k}{\rho}A$$

$$- A^2 \left\{ 1 + \frac{P_\rho \rho^2}{V^2} + \frac{(V^2 - U_M^2)}{V^2}\left(1 + \frac{2\rho\psi_{\rho\rho\theta}}{\psi_{\rho\theta}}\right) \right. \qquad (8.175)$$

$$\left. + \frac{(V^2 - U_M^2)^2}{V^4 p_\theta \psi_{\rho\theta}}(\rho V^2 \psi_{\rho\theta\theta} + \rho\psi_{\rho\alpha_x\alpha_x} + 2\psi_{\alpha_x\alpha_x}) \right\},$$

and

$$2V\psi_{\theta\theta}\frac{\delta C}{\delta t} - \psi_{\theta\rho}\frac{\delta B}{\delta t} - \psi_{\theta\theta}V^2[\alpha_{xxx}] + \psi_{\rho\theta}V[\rho_{xx}]$$

$$- \psi_{\alpha_x\alpha_x}[\alpha_{xxx}] = \left(2\psi_{\theta\theta\rho}V^2 + \frac{1}{\rho}\psi_{\alpha_x\alpha_x} + 2\psi_{\rho\alpha_x\alpha_x}\right)BC \qquad (8.176)$$

$$- (\psi_{\theta\theta\theta}V^3 + 3V\psi_{\theta\alpha_x\alpha_x})C^2 - 3\psi_{\theta\theta}VAC + \psi_{\rho\theta}AB - V\psi_{\rho\rho\theta}B^2.$$

We now wish to remove the terms in $[u_{xx}]$, $[\rho_{xx}]$ and $[\alpha_{xxx}]$. To do this we form $(8.174) + \mu_1 \times (8.175) + \mu_2 \times (8.176)$. Upon inspection of the terms involving $[u_{xx}]$, $[\rho_{xx}]$ and $[\alpha_{xxx}]$ we see that the choices

$$\mu_1 = \frac{\rho}{V}, \qquad \mu_2 = \frac{(V^2 - p_\rho)}{V^2 \psi_{\rho\theta}}$$

remove the terms in $[u_{xx}]$ and $[\rho_{xx}]$. However, it then results that the coefficient of the term in $[\alpha_{xxx}]$ is also zero due to the wavespeed equation (8.165). After further calculations, eliminating B and C one then arrives at the amplitude equation

$$
\begin{aligned}
2&\left(\frac{U_M^2}{V^2} - \frac{(V^2 - U_M^2)^2}{\kappa V^2}\right)\frac{\delta A}{\delta t} + \frac{k}{\rho}A \\
&+ A^2\left\{3 + \frac{P_\rho \rho^2}{V^2} + 3(V^2 - U_M^2)\left(\frac{\rho\psi_{\rho\rho\theta}}{V^2\psi_{\rho\theta}} - \frac{\psi_{\theta\theta}}{\rho^2\psi_{\rho\theta}}\right)\right. \\
&+ \frac{3(V^2 - U_M^2)^2}{V^4\rho_\theta\psi_{\rho\theta}}(\rho V^2\psi_{\rho\theta\theta} + \rho\psi_{\rho\alpha_x\alpha_x} + \psi_{\alpha_x\alpha_x} - V^2\psi_{\theta\theta}) \\
&+ \left.\frac{(V^2 - U_M^2)^3}{V^4 p_\theta^2\psi_{\rho\theta}}(V^2\psi_{\theta\theta\theta} + 3\psi_{\theta\alpha_x\alpha_x})\right\} = 0.
\end{aligned}
\tag{8.177}
$$

The solution to this equation is easily written down, cf. (8.31). We see that as the thermal terms vanish, $V^2 \to U_M^2$ where $U_M^2 = dp(\rho)/d\rho$ and then (8.177) reduces to the isothermal amplitude equation of section 8.1.4, as it should. In particular, we note again the strong damping (attenuation) effect of the linear term in (8.177). This is entirely due to the Darcy term involving k and represents the effect the porous medium has on acoustic wave attenuation, although the influence of the thermal (Green-Naghdi) terms is also evident from (8.177).

8.6 Magnetic field effects

We briefly investigate the effect a magnetic field may have on a sound wave propagating through a porous medium. The effects of both electric and magnetic fields could have an important bearing on sound attenuation. Certainly ceramics are often piezoelectric materials and we do expect a study of electromagnetic effects to be worthwhile.

The idea is to generalize the Jordan-Darcy theory of section 8.1.2 and add to this the theory for a nonlinear magnetic fluid, cf. (Roberts, 1981), (Straughan, 1986). The equations under consideration are then the continuity equation

$$
\dot{\rho} + \rho v_{i,i} = 0,
\tag{8.178}
$$

the equation for the mean magnetic field B_i,

$$
\dot{B}_i = v_{i,j}B_j - v_{j,j}B_i,
\tag{8.179}
$$

and the momentum equation with a Darcy term,

$$
\rho\dot{v}_i = \sigma_{ij,j} - kv_i.
\tag{8.180}
$$

The constitutive equation for σ_{ij} is

$$\sigma_{ij} = -(p + H_m B_m)\delta_{ij} + H_i B_j \tag{8.181}$$

where H_i is the magnetizing force and p is the fluid pressure. The functions H_i and p are given in terms of a Helmholtz free energy function $\psi = \psi(\rho, B)$, $B = |\mathbf{B}|$, by

$$p = \rho^2 \frac{\partial \psi}{\partial \rho}, \qquad H_i = \rho \frac{\partial \psi}{\partial B_i}. \tag{8.182}$$

An acceleration wave for equations (8.178) – (8.180) is a singular surface S across which $v_{i,t}, v_{i,j}, B_{i,t}, B_{i,j}, \rho_t, \rho_{,i}$ and their higher derivatives suffer at most a finite discontinuity, but v_i, B_i and ρ are continuous everywhere. The wave amplitudes are defined by

$$a_i = [B_{i,j} n_j], \qquad b = [\rho_{,i} n_i], \qquad c_i = [v_{i,j} n_j].$$

The analysis of the wavespeeds follows exactly that given by (Straughan, 1986) and the magnetic field does play a key role. One needs to introduce orthogonal surface coordinates u^α ($\alpha = 1, 2$) on S. Then, with $x^i_{;\alpha} = \partial x^i / \partial u^\alpha$ being the tangential vectors to S one resolves a_i, B_i and c_i into components in the direction of the normal to S and its tangential directions. One writes

$$a^i = a_n n^i + a^\alpha x^i_{;\alpha}, \qquad c^i = c_n n^i + c^\alpha x^i_{;\alpha}, \qquad B^i = B_\| n^i + B^\alpha_\perp x^i_{;\alpha},$$

where $B_\|$ and $B^\alpha_\perp$ are the components of B^i in the normal and tangential directions to S, with a_n, c_n the normal components of a_i and c_i, a^α, c^α the tangential ones. The analysis takes the jumps of (8.178) – (8.180) and one may show $a_n = 0$. The remaining equations are written in terms of the six variables b, c_n, a^1, a^2, c^1 and c^2. Details are given in (Straughan, 1986) although he does not have the $-kv_i$ term in equation (8.180). The wavespeed equation is found to be a sixth order equation which factorizes to yield an Alfven wave with speed

$$V^2 = B_\|^2 \frac{1}{B} \frac{\partial \psi}{\partial B}$$

and a fast and a slow wave with speeds which satisfy the quadratic in V^2, namely

$$V^4 - V^2 \left(p_\rho + B \frac{\partial \psi}{\partial B} + B_T^2 \left\{ \frac{2\rho}{B} \psi_{\rho B} + \frac{1}{B} \psi_B + \psi_{BB} \right\} \right)$$

$$+ \frac{p_\rho B_\|^2}{B} \left(\psi_B + B_T^2 \left(\frac{1}{B} \psi_B \right)_B \right) - \frac{B_\|^2 B_T^2}{B^2} (\rho \psi_B)_\rho = 0,$$

where $B_T^2 = B_\alpha^\perp B^\alpha_\perp$. Conditions necessary for real and positive wavespeeds are derived by (Straughan, 1986), and the strong effect of the magnetic field on the wavespeeds is evident. As with the basic theory of section 8.1.2 one has to proceed to the amplitude equation to see the effect of the porous medium via the k coefficient.

9
Numerical Solution of Eigenvalue Problems

9.1 The compound matrix method

9.1.1 The shooting method

The purpose of this chapter is to describe three very efficient methods for solving eigenvalue problems of the type encountered in linear and nonlinear stability problems associated with porous media. The techniques referred to are the compound matrix method, the Chebyshev tau technique, and a Legendre - Galerkin method. The chapter is intended to be a practical guide as to how to solve relevant eigenvalue problems. Several examples from porous convection are included. In order to introduce the compound matrix method we commence by briefly describing a standard shooting method.

We begin with an investigation of the second order eigenvalue problem,

$$\frac{d^2u}{dx^2} + \lambda u = 0, \qquad 0 < x < 1,$$
$$u(0) = u(1) = 0. \tag{9.1}$$

It is easy to see that the solution of this problem is given by $u = \sin n\pi x$, and the associated eigenvalues are

$$\lambda_n = n^2\pi^2, \qquad n = 1, 2, \ldots. \tag{9.2}$$

For stability studies it is often only the smallest eigenvalue that is of interest, i.e., λ_1.

B. Straughan, *Stability and Wave Motion in Porous Media*,
DOI: 10.1007/978-0-387-76543-3_9, © Springer Science+Business Media, LLC 2008

To determine λ_1 numerically we retain the boundary condition $u(0) = 0$, but replace $u(1) = 0$ by a prescribed condition on $u'(0)$, a convenient choice being $u'(0) = 1$, thereby converting the *boundary value problem* to an *initial value* one. Two values of λ_1 are selected, say $0 < \lambda_1^{(1)} < \lambda_1^{(2)}$, and then the initial value problem is integrated numerically to find $u_1(1)$, $u_2(1)$, where u_i denotes the solution corresponding to $\lambda_1^{(i)}$: (a high order, variable step Runge-Kutta-Verner technique is often adequate for the numerical integration, although I frequently use an extrapolation method, see (Stoer and Bulirsch, 1993), p. 484). The idea is to use $u_1(1)$, $u_2(1)$ so found to ensure $u(1)$ is as close to zero as required by some pre-specified degree of accuracy. An iteration technique is then employed to find a sequence $u_k(1)$, corresponding to $\lambda_1^{(k)}$, such that

$$|u_k(1)| < \epsilon, \qquad k \text{ large enough}, \tag{9.3}$$

where ϵ is a user specified tolerance. I have found the secant method, see e.g., (Cheney and Kincaid, 1985) p. 97, is a suitable routine for this purpose. Once a $u_k(1)$ is determined to satisfy (9.3), $\lambda_1^{(k)}$ is then the required numerical estimate of the first eigenvalue to (9.1).

For practical purposes it is usually necessary to have some guide as to what values to select for $\lambda_1^{(1)}$, $\lambda_1^{(2)}$. However, in nonlinear energy stability theory one can usually use linear stability theory as a guide for this. For linear stability theory it is often possible to use known results from related problems that can be solved analytically.

9.1.2 A fourth order equation

The eigenvalue problems encountered in porous stability problems are more complicated than that discussed in section 9.1.1, but the basic numerical shooting method is the same. To illustrate how the shooting method works on a system we use the following fourth order eigenvalue problem

$$\frac{d^2}{dx^2}\left[(1 - \theta x)^3 \frac{d^2u}{dx^2}\right] - \lambda(1 - \theta x)u = 0, \quad x \in (0, 1),$$

$$u = \frac{d^2u}{dx^2} = 0, \qquad\qquad\qquad x = 0, 1, \tag{9.4}$$

where the constant θ satisfies $0 \le \theta < 1$.

To solve (9.4) for the first (lowest) eigenvalue λ_1 does not appear possible analytically, in general, i.e., for $\theta \ne 0$. To solve (9.4) by a shooting method we first write it as a system of four first order differential equations in the vector $\mathbf{u} = (u, u', u'', u''')$, where $u' = du/dx$, $u'' = d^2u/dx^2$, etc. The boundary conditions at $x = 1$ on u, u'' are replaced in turn by $u' = 1$, $u''' = 0$, and then $u' = 0$, $u''' = 1$, at $x = 0$. The two initial value problems thereby obtained are then integrated numerically. Let the solution so found be written as a linear combination of the two solutions so obtained, say

$u = \alpha v + \beta w$. Then, the correct boundary condition $u = u'' = 0$ at $x = 1$ is imposed, and this requires

$$\det \begin{pmatrix} v & v'' \\ w & w'' \end{pmatrix} = 0 \qquad (9.5)$$

to hold at $x = 1$.

While the shooting method is easy to understand and implement, it suffers from a serious drawback. This is that one has numerically to locate the zero of a determinant; e.g., in (9.5), one has to locate

$$v(1)w''(1) - w(1)v''(1) = 0. \qquad (9.6)$$

The two quantities $v(1)w''(1)$ and $w(1)v''(1)$ must, therefore, be very close although neither need be close to zero (and generally will not). One is thus faced with subtracting two nearly identical quantities, and this can lead to very large round off errors and significant error build up during the solution of a convection eigenvalue problem. There are many ways to overcome this: we describe only one. This is the compound matrix technique, which has distinct advantages for energy eigenvalue problems. Basically the idea is to remove the troublesome location of the zero of a determinant by converting to a system of ordinary differential equations in the determinants themselves.

9.1.3   The compound matrix method

We begin this section with an accurate numerical calculation of eigenvalues to (9.4). For $0 \le \theta < 0.9$ accurate results are evidently easily found by the standard shooting method described in section 9.1.1. However, for the case where $\theta \to 1^-$ for which (9.4) becomes singular, other methods must be employed. Some useful information may be gleaned in this case with the compound matrix technique.

To solve (9.4) by the compound matrix method we let $\mathbf{U} = (u, u', u'', u''')^T$, and then suppose $\mathbf{U}_1$ and $\mathbf{U}_2$ are solutions to (9.4) with values at $x = 0$ of $(0, 1, 0, 0)^T$ and $(0, 0, 0, 1)^T$, respectively. A new six vector

$$\mathbf{Y} = (y_1, y_2, y_3, y_4, y_5, y_6)^T$$

is defined as the 2×2 minors of the 4×2 solution matrix whose first column is $\mathbf{U}_1$ and second $\mathbf{U}_2$. So,

$$\begin{aligned}
y_1 &= u_1 u_2' - u_1' u_2, \\
y_2 &= u_1 u_2'' - u_1'' u_2, \\
y_3 &= u_1 u_2''' - u_1''' u_2, \\
y_4 &= u_1' u_2'' - u_1'' u_2', \\
y_5 &= u_1' u_2''' - u_1''' u_2', \\
y_6 &= u_1'' u_2''' - u_1''' u_2''.
\end{aligned} \qquad (9.7)$$

The 2×2 minors are the determinants we refer to at the end of section 9.1.2. The variable y_2 corresponds to the quantity in (9.6). With the compound matrix method we replace (9.6) by $y_2(1) = 0$ and thus avoid the problem of round off error due to subtraction.

By direct calculation from (9.4) the initial value problem for the y_i is found to be

$$
\begin{aligned}
y_1' &= y_2, \\
y_2' &= y_3 + y_4, \\
y_3' &= y_5 + 6\frac{\theta}{M}y_3 - 6\left(\frac{\theta}{M}\right)^2 y_2, \\
y_4' &= y_5, \\
y_5' &= y_6 + 6\frac{\theta}{M}y_5 - 6\left(\frac{\theta}{M}\right)^2 y_4 - \frac{\lambda}{M^2}y_1, \\
y_6' &= 6\frac{\theta}{M}y_6 - \frac{\lambda}{M^2}y_2,
\end{aligned}
\tag{9.8}
$$

where $M = 1 - \theta x$. From the initial conditions on $\mathbf{U}_1$ and $\mathbf{U}_2$ we see that system (9.8) is to be integrated numerically subject to the initial condition

$$
y_5(0) = 1 \tag{9.9}
$$

and the final condition

$$
y_2(1) = 0. \tag{9.10}
$$

Again the zero in (9.10) is located to a pre-assigned degree of accuracy.

We have already noted that the compound matrix method is designed to avoid round off error, and this technique works well if the system of differential equations is stiff.

In stability problems in porous media one typically encounters an eigenvalue problem in which the system of equations appears as a system of coupled second order equations. In the light of this and to describe the compound matrix method further we consider the general linear system

$$
\begin{aligned}
w'' &= \alpha_1 w' + \alpha_2 w + \alpha_3 \theta' + \alpha_4 \theta, \\
\theta'' &= \beta_1 w' + \beta_2 w + \beta_3 \theta' + \beta_4 \theta,
\end{aligned}
\tag{9.11}
$$

where a prime denotes differentiation with respect to z, $\alpha_1, \ldots, \alpha_4$, $\beta_1, \ldots, \beta_4$, are known coefficients which may depend on z, they may be complex, and $z \in (0,1)$. One or more of the coefficients contains an eigenvalue, σ say. The boundary conditions we consider are

$$
w = \theta = 0 \quad \text{at} \quad z = 0, 1. \tag{9.12}
$$

Other boundary conditions are easily incorporated. System (9.11) is typical of the eigenvalue problems which occur in porous convection stability problems.

We introduce the variables $y_1, \ldots, y_6$, being the 2×2 minors arising from w and θ, i.e.

$$
\begin{aligned}
y_1 &= w_1 w_2' - w_2 w_1', & y_4 &= w_1' \theta_2 - w_2' \theta_1, \\
y_2 &= w_1 \theta_2 - w_2 \theta_1, & y_5 &= w_1' \theta_2' - w_2' \theta_1', \\
y_3 &= w_1 \theta_2' - w_2 \theta_1', & y_6 &= \theta_1 \theta_2' - \theta_2 \theta_1'.
\end{aligned}
\tag{9.13}
$$

The y_i variables satisfy the matrix equation

$$
\mathbf{y}' = A\mathbf{y}, \tag{9.14}
$$

where A is the 6×6 matrix

$$
A = \begin{pmatrix}
\alpha_1 & \alpha_4 & \alpha_3 & 0 & 0 & 0 \\
0 & 0 & 1 & 1 & 0 & 0 \\
\beta_1 & \beta_4 & \beta_3 & 0 & 1 & 0 \\
0 & \alpha_2 & 0 & \alpha_1 & 1 & -\alpha_3 \\
-\beta_2 & 0 & \alpha_2 & \beta_4 & \alpha_1 + \beta_3 & \alpha_4 \\
0 & -\beta_2 & 0 & -\beta_1 & 0 & \beta_3
\end{pmatrix}
$$

Due to the conditions on w, θ at $z = 0$, we take $w_1'(0) = 1$, $\theta_2'(0) = 1$, and then the eigenvalues σ are found by integrating (9.14) from 0 to 1 employing the initial condition

$$
y_5(0) = 1. \tag{9.15}
$$

To satisfy the final conditions $w(1) = \theta(1) = 0$, we must iterate on the condition on y_2,

$$
y_2(1) = 0. \tag{9.16}
$$

By having to satisfy condition (9.16) on the single variable y_2 we avoid the problem inherent in the standard shooting method where one has to subtract nearly equal quantities.

The determination of σ_i is relatively straightforward. The procedure whereby one calculates the corresponding eigenfunctions (w_i, θ_i) is described in (Straughan and Walker, 1996b). Rigorous mathematical results involving the compound matrix method for application to eigenvalue problems, and related results, may be found in the interesting papers of (Allen and Bridges, 2002), (Brown and Marletta, 2003), (Davies, 1999), (Dorfmann and Haughton, 2006), (Greenberg and Marletta, 2000; Greenberg and Marletta, 2001; Greenberg and Marletta, 2004), (Ivansson, 2003), (Shubov and Balogh, 2005) and (Theofilis, 2003; Theofilis et al., 2004).

9.1.4 Penetrative convection in a porous medium

A good problem to use as a test for a numerical method is the one which arises in the situation of penetrative convection in a porous medium. The physics of this problem is presented in sections 17.6 and 19.7 of (Straughan, 2004a). The perturbation equations for linearised instability for penetrative convection in an isotropic porous medium are

$$p_{,i} = -u_i - 2R\theta(\xi - z)\delta_{i3}, \qquad u_{i,i} = 0,$$
$$\theta_{,t} = -Rw + \Delta\theta, \tag{9.17}$$

for $\mathbf{x} \in \mathbb{R}^2 \times (0,1)$, where p, u_i, θ are perturbations of pressure, velocity, temperature, $w = u_3$, R^2 is the Rayleigh number, $\xi = 4/T_u$, with T_u being the temperature of the upper surface. We here restrict attention to prescribed temperature and normal component of velocity so that

$$w = 0, \quad \theta = 0 \quad \text{on} \quad z = 0, 1. \tag{9.18}$$

Upon representing the time dependency by $e^{\sigma t}$, σ being the growth rate, and employing normal modes system (9.17), (9.18) reduces to

$$(D^2 - a^2)W - 2a^2 R(\xi - z)\Theta = 0,$$
$$(D^2 - a^2)\Theta - RW - \sigma\Theta = 0, \tag{9.19}$$

$z \in (0,1)$, where a is the wavenumber. The boundary conditions are $W = \Theta = 0$, $z = 0, 1$.

This is an example of a system where the coefficients are functions of the spatial variable z. Also, as T_u increases the coefficient involving $(\xi - z)$ has a strong effect leading to a stiff system and the eigenfunctions vary strongly.

In general, one must solve (9.17), (9.18) for the eigenvalue σ with R given. However, one can show exchange of stabilities holds for (9.17), (9.18), and then to find the instability boundary it is sufficient to take $\sigma = 0$ in (9.19). The compound matrix equations and boundary conditions are then (9.14) – (9.16) with A here given by

$$A = \begin{pmatrix} 0 & 2Ra^2M & 0 & 0 & 0 & 0 \\ 0 & 0 & 1 & 1 & 0 & 0 \\ 0 & a^2 & 0 & 0 & 1 & 0 \\ 0 & a^2 & 0 & 0 & 1 & 0 \\ -R & 0 & a^2 & a^2 & 0 & 2a^2RM \\ 0 & -R & 0 & 0 & 0 & 0 \end{pmatrix}$$

To see this, one sets $\sigma = 0$ in (9.19) and solves for the eigenvalue R.

9.2 The Chebyshev tau method

9.2.1 The D^2 Chebyshev tau method

We describe the Chebyshev tau method in the context of system (9.11). Thus, we define the operators L_1 and L_2 by

$$L_1(u,v) = u'' - \alpha_2 u - \alpha_4 v - \alpha_1 u' - \alpha_3 v',$$
$$L_2(u,v) = v'' - \beta_2 u - \beta_4 v - \beta_1 u' - \beta_3 v'. \tag{9.20}$$

The Chebyshev tau method is very general, and the coefficients α_i, β_i may depend on z and may also be complex. The eigenvalue σ appears in one or more coefficients. System (9.11) is equivalent to

$$L_1(u,v) = 0, \qquad L_2(u,v) = 0, \tag{9.21}$$

on the domain (-1,1) together with the boundary conditions

$$u = v = 0 \qquad \text{at} \quad z = \pm 1. \tag{9.22}$$

The system (9.11) has been transformed from (0,1) to (-1,1) as this is the natural domain in which to use Chebyshev polynomials. The above choice of boundary conditions is not necessary and other boundary conditions may be handled, cf. (Straughan and Walker, 1996b), and sections 9.2.4, 9.2.5 of this book.

We now describe the procedure for finding eigenvalues and eigenfunctions to (9.21), (9.22). It is important to realise that other boundary conditions may be handled, and, in particular, higher order systems of differential equations are naturally dealt with by the same technique.

The key idea is to write u, v as a finite series of Chebyshev polynomials

$$u = \sum_{k=0}^{N+2} a_k T_k(z), \qquad v = \sum_{k=0}^{N+2} b_k T_k(z). \tag{9.23}$$

The exact solution to the differential equation is an infinite series, i.e. let $N \to \infty$ in (9.23). Due to the truncation, to solve (9.21) with the approximate form (9.23), we solve

$$L_1(u,v) = \tau_1 T_{N+1} + \tau_2 T_{N+2}, \qquad L_2(u,v) = \hat{\tau}_1 T_{N+1} + \hat{\tau}_2 T_{N+2}. \tag{9.24}$$

In (9.24) the parameters $\tau_1, \tau_2, \hat{\tau}_1, \hat{\tau}_2$ are effectively error indicators for the truncation in (9.23).

To determine the unknown coefficients a_i and b_i we take the inner product with T_i of (9.24) in the weighted $L^2(-1,1)$ space with inner product

$$(f,g) = \int_{-1}^{1} \frac{fg}{\sqrt{1-z^2}} \, dz \, .$$

Let us denote the associated norm by $\|\cdot\|$. The Chebyshev polynomials are orthogonal in this space, and thus (9.24) leads to the $2(N+1)$ equations

$$(L_1(u,v), T_i) = 0 \quad (L_2(u,v), T_i) = 0 \qquad i = 0, 1, \ldots, N. \qquad (9.25)$$

There are four more conditions which arise by taking inner products, and these are

$$(L_1(u,v), T_{N+j}) = \tau_j \|T_{N+j}\|^2, \quad (L_2(u,v), T_{N+j}) = \hat{\tau}_j \|T_{N+j}\|^2, \quad j = 1, 2.$$

These four equations yield the tau coefficients $\tau_1, \tau_2, \hat{\tau}_1, \hat{\tau}_2$, which in turn are measures of the error involved in the truncation (9.23). To derive four more equations for a_i and b_i to add to (9.25) we employ the boundary conditions. The Chebyshev polynomials $T_n(z)$ satisfy $T_n(\pm 1) = (\pm 1)^n$, and this together with (9.22) and (9.23) yield

$$\begin{aligned}
\sum_{n=0}^{N+2} (-1)^n a_n = 0, \quad & \sum_{n=0}^{N+2} a_n = 0, \\
\sum_{n=0}^{N+2} (-1)^n b_n = 0, \quad & \sum_{n=0}^{N+2} b_n = 0.
\end{aligned} \qquad (9.26)$$

Equations (9.25) and (9.26) yield a system of $2(N+3)$ equations for the $2(N+3)$ unknowns a_i, b_i, $i = 0, \ldots, N+2$. We now suppose α_i, β_i are constant. If they are functions of z then they must be expanded in a series of Chebyshev polynomials, cf. (Orszag, 1971), p. 702. One then uses the relation $2T_m T_n = (T_{m+n} + T_{|m-n|})$ to write expressions as a linear combination of the T_i. For many convection problems the coefficients are linear, quadratic or third order polynomials and these are easily handled.

To calculate the coefficients in (9.25) we observe that the derivative of a Chebyshev polynomial is a linear combination of lower order Chebyshev polynomials and it may be shown that

$$T_n' = \begin{cases} 2n(T_{n-1} + \ldots + T_1), & n \text{ even}, \\ 2n(T_{n-1} + \ldots + T_2) + nT_0, & n \text{ odd}. \end{cases} \qquad (9.27)$$

By recalling (9.20) and using (9.23) and (9.27), equations (9.25) are reduced to the $2(N+1)$ algebraic equations

$$\begin{aligned}
a_i^{(2)} - \alpha_2 a_i - \alpha_4 b_i - \alpha_1 a_i^{(1)} - \alpha_3 b_i^{(1)} = 0, \quad i = 0, \ldots, N, \\
b_i^{(2)} - \beta_2 a_i - \beta_4 b_i - \beta_1 a_i^{(1)} - \beta_3 b_i^{(1)} = 0, \quad i = 0, \ldots, N.
\end{aligned} \qquad (9.28)$$

The coefficients $a_i^{(1)}$ and $a_i^{(2)}$ are given by

$$a_i^{(1)} = \frac{2}{c_i} \sum_{\substack{p=i+1 \\ p+i \text{ odd}}}^{p=N+2} p a_p, \quad a_i^{(2)} = \frac{2}{c_i} \sum_{\substack{p=i+2 \\ p+i \text{ even}}}^{p=N+2} p(p^2 - i^2) a_p. \qquad (9.29)$$

A similar representation holds for $b_i^{(1)}$, $b_i^{(2)}$. The coefficients c_i have form $c_0 = 2, c_i = 1, i = 1, 2, \ldots$. The $2(N+1)$ equations (9.28) together with the four equations (9.26) form a system of simultaneous linear equations for the $2(N+3)$ unknowns (a_i, b_i). This may be written in matrix form as

$$A\mathbf{x} = \sigma B\mathbf{x}, \qquad (9.30)$$

where $\mathbf{x} = (a_0, \ldots, a_{N+2}, b_0, \ldots, b_{N+2})^T$.

The matrices involved in the definition of $a_i^{(1)}$ and $a_i^{(2)}$ may alternatively be derived as follows. We know that

$$u' = \sum_{s=0}^{N+2} a_s T_s'(z)$$

$$= \sum_{s=0}^{N+2} a_s \left(\sum_{r=0}^{N+2} D_{rs} T_r \right)$$

$$= \sum_{r=0}^{N+2} \left(\sum_{s=0}^{N+2} D_{rs} a_s \right) T_r$$

and so

$$a_r^{(1)} = \sum_{s=0}^{N+2} D_{rs} a_s.$$

In addition,

$$u'' = \sum_{r=0}^{N+2} \left(\sum_{s=0}^{N+2} D_{rs} a_s^{(1)} \right) T_r.$$

Therefore,

$$a_r^{(2)} = \sum_{s=0}^{N+2} D_{rs} a_s^{(1)}$$

$$= \sum_{s=0}^{N+2} D_{rs} \sum_{k=0}^{N+2} D_{sk} a_k$$

$$= \sum_{s=0}^{N+2} \sum_{k=0}^{N+2} D_{rs} D_{sk} a_k .$$

The differentiation matrix D, and second differentiation matrix D^2 thus arise naturally. These matrices and their coefficients take the form

$$
\begin{aligned}
D_{0,2j-1} &= 2j - 1, & j &\geq 1, \\
D_{i,i+2j-1} &= 2(i + 2j - 1), & i &\geq 1, j \geq 1, \\
D_{0,2j}^2 &= \frac{1}{2}(2j)^3, & j &\geq 1, \\
D_{i,i+2j}^2 &= (i + 2j)4j(i + j), & i &\geq 1, j \geq 1,
\end{aligned}
\qquad (9.31)
$$

or in matrix form

$$D = \begin{pmatrix} 0 & 1 & 0 & 3 & 0 & 5 & 0 & 7 & 0 & 9 & \dots \\ 0 & 0 & 4 & 0 & 8 & 0 & 12 & 0 & 16 & 0 & \dots \\ 0 & 0 & 0 & 6 & 0 & 10 & 0 & 14 & 0 & 18 & \dots \\ 0 & 0 & 0 & 0 & 8 & 0 & 12 & 0 & 16 & 0 & \dots \\ 0 & 0 & 0 & 0 & 0 & 10 & 0 & 14 & 0 & 18 & \dots \\ \dots & \dots & \dots & \dots & \dots & \dots & \dots & \dots & \dots & \dots & \dots \end{pmatrix}$$

$$D^2 = \begin{pmatrix} 0 & 0 & 4 & 0 & 32 & 0 & 108 & \dots \\ 0 & 0 & 0 & 24 & 0 & 120 & 0 & \dots \\ 0 & 0 & 0 & 0 & 48 & 0 & 192 & \dots \\ \dots & \dots & \dots & \dots & \dots & \dots & \dots & \dots \end{pmatrix}$$

Note that in the matrix sense $D^2 = D \cdot D$. The B matrix in (9.30) is singular due to the way the boundary condition rows are added to A. When it is possible, it is usually best to remove the singular behaviour since this can result in the formation of spurious eigenvalues (i.e. numbers which appear in the eigenvalue list, but which are not eigenvalues).

For the boundary conditions (9.22) we may easily eliminate a_{N+1}, a_{N+2}, b_{N+1}, b_{N+2}. Suppose N is odd, then

$$a_{N+1} = -(a_0 + a_2 + \dots + a_{N-1}), \qquad a_{N+2} = -(a_1 + a_3 + \dots + a_N). \quad (9.32)$$

Similar forms hold for the b's. This allows us to remove the $N+1$ and $N+2$ rows of D^2 and eliminate the $N+1$, $N+2$ columns. This yields $(N+1) \times (N+1)$ matrices D^2, and the matrix problem resulting from (9.30) does not suffer from B being singular because of zero boundary condition rows. Further analyses of singularities due to boundary conditions are contained in (Bourne, 2003), (Straughan, 2001a), (Straughan and Walker, 1996b).

The equation which results from (9.25) has again form (9.30) but now A and B are $(N+1) \times (N+1)$ matrices and $\mathbf{x} = (a_0, \dots, a_N, b_0, \dots, b_N)$. Explicit details of A, B are given for the problem of penetrative convection in section 9.2.2. The eigenvalues of the generalised eigenvalue problem (9.30) are found efficiently using the QZ algorithm. This algorithm is available in many standard libraries, e.g. in the routines F02BJF, F02GJF of the NAG library. Since u and v have the forms (9.23) the calculation of the eigenfunctions using the Chebyshev tau method is really efficient. As soon as we know the coefficients a_k and b_k, u and v follow immediately from (9.23).

9.2.2 Penetrative convection

The Chebyshev tau method requires solution of (9.19) with $\sigma = 0$, which is approximated by an equation of form (9.30). Now $\mathbf{x} =$

$(W_0, \ldots, W_N, \Theta_0, \ldots, \Theta_N)$, and the matrices A, B are given by

$$A = \begin{pmatrix} D^2 - a^2 I & 0 \\ 0 & D^2 - a^2 I \end{pmatrix} \qquad B = \begin{pmatrix} 0 & 2a^2(\xi I - M) \\ I & 0 \end{pmatrix}.$$

For coding purposes we work with $N \times N$ matrices and with $i = 1, \ldots, N$, and then in the above M is the $N \times N$ matrix arising from the Chebyshev representation of z, i.e.

$$M_{i,i+1} = \frac{1}{2}, \qquad\qquad i = 1, \ldots, N-1,$$
$$M_{21} = 1; \quad M_{i+1,i} = \frac{1}{2}, \quad i = 2, \ldots, N-1; \qquad \text{rest } 0. \tag{9.33}$$

The matrix equation (9.30) is conveniently solved by the QZ algorithm of (Moler and Stewart, 1973). This routine yields all eigenvalues and eigenfunctions with no trouble.

9.2.3   Fluid overlying a porous layer

We further illustrate the Chebyshev tau method by using the convection problem of section 6.2. For completeness we recollect the perturbation equations of section 6.2.2 and the boundary conditions of section 6.2.3. The differential equations are, in $z \in (0,1)$,

$$(D^2 - a^2)^2 W - a^2 Ra\Theta = \frac{\sigma}{Pr}(D^2 - a^2)W, \tag{9.34}$$
$$(D^2 - a^2)\Theta - W = \sigma\Theta, \tag{9.35}$$

in $z_m \in (-1, 0)$,

$$(D^2 - a_m^2)W_m + a_m^2 Ra_m \Theta_m = -\frac{\sigma_m \delta^2}{\phi Pr_m}(D^2 - a_m^2)W_m, \tag{9.36}$$
$$(D^2 - a_m^2)\Theta_m - W_m = \sigma_m G_m \Theta_m, \tag{9.37}$$

and we recall that

$$\sigma_m = \frac{\hat{d}^2}{\epsilon_T}\sigma, \qquad Ra_m = Ra\frac{(\delta\epsilon_T)^2}{\hat{d}^4}. \tag{9.38}$$

The boundary conditions are, on $z = -1$,

$$W_m = 0, \tag{9.39}$$
$$\Theta_m = 0, \tag{9.40}$$

on $z = 1$,

$$W = 0, \tag{9.41}$$
$$D\Theta + L\Theta = 0, \tag{9.42}$$
$$D^2 W = Ma\Delta^*\theta = -Ma.a^2\Theta, \tag{9.43}$$

at the interface $z = 0$,

$$W = \hat{d}W_m, \tag{9.44}$$

$$\hat{d}\Theta = \epsilon_T^2 \Theta_m, \tag{9.45}$$

$$D\Theta = \epsilon_T D_p \Theta_m, \tag{9.46}$$

$$D^2 W - \frac{\alpha \hat{d}}{\delta} DW + \frac{\alpha \hat{d}^3}{\delta} D_p W_m = 0, \tag{9.47}$$

$$\frac{\hat{d}^4}{\phi Pr_m} \sigma_m D_p W_m + \frac{\hat{d}^4}{\delta^2} D_p W_m = \frac{\sigma}{Pr} DW - D^3 W - 3\Delta^* DW. \tag{9.48}$$

We first write equation (9.34) as a system of second order equations, and so write

$$(D^2 - a^2)W - A = 0,$$
$$(D^2 - a^2)A - a^2 Ra\Theta = \frac{\sigma}{Pr} A. \tag{9.49}$$

The idea is to treat $W, A, \Theta, W_m, \Theta_m$ as independent variables.

We now transform the intervals (0,1) and (-1,0) to the interval (-1,1) and ensure that the interface $z = 0$ is common to both sets of equations when defined on (-1,1). We here select the variables $\hat{z} = 2z-1$ and $\hat{z}_m = -2z_m-1$. This maps the boundary $z = 1$ to $\hat{z} = 1$, $z_m = -1$ to $\hat{z}_m = 1$, and the interface $z = 0$ to $\hat{z} = \hat{z}_m = -1$. Note that

$$D = \frac{d}{dz} = 2\frac{d}{d\hat{z}}, \qquad D = \frac{d}{dz_m} = -2\frac{d}{d\hat{z}_m}.$$

Thus, equations (9.49), (9.35) – (9.37) are transformed to, on $(-1,1)$, employing D to denote $d/d\hat{z}$ or $d/d\hat{z}_m$, as appropriate,

$$(4D^2 - a^2)W - A = 0, \tag{9.50}$$

$$(4D^2 - a^2)A - a^2 Ra\Theta = \frac{\sigma}{Pr} A, \tag{9.51}$$

$$(4D^2 - a^2)\Theta - W = \sigma\Theta, \tag{9.52}$$

$$(4D^2 - a_m^2)W_m + a_m^2 Ra_m \Theta_m = -\sigma \frac{\hat{d}^2 \delta^2}{\epsilon_T \phi Pr_m} (4D^2 - a_m^2)W_m, \tag{9.53}$$

$$(4D^2 - a_m^2)\Theta_m - W_m = \sigma \frac{\hat{d}^2 G_m}{\epsilon_T} \Theta_m. \tag{9.54}$$

For given values of the parameters we have to solve equations (9.50) – (9.54) for σ on $(-1,1)$ subject to boundary conditions (9.39) – (9.48), which transform to, on $\hat{z}_m = 1$,

$$W_m = 0, \tag{9.55}$$

$$\Theta_m = 0, \tag{9.56}$$

on $\hat{z} = 1$,

$$W = 0, \qquad (9.57)$$
$$2D\Theta + L\Theta = 0, \qquad (9.58)$$
$$a^2 W + A + Ma.a^2\Theta = 0, \qquad (9.59)$$

and at (what is now the interface) $\hat{z} = \hat{z}_m = -1$,

$$W = \hat{d}W_m, \qquad (9.60)$$
$$\hat{d}\Theta = \epsilon_T^2 \Theta_m, \qquad (9.61)$$
$$D\Theta + \epsilon_T D\Theta_m = 0, \qquad (9.62)$$
$$A + a^2 W - \frac{2\hat{a}\hat{d}}{\delta} DW - \frac{2\hat{a}\hat{d}^3}{\delta} DW_m = 0, \qquad (9.63)$$
$$DA - 2a^2 DW - \frac{\hat{d}^4}{\delta^2} DW_m = \frac{\sigma}{Pr} DW + \sigma \frac{\hat{d}^6}{\phi Pr_m \epsilon_T} DW_m. \qquad (9.64)$$

We now write $W, A, \Theta, W_m, \Theta_m$ in the form of a series of Chebyshev polynomials (really an infinite series, but we truncate) so

$$W = \sum_{i=0}^{N+2} W_i T_i(\hat{z}), \qquad A = \sum_{i=0}^{N+2} A_i T_i(\hat{z}), \qquad \Theta = \sum_{i=0}^{N+2} \Theta_i T_i(\hat{z}),$$
$$W^m = \sum_{i=0}^{N+2} W_i^m T_i(\hat{z}_m), \qquad \Theta^m = \sum_{i=0}^{N+2} \Theta_i^m T_i(\hat{z}_m).$$

The D^2 matrices have the form as in section 9.2.1. Let us denote the boundary conditions (9.55) – (9.64) by BC1 to BC10. We must use the relations $T_n(\pm 1) = (\pm 1)^n$, $T_n'(\pm 1) = (\pm 1)^{n-1} n^2$, to discretize the boundary conditions (9.55) – (9.64). For example, conditions (9.55), (9.58), (9.62), (9.64) become

$$W_0^m + W_1^m + \ldots + W_{N+2}^m = 0,$$
$$\sum_{i=0}^{N+2} (2i^2 + L)\Theta_i = 0,$$
$$\sum_{i=0}^{N+2} (-1)^{i-1} i^2 \Theta_i + \epsilon_T \sum_{i=0}^{N+2} (-1)^{i-1} i^2 \Theta_i^m = 0,$$

and

$$- 2a^2 \sum_{i=0}^{N+2} (-1)^{i-1} i^2 W_i + \sum_{i=0}^{N+2} (-1)^{i-1} i^2 A_i$$

$$- \frac{\hat{d}^4}{\delta^2} \sum_{i=0}^{N+2} (-1)^{i-1} i^2 W_i^m$$

$$= \frac{\sigma}{Pr} \sum_{i=0}^{N+2} (-1)^{i-1} i^2 W_i + \sigma \frac{\hat{d}^6}{\epsilon_T \phi Pr_m} \sum_{i=0}^{N+2} (-1)^{i-1} i^2 W_i^m .$$

In this way we reduce the eigenvalue problem (9.50) – (9.64) to solving the matrix equation

$$A\mathbf{x} = \sigma B\mathbf{x},$$

where $\mathbf{x}$ is the $5(N + 3)$ vector with components $W_0, \ldots, W_{N+2}$, $A_0, \ldots, A_{N+2}$, $\Theta_0, \ldots, \Theta_{N+2}$, $W_0^m, \ldots, W_{N+2}^m$, $\Theta_0^m, \ldots, \Theta_{N+2}^m$, and the matrices A and B are given by

$$A = \begin{pmatrix}
4D^2 - a^2 I & -I & 0 & 0 & 0 \\
BC3 & 0\ldots0 & 0\ldots0 & 0\ldots0 & 0\ldots0 \\
BC6 & 0\ldots0 & 0\ldots0 & BC6 & 0\ldots0 \\
0 & 4D^2 - a^2 I & -a^2 Ra I & 0 & 0 \\
BC5 & BC5 & BC5 & 0\ldots0 & 0\ldots0 \\
BC9 & BC9 & 0\ldots0 & BC9 & 0\ldots0 \\
-I & 0 & 4D^2 - a^2 I & 0 & 0 \\
0\ldots0 & 0\ldots0 & BC4 & 0\ldots0 & 0\ldots0 \\
0\ldots0 & 0\ldots0 & BC7 & 0\ldots0 & BC7 \\
0 & 0 & 0 & 4D^2 - a_m^2 I & a_m^2 Ra_m I \\
0\ldots0 & 0\ldots0 & 0\ldots0 & BC1 & 0\ldots0 \\
BC10 & BC10 & 0\ldots0 & BC10 & 0\ldots0 \\
0 & 0 & 0 & -I & 4D^2 - a_m^2 I \\
0\ldots0 & 0\ldots0 & 0\ldots0 & 0\ldots0 & BC2 \\
0\ldots0 & 0\ldots0 & BC8 & 0\ldots0 & BC8
\end{pmatrix}$$

and

$$
B = \begin{pmatrix}
0 & 0 & 0 & 0 & 0 \\
0\ldots0 & 0\ldots0 & 0\ldots0 & 0\ldots0 & 0\ldots0 \\
0\ldots0 & 0\ldots0 & 0\ldots0 & 0\ldots0 & 0\ldots0 \\
0 & Pr^{-1}I & 0 & 0 & 0 \\
0\ldots0 & 0\ldots0 & 0\ldots0 & 0\ldots0 & 0\ldots0 \\
0\ldots0 & 0\ldots0 & 0\ldots0 & 0\ldots0 & 0\ldots0 \\
0 & 0 & I & 0 & 0 \\
0\ldots0 & 0\ldots0 & 0\ldots0 & 0\ldots0 & 0\ldots0 \\
0\ldots0 & 0\ldots0 & 0\ldots0 & 0\ldots0 & 0\ldots0 \\
0 & 0 & 0 & -\frac{\delta^2 \hat{d}^2}{\epsilon_T \phi Pr_m}(4D^2 - a_m^2 I) & 0 \\
0\ldots0 & 0\ldots0 & 0\ldots0 & 0\ldots0 & 0\ldots0 \\
BC10 & 0\ldots0 & 0\ldots0 & BC10 & 0\ldots0 \\
0 & 0 & 0 & 0 & (\hat{d}^2 G_m/\epsilon_T)I \\
0\ldots0 & 0\ldots0 & 0\ldots0 & 0\ldots0 & 0\ldots0 \\
0\ldots0 & 0\ldots0 & 0\ldots0 & 0\ldots0 & 0\ldots0
\end{pmatrix}
$$

In these expressions BC1 – BC10 denote the discrete form of the boundary conditions.

9.2.4 The D Chebyshev tau method

We illustrate this technique by direct application to the penetrative convection system (9.19). Unlike section 9.2.2 we do not assume $\sigma \in \mathbb{R}$ in (9.19) and solve instead for the eigenvalue σ. The idea is to write (9.19) as a system of first order equations. Thus, we introduce variables U and V via $DW = U, D\Theta = V$. Then, we may alternatively write (9.19) in the form

$$
\begin{aligned}
DW - U &= 0, \\
DU - a^2 W - 2a^2 R(\xi - z)\Theta &= 0, \\
D\Theta - V &= 0, \\
DV - a^2 \Theta - RW - \sigma\Theta &= 0.
\end{aligned}
\tag{9.65}
$$

The idea is that W, U, Θ and V are regarded as independent variables. The boundary conditions are (9.18), i.e.

$$W = 0, \quad \Theta = 0, \quad z = 0, 1.$$

Equations (9.65) and the boundary conditions are rewritten on the domain $z \in (-1, 1)$ and then W, U, Θ, V are written as a series of Chebyshev polynomials in the form (9.23). In this manner we find the Chebyshev tau method requires solution of the matrix equation

$$A\mathbf{x} = \sigma B\mathbf{x},$$

where now $\mathbf{x} = (W_0, \ldots, W_N, U_0, \ldots, U_N, \Theta_0, \ldots, \Theta_N, V_0, \ldots, V_N)$ and A and B are given by

$$A = \begin{pmatrix} D & -I & 0 & 0 \\ -a^2 I & D & -2a^2(\xi I - M) & 0 \\ 0 & 0 & D & -I \\ -RI & 0 & -a^2 I & D \end{pmatrix} \qquad B = \begin{pmatrix} 0 & 0 & 0 & 0 \\ 0 & 0 & 0 & 0 \\ 0 & 0 & 0 & 0 \\ 0 & 0 & I & 0 \end{pmatrix}$$

with M being the matrix whose entries are given by (9.33).

Again, we note that we have described how to solve equations (9.19) directly for σ, assuming R is given. This is different from section 9.2.2 where σ is set equal to 0 and we solve for R. In general, for a porous medium stability problem we do not know whether σ will be real or not and we have to solve as above.

For the system above, the boundary conditions are added to the $i(N + 1)$th rows, $i = 1, 2, 3, 4$. Thus, we add the conditions $W = 0$ to the rows $N + 1$ and $2(N + 1)$, and the conditions $\Theta = 0$ to the rows $3(N + 1)$ and $4(N + 1)$. The conditions $W = 0, \Theta = 0$ are added by using the relations (9.26).

The question of removing the boundary condition rows in the D-Chebyshev tau method is addressed in (Payne and Straughan, 2000a), (Straughan, 2001a), (Bourne, 2003).

9.2.5 Natural variables

To illustrate this method we again employ the equations for penetrative convection in a porous medium, equations (9.17). So, the linearized perturbation equations are

$$\begin{aligned} \pi_{,i} &= -u_i - 2RM(z)\theta k_i, \\ u_{i,i} &= 0, \\ \theta_t &= -Rw + \Delta\theta, \end{aligned} \qquad (9.66)$$

$(x, y) \in \mathbb{R}^2$, $z \in (0, 1)$, $t > 0$. On the lower boundary $z = 0$ we assume

$$\theta = w = 0. \qquad (9.67)$$

However, on the upper surface we suppose the temperature satisfies a mixed boundary condition of the form

$$\frac{\partial \theta}{\partial z} + \mu\theta = 0, \qquad z = 1, \qquad (9.68)$$

while we also suppose the pressure is constant there. In this case, the appropriate boundary condition is not $w = 0$. Instead, since the pressure is constant, the pressure perturbation, π, is zero at $z = 1$. Because of this $\pi_x = 0, \pi_y = 0$ on $z = 1$, and then from $(9.66)_1$, $u = v = 0$ on $z = 1$. Then equation $(9.66)_2$, namely, $u_x + v_y + w_z = 0$, implies that $w_z = 0$ on $z = 1$.

Thus, the correct boundary conditon is

$$\frac{\partial w}{\partial z} = 0, \qquad z = 1. \tag{9.69}$$

Now, write $u_i = U_i(z)\, g(x, y)\, e^{\sigma t}$, where g is a planform such that $\Delta^* g = -a^2 g$, a being a wavenumber, with similar representations for θ and π. We eliminate u and v from (9.66) – (9.69). We must then solve the system of equations

$$0 = (D^2 - a^2)W - 2RMa^2\Theta,$$
$$\sigma\Theta = -RW + (D^2 - a^2)\Theta, \tag{9.70}$$

subject to the boundary conditions

$$\Theta = W = 0, \qquad z = 0,$$
$$DW = 0, \quad D\Theta + \mu\Theta = 0, \qquad z = 1. \tag{9.71}$$

(Payne and Straughan, 2000a) suggest a natural way to solve (9.70) and (9.71) using a D - Chebyshev tau method. Instead of introducing $U = DW, V = D\Theta$ as in section 9.2.4, they advocate using the structure of the boundary conditions (9.71) to suggest natural variables U, V. So, we select $U = DW$, $V = D\Theta + \mu\Theta$. Equivalently to solving (9.70), (9.71), we must now solve the system

$$DW - U = 0,$$
$$DU - a^2W - 2RMa^2\Theta = 0,$$
$$D\Theta + \mu\Theta - V = 0,$$
$$DV - \mu V + (\mu^2 - a^2)\Theta - RW = \sigma\Theta, \tag{9.72}$$

together with the easily implementable boundary conditions

$$\Theta = W = 0, \quad z = 0; \qquad U = 0, V = 0, \quad z = 1. \tag{9.73}$$

One may easily employ a Chebyshev polynomial expansion of W, U, Θ, V and write (9.72) in the approximate form $A\mathbf{x} = \sigma B\mathbf{x}$ where A, B are square matrices. Due to the simple form of boundary conditions, (9.73), one may remove boundary condition rows and thereby incorporate the boundary conditions in the matrices A and B. Details of this procedure are given in (Payne and Straughan, 2000a), pp. 824, 825.

9.3 Legendre-Galerkin method

9.3.1 Fourth order system

We describe the Legendre - Galerkin technique by starting with an analysis of system (9.11). In this section we suppose the coefficients $\alpha_1, \ldots, \alpha_4, \beta_1, \ldots, \beta_4$ are constants and the eigenvalue σ appears linearly in one or more of these coefficients. The extension of the technique to the situation in which α_i or β_i depend on the variable z is discussed in section 9.3.2. To solve equations (9.11) we assume they have been transformed to the interval $(-1, 1)$ although we still keep the same form for the coefficients. Thus, our interest is to solve the system

$$
\begin{aligned}
w'' &= \alpha_1 w' + \alpha_2 w + \alpha_3 \theta' + \alpha_4 \theta, \\
\theta'' &= \beta_1 w' + \beta_2 w + \beta_3 \theta' + \beta_4 \theta,
\end{aligned}
\tag{9.74}
$$

$' = d/dz$, with $z \in (-1, 1)$. The boundary conditions are

$$
w = \theta = 0 \qquad \text{at} \quad z = \pm 1.
\tag{9.75}
$$

Before describing the Legendre - Galerkin method we refer to some basic properties of Legendre polynomials. We denote the Legendre polynomial of order n by $P_n(z)$. Further details of its properties may be found in the book by (Sneddon, 1980), chapter 3. The $P_n(z)$ are a system of orthogonal polynomials which satisfy

$$
P_n(\pm 1) = (\pm 1)^n,
\tag{9.76}
$$

together with

$$
(2n + 1)P_n = P'_{n+1} - P'_{n-1},
\tag{9.77}
$$

and

$$
(P_i, P_j) = \int_{-1}^{1} P_i P_j \, dz = \begin{cases} 2/(2i + 1), & i = j \\ 0, & i \neq j, \end{cases}
\tag{9.78}
$$

and

$$
z P_n = \left(\frac{n + 1}{2n + 1} \right) P_{n+1} + \left(\frac{n}{2n + 1} \right) P_{n-1}.
\tag{9.79}
$$

Proofs of these relations are given by (Sneddon, 1980), equations (13.5a,b), (14.6), (15.8) and (14.7), respectively.

The Legendre - Galerkin method we now describe was effectively used by (Shen, 1994) and a modification by (Kirchner, 2000). The treatment for finding eigenvalues for porous stability problems and related issues was given by (Hill and Straughan, 2005; Hill and Straughan, 2006). The key is to introduce the basis function

$$
\phi_i(z) = \int_{-1}^{z} P_i(s) \, ds, \quad i = 1, \dots, p - 1,
\tag{9.80}
$$

$p \geq 2$. The reason why the method is so attractive is the relation (9.77) which allows us to also write ϕ_i in the form

$$
\phi_i(z) = \frac{1}{(2i + 1)} \left(P_{i+1}(z) - P_{i-1}(z) \right).
\tag{9.81}
$$

Note that $\phi_i(-1) = 0$ and $\phi_i(1) = 0$, relations which are useful for homogeneous boundary conditions. Thus, we write w and θ in (9.74) as a series in the basis ϕ_i. So, we write

$$w = \sum_{k=1}^{N} w_k \phi_k \,, \qquad \theta = \sum_{k=1}^{N} \theta_k \phi_k \,. \tag{9.82}$$

We next multiply $(9.74)_1$ by ϕ_i, $(9.74)_2$ by ϕ_i and integrate by parts, recalling $\phi_i(\pm 1) = 0$, to find

$$\begin{aligned} (w', \phi_i') + \alpha_1(w', \phi_i) + \alpha_2(w, \phi_i) + \alpha_3(\theta', \phi_i) + \alpha_4(\theta, \phi_i) &= 0, \\ (\theta', \phi_i') + \beta_1(w', \phi_i) + \beta_2(w, \phi_i) + \beta_3(\theta', \phi_i) + \beta_4(\theta, \phi_i) &= 0. \end{aligned} \tag{9.83}$$

The brackets $(\cdot, \cdot)$ in (9.83) denote the inner product on $L^2(-1, 1)$, i.e. $(f, g) = \int_{-1}^{1} f(s) g(s) ds$. It is now necessary to evaluate the terms $-(w'', \phi_i) = (w', \phi_i')$, (w', ϕ_i) and (w, ϕ_i). The remaining terms in (9.83) all fall into these three categories. In this way, we reduce the solution of (9.74), (9.75) to the solution of a matrix eigenvalue problem.

Firstly,

$$\begin{aligned} -(w'', \phi_i) &= (w', \phi_i') \\ &= \sum_{k=1}^{N} w_k(\phi_k', \phi_i') \\ &= \sum_{k=1}^{N} w_k(P_k, P_i) \end{aligned}$$

since from (9.80), $\phi_i' = P_i$. Then, employing (9.78) we obtain

$$(w', \phi_i') = w_i \|P_i\|^2 = \frac{2}{(2i+1)} \, w_i \,, \tag{9.84}$$

where $\|P_i\|^2 = \int_{-1}^{1} P_i^2 ds$. Thus, use of the ϕ_i basis has the desirable property that the second derivative operator effectively becomes a diagonal matrix.

The next class of term has form

$$(w', \phi_i) = \int_{-1}^{1} w'\phi_i dz$$

$$= \sum_{k=1}^{N} w_k(\phi'_k, \phi_i)$$

$$= \sum_{k=1}^{N} w_k(P_k, \phi_i)$$

$$= \sum_{k=1}^{N} \frac{w_k}{(2i+1)} (P_k, P_{i+1} - P_{i-1})$$

$$= \frac{w_{i+1}}{(2i+1)} \|P_{i+1}\|^2 - \frac{w_{i-1}}{(2i+1)} \|P_{i-1}\|^2$$

$$= \frac{2w_{i+1}}{(2i+3)(2i+5)} - \frac{2w_{i-1}}{(2i-1)(2i+1)}, \tag{9.85}$$

where (9.78), (9.80) and (9.81) have been employed. Let us note that (9.85) contributes to terms either side of the diagonal in a banded structure of matrix.

Finally we consider terms of the form

$$(w, \phi_i) = \int_{-1}^{1} w\phi_i dz$$

$$= \sum_{k=1}^{N} w_k(\phi_k, \phi_i)$$

$$= \sum_{k=1}^{N} w_k \left(\frac{P_{k+1} - P_{k-1}}{2k+1}, \frac{P_{i+1} - P_{i-1}}{2i+1} \right)$$

$$= \frac{w_i}{(2i+1)^2} \|P_{i+1}\|^2 - \frac{w_{i+2}}{(2i+1)(2i+3)} \|P_{i+1}\|^2$$

$$+ \frac{w_i}{(2i+1)^2} \|P_{i-1}\|^2 - \frac{w_{i-2}}{(2i-1)(2i+1)} \|P_{i-1}\|^2$$

$$= -\frac{2w_{i+2}}{(2i+3)^2(2i+1)} + \frac{4w_i}{(2i-1)(2i+1)(2i+3)}$$

$$- \frac{2w_{i-2}}{(2i-1)^2(2i+1)}. \tag{9.86}$$

By using relations like (9.84), (9.85) and (9.86) in equations (9.83) we arrive at a matrix eigenvalue problem of form

$$A\mathbf{x} = \sigma B\mathbf{x} \tag{9.87}$$

where $\mathbf{x} = (w_1, \ldots, w_N, \theta_1, \ldots, \theta_N)^T$. It is important to note that the matrices A and B are of form

$$A = \begin{pmatrix} A_{11} & A_{12} \\ A_{21} & A_{22} \end{pmatrix} \tag{9.88}$$

and

$$B = \begin{pmatrix} B_{11} & B_{12} \\ B_{21} & B_{22} \end{pmatrix} \tag{9.89}$$

where each A_{ij}, B_{ij} is an $n \times n$ matrix which has the structure

$$A_{ij} = \begin{pmatrix} a_{11} & a_{12} & a_{13} & 0 & \cdots & & 0 \\ a_{21} & \ddots & \ddots & \ddots & & \ddots & \vdots \\ a_{31} & \ddots & \ddots & \ddots & & \ddots & 0 \\ 0 & \ddots & \ddots & \ddots & & \ddots & a_{(N-2)N} \\ \vdots & \ddots & \ddots & \ddots & & \ddots & a_{(N-1)N} \\ 0 & \cdots & 0 & a_{N(N-2)} & a_{N(N-1)} & a_{NN} \end{pmatrix}$$

i.e. the matrices A and B are block 5-banded. They are thus sparse. This leads to a major difference with the Chebyshev tau method of section 9.2. For larger matrices A and B we are able to use an iterative solver like the Arnoldi method which may be found in the ARPACK package, see (Lehoucq et al., 1998). This leads to a much more efficient solver for equation (9.87) than the QZ algorithm when many basis functions are required. The last comment is particularly important when 2 and 3-D eigenvalue problems are considered as is pointed out in section 9.3.3. Details of the speed up achieved and performance of the Arnoldi algorithm may be found in (Hill and Straughan, 2005).

9.3.2 *Penetrative convection*

We further illustrate the Legendre - Galerkin method by application to the penetrative convection system (9.19), (9.18). We recast this problem in the interval $(-1, 1)$ and so have to solve

$$\begin{aligned} (4D^2 - a^2)W - a^2 R[2\xi - (z+1)]\Theta &= 0, \\ (4D^2 - a^2)\Theta - RW &= \sigma\Theta, \end{aligned} \tag{9.90}$$

together with

$$W = \Theta = 0 \quad \text{at} \quad z = \pm 1. \tag{9.91}$$

With ϕ_i defined as in (9.80) we write W, Θ as

$$W = \sum_{k=1}^{N} W_k \phi_k, \qquad \Theta = \sum_{k=1}^{N} \Theta_k \phi_k. \tag{9.92}$$

We then multiply each of equations (9.90) by ϕ_i and integrate over $(-1, 1)$ and use the boundary conditions to find

$$
\begin{aligned}
&4(W', \phi_i') + a^2(W, \phi_i) + a^2 R(2\xi - 1)(\Theta, \phi_i) - a^2 R(z\Theta, \phi_i) = 0, \\
&4(\Theta', \phi_i') + a^2(\Theta, \phi_i) + R(W, \phi_i) = -\sigma(\Theta, \phi_i).
\end{aligned}
\tag{9.93}
$$

The terms in (9.93) are all of form (9.84), (9.85) or (9.86) apart from the term $-(z\Theta, \phi_i)$. The terms not involving z directly are handled as in section 9.3.1. To handle the z term we use relation (9.79). Thus,

$$
\begin{aligned}
(z\Theta, \phi_i) &= \sum_{k=1}^{N} \Theta_k (z\phi_k, \phi_i) \\
&= \sum_{k=1}^{N} \Theta_k \left(z \frac{(P_{k+1} - P_{k-1})}{2k+1}, \phi_i \right) \\
&= \sum_{k=1}^{N} \Theta_k \left(\frac{1}{(2k+1)} \left\{ \frac{(k+2)}{(2k+3)} P_{k+2} + \frac{(k+1)}{(2k+3)} P_k \right\} \right. \\
&\quad \left. - \frac{1}{(2k+1)} \left\{ \frac{k}{(2k-1)} P_k + \frac{(k-1)}{(2k-1)} P_{k-2} \right\}, \phi_i \right) \\
&= \sum_{k=1}^{N} \Theta_k \left(\frac{k+2}{(2k+1)(2k+3)} P_{k+2} - \frac{P_k}{(2k+3)(2k-1)} \right. \\
&\quad \left. - \frac{(k-1)}{(2k+1)(2k-1)} P_{k-2}, \frac{P_{i+1} - P_{i-1}}{(2i+1)} \right) \\
&= \Theta_{i-1} \frac{(i+1)}{(2i-1)(2i+1)^2} \|P_{i+1}\|^2 - \Theta_{i-3} \frac{(i-1)}{(2i-5)(2i-3)(2i+1)} \|P_{i-1}\|^2 \\
&\quad - \Theta_{i+1} \frac{1}{(2i+5)(2i+1)^2} \|P_{i+1}\|^2 + \Theta_{i-1} \frac{1}{(2i-3)(2i+1)^2} \|P_{i-1}\|^2 \\
&\quad - \Theta_{i+3} \frac{(i+2)}{(2i+7)(2i+5)(2i+1)} \|P_{i+1}\|^2 + \Theta_{i+1} \frac{i}{(2i+3)(2i+1)^2} \|P_{i-1}\|^2
\end{aligned}
$$

with i taking appropriate values

$$
\begin{aligned}
&= \Theta_{i-1} \frac{2(i+1)}{(2i-1)(2i+1)^2(2i+3)} - \Theta_{i-3} \frac{2(i-1)}{(2i-5)(2i-3)(2i-1)(2i+1)} \\
&\quad - \Theta_{i+1} \frac{2}{(2i+1)^2(2i+3)(2i+5)} + \Theta_{i-1} \frac{2}{(2i-3)(2i-1)(2i+1)^2} \\
&\quad - \Theta_{i+3} \frac{2(i+2)}{(2i+1)(2i+3)(2i+5)(2i+7)} \\
&\quad + \Theta_{i+1} \frac{2i}{(2i+3)(2i-1)(2i+1)^2}.
\end{aligned}
\tag{9.94}
$$

In this way, we see that equations (9.93) lead to a matrix eigenvalue problem of form (9.87) where $\mathbf{x} = (W_1, \dots, W_N, \Theta_1, \dots, \Theta_N)^T$ and A and B have a banded structure like (9.88), (9.89), except the $(1,2)$ block of A is 7 banded. However, the Arnoldi method may again be used to solve the matrix eigenvalue problem efficiently. Details of the performance of the Arnoldi method for several examples similar to that of this section are contained in (Hill and Straughan, 2005).

9.3.3   Extension of the method

The Legendre - Galerkin technique has other desirable features. We may, for example, deal with other coefficients of z in the equations rather than just z itself. For a polynomial function, $h(z)$ say, repeated use of relation (9.79) allows us to incorporate $h(z)$ quickly. For example,

$$
\begin{aligned}
z^2 Pn =& z\left(\left[\frac{n+1}{2n+1}\right]P_{n+1} + \left[\frac{n}{2n+1}\right]P_{n-1}\right) \\
=& \left(\frac{n+1}{2n+1}\right)\left(\frac{n+2}{2n+3}\right)P_{n+2} + \left(\frac{n+1}{2n+1}\right)\left(\frac{n+1}{2n+3}\right)P_n \\
& + \left(\frac{n}{2n+1}\right)\left(\frac{n}{2n-1}\right)P_n + \left(\frac{n}{2n+1}\right)\left(\frac{n-1}{2n-1}\right)P_{n-2} \\
=& \frac{(n+1)(n+2)}{(2n+1)(2n+3)}P_{n+2} + \frac{(4n^3+6n^2-1)}{(2n-1)(2n+1)(2n+3)}P_n \\
& + \frac{n(n-1)}{(2n-1)(2n+1)}P_{n-2}.
\end{aligned}
\tag{9.95}
$$

By using (9.95) we may easily account for a coefficient z^2 in the equations. The effect is to increase the bandwidth of an appropriate block of the A or B matrices. Further details of incorporating a general function $g(z)$ may be found in (Hill and Straughan, 2005).

Another very important advantage of the Legendre - Galerkin method is its efficient applicability to two or three - dimensional eigenvalue problems. Here, one works with the primitive variables. For example, in the penetrative convection problem one would use $u_1 = u$, $u_2 = v$, $u_3 = w$, θ and π, and one writes each as an expansion of the tensor product of the ϕ_i (π is expanded in P_i). For example, in 2-D,

$$
u = \sum_{k=1}^{N}\sum_{j=1}^{N} U_{kj}\alpha_{kj}(x,y)
$$

where

$$
\alpha_{kj}(x,y) = \phi_k(x)\phi_j(y) = \int_{-1}^{x} P_k(s)ds \int_{-1}^{y} P_j(r)dr.
$$

Such a procedure leads to a matrix equation of form (9.87) where A, B have a more complicated (larger) structure. However, they are usually sparse (block banded) and the Arnoldi technique is an efficient solver for the matrix eigenvalue problem. Further details of this procedure are given by (Hill and Straughan, 2006).

Finally, we briefly remark on the use of other orthogonal polynomials in a Galerkin - like idea. For example, why do we not consider basis functions like $\int_{-1}^{z} T_n(s)ds$ with T_n being a Chebyshev polynomial? Such an idea would be very appealing due to the fact that

$$T_m T_n = \frac{1}{2}(T_{m+n} + T_{|m-n|}) \qquad (9.96)$$

a relation which is very useful when dealing with coefficients which depend on z. (The other orthogonal polynomials do not satisfy such a simple relation for their product as (9.96).) However, what makes the Legendre polynomials work so well in the Galerkin formulation is relation (9.77). For other orthogonal polynomials, the analogous relations do not appear to be so useful. For example, for the Chebyshev polynomials we have

$$2T_n(z) = \frac{T'_{n+1}(z)}{(n+1)} - \frac{T'_{n-1}(z)}{(n-1)}. \qquad (9.97)$$

A proof of this follows by differentiating the relation

$$(1 - x^2)T'_n = n(T_{n-1} - xT_n), \qquad n \geq 1,$$

see (Gardner et al., 1989), p. 165, to find

$$(1 - x^2)T''_n - 2xT'_n = nT'_{n-1} - nT_n - nxT'_n, \qquad (9.98)$$

where we momentarily use x as the independent variable in place of z. Substitute for $(1 - x^2)T''_n$ from (A5) of (Gardner et al., 1989), i.e. use

$$(1 - x^2)T''_n = xT'_n - n^2 T_n,$$

to obtain from (9.98)

$$xT'_n = \frac{nT'_{n-1}}{(n-1)} + nT_n. \qquad (9.99)$$

Now, differentiate the recursion relation (A3) of (Gardner et al., 1989) to find

$$T'_{n+1} - 2xT'_n - 2T_n + T'_{n-1} = 0.$$

Substitute for xT'_n from (9.99) and we derive (9.97).

The fact that the coefficients of T'_{n+1} and T'_{n-1} are not the same in (9.97) means application of a Chebyshev - Galerkin technique is not so straightforward as the Legendre - Galerkin method outlined here. Nevertheless, such application is very interesting, and details of such methods may be found in (Pop, 1997; Pop, 2000), (Pop and Gheorghiu, 1996), (Gheorghiu and Pop, 1996), and in (Hill, 2005a), chapter 8.

References

Akyildiz, F. T. (2007). Stokes' first problem for a Newtonian fluid in a non-Darcian porous half-space using a Laguerre-Galerkin method. *Math. Meth. Appl. Sci.*, 30:2263–2277.

Akyildiz, F. T. and Bellout, H. (2004). The extended Graetz problem for dipolar fluids. *Int. J. Heat Mass Transfer*, 47:2747–2753.

Akyildiz, F. T. and Bellout, H. (2005). Chaos in the thermal convection of a Newtonian fluid with a temperature dependent viscosity. *Appl. Math. Computation*, 162:1103–1118.

Albers, B. (2003). Relaxation analysis and linear stability vs. adsorption in porous materials. *Continuum Mech. Thermodyn.*, 15:73–95.

Albers, B. and Wilmansky, K. (2005). Modelling acoustic waves in saturated poroelastic media. *J. Engng. Mech. (ASCE)*, 131:974–985.

Alex, S. M., Patil, P. R., and Venkatakrishnan, K. S. (2001). Variable gravity effects on thermal instability in a porous medium with internal heat source and inclined temperature gradient. *Fluid Dyn. Res.*, 29:1–6.

Alikakos, N. D. and Rostamian, R. (1981). Large time behaviour of solutions of Neumann boundary value problem for the porous medium equation. *Indiana Univ. Math. J.*, 30:749–785.

Allen, L. and Bridges, T. J. (2002). Numerical exterior algebra and the compound matrix method. *Numer. Math.*, 92:197–232.

Allen, M. B. and Khosravani, A. (1992). Solute transport via alternating direction collocation using the method of modifed characteristics. *Adv. Water Resources*, 15:125–132.

Alvarez-Ramirez, J., Fernandez-Anaya, G., Valdes-Parada, F. J., and Ochoa-Tapia, J. A. (2006). A high order extension for Cattaneo's diffusion equation. *Physica A - Statistical Mechanics and its Applications*, 368:345–354.

Alvarez-Ramirez, J., Puebla, H., and Ochoa-Tapia, J. A. (2001). Linear boundary control for a class of nonlinear pde processes. *Systems and Control Letters*, 44:395–403.

Ames, K. A., Clark, G. W., Epperson, J. F., and Oppenheimer, S. F. (1998). A comparison of regularizations for an ill-posed problem. *Mathematics of Computation*, 67:1451–1471.

Ames, K. A. and Payne, L. E. (1989). Decay estimates in steady pipe flow. *SIAM J. Math. Anal.*, 20:789–815.

Ames, K. A. and Payne, L. E. (1994). On stabilizing against modeling errors in a penetrative convection problem for a porous medium. *Math. Models Meth. Appl. Sci.*, 4:733–740.

Ames, K. A. and Payne, L. E. (1997). Continuous dependence results for a problem in penetrative convection. *Quart. Appl. Math.*, 55:769–790.

Ames, K. A. and Payne, L. E. (1999). Continuous dependence on modelling for some well-posed perturbations of the backward heat equation. *J. Inequalities and Applications*, 3:51–64.

Ames, K. A., Payne, L. E., and Schaefer, P. W. (1993). Spatial decay estimates in time - dependent Stokes flow. *SIAM J. Math. Anal.*, 24:1395–1413.

Ames, K. A., Payne, L. E., and Schaefer, P. W. (2004a). Energy and pointwise bounds in some nonstandard parabolic problems. *Proc. Royal Soc. Edinburgh A*, 134:1–9.

Ames, K. A., Payne, L. E., and Schaefer, P. W. (2004b). On a nonstandard problem for heat conduction in a cylinder. *Applicable Analysis*, 83:125–133.

Ames, K. A., Payne, L. E., and Song, J. C. (2001). Spatial decay in pipe flow of a viscous fluid interfacing a porous medium. *Math. Models Meth. Appl. Sci.*, 11:1547–1562.

Ames, K. A. and Straughan, B. (1992). Continuous dependence results for initially prestressed thermoelastic bodies. *Int. J. Engng. Sci.*, 30:7–13.

Ames, K. A. and Straughan, B. (1997). *Non-standard and improperly posed problems*, volume 194 of *Mathematics in Science and Engineering*. Academic Press, San Diego.

Amili, P. and Yortsos, Y. C. (2004). Stability of heat pipes in vapor-dominated systems. *Int. J. Heat Mass Transfer*, 47:1233–1246.

Armandi, M., Bonelli, B., Bottero, I., Otero Areán, C., and Garrone, E. (2007). Synthesis and characterization of ordered porous carbons with potential applications as hydrogen storage media. *Microporous and Mesoporous Materials*, 103:150–157.

Aronson, D. G. and Caffarelli, L. A. (1983). The initial trace of a solution of the porous medium equation. *Trans. Amer. Math. Soc.*, 280:351–366.

Aronson, D. G. and Peletier, L. A. (1981). Large time behaviour of solutions of the porous medium equation in bounded domains. *J. Differential Equations*, 39:378–412.

Avramenko, A. A. and Kuznetsov, A. V. (2004). Stability of a suspension of gyrotatic micoorganisms in superimposed fluid and porous layers. *Int. Commun. Heat Mass. Transfer*, 31:1057–1066.

Ayrault, C., Moussatov, A., Castagnéde, B., and Lafarge, D. (1999). Ultrasonic characterization of plastic foams via measurements with static pressure variations. *Appl. Phys, Lett.*, 74:3224–3226.

Ball, J. M. (1977). Remarks on blow-up and nonexistence theorems for nonlinear evolution equations. *Quart. J. Math.*, 28:473–486.

Banu, N. and Rees, D. A. S. (2002). Onset of Darcy-Bénard convection using a thermal non-equilibrium model. *Int. J. Heat Mass Transfer*, 45:2221–2228.

Bardan, G., Bergeon, A., Knobloch, E., and Mojtabi, A. (2000). Nonlinear doubly diffusive convection in vertical enclosures. *Physica D*, 138:91–113.

Bardan, G., Knobloch, E., Mojtabi, A., and Khallouf, H. (2001). Nonlinear doubly diffusive convection with vibration. *Fluid Dyn. Res.*, 28:159–187.

Bardan, G. and Mojtabi, A. (1998). Theoretical stability study of double-diffusive convection in a square cavity. *Comptes Rendues de l'Academie des Sciences*, 326:851–857.

Barenblatt, G. I. and Vazquez, J. L. (2004). Nonlinear diffusion and image contour enhancement. *Interfaces and Free Boundaries*, 6:31–54.

Basu, S., Chowdhury, S., and Chakraborty, S. (2007). Influences of pressure gradients in freezing Poiseuille-Couette flows. *Int. J. Heat Mass Transfer*, 50:4493–4498.

Batchelor, G. K. (1967). *An introduction to fluid dynamics*. Cambridge University Press, Cambridge.

Bear, J. and Gilman, A. (1995). Migration of salts in the unsaturated zone caused by heating. *Transport in Porous Media*, 19:139–156.

Beavers, G. S. and Joseph, D. D. (1967). Boundary conditions at a naturally impermeable wall. *J. Fluid Mech.*, 30:197–207.

Bellomo, N. and Preziosi, L. (1995). *Modelling mathematical methods and scientific computation*. CRC Press, Boca Raton.

Bennethum, L. S. and Giorgi, T. (1997). Generalized Forchheimer equation for two - phase flow based on hybrid mixture theory. *Trans. Porous Media*, 26:261–275.

Bhadauria, B. S. (2007). Double diffusive convection in a porous medium with modulated temperature on the boundaries. *Trans. Porous Media*, 70:191–211.

Bhandar, A. S., Vogel, M. J., and Steen, P. H. (2004). Energy landscapes and bistability to finite-amplitude disturbances for the capillary bridge. *Phys. Fluids*, 16:3063–3069.

Biot, M. A. (1956a). The theory of propagation of elastic waves in fluid-saturated porous solid. I. Low frequency range. *J. Acoust. Soc. Am.*, 28:168–178.

Biot, M. A. (1956b). The theory of propagation of elastic waves in fluid-saturated porous solid. II. Higher frequency range. *J. Acoust. Soc. Am.*, 28:179–191.

Blest, D. C., McKee, S., Zulkifle, A. K., and Marshall, P. (1999). Curing simulation by autoclave resin infusion. *Composites Science and Technology*, 59:2297–2313.

Bleustein, J. L. and Green, A. E. (1967). Dipolar fluids. *Int. J. Engng. Sci.*, 5:323–340.

Blyth, M. G., Luo, H., and Pozrikidis, C. (2006). Stability of axisymmetric core-annular flow in the presence of an insoluble surfactant. *J. Fluid. Mech.*, 548:207–235.

Blyth, M. G. and Pozrikidis, C. (2004a). Effect of inertia on the Marangoni instability of two-layer channel flow. Part II: normal mode analysis. *J. Engng. Math.*, 50:329–341.

Blyth, M. G. and Pozrikidis, C. (2004b). Effect of surfactant on the stability of film flow down an inclined plane. *J. Fluid. Mech.*, 521:241–250.

Boano, F., Revelli, R., and Ridolfi, L. (2007). Bedform-induced hyporheic exchange with unsteady flows. *Advances in Water Resources*, 30:148–156.

Bofill, F. and Quintanilla, R. (2003). On the uniqueness and spatial behaviour of anti-plane shear deformations of swelling porous elastic soils backward in time. *Int. J. Engng. Sci.*, 15:1815–1826.

Bogorodskii, P. V. and Nagurnyi, A. P. (2000). Under - ice meltwater puddles: a factor of fast sea ice melting in the Arctic. *Doklady Earth Sciences*, 373: 885–887.

Borja, R. I. (2004). Cam - Clay plasticity. Part V: A mathematical framework for three -phase deformation and strain localization analysis of partially saturated porous media. *Comput. Methods Appl. Mech.*, 193:5301–5338.

Bourne, D. (2003). Hydrodynamic stability, the Chebshev tau method, and spurious eigenvalues. *Continuum Mech. Thermodyn.*, 15:571–579.

Boutat, M., d'Angelo, Y., Hilout, S., and Lods, V. (2004). Existence and finite-time blow-up for the solution to a thin-film surface evolution problem. *Asymptotic Analysis*, 38:93–128.

Bowen, R. M. (1982). Compressible porous media models by use of the theory of mixtures. *Int. J. Engng. Sci.*, 20:697–735.

Bresch, D., Essoufi, E. H., and Sy, M. (2007). Effect of density dependent viscosities on multiphasic incompressible fluid models. *J. Math. Fluid Mech.*, 9:377–397.

Bresch, D. and Sy, M. (2003). Convection in rotating porous media: The planetary geostrophic equations, used in geophysical fluid dynamics, revisited. *Continuum Mech. Thermodyn.*, 15:247–263.

Brinkman, H. C. (1947). A calculation of viscous force exerted by a flowing fluid on a dense swarm of particles. *Appl. Sci. Res.*, 1:27–34.

Brown, M. and Marletta, M. (2003). Spectral inclusion and spectral exactness for singular non - self - adjoint hamiltonian systems. *Proc. Roy. Soc. London A*, 459:1987–2009.

Brusov, P., Brusov, P., Lawes, G., Lee, C., Matsubara, A., Ishikawa, O., and Majumdar, P. (2003). Novel sound phenomena in superfluid helium in aerogel and other impure superfluids. *Physics Letters A*, 310:311–321.

Budu, P. (2002). *Conditional and unconditional nonlinear stability in fluid dynamics*. PhD thesis, University of Durham.

Buishvili, T., Kekutia, S., Tkeshelashvili, O., and Tkeshelashvili, L. (2002). Theory of sound propagation in superfluid - filled porous media. *Physics Letters A*, 300:672–686.

Butler, K. M. and Farrell, B. F. (1992). Three-dimensional optimal perturbations in viscous shear flow. *Phys. Fluids A*, 4:1637–1650.

Butler, K. M. and Farrell, B. F. (1994). Nonlinear equilibration of two-dimensional optimal perturbations in viscous shear flow. *Phys. Fluids A*, 6:2011–2020.

Calmidi, V. V. and Mahajan, R. L. (2000). Forced convection in high porosity foams. *Trans. ASME. J. Heat Transfer*, 122:557–565.

Capone, F. and Rionero, S. (2000). Thermal convection with horizontally periodic temperature gradient. *Rend. Accad. Sci. Fis. Matem. Napoli*, 67:119–128.

Carr, M. (2003a). *Convection in porous media flows.* PhD thesis, University of Durham.

Carr, M. (2003b). A model for convection in the evolution of under-ice melt ponds. *Continuum Mech. Thermodyn.*, 15:45–54.

Carr, M. (2003c). Unconditional nonlinear stability for temperature dependent density flow in a porous medium. *Math. Models Meth. Appl. Sci.*, 13:207–220.

Carr, M. (2004). Penetrative convection in a superposed porous - medium - fluid layer via internal heating. *J. Fluid Mech.*, 509:305–329.

Carr, M. and de Putter, S. (2003). Penetrative convection in a horizontally isotropic porous layer. *Continuum Mech. Thermodyn.*, 15:33–43.

Carr, M. and Straughan, B. (2003). Penetrative convection in fluid overlying a porous layer. *Advances in Water Resources*, 26:263–276.

Cattaneo, C. (1948). Sulla conduzione del calore. *Atti Sem. Mat. Fis. Modena*, 3:83–101.

Caviglia, G., Morro, A., and Straughan, B. (1992a). Reflection and refraction at a variable porosity interface. *J. Acoustical Soc. Amer.*, 1992:1113–1119.

Caviglia, G., Morro, A., and Straughan, B. (1992b). Thermoelasticity at cryogenic temperatures. *Int. J. Nonlinear Mech.*, 27:251–263.

Celebi, A. O., Kalantarov, V. K., and Ugurlu, D. (2006). On continuous dependence on coefficients of the Brinkman-Forchheimer equations. *Appl. Math. Letters*, 2006:801–807.

Chadwick, P. and Currie, P. K. (1974). Intrinsically characterized acceleration waves in heat-conducting elastic materials. *Proc. Camb. Phil. Soc.*, 76: 481–491.

Chadwick, P. and Currie, P. K. (1975). On the existence of transverse elastic acceleration waves. *Math. Proc. Camb. Phil. Soc.*, 77:405–413.

Chamkha, A. J., Pop, I., and Takhar, H. S. (2006). Marangoni mixed convection boundary layer flow. *Meccanica*, 41:219–232.

Chandesris, M. and Jamet, D. (2006). Boundary conditions at a planar fluid-porous interface for Poiseuille flow. *Int. J. Heat Mass Transfer*, 49:2137–2150.

Chang, M. H. (2004). Stability of convection induced by selective absorption of radiation in a fluid overlying a porous layer. *Phys. Fluids*, 16:3690–3698.

Chang, M. H. (2005). Thermal convection in superposed fluid and porous layers subjected to a horizontal plane Couette flow. *Phys. Fluids*, 17:064106-1–064106-7.

Chang, M. H. (2006). Thermal convection in superposed fluid and porous layers subjected to a plane Poiseuille flow. *Phys. Fluids*, 18:035104-1–035104-10.

Chang, M. H., Chen, F., and Straughan, B. (2006). Instability of Poiseuille flow in a fluid overlying a porous layer. *J. Fluid Mech.*, 564:287–303.

Charrier-Mojtabi, M. C., Karimi-Fard, M., Azaiez, M., and Mojtabi, A. (1998). Onset of a double-diffusive convective regime in a rectangular porous cavity. *J. Porous Media*, 1:107–121.

Chasnov, J. R. and Tse, K. L. (2001). Turbulent penetrative convection with an internal heat source. *Fluid Dyn. Res.*, 28:397–421.

Chen, F. (1990). Throughflow effects on convective instability in superposed fluid and porous layers. *J. Fluid Mech.*, 231:113–133.

Chen, F. (1993). Stability of Taylor-Dean flow in an annulus with arbitrary gap spacing. *Phys. Rev. E*, 48:1036–1045.

Chen, F. (2003). Personal communication.

Chen, F. and Chang, M. H. (1992). Stability of Taylor-Dean flow in a small gap between rotating cylinders. *J. Fluid Mech.*, 243:443–455.

Chen, F. and Chen, C. F. (1988). Onset of finger convection in a horizontal porous layer underlying a fluid layer. *J. Heat Transfer*, 3:403–409.

Chen, F. and Chen, C. F. (1992). Convection in superposed fluid and porous layers. *J. Fluid Mech.*, 234:97–119.

Chen, P. J. (1973). Growth and decay of waves in solids. In Flügge, S. and Truesdell, C., editors, *Handbuch der Physik*, volume VIa/3, pages 303–402. Springer, Berlin.

Chen, P. J. and Gurtin, M. E. (1968). On a theory of heat conduction involving two temperatures. *ZAMP*, 19:614–627.

Cheney, W. and Kincaid, D. (1985). *Numerical mathematics and computing.* Brooks-Cole, Monterey.

Chirita, S. and Ciarletta, M. (2003). Spatial behaviour of solutions in the plane Stokes flow. *J. Math. Anal. Appl.*, 277:571–588.

Chirita, S. and Ciarletta, M. (2006). Spatial estimates for the constrained anisotropic elastic cylinder. *J. Elasticity*, 85:189–213.

Chirita, S. and Ciarletta, M. (2008). On the structural stability of a thermoelastic model of porous media. *Math. Meth. Appl. Sci.*, 31:19–34.

Chirita, S., Ciarletta, M., and Fabrizio, M. (2001). Some spatial decay estimates in time-dependent Stokes slow flows. *Appl. Anal.*, 77:211–231.

Chirita, S., Ciarletta, M., and Straughan, B. (2006). Structural stability in porous elasticity. *Proc. Roy. Soc. London A*, 462:2593–2605.

Choi, C. K., Chung, T. J., and Kim, M. C. (2004). Buoyancy effects in plane Couette flow heated uniformly from below with constant heat flux. *Int. J. Heat Mass Transfer*, 47:2629–2636.

Christopherson, D. G. (1940). Note on the vibration of membranes. *Quart. J. Math.*, 11:63–65.

Christov, C. I. and Jordan, P. M. (2005). Heat conduction paradox involving second sound propagation in moving media. *Physical Review Letters*, 94:154301-1–154301-4.

Christov, I., Jordan, P. M., and Christov, C. I. (2006). Nonlinear acoustic propagation in homentropic perfect gases: a numerical study. *Physics Letters A*, 353:273–280.

Christov, I., Jordan, P. M., and Christov, C. I. (2007). Modelling weakly nonlinear acoustic wave propagation. *Quart. Jl. Mech. Appl. Math.*, 60:473–495.

Chung, C. A. and Chen, F. (2000a). Convection in directionally solidifying alloys under inclined rotation. *J. Fluid Mech.*, 412:93–123.

Chung, C. A. and Chen, F. (2000b). Onset of plume convection in mushy layers. *J. Fluid Mech.*, 408:53–82.

Ciarletta, M. and Iesan, D. (1993). *Non-classical elastic solids.* Longman, New York.

Ciarletta, M. and Straughan, B. (2006). Poroacoustic acceleration waves. *Proc. Roy. Soc. London A*, 462:3493–3499.

Ciarletta, M. and Straughan, B. (2007a). Poroacoustic acceleration waves with second sound. *J. Sound Vibration*, 306:725–731.

Ciarletta, M. and Straughan, B. (2007b). Thermo-poroacoustic acceleration waves in elastic materials with voids. *J. Math. Anal. Appl.*, 333:142–150.

Ciarletta, M., Straughan, B., and Zampoli, V. (2007). Thermo-poroacoustic acceleration waves in elastic materials with voids without energy dissipation. *Int. J. Engng. Sci.*, 45:736–743.

Ciarletta, M. and Zampoli, A. (2006). Personal communication.

Ciesjko, M. and Kubik, J. (1999). Derivation of matching conditions at the contact surface between fluid-saturated porous solid and bulk fluid. *Transport in Porous Media*, 34:319–336.

Clark, P. U., Pisias, N. G., Stocker, T. F., and Weaver, A. J. (2002). The role of thermohaline circulation in abrupt climate change. *Nature*, 415:863–869.

Coleman, B. D., Fabrizio, M., and Owen, D. R. (1982). On the thermodynamics of second sound in dielectric crystals. *Arch. Rational Mech. Anal.*, 80:135–158.

Coleman, B. D. and Noll, W. (1963). The thermodynamics of elastic materials with heat conduction and viscosity. *Arch. Rational Mech. Anal.*, 13:167–178.

Cortis, A. and Ghezzehei, T. A. (2007). On the transport of emulsions in porous media. *J. Colloid and Interface Science*, 313:1–4.

Crisciani, F. (2004). Nonlinear stability of the Sverdrup flow against mesoscale disturbances. *Nuovo Cimento C*, 27:7–16.

Darcy, H. (1856). *Les fontaines publiques de la ville de Dijon*. Dalmont, Paris.

Das, D. B. and Lewis, M. (2007). Dynamics of fluid circulation in coupled free and heterogeneous porous domains. *Chem. Engng.Sci.*, 62:3549–3573.

Das, D. B., Nassehi, V., and Wakeman, R. J. (2002). A finite volume model for the hydrodynamics of combined free and porous flow in sub-surface regions. *Advances in Environmental Research*, 7:35–58.

Davies, E. B. (1999). Pseudo - spectra, the harmonic oscillator and complex resonances. *Proc. Roy. Soc. London A*, 455:585–599.

de Boer, R. (1999). *Theory of Porous Media: Highlights in Historical Development and Current State*. Springer, Berlin.

de Boer, R., Ehlers, W., and Liu, Z. F. (1993). One - dimensional transient wave propagation in fluid - saturated incompressible porous media. *Arch. Appl. Mech.*, 63:59–72.

De Cicco, S. and Diaco, M. (2002). A theory of thermoelastic materials with voids without energy dissipation. *J. Thermal Stresses*, 25:493–503.

De Ville, A. (1996). On the properties of compressible gas flow in a porous media. *Trans. Porous Media*, 22:287–306.

dell'Isola, F. and Hutter, K. (1998). A qualitative analysis of the dynamics of a sheared and pressurized layer of saturated soil. *Proc. Roy. Soc. London A*, 454:3105–3120.

dell'Isola, F. and Hutter, K. (1999). Variations of porosity in a sheared pressurized layer of saturated soil induced by vertical drainage of water. *Proc. Roy. Soc. London A*, 455:2841–2860.

Di Benedetto, E. (1983). Continuity of weak solutions to a general porous medium equation. *Indiana Univ. Math. J.*, 32:83–118.

Diebold, A. C. (2005). Subsurface imaging with scanning ultrasound holography. *Science*, 310:61–62.

Dincov, D. D., Parrott, K. A., and Pericleous, K. A. (2004). Heat and mass transfer in two - phase porous materials under intensive micorwave heating. *J. Food Engng.*, 65:403–412.

Discacciati, M., Miglio, E., and Quarteroni, A. (2002). Mathematical and numerical models for coupling surface and groundwater flows. *Appl. Numer. Math.*, 43:57–74.

406   References

Doering, C. R. and Constantin, P. (1998). Bounds for heat transport in a porous layer. *J. Fluid Mech.*, 376:263–296.

Doering, C. R., Eckhardt, B., and Schumacher, J. (2006a). Failure of energy stability in Oldroyd-B fluids at arbitrarily low Reynolds numbers. *J. Non-Newtonian Fluid Mech.*, 135:92–96.

Doering, C. R. and Foias, C. (2002). Energy dissipation in body - forced turbulence. *J. Fluid Mech.*, 467:289–306.

Doering, C. R. and Gibbon, J. D. (1995). *Applied Analysis of the Navier-Stokes Equations*. Cambridge University Press, Cambridge.

Doering, C. R., Otto, F., and Reznikoff, M. G. (2006b). Bounds on vertical heat transport for infinite - Prandtl - number Rayleigh - Bénard convection. *J. Fluid Mech.*, 560:229–241.

Doering, C. R., Spiegel, E. A., and Worthing, R. A. (2000). Energy dissipation in a shear layer with suction. *Phys. Fluids*, 12:1955–1968.

Dongarra, J. J., Straughan, B., and Walker, D. W. (1996). Chebyshev tau - QZ algorithm methods for calculating spectra of hydrodynamic stability problems. *Appl. Numer. Math.*, 22:399–435.

Dorfmann, A. and Haughton, D. M. (2006). Stability and bifurcation of compressed elastic cylindrical tubes. *Int. J. Engng. Sci.*, 44:1353–1365.

Drazin, P. G. and Reid, W. H. (1981). *Hydrodynamic stability*. Cambridge University Press.

Du, N., Fan, J., Wu, H., Chen, S., and Liu, Y. (2007). An improved model of heat transfer through penguin feathers and down. *J. Theoretical Biology*, 248:727–735.

Duka, B., Ferrario, C., Passerini, A., and Piva, S. (2007). Non-linear approximations for natural convection in a horizontal annulus. *Int. J. Nonlinear Mech.*, 42:1055–1061.

Dupuit, J. (1863). *Etudes thèoriques et pratiques sur le mouvement des eaux.* Dunod, Paris.

El-Habel, F., Mendoza, C., and Bagtzoglou, A. C. (2002). Solute transport in open channel flows and porous streambeds. *Advances in Water Resources*, 25:455–469.

El-Hakiem, M. A. (1999). Effect of transverse magnetic field on natural convection in boundary layer flow field of micropolar fluids in a porous medium. *Appl. Mech. Engng. Poland*, 4:508–509.

Elbarbary, E. M. E. (2007). Pseudospectral integration matrix and boundary value problems. *Int. J. Computer Math.*, 84:1851–1861.

Ennis-King, J., Preston, I., and Paterson, L. (2005). Onset of convection in anisotropic porous media subject to a rapid change in boundary conditions. *Phys. Fluids*, 17:084107–1–084107–15.

Eremeyev, V. A. (2005). Acceleration waves in incompressible elastic media. *Doklady Physics*, 50:204–206.

Eringen, A. C. (1990). Theory of microstretch elastic solids. *Int. J. Engng. Sci.*, 28:1291–1301.

Eringen, A. C. (1994). A continuum theory of swelling porous elastic soils. *Int. J. Engng. Sci.*, 32:1337–1349.

Eringen, A. C. (2004a). Corrigendum to "A continuum theory of swelling porous elastic soils". *Int. J. Engng. Sci.*, 42:949–949.

Eringen, A. C. (2004b). Electromagnetic theory of microstretch elasticity and bone modeling. *Int. J. Engng. Sci.*, 42:231–242.

Ewing, R. E., Iliev, O. P., and Lazarov, R. D. (1994). Numerical simulation of contaminant transport due to flow in liquid and porous media. Manuscript, Texas A & M University.

Ewing, R. E. and Weekes, S. (1998). Numerical methods for contaminant transport in porous media. *Computational Mathematics*, 202:75–95.

Fabrizio, M. (1994). *Introduzione alla meccanica razionale e ai suoi metodi matematici*. Zanichelli, Bologna, second edition.

Fabrizio, M. and Morro, A. (2003). *Electromagnetism of continuous media*. Oxford University Press, Oxford.

Farrell, B. F. (1988a). Optimal excitation of neutral Rossby waves. *J. Atmos. Sci.*, 45:163–172.

Farrell, B. F. (1988b). Optimal excitation of perturbations in viscous shear flow. *Phys. Fluids*, 31:2093–2102.

Farrell, B. F. (1989). Optimal excitation of baroclinic waves. *J. Atmos. Sci.*, 46:1193–1206.

Fellah, Z. E. A. and Depollier, C. (2000). Transient acoustic wave propagation in rigid porous media: a time-domain approach. *J. Acoust. Soc. Am.*, 107: 683–688.

Fellah, Z. E. A., Depollier, C., Berger, S., Lauriks, W., Trompette, P., and Chapelon, J. Y. (2003). Determination of transport parameters in airsaturated porous materials via reflected ultrasonic waves. *J. Acoust. Soc. Am.*, 114:2561–2569.

Feuillade, C. (2007). Personal communication.

Fichera, G. (1992). Is the Fourier theory of heat propagation paradoxical? *Rend. Circolo Matem. Palermo*, 41:5–28.

Fila, M. and Winkler, M. (2008). Single - point blow - up on the boundary where the zero Dirichlet boundary condition is imposed. *J. European Math. Soc.*, 10:105–132.

Firdaouss, M., Guermond, J. L., and Le Quére, P. (1997). Nonlinear corrections to Darcy's law at low Reynolds numbers. *J. Fluid Mech.*, 343:331–350.

Flavin, J. N. (2006). The evolution to a steady state for a porous medium model. *J. Math. Anal. Appl.*, 322:393–402.

Flavin, J. N. and Rionero, S. (1995). *Qualitative Estimates for Partial Differential Equations*. CRC Press, Boca Raton.

Flavin, J. N. and Rionero, S. (1998). Asymptotic and other properties for a nonlinear diffusion model. *J. Math. Anal. Appl.*, 228:119–140.

Flavin, J. N. and Rionero, S. (2003). Stability properties for nonlinear diffusion in porous and other media. *J. Math. Anal. Appl.*, 281:221–232.

Forchheimer, P. (1901). Wasserbewegung durch boden. *Z. Vereines Deutscher Ingnieure*, 50:1781–1788.

Ford, R. M. and Harvey, R. W. (2007). Role of chemotaxis in the transport of bacteria through saturated porous media. *Advances in Water Resources*, 30:1608–1617.

Franchi, F. and Straughan, B. (1993a). Continuous dependence on the body force for solutions to the Navier-Stokes equations and on the heat supply in a model for double diffusive porous convection. *J. Math. Anal. Appl.*, 172:117–129.

Franchi, F. and Straughan, B. (1993b). Stability and nonexistence results in the generalized theory of a fluid of second grade. *J. Math. Anal. Appl.*, 180:122–137.

Franchi, F. and Straughan, B. (1994). Spatial decay estimates and continuous depedence on modelling for an equation from dynamo theory. *Proc. Roy. Soc. London A*, 445:437–451.

Franchi, F. and Straughan, B. (1996). Structural stability for the Brinkman equations of porous media. *Math. Meth. Appl. Sci.*, 19:1335–1347.

Franchi, F. and Straughan, B. (2003). Continuous dependence and decay for the Forchheimer equations. *Proc. Roy. Soc. London A*, 459:3195–3202.

Fu, Y. B. and Scott, N. H. (1988). Acceleration wave propagation in an inhomogeneous heat conducting elastic rod of slowly varying cross section. *J. Thermal Stresses*, 15:253–264.

Fu, Y. B. and Scott, N. H. (1991). The transistion from acceleration wave to shock wave. *Int. J. Engng. Sci.*, 29:617–624.

Galakov, E. (2007). Some nonexistence results for quasilinear pdes. *Communcations on Pure and Applied Analysis*, 6:141–161.

Galdi, G. P. and Straughan, B. (1985). Exchange of stabilities, symmetry and nonlinear stability. *Arch. Rational Mech. Anal.*, 89:211–228.

Gales, C. (2003). On the asymptotic partition of energy in the theory of swelling porous elastic soils. *Arch. Mech.*, 55:91–107.

Garai, M. and Pompoli, F. (2005). A simple empirical model of polyester fibre materials for acoustical applications. *Applied Acoustics*, 66:1383–1398.

Gardner, D. R., Trogdon, S. A., and Douglas, R. W. (1989). A modified tau spectral method that eliminates spurious eigenvalues. *J. Computational Physics*, 80:137–167.

Gheorghiu, C. I. and Pop, I. S. (1996). A modified Chebyshev - tau method for a hydrodynamic stability problem. In *Approximation and Optimization*, volume 2, pages 119–126.

Ghorai, S. and Hill, N. A. (2005). Penetrative phototactic bioconvection. *Phys. Fluids*, 17:Art. No. 074101.

Giacobbo, F. and Patelli, E. (2007). Monte Carlo simulation of nonlinear reactive contaminant transport in unsaturated porous media. *Annals of Nuclear Energy*, 34:51–63.

Gigler, J. K., van Loon, W. K. P., van den Berg, J. V., Sonneveld, C., and Meerdink, G. (2000a). Natural wind drying of willow stems. *Biomass and Bioenergy*, 19:153–163.

Gigler, J. K., van Loon, W. K. P., Vissers, M. M., and Bot, G. P. A. (2000b). Forced convective drying of willow chips. *Biomass and Bioenergy*, 19: 259–270.

Gilman, A. and Bear, J. (1996). The influence of free convection on soil salinization in arid regions. *Trans. Porous Media*, 23:275–301.

Giorgi, T. (1997). Derivation of the Forchheimer law via matched asymptotic expansions. *Trans. Porous Media*, 29:191–206.

Givler, R. C. and Altobelli, S. A. (1994). A determination of effective viscosity for the Brinkman-Forchheimer flow model. *J. Fluid Mech.*, 258:355–370.

Goharzadeh, A., Khalili, A., and Jorgensen, B. B. (2005). Transistion layer thickness at a fluid-porous interface. *Phys. Fluids*, 17:057102-1–057102-10.

Goodman, M. A. and Cowin, S. C. (1972). A continuum theory for grnaular materials. *Arch. Rational Mech. Anal.*, 44:249–266.

Govender, S. (2006a). Effect of Darcy-Prandtl number on the stability of solutal convection in solidifying binary alloy systems. *J. Porous Media*, 9:523–539.

Govender, S. (2006b). Stability of gravity driven convection in a cylindrical porous layer subjected to vibration. *Trans. Porous Media*, 63:489–502.

Graffi, D. (1960). Ancora sul teorema di unicità per le equazioni del moto dei fluidi. *Atti della Accademia delle Scienze dell'Instituto di Bologna*, 8:7–14. (Series 11).

Graffi, D. (1999). Ancora sul teorema di unicità per le equazioni del moto dei fluidi. In Fabrizio, M., Grioli, G., and Renno, P., editors, *Opere Scelte*, pages 273–280. C.N.R. Gruppo Nazionale per la Fisica Matematica, Bologna.

Green, A. E. (1972). A note on linear thermoelasticity. *Mathematika*, 19:69–75.

Green, A. E. and Laws, N. (1972). On the entropy production inequality. *Arch. Rational Mech. Anal.*, 45:47–53.

Green, A. E. and Naghdi, P. M. (1968). A note on simple dipolar stresses. *J. Mécanique*, 7:465–474.

Green, A. E. and Naghdi, P. M. (1970). A note on dipolar inertia. *Q. Appl. Math.*, 28:458–460.

Green, A. E. and Naghdi, P. M. (1991). A re-examination of the basic postulates of thermomechanics. *Proc. Roy. Soc. London A*, 432:171–194.

Green, A. E. and Naghdi, P. M. (1992). On undamped heat waves in an elastic solid. *J. Thermal Stresses*, 15:253–264.

Green, A. E. and Naghdi, P. M. (1993). Thermoelasticity without energy-dissipation. *J. Elasticity*, 31:189–208.

Green, A. E. and Naghdi, P. M. (1995). A unified procedure for construction of theories of deformable media. I. Classical continuum physics. *Proc. Roy. Soc. London A*, 448:335–356.

Green, A. E. and Naghdi, P. M. (1996). An extended theory for incompressible viscous fluid flow. *J. Non-Newtonian Fluid Mech.*, 66:233–255.

Green, A. E., Naghdi, P. M., and Rivlin, R. S. (1965). Directors and multipolar displacements in continuum mechanics. *Int. J. Engng. Sci.*, 2:611–620.

Green, A. E. and Rivlin, R. S. (1967). The relation between director and multipolar theories in continuum mechanics. *ZAMP*, 18:208–218.

Greenberg, L. and Marletta, M. (2000). Numerical methods for higher order Sturm-Liouville problems. *J. Computational and Applied Math.*, 125: 367–383.

Greenberg, L. and Marletta, M. (2001). Numerical solution of non self-adjoint Sturm-Liouville problems and related systems. *SIAM J. Numer. Anal.*, 38:1800–1845.

Greenberg, L. and Marletta, M. (2004). The Ekman flow and related problems: spectral theory and numerical analysis. *Math. Proc. Camb. Phil. Soc.*, 136:719–764.

Gultop, T. (2006). On the propagation of acceleration waves in incompressible hyperelastic solids. *J. Sound Vibration*, 462:409–418.

Guo, J. and Kaloni, P. N. (1995a). Doubly diffusive convection in a porous medium, nonlinear stability, and the Brinkman effect. *Stud. Appl. Math.*, 94:341–358.

Guo, J. and Kaloni, P. N. (1995b). Nonlinear stability of convection induced by inclined thermal and solutal gradients. *ZAMP*, 46:645–654.

Guo, J. and Kaloni, P. N. (1995c). Nonlinear stability problem of a rotating doubly diffusive porous layer. *J. Math. Anal. Appl.*, 190:373–390.

Guo, J., Qin, Y., and Kaloni, P. N. (1994). Nonlinear stability problem of a rotating doubly diffusive fluid layer. *Int. J. Engng. Sci.*, 32:1207–1219.

Gurtin, M. E. and Pipkin, A. C. (1968). A general theory of heat conduction with finite wavespeeds. *Arch. Rational Mech. Anal.*, 31:113–126.

Gustavsson, L. H. (1981). Resonant growth of three-dimensional disturbances in plane Poiseuille flow. *J. Fluid Mech.*, 112:253–264.

Gustavsson, L. H. (1986). Excitation of direct resonances in plane Poiseuille flow. *Stud. Appl. Math.*, 75:227–248.

Gustavsson, L. H. (1991). Energy growth of three-dimensional disturbances in plane Poiseuille flow. *J. Fluid Mech.*, 224:241–260.

Gustavsson, L. H. and Hultgren, L. S. (1980). A resonance mechanism in plane Couette flow. *J. Fluid Mech.*, 98:149–159.

Han, P., Tang, D., and Zhou, L. (2006). Numerical analysis of two-dimensional lagging thermal behaviour under short-pulse-laser heating on surface. *Int. J. Engng. Sci.*, 44:1510–1519.

Harrington, G. A., Hendry, M. J., and Robinson, N. I. (2007). Impact of permeable conduits on solute transport in aquitards: mathematical models and their application. *Water Resources Research*, 43:W05441.

Hassanizadeh, S. M. and Gray, W. G. (1990). Mechanics and thermodynamics of multiphase flow in porous media including interphase boundaries. *Adv. Water Resources*, 13:169–186.

Hassanizadeh, S. M. and Gray, W. G. (1993). Thermodynamic basis of capillary pressure in porous media. *Water Resources Research*, 29:3389–3405.

Hayat, T., Abbas, Z., and Ashgar, S. (2005). Heat transfer analysis on rotating flow of a second grade fluid past a porous plate with variable suction. *Mathematical Problems in Engineering*, 5:555–582.

Hayat, T. and Khan, M. (2005). Homotopy solutions for a generalized second grade fluid past a porous plate. *Nonlinear Dynamics*, 42:395–405.

Hetnarsky, R. B. and Ignaczak, J. (1999). Generalized thermoelasticity. *J. Thermal Stresses*, 22:451–461.

Hifdi, A., Touhami, M. O., and Naciri, J. K. (2004a). Channel entrance flow and its linear stability. *J. Statistical Mechanics - Theory and Experiment*, page P06003.

Hifdi, A., Touhami, M. O., and Naciri, J. K. (2004b). Linear stability of nearly parallel symmetric flows in a channel. *Comptes Rendues Mécanique*, 332:859–866.

Hill, A. A. (2003). Convection due to the selective absorption of radiation in a porous medium. *Continuum Mech. Thermodyn.*, 15:275–285.

Hill, A. A. (2004a). Conditional and unconditional nonlinear stability for convection induced by absorption of radiation in a porous medium. *Continuum Mech. Thermodyn.*, 16:305–318.

Hill, A. A. (2004b). Convection induced by the selective absorption of radiation for the Brinkman model. *Continuum Mech. Thermodyn.*, 16:43–52.

Hill, A. A. (2004c). Penetrative convection induced by the absorption of radiation with a nonlinear internal heat source. *Dynam. Atmospheres Oceans*, 38: 57–67.

Hill, A. A. (2005a). *Convection problems in porous media*. PhD thesis, University of Durham.

Hill, A. A. (2005b). Double-diffusive convection in a porous medium with a concentration based internal heat source. *Proc. Roy. Soc. London A*, 461:561–574.

Hill, A. A., Rionero, S., and Straughan, B. (2007). Global stability for penetrative convection with throughflow in a porous material. *IMA J. Appl. Math.*, 72:635–643.

Hill, A. A. and Straughan, B. (2005). A Legendre spectral element method for eigenvalues in hydrodynamic stability. *J. Computational Appl. Math.*, 193:363–381.

Hill, A. A. and Straughan, B. (2006). Linear and nonlinear stability thresholds for thermal convection in a box. *Math. Meth. Appl. Sci.*, 29:2123–2132.

Hill, A. A. and Straughan, B. (2008). Poiseuille flow of a fluid overlying a porous layer. *J. Fluid Mech.*, 603:137–149.

Hirata, S. C., Goyeau, B., Gobin, D., Carr, M., and Cotta, R. M. (2007). Linear stability of natural convection in superposed fluid and porous layers:Influence of the interfacial modelling. *Int. J. Heat Mass Transfer*, 50:1356–1367.

Hirata, S. C., Goyeau, B., Gobin, D., and Cotta, R. M. (2006). Stability of natural convection in superposed fluid and porous layers using integral transforms. *Numer. Heat Transfer, Part B, Fundamentals*, 50:409–424.

Hirota, C. and Ozawa, K. (2006). Numerical method of estimating the blow-up time and rate of the solution of ordinary differential equations - An application to the blow-up problems of partial differential equations. *J. Comp. Appl. Math.*, 193:614–637.

Hirsch, M. W. and Smale, S. (1974). *Differential equations, dynamical systems, and linear algebra*. Academic Press, New York.

Hoffmann, K. H., Marx, D., and Botkin, N. D. (2007). Drag on spheres in micropolar fluids with non-zero boundary conditions for microrotations. *J. Fluid Mech.*, 590:319–330.

Holzapfel, G. A. (2000). *Nonlinear solid mechanics*. Wiley, Chichester.

Holzbecher, E. (2004). Free convection in open-top enclosures filled with a porous medium heated from below. *Numer. Heat Trans. Part A-Applications*, 46:241–254.

Hooper, A. P. and Grimshaw, R. (1996). Two-dimensional disturbance growth of linearly stable viscous shear flows. *Phys. Fluids*, 8:1424–1432.

Hoppe, R. H. W., Porta, P., and Vassilevski, Y. (2007). Computational issues related to iterative coupling of subsurface and channel flows. *Calcolo*, 44: 1–20.

Horgan, C. O. (1978). Plane entry flows and energy estimates for the Navier - Stokes equations. *Arch. Rational Mech. Anal.*, 68:359–381.

Horgan, C. O. and Payne, L. E. (1983). On inequalities of Korn, Friedrichs and Babuska - Aziz. *Arch. Rational Mech. Anal.*, 82:165–179.

Horgan, C. O. and Payne, L. E. (1992). A Saint-Venant principle for a theory of nonlinear plane elasticity. *Quart. Appl. Math.*, 50:641–675.

Horgan, C. O., Payne, L. E., and Wheeler, L. T. (1984). Spatial decay estimates in transient heat conduction. *Quart. Appl. Math.*, 42:119–127.

412   References

Horgan, C. O. and Quintanilla, R. (2005). Spatial behaviour of solutions of the dual - phase - lag heat equation. *Math. Meth. Appl. Sci.*, 28:43–57.

Horgan, C. O. and Wheeler, L. T. (1978). Spatial decay estimates for the Navier - Stokes equations with application to the problem of entry flow. *SIAM J. Appl. Math.*, 35:97–116.

Hurle, D. T. J. and Jakeman, E. (1971). Soret driven thermo - solutal convection. *J. Fluid Mech.*, 47:667–687.

Iesan, D. (2004). *Thermoelastic models of continua*. Kluwer, Dordrecht.

Iesan, D. (2005). Second-order effects in the torsion of elastic materials with voids. *ZAMM*, 85:351–365.

Iesan, D. (2006). Nonlinear plane strain of elastic materials with voids. *Math. Mech. Solids*, 11:361–384.

Iesan, D. and Quintanilla, R. (1995). Decay estimates and energy bounds for porous elastic cylinders. *ZAMP*, 46:268–281.

Iesan, D. and Scalia, A. (2006). Propagation of singular surfaces in thermo-microstretch continua with memory. *Int. J. Engng. Sci.*, 44:845–858.

Itagaki, S., McGeer, P. L., Akiyama, H., Zhu, S., and Selkoe, D. (1989). Relationship of microglia and astrocytes to amyloid deposits of alzheimer's disease. *J. Neuroimmunol.*, 24:173–182.

Ivansson, S. (2003). Compound matrix Riccati method for solving boundary - value problems. *ZAMM*, 83:535–548.

Jaballah, S., Sammouda, H., and Belghith, A. (2007). Effect of surface radiation on the natural - convection stability in a two - dimensional enclosure with diffusely emitting boundary walls. *Numerical Heat Transfer Part A - Applications*, 51:495–516.

Jäger, W. and Mikelic, A. (1998). On the interface boundary condition by Beavers, Joseph and Saffman. Interdisziplinäres Zentrum für Wissenschaftliches Rechnen der Universität Heidelberg, Preprint 98-12.

Jäger, W., Mikelic, A., and Neuß (1999). Asymptotic analysis of the laminar viscous flow over a porous bed. Interdisziplinäres Zentrum für Wissenschaftliches Rechnen der Universität Heidelberg, Preprint 99-33.

Jaisaardsuetrong, J. and Straughan, B. (2007). Thermal waves in a rigid heat conductor. *Physics Letters A*, 366:433–436.

Jiang, Q. and Rajapakse, R. K. N. D. (1994). On coupled heat - moisture transfer in deformable porous media. *Q. Jl. Mech. Appl. Math.*, 47:53–68.

Johnson, D. and Narayanan, R. (1996). Experimental observation of dynamic mode switching in interfacial - tension - driven convection near a codimension - two point. *Phys. Rev E*, 54:R3102–R3104.

Johnson, D. L., Plona, T. J., and Kajima, H. (1994). Probing porous media with first and second sound. II. Acoustic properties of water saturated porous media. *J. Appl. Phys.*, 76:115–125.

Jones, I. P. (1973). Low Reynolds number flow past a porous spherical shell. *Proc. Camb. Phil. Soc.*, 73:231–238.

Jordan, P. M. (2005a). Growth and decay of acoustic acceleration waves in Darcy-type porous media. *Proc. Roy. Soc. London A*, 461:2749–2766.

Jordan, P. M. (2005b). Growth and decay of shock and acceleration waves in a traffic flow model with relaxation. *Physica D*, 207:220–229.

Jordan, P. M. (2006). Finite amplitude acoustic travelling waves in a fluid that saturates a porous medium: Acceleration wave formation. *Phys. Letters A*, 355:216–221.

Jordan, P. M. (2007). Growth, decay and bifurcation of shock amplitudes under the type-II flux law. *Proc. Roy. Soc. London A*, 463:2783–2798.

Jordan, P. M. and Christov, C. I. (2005). A simple finite difference scheme for modelling the finite-time blow-up of acoustic acceleration waves. *J. Sound Vibration*, 281:1207–1216.

Jordan, P. M. and Puri, A. (2005). Growth/decay of transverse acceleration waves in nonlinear elastic media. *Phys. Letters A*, 341:427–434.

Jordan, P. M. and Puri, P. (1999). Exact solutions for the flow of a dipolar fluid on a suddenly accelerated flat plate. *Acta Mechanica*, 137:183–194.

Jordan, P. M. and Puri, P. (2002). Exact solutions for the unsteady plane Couette flow of a dipolar fluid. *Proc. Roy. Soc. London A*, 458:1245–1272.

Jordan, P. M. and Puri, P. (2003). Stokes' first problem for a Rivlin - Ericksen fluid of second grade in a porous half space. *Int. J. Nonlinear Mech.*, 38:1019–1025.

Jordan, P. M. and Straughan, B. (2006). Acoustic acceleration waves in homentropic Green and Naghdi gases. *Proc. Roy. Soc. London A*, 462:3601–3611.

Joseph, D. D. (1965). On the stability of the Boussinesq equations. *Arch. Rational Mech. Anal.*, 20:59–71.

Joseph, D. D. (1966). Nonlinear stability of the Boussinesq equations by the method of energy. *Arch. Rational Mech. Anal.*, 22:163–184.

Joseph, D. D. (1970). Global stability of the conduction-diffusion solution. *Arch. Rational Mech. Anal.*, 36:285–292.

Joseph, D. D. (1976a). *Stability of fluid motions*, volume 2. Springer.

Joseph, D. D. (1976b). *Stability of fluid motions*, volume 1. Springer.

Jou, D. and Criado-Sancho, M. (1998). Thermodynamic stability and temperature overshooting in dual-phase-lag heat transfer. *Phys. Letters A*, 248:172–178.

Jovanovic, B. S. and Vulkov, L. G. (2005). Energy stability for a class of two-dimensional interface linear parabolic problems. *J. Math. Anal. Appl.*, 311:120–138.

Jugjai, S. and Phothiya, C. (2007). Liquid fuels-fired porous combustor-heater. *Fuel*, 86:1062–1068.

Jung, W. Y. (2008). A combined honeycomb and solid viscoelastic material for structural damping applications. http://mceer.buffalo.edu/publications/resaccom/02-SP09/pdf_screen/08_Jung.pdf.

Kaiser, R. Tilgner, A. and von Wahl, W. (2005). A generalized energy functional for plane Couette flow. *SIAM J. Math. Anal.*, 37:438–454.

Kaiser, R. and Mulone, G. (2005). A note on nonlinear stability of plane parallel shear flows. *J. Math. Anal. Appl.*, 302:543–556.

Kaloni, P. N. and Qiao, Z. C. (2001). Nonlinear convection in a porous medium with inclined temperature gradient and variable gravity effects. *Int. J. Heat Mass Transfer*, 44:1585–1591.

Kameyama, N. and Sugiyama, M. (1996). Analysis of acceleration waves in crystalline solids based on a continuum model incorporating microscopic thermal vibration. *Continuum Mech. Thermodyn.*, 8:351–359.

Kaminski, W. (1990). Hyperbolic heat conduction equation for materials with a non-homogeneous inner structure. *ASME J. Heat Transfer*, 112:555–560.

Kapoor, V. (1996). Criterion for instability of steady - state unsaturated flows. *Trans. Porous Media*, 25:313–334.

Karimi-Fard, M., Charrier-Mojtabi, M. C., and Mojtabi, A. (1999). Onset of stationary and oscillatory convection in a tilted porous cavity saturated with a binary fluid: linear stability analysis. *Phys. Fluids*, 11:1346–1358.

Kato, Y., Hashiba, M., and Fujimura, K. (2003). The onset of penetrative double-diffusive convection. *Fluid Dynamics Research*, 32:295–316.

Kekutia, S. E. and Chkhaidze, N. D. (2005). Equations for superfluid ^{3}He - ^{4}He solution filled porous media. *J. Statistical Mechanics: Theory and Experiment*, page P12008.

Khadra, A., Liu, X. Z., and Shen, X. M. (2005). Impulsive control and synchronization of spatiotemporal chaos. *Chaos Solitons and Fractals*, 26:615–636.

Khadrawi, A. and Al-Nimr, M. (2005). The effect of the local inertia term on the transient free convection from a vertical plate inserted in a semi-infinite domain partly filled with porous material. *Trans. Porous Media*, 59:139–153.

Khaled, A. R. A. and Vafai, K. (2003). The role of porous media in modelling flow and heat transfer in biological tissues. *Int. J. Heat Mass Transfer*, 46:4989–5003.

Kielhöfer, H. (1975). Existenz und regularität von lösungen semilinearen parabolischer anfangs - randwertprobleme. *Math. Zeit.*, 142:131–160.

Kim, M. C. and Choi, C. K. (2004). The onset of Taylor-Görtler vortices in impulsively decelerating swirl flow. *Korean J. Chem. Engng.*, 21:767–772.

Kim, M. C. and Choi, C. K. (2006). The onset of Taylor-Görtler vortices during impulsive spin - down to rest. *Chem. Engng. Sci.*, 61:6478–6485.

Kim, M. C., Kim, S., and Choi, C. K. (2006). The convective stability of circular Couette flow induced by a linearly accelerated inner cylinder. *European J. Mech. B - Fluids*, 25:74–82.

Kim, M. C., Lee, S. B., Kim, S., and Chung, B. J. (2003). Thermal instability of viscoelastic fluids in porous media. *Int. J. Heat Mass Transfer*, 46:5065–5072.

Kim, M. C., Park, J. H., and Choi, C. K. (2005). Onset of buoyancy - driven convection in the horizontal fluid layer subjected to ramp heating from below. *Chem. Engng. Sci.*, 60:5363–5371.

Kirane, M., Laskri, Y., and Tatar, N. E. (2005). Critical exponents of Fujita type for certain evolution equations and systems with spatio-temporal fractional derivatives. *J. Math. Anal. Appl.*, 312:488–501.

Kirchner, N. P. (2000). Computational aspects of the spectral Galerkin FEM for the Orr-Sommerfeld equation. *Int. J. Numer. Meth. Fluids*, 32:119–137.

Kladias, N. and Prasad, V. (1991). Experimental verification of Darcy - Brinkman - Forchheimer flow model for natural convection in porous media. *J. Thermophys.*, 5:560–576.

Knutti, R. and Stocker, T. F. (2000). Influence of thermohaline convection on projected sea level rise. *J. Climatology*, 13:1997–2001.

Kolymbas, D. (1998). Behaviour of liquified sand. *Phil. Trans. Roy. Soc. London A*, 356:2609–2622.

Krishnamurti, R. (1975). On cellular cloud patterns. Part I: Mathematical model. *J. Atmos. Sci.*, 32:1353–1363.

Krishnamurti, R. (1997). Convection induced by selective absorbtion of radiation: A laboratory model of conditional instability. *Dynamics of Atmospheres and Oceans*, 27:367–382.

Kwok, L. P. and Chen, C. F. (1987). Stability of thermal convection in a vertical porous layer. *J. Heat Transfer*, 109:889–893.

Lakes, R. S. (2008). Negative Poisson's ratio materials. http://silver.neep.wisc.edu/ lakes/Poisson.html.

Larson, V. E. (2000). Stability properties of and scaling laws for a dry radiative-convective atmosphere. *Q. J. Royal Meteorological Soc.*, 126:145–171.

Larson, V. E. (2001). The effects of thermal radiation on dry convective instability. *Dynamics of Atmospheres and Oceans*, 34:45–71.

Layton, W. J., Schieweck, F., and Yotov, I. (2003). Coupling fluid flow with porous media flow. *SIAM J. Numer. Anal.*, 40:2195–2218.

Le Bars, M. and Worster, M. G. (2006). Interfacial conditions between a pure fluid and a porous medium: implications for binary alloy solidification. *J. Fluid Mech.*, 550:149–173.

Lehoucq, R. B., Sorenson, D. C., and Yang, C. (1998). *ARPACK user's guide: solution of large scale eigenvalue problems with implicitly restarted Arnoldi methods*. SIAM, Philadelphia.

Lewis, R. W. and Schrefler, B. A. (1998). *The finite element method in the static and dynamic deformation and consolidation of porous media*. Wiley, Chichester, second edition.

Li, F., Ozen, O., Aubry, N., Papageorgiou, D. T., and Petropoulos, P. G. (2007). Linear stability of a two-fluid interface for electrohydrodynamic mixing in a channel. *J. Fluid Mech.*, 583:347–377.

Lide, D. R., editor (1991). *Handbook of Chemistry and Physics*. CRC Press, Boca Raton.

Lin, C. and Payne, L. E. (1993). On the spatial decay of ill - posed parabolic problems. *Math. Models Meth. Appl. Sci.*, 3:563–575.

Lin, C. and Payne, L. E. (1994). The influence of domain and diffusivity perturbations on the decay of end effects in heat conduction. *SIAM J. Math. Anal.*, 25:1241–1258.

Lin, C. and Payne, L. E. (2004). Continuous dependence of heat flux on spatial geometry for the generalized Maxwell-Cattaneo system. *ZAMP*, 55:575–591.

Lin, C. and Payne, L. E. (2007a). Structural stability for a Brinkman fluid. *Math. Meth. Appl. Sci.*, 30:567–578.

Lin, C. and Payne, L. E. (2007b). Structural stability for the Brinkman equations of flow in double diffusive convection. *J. Math. Anal. Appl.*, 325:1479–1490.

Lindsay, K. A. and Straughan, B. (1978). Acceleration waves and second sound in a perfect fluid. *Arch. Rational Mech. Anal.*, 68:53–87.

Lindsay, K. A. and Straughan, B. (1979). Propagation of mechanical and temperature acceleration waves in thermoelastic materials. *ZAMP*, 30:477–490.

Lombardo, S. and Mulone, G. (2002a). Double-diffusive convection in porous media: The Darcy and Brinkman models. In Monaco, R., Pandolfi-Bianchi, M., and Rionero, S., editors, *Proc. Wascom 2001*.

Lombardo, S. and Mulone, G. (2002b). Necessary and sufficient conditions for global nonlinear stability for rotating double-diffusive convection in a porous medium. *Continuum Mech. Thermodyn.*, 14:527–540.

Lombardo, S. and Mulone, G. (2003). Nonlinear stability and convection for laminar flows in a porous medium with the Brinkman law. *Math. Meth. Appl. Sci.*, 26:453–462.

Lombardo, S. and Mulone, G. (2005). Necessary and sufficient conditions via the eigenvalues - eigenvectors method: an application to the magnetic Bénard problem. *Nonlinear Analysis*, 63:e2091–e2101.

Lombardo, S., Mulone, G., and Straughan, B. (2001). Stability in the Bénard problem for a double-diffusive mixture in a porous medium. *Math. Meth. Appl. Sci.*, 24:1229–1246.

Lopez de Haro, M., del Rio, J. A., and Whitaker, S. (1996). Flow of Maxwell fluids in porous media. *Trans. Porous Media*, 25:167–192.

Lu, J. W. and Chen, F. (1997). Assessment of mathematical models for the flow in directional solidification. *J. Crystal Growth*, 171:601–613.

Lu, W. S. and Shao, H. Y. (2003). Generalized nonlinear subcritical symmetric instability. *Adv. Atmos. Sci.*, 20:623–630.

Luca, M., Chavez-Ross, A., Edelstein-Keshet, L., and Mogilner, A. (2003). Chemotactic signalling, microglia, and Alzheimer's disease senile plaques: is there a connection. *Bull. Math. Biol.*, 65:693–730.

Luo, H. X. and Pozrikidis, C. (2006). Effect of inertia on film flow over oblique and three-dimensional corrugations. *Phys. Fluids*, 18:078107–078107.

Ly, H. V. and Titi, E. S. (1999). Global Gevrey regularity for Bénard convection in a porous medium with zero Darcy-Prandtl number. *J. Nonlinear Sci.*, 9:333–362.

Mack, L. M. (1976). A numerical study of the temporal eigenvalue spectrum of the Blasius boundary layer. *J. Fluid Mech.*, 73:497–520.

Mahidjiba, A., Robillard, L., and Vasseur, P. (2003). Linear stability of cold water saturating an anisotropic porous medium - effect of confinement. *Int. J. Heat Mass Transfer*, 46:323–332.

Malashetty, M. S., Gaikwad, S. N., and Swamy, M. (2006). An analytical study of linear and non-linear double diffusive convection with Soret effect in couple stress liquids. *Int. J. Thermal Sci.*, 45:897–907.

Malashetty, M. S., Shivakumara, I. S., and Kulkarni, S. (2005). The onset of Lapwood-Brinkman convection using a thermal non-equilibrium model. *Int. J. Heat Mass Transfer*, 48:1155–1163.

Malashetty, M. S. and Swamy, M. (2007a). Combined effect of thermal modulation and rotation on the onset of stationary convection in a porous layer. *Trans. Porous Media*, 69:313–330.

Malashetty, M. S. and Swamy, M. (2007b). The effect of rotation on the onset of convection in a horizontal anisotropic porous layer. *Int. J. Thermal Sci.*, 46:1023–1032.

Malashetty, M. S., Swamy, M., and Kulkarni, S. (2007). Thermal convection in a rotating porous layer using a thermal nonequilibrium model. *Phys. Fluids*, 19:052102–052102.

Malik, S. V. and Hooper, A. P. (2007). Three-dimensional disturbances in channel flows. *Phys. Fluids*, 19:052102–052102.

Man, C. S. (1992). Nonsteady channel flow of ice as a modified second - order fluid with power - law viscosity. *Arch. Rational Mech. Anal.*, 119:35–57.

Man, C. S. and Sun, Q. K. (1987). On the significance of normal stress effects in the flow of glaciers. *J. Glaciology*, 33:268–273.

Manole, D. M., Lage, J. L., and Nield, D. A. (1994). Convection induced by inclined thermal and solutal gradients with horizontal mas flow, in a shallow horizontal layer of a porous medium. *Int. J. Heat Mass Transfer*, 37:2047–2057.

Mariano, P. M. and Sabatini, L. (2000). Homothermal acceleration waves in multifield theories of continua. *Int. J. Non-Linear Mech.*, 35:963–977.

Martin, S. and Kauffman, P. (1974). The evolution of under-ice melt ponds, or double diffusion at the freezing point. *J. Fluid Mech.*, 64:507–527.

Martins, R. C. and Silva, C. L. M. (2004). Computational design of accelerated life testing applied to frozen green beans. *J. Food Engineering*, 64:455–464.

Massoudi, M. (2005). An anisotropic constitutive relation for the stress tensor of a rod-like (fibrous-type) granular material. *Mathematical Problems in Engineering*, 6:679–702.

Massoudi, M. (2006a). On the heat flux vector for flowing granular materials. Part I: Effective thermal conductivity and background. *Math. Meth. Appl. Sci.*, 29:1585–1598.

Massoudi, M. (2006b). On the heat flux vector for flowing granular materials. Part II: Derivation and special cases. *Math. Meth. Appl. Sci.*, 29:1599–1613.

Massoudi, M. and Phuoc, T. X. (2004). Flow of a generalized second grade non-Newtonian fluid with variable viscosity. *Cont. Mech. Thermodyn.*, 16: 529–538.

Massoudi, M. and Phuoc, T. X. (2007). Conduction and dissipation in the shearing flow of granular materials modelled as non-Newtonian fluids. *Powder Technology*, 175:146–162.

Massoudi, M., Vaidya, A., and Wulandana, R. (2008). Natural convection flow of a generalized second grade fluid between two vertical walls. *Nonlinear Analysis: Real World Applications*, 9:80–93.

Maysenhölder, W., Berg, A., and Leistner, P. (2004). Acoustic properties of aluminium foams - measurements and modelling. CFA/DAGA'04, Strasbourg, 22–25/03/2004. See: www.ibp.fhg.de/ba/forschung/aluschaum/aluschaum.pdf.

McKay, G. (1998). Onset of buoyancy-driven convection in superposed reacting fluid and porous layers. *J. Engng. Math.*, 33:31–46.

McKay, G. (2001). The Beavers and Joseph condition for velocity slip at the surface of a porous medium. In Straughan, B., Greve, R., Ehrentraut, H., and Wang, Y., editors, *Continuum Mechanics and Applications in Geophysics and the Environment*. Springer.

McKay, G. and Straughan, B. (1993). Patterned ground formation under water. *Continuum Mech. Thermodyn.*, 5:145–162.

Mehta, P. G. (2004). A unified well-posed computational approach for the 2D Orr-Sommerfeld problem. *J. Computational Physics*, 199:541–557.

Meyer, R. J. (2006). Ultrasonic drying of saturated porous solids via second sound. http://www.freepatentsonline.com/6376145.html.

Mielke, A. (1997). Mathematical analysis of sideband instabilities with application to Rayleigh - Bénard convection. *J. Nonlinear Science*, 7:57–99.

Miglio, E., Quarteroni, A., and Saleri, F. (2003). Coupling of free surface and groundwater flows. *Computers and Fluids*, 32:73–83.

Mitra, K., Kumar, S., Vedavarz, A., and Moallemi, M. K. (1995). Experimental evidence of hyperbolic heat conduction in processed meat. *Trans. ASME, J. Heat Transfer*, 117:568–573.

Moler, C. B. and Stewart, G. W. (1973). An algorithm for generalized matrix eigenproblems. *SIAM. J. Numerical Anal.*, 10:241–256.

Morro, A. and Ruggeri, T. (1988). Non-equilibrium properties of solids obtained from second - sound measurements. *J. Phys. C: Solid State Phys.*, 21: 1743–1752.

Mouraille, O., Mulder, W. A., and Luding, S. (2006). Sound wave acceleration in granular materials. *J. Statistical Mech., Theory and Experiment*, pages 1–15. P07023.

Moussatov, A., Ayrault, C., and Castagnéde, B. (2001). Porous material characterization - ultrasonic method for estimation of tortuosity and characteristic length using a barometric chamber. *Ultrasonics*, 39:195–202.

Mu, M. and Xu, J. C. (2007). A two-grid method of a mixed Stokes - Darcy model for coupling fluid flow with porous medium flow. *SIAM J. Numer. Anal.*, 45:1801–1813.

Mulone, G. (2004). Stabilizing effects in dynamical systems: linear and nonlinear stability conditions. *Far East J. Applied Math.*, 15:117–135.

Mulone, G. and Straughan, B. (2006). An operative method to obtain necessary and sufficient stability conditions for double diffusive convection in porous media. *ZAMM*, 86:507–520.

Mulone, G., Straughan, B., and Wang, W. (2007). Stability of epidemic models with evolution. *Studies in Applied Math.*, 118:117–132.

Murdoch, A. J. and Soliman, A. (1999). On the slip-boundary condition for liquid flow over planar porous boundaries. *Proc. Roy. Soc. London A*, 455:1315–1340.

Nakanishi, K. and Tanaka, N. (2007). Sol-gel with phase separation. hierarchically porous materials optimized for high-performance liquid chromatography separations. *Accounts of Chemical Research*, 40:863–873.

Naulin, V., Nielsen, A. H., and Rasmussen, J. J. (2005). Turbulence spreading, anomalous transport, and pinch effect. *Physics of Plasmas*, 12:122306–122306.

Nerli, A., Camarri, S., and Salvetti, M. V. (2007). A conditional stability criterion based on generalized energies. *J. Fluid Mech.*, 581:277–286.

Nicodemus, R., Grossmann, S., and Holthaus, M. (1997a). Improved variational principle for bounds on energy dissipaton in turbulent shear flow. *Physica D*, 101:178–190.

Nicodemus, R., Grossmann, S., and Holthaus, M. (1997b). Variational bound on energy dissipaton in turbulent shear flow. *Phys. Rev. Lett.*, 79:4170–4173.

Nicodemus, R., Grossmann, S., and Holthaus, M. (1998). The background flow method. Part 1. Constructive approach to bounds on energy dissipation. *J. Fluid Mech.*, 363:281–300.

Nicodemus, R., Grossmann, S., and Holthaus, M. (1999). Towards lowering dissipation bounds for turbulent flows. *European Physical J. B*, 10:385–396.

Nield, D. A. (1977). Onset of convection in a fluid layer overlying a layer of a porous medium. *J. Fluid Mech.*, 81:513–522.

Nield, D. A. (1983). The boundary correction for the Rayleigh - Darcy problem: limitations of the Brinkman equation. *J. Fluid Mech.*, 128:37–46.

Nield, D. A. (1987a). Convective instability in porous media with throughflow. *AIChE Journal*, 33:1222–1224.

Nield, D. A. (1987b). Throughflow effects in the Rayleigh-Bénard convective instability problem. *J. Fluid Mech.*, 185:353–360.

Nield, D. A. (1991a). Convection in a porous medium with inclined temperature gradient. *Int. J. Heat Mass Transfer*, 34:87–92.

Nield, D. A. (1991b). The limitations of the Brinkman - Forchheimer equation in modeling flow in a saturated porous medium and at an interface. *Int. J. Heat Fluid Flow*, 12:269–272.

Nield, D. A. (1994). Convection in a porous medium with inclined temperature gradient: additional results. *Int. J. Heat Mass Transfer*, 37:3021–3025.

Nield, D. A. (1998). Instability and turbulence in convective flows in porous media. In Debnath, L. and Riahi, D., editors, *Nonlinear instability, chaos and turbulence*, pages 225–276. WIT Press, Boston.

Nield, D. A. (2003). The stability of flow in a channel or duct occupied by a porous medium. *Int. J. Heat Mass Transfer*, 46:4351–4354.

Nield, D. A. (2007). Comment on the effect of anisotropy on the onset of convection in a porous medium. *Advances Water Res.*, 30:696–697.

Nield, D. A. and Bejan, A. (2006). *Convection in Porous Media*. Springer, New York, third edition.

Nield, D. A. and Kuznetsov, A. V. (2001). The interaction of thermal non-equilibrium and heterogeneous conductivity effects in forced convection in layered porous channels. *Int. J. Heat Mass Transfer*, 44:4369–4373.

Nield, D. A. and Kuznetsov, A. V. (2006). The onset of convection in a bidisperse porous medium. *Int. J. Heat Mass Transfer*, 49:3068–3074.

Nield, D. A. and Kuznetsov, A. V. (2007). The effect of combined vertical and horizontal heterogeneity on the onset of convection in a bidisperse porous medium. *Int. J. Heat Mass Transfer*, 50:3329–3339.

Nield, D. A. and Kuznetsov, A. V. (2008). Natural convection about a vertical plate embedded in a bidisperse porous medium. *Int. J. Heat Mass Transfer*, 51:1658–1664.

Nield, D. A., Manole, D. M., and Lage, J. L. (1993). Convection induced by inclined thermal and solutal gradients in a shallow horizontal layer of a porous medium. *J. Fluid Mech.*, 257:559–574.

Normand, C. and Azouni, A. (1992). Penetrative convection in an internally heated layer of water near the maximum density point. *Phys. Fluids A*, 4:243–253.

Nunziato, J. W. and Cowin, S. C. (1979). A nonlinear theory of elastic materials with voids. *Arch. Rational Mech. Anal.*, 72:175–201.

Nunziato, J. W. and Walsh, E. K. (1977). On the influence of void compaction and material non-uniformity on the propagation of one-dimensional acceleration waves in granular materials. *Arch. Rational Mech. Anal.*, 64:299–316.

Nunziato, J. W. and Walsh, E. K. (1978). Addendum, On the influence of void compaction and material non-uniformity on the propagation of one-dimensional acceleration waves in granular materials. *Arch. Rational Mech. Anal.*, 67:395–398.

Ochoa-Tapia, J. A. and Whitaker, S. (1995a). Momentum transfer at the boundary between a porous medium and a homogeneous fluid. I. Theoretical development. *Int. J. Heat Mass Transfer*, 38:2635–2646.

Ochoa-Tapia, J. A. and Whitaker, S. (1995b). Momentum transfer at the boundary between a porous medium and a homogeneous fluid. II. Comparison with experiment. *Int. J. Heat Mass Transfer*, 38:2647–2655.

Ochoa-Tapia, J. A. and Whitaker, S. (1997). Heat transfer at the boundary between a porous medium and a heterogeneous fluid. *Int. J. Heat Mass Transfer*, 40:2691–2707.

Ogden, R. W. (1997). *Non-linear elastic deformations*. Dover, Mineola, New York.

Orszag, S. A. (1971). Accurate solution of the Orr-Sommerfeld stability equation. *J. Fluid Mech.*, 50:689–703.

Ostoja-Starzewski, M. and Trebicki, J. (1999). On the growth and decay of acceleration waves in random media. *Proc. Roy. Soc. London A*, 455:2577–2614.

Otero, J., Dontcheva, L. A., Johnston, H., Worthing, R. A., Kurganov, A., Petrova, G., and Doering, C. R. (2004). High - Rayleigh - number convection in a fluid - saturated porous layer. *J. Fluid Mech.*, 500:263–281.

Otto, J. (1929). Thermische Zustandsgrößen Gase bei mittleren und kleinen Drucken. In *Handbuch der Experimental Physik*, volume VIII/2. Springer.

Ouellette, J. (2004). Seeing with sound. Acoustic microscopy advances beyond failure analysis. American Institute of Physics Website http://www.aip.org/tip/INPHFA/vol-10/iss-3/p14.html.

Payne, L. E. (1964). Uniqueness criteria for steady state solutions of the Navier-Stokes equations. In *Atti del Simposio Internazionale sulle Applicazioni dell'Analisi alla Fisica Matematica, Cagliari - Sassari*, pages 130–153, Roma. Cremonese.

Payne, L. E. (1975). *Improperly Posed Problems in Partial Differential Equations*, volume 22 of *CBMS-NSF Regional Conference Series in Applied Mathematics*. SIAM.

Payne, L. E. and Philippin, G. A. (1995). Pointwise bounds and spatial decay estimates in heat conduction problems. *Math. Models Meth. Appl. Sci.*, 5:755–775.

Payne, L. E., Rodrigues, J. F., and Straughan, B. (2001). Effect of anisotropic permeability on Darcy's law. *Math. Meth. Appl. Sci.*, 24:427–438.

Payne, L. E. and Schaefer, P. W. (2002). Energy bounds for some nonstandard problems in partial differential equations. *J. Math. Anal. Appl.*, 273:75–92.

Payne, L. E. and Schaefer, P. W. (2006). Lower bounds for blow-up time in parabolic problems under Neumann conditions. *Applicable Anal.*, 85:1301–1311.

Payne, L. E. and Schaefer, P. W. (2007). Lower bounds for blow-up time in parabolic problems under Dirichlet conditions. *J. Math. Anal. Appl.*, 328:1196–1205.

Payne, L. E., Schaefer, P. W., and Song, J. C. (2004). Some nonstandard problems in viscous flow. *Math. Meth. Appl. Sci.*, 27:2045–2053.

Payne, L. E. and Song, J. C. (1997). Spatial decay estimates for the Brinkman and Darcy flows in a semi-infinite cylinder. *Continuum Mech. Thermodyn.*, 9:175–190.

Payne, L. E. and Song, J. C. (2000). Spatial decay for a model of double diffusive convection in Darcy and Brinkman flows. *ZAMP*, 51:867–889.

Payne, L. E. and Song, J. C. (2002). Spatial decay bounds for the Forchheimer equations. *Int. J. Engng. Sci.*, 40:943–956.

Payne, L. E. and Song, J. C. (2004a). Spatial decay estimates for the Maxwell-Cattaneo equations with mixed boundary conditions. *ZAMP*, 55:962–973.

Payne, L. E. and Song, J. C. (2004b). Temporal decay bounds in generalized heat conduction. *J. Math. Anal. Appl.*, 294:82–95.

Payne, L. E. and Song, J. C. (2007a). Lower bounds for the blow-up time in a temperature dependent Navier-Stokes flow. *J. Math. Anal. Appl.*, 335: 371–376.

Payne, L. E. and Song, J. C. (2007b). Spatial decay in a double diffusive convection problem in Darcy flow. *J. Math. Anal. Appl.*, 330:864–875.

Payne, L. E., Song, J. C., and Straughan, B. (1999). Continuous dependence and convergence results for Brinkman and Forchheimer models with variable viscosity. *Proc. Roy. Soc. London A*, 455:2173–2190.

Payne, L. E. and Straughan, B. (1987). Unconditional nonlinear stability in penetrative convection. *Geophys. Astrophys. Fluid Dyn.*, 39:57–63.

Payne, L. E. and Straughan, B. (1989). Critical Rayleigh numbers for oscillatory and nonlinear convection in an isotropic thermomicropolar fluid. *Int. J. Engng. Sci.*, 27:827–836.

Payne, L. E. and Straughan, B. (1996). Stability in the initial-time geometry problem for the Brinkman and Darcy equations of flow in porous media. *J. Math. Pures et Appl.*, 75:225–271.

Payne, L. E. and Straughan, B. (1998a). Analysis of the boundary condition at the interface between a viscous fluid and a porous medium and related modelling questions. *J. Math. Pures et Appl.*, 77:317–354.

Payne, L. E. and Straughan, B. (1998b). Structural stability for the Darcy equations of flow in porous media. *Proc. Roy. Soc. London A*, 454:1691–1698.

Payne, L. E. and Straughan, B. (1999a). Convergence and continuous dependence for the Brinkman - Forchheimer equations. *Stud. Appl. Math.*, 102:419–439.

Payne, L. E. and Straughan, B. (1999b). Convergence for the equations for a Maxwell fluid. *Stud. Appl. Math.*, 103:267–278.

Payne, L. E. and Straughan, B. (1999c). Effect of errors in the spatial geometry for temperature dependent Stokes flow. *J. Math. Pures et Appl.*, 78:609–632.

Payne, L. E. and Straughan, B. (2000a). A naturally efficient numerical technique for porous convection stability with non-trivial boundary conditions. *Int. J. Num. Anal. Meth. Geomech.*, 24:815–836.

Payne, L. E. and Straughan, B. (2000b). Unconditional nonlinear stability in temperature - dependent viscosity flow in a porous medium. *Stud. Appl. Math.*, 105:59–81.

Payne, L. E. and Weinberger, H. F. (1958). New bounds for solutions of second order elliptic partial differential equations. *Pacific J. Math.*, 8:551–573.

Pestov, I. (1998). Stability of vapour - liquid counterflow in porous media. *J. Fluid Mech.*, 364:273–295.

Pieters, G. J. M. (2004). *Stability and evolution of gravity - driven flow in porous media applied to hydrological and ecological problems.* PhD thesis, Technische Universiteit Eindhoven. ISBN 90-386-0932-9.

Pieters, G. J. M., Pop, I. S., and van Duijn, C. J. (2004). A note on the solution of a coupled parabolic - elliptic system arising in linear stability analysis of gravity - driven porous media flow. CASA Report 04-30, Technische Universiteit Eindhoven.

Pieters, G. J. M. and Schuttelaars, H. M. (2007). On the nonlinear dynamics of a saline boundary layer formed by throughflow near the surface of a porous medium. Submitted.

Pieters, G. J. M. and van Duijn, C. J. (2006). Transient growth in linearly stable gravity - driven flow in porous media. *Eur. J. Mech. B/Fluids*, 25:83–94.

Pop, I. S. (1997). A stabilized approach for the Chebyshev tau method. *Studia Univ. Babes-Bolyai, Mathematica*, 42:67–79.

Pop, I. S. (2000). A stabilized Chebyshev - Galerkin approach for the biharmonic operator. *Bul. Stint. Univ. Baia Mare*, 16:335–344.

Pop, I. S. and Gheorghiu, C. I. (1996). A Chebyshev - Galerkin method for fourth order problems. In *Approximation and Optimization*, volume 2, pages 217–220.

Potherat, A. (2007). Quasi-two-dimensional perturbations in duct flows under transverse magnetic field. *Phys. Fluids*, 19:074104.

Potter, H. (1992). The involvement of astrocytes and an acute phase response in the amyloid deposition of Alzheimer's disease. *Prog. Brain Res.*, 94:447–458.

Proctor, M. R. E. (1981). Steady subcritical thermohaline convection. *J. Fluid Mech.*, 105:507–521.

Protter, M. H. and Weinberger, H. F. (1967). *Maximum principles in differential equations*. Prentice-Hall, Englewood Cliffs.

Puri, P. (2005). Stability and eigenvalue bounds of the flow of a dipolar fluid between two parallel plates. *Proc. Roy. Soc. London A*, 461:1401–1421.

Puri, P. and Jordan, P. M. (1999a). Stokes's first problem for a dipolar fluid with nonclassical heat conduction. *J. Engng. Math.*, 36:219–240.

Puri, P. and Jordan, P. M. (1999b). Wave structure in Stokes' second problem for a dipolar fluid with nonclassical heat conduction. *Acta Mechanica*, 133: 145–160.

Puri, P. and Jordan, P. M. (2004). On the propagation of plane waves in type-III thermoelastic media. *Proc. Roy. Soc. London A*, 460:3203–3221.

Puri, P. and Jordan, P. M. (2006). On the steady shear flow of a dipolar fluid in a porous half-space. *Int. J. Engng. Sci.*, 44:227–240.

Qiao, Z. C. and Kaloni, P. N. (1998). Nonlinear convection in a porous medium with inclined temperature gradient and vertical throughflow. *Int. J. Heat Mass Transfer*, 41:2549–2552.

Qin, Y. and Chadam, J. (1996). Nonlinear convective stability in a porous medium with temperature - dependent viscosity and inertial drag. *Stud. Appl. Math.*, 96:273–288.

Qin, Y., Guo, J., and Kaloni, P. N. (1995). Double diffusive penetrative convection in porous media. *Int. J. Engng. Sci.*, 33:303–312.

Qin, Y. and Kaloni, P. N. (1992). Steady convection in a porous medium based upon the Brinkman model. *IMA J. Appl. Math.*, 35:85–95.

Qin, Y. and Kaloni, P. N. (1993). Creeping flow past a porous spherical shell. *Z. Angew. Math. Mech.*, 73:77–84.

Qin, Y. and Kaloni, P. N. (1994). Convective instabilities in anisotropic porous media. *Stud. Appl. Math.*, 91:189–204.

Qin, Y. and Kaloni, P. N. (1998). Spatial decay estimates for plane flow in the Brinkman - Forchheimer model. *Quart. Appl. Math.*, 56:71–87.

Quinlan, R. A. and Straughan, B. (2005). Decay bounds in a model for aggregation of microglia: application to Alzheimer's disease senile plaques. *Proc. Roy. Soc. London A*, 461:2887–2897.

Quintanilla, R. (2002a). Exponential stability for one-dimensional problem of swelling porous elastic soils with fluid saturation. *J. Comput. Appl. Math.*, 145:525–533.

Quintanilla, R. (2002b). Exponential stability in the dual - phase - lag heat conduction theory. *J. Non-Equilib. Thermodyn.*, 27:217–227.

Quintanilla, R. (2002c). On the linear problem of swelling porous elastic soils. *J. Math. Anal. Appl.*, 269:50–72.

Quintanilla, R. (2002d). On the linear problem of swelling porous elastic soils with incompressible fluid. *Int. J. Engng. Sci.*, 40:1485–1494.

Quintanilla, R. (2003). On existence and stability in the theory of swelling porous elastic soils. *IMA J. Appl. Math.*, 68:491–506.

Quintanilla, R. (2004). Exponential stability in porous media problem saturated by multiple immiscible fluids. *Appl. Math. Computation*, 150:661–668.

Quintanilla, R. and Racke, R. (2003). Stability in thermoelasticity of type III. *Discrete and Continuous Dynamical Systems B*, 1:383–400.

Quintanilla, R. and Racke, R. (2006). A note on stability in dual-phase-lag heat conduction. *Int. J. Heat Mass Transfer*, 49:1209–1213.

Quintanilla, R. and Racke, R. (2007). Qualitative aspects in dual-phase-lag heat conduction. *Proc. Roy. Soc. London A*, 463:659–674.

Quintanilla, R. and Straughan, B. (2000). Growth and uniqueness in thermoelasticity. *Proc. Roy. Soc. London A*, 456:1419–1429.

Quintanilla, R. and Straughan, B. (2002). Explosive instabilities in heat transmission. *Proc. Roy. Soc. London A*, 458:2833–2837.

Quintanilla, R. and Straughan, B. (2004). Discontinuity waves in type III thermoelasticity. *Proc. Roy. Soc. London A*, 460:1169–1175.

Quintanilla, R. and Straughan, B. (2005a). Bounds for some non-standard problems in porous flow and viscous Green-Naghdi fluids. *Proc. Roy. Soc. London A*, 461:3159–3168.

Quintanilla, R. and Straughan, B. (2005b). Energy bounds for some non-standard problems in thermoelasticity. *Proc. Roy. Soc. London A*, 461:1147–1162.

Quintanilla, R. and Straughan, B. (2008). Nonlinear waves in a Green-Naghdi dissipationless fluid. *J. Non-Newtonian Fluid Mech.* doi:10.1016/j-jnnfm.2008.04.006.

Rai, A. (2003). Breakdown of acceleration waves in radiative magnetic fluids. *Defence Science Journal*, 53:425–430.

Raiser, G. F., Wise, J. L., Clifton, R. J., Grady, D. E., and Cox, D. E. (1994). Plate impact response of ceramics and glasses. *J. Appl. Phys.*, 75:3862–3869.

Rajagopal, K. R. and Truesdell, C. (1999). *An introduction to the mechanics of fluids*. Birkhäuser, Basel.

Rajapakse, R. K. N. D. and Senjuntichai, T. (1995). Dynamic response of a multi-layered poroelastic medium. *Earthquake Engineering and Structural Dynamics*, 24:703–722.

Ramos, E. and Vargas, M. (2005). The Boussinesq approximation in a rotating frame of reference. *J. Non-Equilibrium Thermodyn.*, 30:21–37.

Rappoldt, C., Pieters, G. J. M., Adema, E. B., Baaijens, G. J., Grootjans, A. P., and van Duijn, C. J. (2003). Buoyancy driven flow in a peat moss layer as a mechanism for solute transport. *Proc. Nat. Acad. Sci.*, 100:14937–14942.

Reddy, S. C. and Henningson, D. S. (1993). Energy growth in viscous channel flows. *J. Fluid Mech.*, 252:209–238.

Rees, D. A. S. (2002). The onset of Darcy - Brinkman convection in a porous layer: an asymptotic analysis. *Int. J. Heat Mass Transfer*, 45:2213–2220.

Rees, D. A. S. and Bassom, A. P. (2000). The onset of Darcy-Bénard convection in an inclined layer heated from below. *Acta Mech.*, 144:103–118.

Rees, D. A. S., Bassom, A. P., and Siddheshwar, P. G. (2008). Local thermal non-equilibrium effects arising from the injection of a hot fluid into a porous medium. *J. Fluid Mech.*, 594:379–398.

Rees, D. A. S. and Postelnicu, A. (2001). The onset of convection in an inclined anisotropic porous layer. *Int. J. Heat Mass Transfer*, 44:4127–4138.

Riahi, D. N. (2007). Inertial effects on rotating flow in a porous layer. *J. Porous Media*, 10:343–356.

Rionero, S. (2001). Asymptotic and other properties of some nonlinear diffusion models. In Straughan, B., Greve, R., Ehrentraut, H., and Wang, Y., editors, *Continuum Mechanics and Applications in Geophysics and the Environment*, pages 56–78. Springer.

Rionero, S. (2004). A rigorous link between the L^2-stability of the solutions to a binary reaction-diffusion system of pde and the stability of the solutions to a binary system of ode. *Rend. Acc. Sc. fis. mat. Napoli*, 71:53–62.

Rionero, S. (2005). L^2-stability of the solutions to a nonlinear binary reaction-diffusion system of pdes. *Rend. Mat. Acc. Lincei*, 16:227–238.

Rionero, S. (2006a). Nonlinear L^2-stability analysis for two species population dynamics in spatial ecology under Neumann boundary data. *Rend. Circolo Matem. Palermo*, 78:273–283.

Rionero, S. (2006b). A nonlinear L^2-stability analysis for two species population dynamics with dispersal. *Mathematical Biosciences and Engineering*, 3: 189–204.

Rionero, S. (2006c). A rigorous reduction of the L^2-stability of the solutions to a nonlinear binary reaction-diffusion system of pdes to the stability of the solutions to a linear binary system of odes. *J. Math. Anal. Appl.*, 319: 377–397.

Rionero, S. and Torcicollo, I. (2000). On an ill-posed problem in nonlinear heat conduction. *Transport Theory and Statistical Physics*, 29:173–186.

Riviere, B. (2005). Analysis of a discontinuous finite element method for the coupled Stokes and Darcy problems. *J. Scientific Computing*, 22:479–500.

Roberts, P. H. (1981). Equilibria and stability of a fluid type II superconductor. *Q. J. Mech. Appl. Math.*, 34:327–343.

Rodrigues, J. F. (1986). A steady state Boussinesq-Stefan problem with continuous extraction. *Annali Matem. Pura Appl.*, 144:203–218.

Rodrigues, J. F. (1992). Weak solutions for thermoconvective flows of Boussinesq-Stefan type. In *Mathematical Topics in Fluid Mechanics*, pages 93–116. Longman.

Rogers, J., Strohmeyer, R., Kovelowski, C. J., and Li, R. (2002). Microglia and inflammatory mechanisms in the clearance of amyloid beta peptide. *Glia*, 40:260–269.

Rosensweig, R. E. (1985). *Ferrohydrodynamics*. Cambridge Univ. Press.

Rothmeyer, M. (1980). The Soret effect and salt-gradient solar ponds. *Solar Energy*, 25:567–568.

Rudraiah, N., Kaloni, P. N., and Radhadevi, P. V. (1989). Oscillatory convection in a viscoelastic fluid through a porous layer heated from below. *Rheologica Acta*, 28:48–53.

Rudraiah, N., Masuoka, T., and Nair, P. (2007). Effect of combined Brinkman - electric boundary layer on the onset of Marangoni electroconvection in a poorly conducting fluid - saturated porous layer cooled from above in the presence of an electric field. *J. Porous Media*, 10:421–433.

Ruggeri, T. and Sugiyama, M. (2005). Hyperbolicity, convexity and shock waves in one-dimensional crystalline solids. *J. Phys. A: Math. Gen.*, 38:4337–4347.

Ruszkowski, M., Garcia-Osorio, V., and Ydstie, B. E. (2005). Passivity based on control of transport reaction systems. *AICHE J.*, 51:3147–3166.

Saffman, P. (1971). On the boundary conditions at the surface of a porous medium. *Stud. Appl. Math.*, 50:93–101.

Saggio-Woyansky, J., Scott, C. E., and Minnear, W. P. (1992). Processing of porous ceramics. *American Ceramic Society Bulletin*, 71:1674–1682.

Salas, K. I. and Waas, A. M. (2007). Convective heat transfer in open cell metal foams. *ASME J. Heat Transfer*, 129:1217–1229.

Sanjuán, N., Simal, S., Bon, J., and Mulet, A. (1999). Modelling of broccoli stems rehydration process. *J. Food Engineering*, 42:27–31.

Saravanan, S. and Kandaswamy, P. (2003). Convection currents in a porous layer with a gravity gradient. *Heat and Mass Transfer*, 39:693–699.

Sattinger, D. H. (1970). The mathematical problem of hydrodynamic stability. *J. Math. Mech.*, 19:797–817.

Scarpa, F. and Tomlinson, G. (2000). Theoretical characteristics of the vibration of sandwich plates with in - plane negative Poisson's ratio values. *J. Sound Vibration*, 230:45–67.

Schaflinger, U. (1994). Interfacial instabilities in a stratified flow of two superposed fluids. *Fluid Dyn. Res.*, 13:299–316.

Schaflinger, U., Acrivos, A., and Stibi, H. (1995). An experimental study of viscous resuspension in a pressure - driven plane channel flow. *Int. J. Multiphase Flow*, 21:693–704.

Schmid, P. J. and Henningson, D. S. (1994). Optimal energy density growth in Hagen - Poiseuille flow. *J. Fluid Mech.*, 277:197–225.

Schmittner, A., Yoshimori, M., and Weaver, A. J. (2002). Instability of glacial climate in a model of the ocean - atmosphere - cryosphere system. *Science*, 295:1489–1493.

Schwingshackl, C. W., Aglietti, G. S., and Cunningham, P. R. (2006). Determination of honeycomb material properties: existing theories and alternative dynamic approach. *J. Aerospace Engng.*, 19:177–183.

Senjuntichai, T. and Rajapakse, R. K. N. D. (1995). Exact stiffness method for quasi-statics of a multi-layered poroelastic medium. *Int. J. Solids Structures*, 32:1535–1553.

Serdyukov, S. I. (2001). A new version of extended irreversible thermodynamics and dual-phase-lag model in heat transfer. *Phys. Letters A*, 281:16–20.

Serdyukov, S. I., Voskresenskii, N. M., Bel'nov, V. K., and Karpov, I. I. (2003). Extended irreversible thermodynamics and generation of the dual-phase-lag model in heat transfer. *J. Non-Equilib. Thermodyn.*, 28:207–219.

Shanthini, R. (1989). Degeneracies of the temporal Orr-Sommerfeld eigenmodes in plane Poiseuille flow. *J. Fluid Mech.*, 201:13–34.

Sharma, R. C. and Gupta, U. (1995). Thermal convection in micropolar fluids in porous medium. *Int. J. Engng. Sci.*, 33:1887–1892.

Sharma, R. C. and Kumar, P. (1997). On micropolar fluids heated from below in hydromagnetics in porous medium. *Czech. J. Phys.*, 47:637–647.

Sharma, R. C. and Kumar, P. (1998). Effect of rotation on thermal convection in micropolar fluids in porous medium. *Indian J. Pure Appl. Math.*, 29:95–104.

Shen, J. (1994). Efficient spectral Galerkin method I. Direct solvers of second and fourth order equations using Legendre polynomials. *SIAM J. Sci. Computing*, 15:1489–1505.

Sheu, L. J. (2006). An autonomous system for chaotic convection in a porous medium using a thermal non-equilibrium model. *Chaos, Solitons and Fractals*, 30:672–689.

Shivakumara, I. S., Suma, S. P., and Chavaraddi, K. B. (2006). Onset of surface - tension - driven convection in superposed layers of fluid and saturated porous medium. *Arch. Mech.*, 58:71–92.

Shubov, M. A. and Balogh, A. (2005). Asymptotic distribution of eigenvalues for the damped string equation: numerical approach. *J. Aerospace Engng.*, 18:69–83.

Siddheshwar, P. G. and Sri Krishna, C. V. (2003). Linear and non - linear analyses of convection in a micropolar fluid occupying a porous medium. *Int. J. Nonlinear Mech.*, 38:1561–1579.

Silva, F. V. E. (2006). On a lemma due to Ladyzhenskaya and Solonnikov and some applications. *Nonlinear Analysis, Theory Methods and Applications*, 64:706–725.

Simal, S., Benedito, J., Sanchez, E. S., and Rosello, C. (1998). Use of ultrasound to increase mass transport rates during osmotic dehydration. *J. Food Engineering*, 36:323–336.

Simal, S., Sanchez, E. S., Bon, J., Femenia, A., and Rosello, C. (2001). Water and salt diffusion during cheese ripening: effect of the external and internal resistances to mass transfer. *J. Food Engineering*, 48:269–275.

Singer, D., Pasierb, F., Ruel, R., and Kojima, H. (1984). Multiple scattering of second sound in superfluid Helium II - filled porous medium. *Phys. Rev. B*, 30:2909–2912.

Sneddon, I. N. (1980). *Special functions of mathematical physics and chemistry.* Longman, London and New York.

Somerville, R. C. J. and Gal-Chen, T. (1979). Numerical simulation of convection with mean vertical motion. *J. Atmos. Sci.*, 36:805–815.

Song, J. C. (2002). Spatial decay estimates in time-dependent double diffusive Darcy plane flow. *J. Math. Anal. Appl.*, 267:76–88.

Song, J. C. (2003). Improved decay estimates in time-dependent Stokes flow. *J. Math. Anal. Appl.*, 288:505–517.

Song, J. C. (2006). A spatial decay bound for an anisotropic penetrative convection Darcy flow. *Appl. Math. Letters*, 19:478–484.

Spencer, A. J. M. (1980). *Continuum Mechanics.* Longman, London.

Stocker, T. F. (2001). The role of simple models in understanding climate change. In Straughan, B., Greve, R., Ehrentraut, H., and Wang, Y., editors, *Continuum Mechanics and Applications in Geophysics and the Environment.* Springer.

Stocker, T. F. and Schmittner, A. (1997). Influence of CO_2 emission rates on the stability of thermohaline convection. *Nature*, 388:862–865.

Stoer, J. and Bulirsch, R. (1993). *Introduction to numerical analysis*. Springer, New York, second edition.

Storesletten, L. (1993). Natural convection in a horizontal porous layer with anisotropic thermal diffusivity. *Trans. Porous Media*, 12:19–29.

Strang, G. (1988). *Linear algebra and its applications*. Harcourt Brace Jovanovich, Orlando.

Straughan, B. (1974). *Qualitative analysis of some equations in contemporary continuum mechanics*. PhD thesis, Heriot-Watt University.

Straughan, B. (1986). Nonlinear waves in a general magnetic fluid. *ZAMP*, 37:274–279.

Straughan, B. (1998). *Explosive instabilities in mechanics*. Springer, Heidelberg.

Straughan, B. (2001a). Porous convection, the Chebyshev tau method and spurious eigenvalues. In Straughan, B., Greve, R., Ehrentraut, H., and Wang, Y., editors, *Continuum Mechanics and Applications in Geophysics and the Environment*. Springer.

Straughan, B. (2001b). A sharp nonlinear stability threshold in rotating porous convection. *Proc. Roy. Soc. London A*, 457:87–93.

Straughan, B. (2001c). Surface tension driven convection in a fluid overlying a porous layer. *J. Computational Phys.*, 170:320–337.

Straughan, B. (2002a). Effect of property variation and modelling on convection in a fluid overlying a porous layer. *Int. J. Num. Anal. Meth. Geomech.*, 26:75–97.

Straughan, B. (2002b). Global stability for convection induced by absorption of radiation. *Dynamics of Atmospheres and Oceans*, 35:351–361.

Straughan, B. (2004a). *The energy method, stability, and nonlinear convection*. Springer, New York, second edition.

Straughan, B. (2004b). Resonant porous penetrative convection. *Proc. Roy. Soc. London A*, 460:2913–2927.

Straughan, B. (2006). Global nonlinear stability in porous convection with a thermal non-equilibrium model. *Proc. Roy. Soc. London A*, 462:409–418.

Straughan, B. and Franchi, F. (1984). Bénard convection and the Cattaneo law of heat conduction. *Proc. Roy. Soc. Edinburgh A*, 96:175–178.

Straughan, B. and Hutter, K. (1999). A priori bounds and structural stability for double diffusive convection incorporating the Soret effect. *Proc. Roy. Soc. London A*, 455:767–777.

Straughan, B. and Tracey, J. (1999). Multi-component convection-diffusion with internal heating or cooling. *Acta Mechanica*, 133:219–238.

Straughan, B. and Walker, D. W. (1996a). Anisotropic porous penetrative convection. *Proc. Roy. Soc. London A*, 452:97–115.

Straughan, B. and Walker, D. W. (1996b). Two very accurate and efficient methods for computing eigenvalues and eigenfunctions in porous convection problems. *J. Computational Phys.*, 127:128–141.

Su, S. and Dai, W. (2006). Comparison of the solutions of a phase-lagging heat transport equation and damped wave equation with a heat source. *Int. J. Heat Mass Transfer*, 49:2793–2801.

Su, S., Dai, W., Jordan, P. M., and Mickens, R. E. (2005). Comparison of the solutions of a phase-lagging heat transport equation and damped wave equation. *Int. J. Heat Mass Transfer*, 48:2233–2241.

Sugiyama, M. (1994). Statistical mechanical study of wave propagation in crystalline solids at finite temperatures. *Current Topics in Acoustics Research*, 1:139–158.

Sun, S. and Wheeler, M. F. (2007). Discontinuous Galerkin methods for simulating bioreactive transport of viruses in porous media. *Advances in Water Resources*, 30:1696–1710.

Sunil, Chand, P., Bharti, P. K., and Mahajan, A. (2008). Thermal convection in micropolar ferrofluid in the presence of rotation. *J. Magnetism and Magnetic Materials*, 320:316–324.

Sunil, Divya, and Sharma, R. (2004). Effect of rotation on a ferromagnetic fluid heated and soluted from below saturating a porous medium. *J. Geophys. Engng.*, 1:116–127.

Sunil, Divya, and Sharma, R. (2005a). Effect of dust particles on thermal convection in a ferromagnetic fluid saturating a porous medium. *J. Magnetism and Magnetic Materials*, 288:183–195.

Sunil, Divya, and Sharma, R. (2005b). The effect of magnetic field dependent viscosity on thermosolutal convection in a ferromagnetic fluid saturating a porous medium. *Trans. Porous Media*, 60:251–274.

Sunil, Divya, and Sharma, V. (2005c). Effect of dust particles on a rotating ferromagnetic fluid heated from below saturating a porous medium. *J. Colloid Interface Sci.*, 291:152–161.

Sunil and Mahajan, A. (2008). A nonlinear stability analysis for magnetized ferrofluid heated from below. *Proc. Roy. Soc. London A*, 464:83–98.

Sunil, Sharma, A., Bharti, P., and Shandil, R. (2006). Effect of rotation on a layer of micropolar ferromagnetic fluid heated from below saturating a porous medium. *Int. J. Engng. Sci.*, 44:683–698.

Suzuki, R. (2006). Universal bounds for quasilinear parabolic equations with convection. *Discrete and Continuous Dynamical Systems*, 16:563–586.

Szczygiel, J. (1999). Diffusion in a bidispersive grain of a reforming catalyst. *Computers and Chemistry*, 23:121–134.

Tabor, H. (1980). Non-convecting solar ponds. *Phil. Trans. Roy. Soc. London A*, 295:423–433.

Tadj, N., Draoui, B., Theodoris, G., Bartzanas, T., and Kittas, C. (2007). Convective heat transfer in a heated greenhouse tunnel. In *VIII International symposium on protected cultivation in mild winter climates: advances in soil and soiless cultivation under protected environment*, volume 747 of *ISHS Acta Horticulturae*.

Taylor, G. I. (1971). A model for the boundary conditions of a porous material. *J. Fluid Mech.*, 49:319–326.

Temam, R. (1988). *Infinite dimensional dynamical systems in Mechanics and Physics*. Springer, New York.

Tersenov, A. S. (2004). The preventive effect of the convection and of the diffusion in the blow-up phenomenon for parabolic equations. *Annales de l'Institut Henri Poincaré - Analyse non Lineaire*, 21:533–541.

Theofilis, V. (2003). Advances in global linear instability analysis of nonparallel and three-dimensional flows. *Progress in Aerospace Sciences*, 39:249–315.

Theofilis, V., Duck, P. W., and Owen, J. (2004). Viscous linear stability analysis of rectangular duct and cavity flows. *J. Fluid Mech.*, 505:249–286.

Truesdell, C. and Noll, W. (1992). *The non-linear field theories of mechanics*. Springer, Berlin, second edition.

Truesdell, C. and Toupin, R. A. (1960). The classical field theories. In Flügge, S., editor, *Handbuch der Physik*, volume III/1, pages 226–793. Springer.

Tse, K. L. and Chasnov, J. R. (1998). A Fourier-Hermite pseudospectral method for penetrative convection. *J. Computational Physics*, 142:489–505.

Tyvand, P. A. and Storesletten, L. (1991). Onset of convection in an anisotropic porous medium with oblique principal axes. *J. Fluid Mech.*, 226:371–382.

Tzou, D. Y. (1995a). The generalized lagging response in small-scale and high-rate heating. *Int. J. Heat Mass Transfer*, 38:3231–3234.

Tzou, D. Y. (1995b). A unified approach for heat conduction from macro to micro-scales. *J. Heat Transfer*, 117:8–16.

Ushijima, T. K. (2000). On the approximation of blow-up time for solutions of nonlinear parabolic equations. *Pub. Res. Inst. Math. Sci., Kyoto*, 36:613–640.

Vadasz, J. J., Govender, S., and Vadasz, P. (2005a). Heat transfer enhancement in nano-fluids suspensions: possible mechanisms and explanations. *Int. J. Heat Mass Transfer*, 48:2673–2683.

Vadasz, J. J., Roy-Aikins, J., and Vadasz, P. (2005b). Sudden or smooth transitions in porous media natural convection. *Int. J. Heat Mass Transfer*, 48:1096–1106.

Vadasz, P. (1995). Coriolis effect on free convection in a long rotating porous box subject to uniform heat generation. *Int. J. Heat Mass Transfer*, 38:2011–2018.

Vadasz, P. (1996). Convection and stability in a rotating porous layer with alternating direction of the centrifugal body force. *Int. J. Heat Mass Transfer*, 39:1639–1647.

Vadasz, P. (1997). Flow in rotating porous media. In Plessis, P. D., editor, *Fluid Transport in Porous Media*. Computational Mechanics Publications.

Vadasz, P. (1998a). Coriolis effect on gravity-driven convection in a rotating porous layer heated from below. *J. Fluid Mech.*, 376:351–375.

Vadasz, P. (1998b). Free convection in rotating porous media. In Ingham, D. and Pop, I., editors, *Transport Phenomena in Porous Media*, pages 285–312. Elsevier.

Vaidya, A. and Wulandana, R. (2006). Non-linear stability for convection with quadratic temperature dependent viscosity. *Math. Meth. Appl. Sci.*, 29:1555–1561.

Valenti, G., Curro, C., and Sugiyama, M. (2004). Acceleration waves analysed by a new continuum model of solids incorporating microscopic thermal vibrations. *Continuum Mech. Thermodyn.*, 16:185–198.

Valerio, J. V., Carvalho, M. S., and Tomei, C. (2007). Filtering the eigenvalues at infinity from the linear stability analysis of incompressible flows. *J. Computational Phys.*, 227:229–243.

van Duijn, C. J., Galiano, G., and Peletier, M. A. (2001). A diffusion - convection problem with drainage arising in the ecology of mangroves. *Interfaces and Free Boundaries*, 3:15–44.

van Duijn, C. J., Pieters, G. J. M., and Raats, P. A. C. (2004). Steady flows in unsaturated soils are stable. *Trans. Porous Media*, 57:215–244.

van Duijn, C. J., Pieters, G. J. M., Wooding, R. A., and van der Ploeg, A. (2002). Stability criteria for the vertical boundary layer formed by throughflow near the surface of a porous medium. In *Environmental Mechanics: Water, Mass and Energy Transfer in the Biosphere, Geophysical Monograph*, volume 192, pages 155–169. American Geophysical Union.

Vasilevich, V. V. and Alexandrovich, R. B. (2008). Glass - fibre - reinforced honeycomb materials. http://www.technologiya.ru/tech/materiale/t0102.html.

Vedavarz, A., Mitra, K., Kumar, S., and Moallemi, M. K. (1992). Effect of hyperbolic heat conduction on temperature distribution in laser irradiated tissue with blood perfusion. *Adv. Biological Heat Mass Transfer, ASME HTD*, 231:7–16.

Wang, Y., Rajapakse, R. K. N. D., and Shah, A. H. (1991). Dynamic interaction between flexible strip foundations. *Earthquake Engineering and Structural Dynamics*, 20:441–454.

Webber, M. (2006). The destabilizing effect of boundary slip on Bénard convection. *Math. Meth. Appl. Sci.*, 29:819–838.

Webber, M. (2007). *Instability of fluid flows, including boundary slip*. PhD thesis, Durham University.

Webber, M. (2008). Instability of thread - annular flow with small characteristic length to three - dimensional disturbances. *Proc. Roy. Soc. London A*, 464:673–690.

Webber, M. and Straughan, B. (2006). Stability of pressure driven flow in a microchannel. *Rend. Circolo Matem. Palermo*, 78:343–357.

Wei, C. and Muraleetharan, K. K. (2007). Linear viscoelastic behaviour of porous media with non-uniform saturation. *Int. J. Engng. Sci.*, 45:698–715.

Wei, Q. (2004). Bounds on convective heat transport in a rotating porous layer. *Mech. Res. Communications*, 31:269–276.

Weingartner, B., Osinov, V. A., and Wu, W. (2006). Acceleration wave speeds in a hypoplastic constitutive model. *Int. J. Non-linear Mech.*, 41:991–999.

Weinstein, T. and Bennethum, L. S. (2006). On the derivation of the transport equation for swelling porous materials with finite deformation. *Int. J. Engng. Sci.*, 44:1408–1422.

Whitaker, S. (1986). Flow in porous media. I: A theoretical derivation of Darcy's law. *Transport in Porous Media*, 1:3–25.

Whitaker, S. (1996). The Forchheimer equation: A theoretical development. *Transport in Porous Media*, 25:27–62.

Whitham, G. B. (1974). *Linear and nonlinear waves*. John Wiley and Sons, New York.

Wilson, D. K. (1997). Simple, relaxational models for the acoustic properties of porous media. *Applied Acoustics*, 50:171–188.

Wood, B. D. and Ford, R. M. (2007). Biological processes in porous media: from the pore scale to the field. *Advances in Water Resources*, 30:1387–1391.

Wood, B. D., Radakovich, K., and Golfier, F. (2007). Effective reaction at a fluid - solid interface: applications to biotransformation in porous media. *Advances in Water Resources*, 30:1630–1647.

Wooding, R. A., Tyler, S. W., and White, I. (1997a). Convection in groundwater below an evaporating salt lake: 1. Onset of instability. *Water Resources Research*, 33:1199–1217.

Wooding, R. A., Tyler, S. W., White, I., and Anderson, P. A. (1997b). Convection in groundwater below an evaporating salt lake: 2. Evolution of fingers or plumes. *Water Resources Research*, 33:1219–1228.

Worster, M. G. (1992). Instabilities of the liquid and mushy regions during solidification of alloys. *J. Fluid Mech.*, 237:649–669.

Xu, X., Chen, S., and Zhang, D. (2006). Convective stability analysis of the long - term storage of carbon dioxide in deep saline aquifers. *Advances in Water Resources*, 29:397–407.

Xu, X., Chen, S., and Zhang, D. (2007). Reply to comment of Nield. *Advances in Water Resources*, 30:698–699.

Ybarra, P. L. G. and Velarde, M. G. (1979). The influence of the Soret and Dufour effects on the stability of a binary gas layer heated from below or above. *Geophys. Astrophys. Fluid Dyn.*, 13:83–94.

Ydstie, B. E. (2002). Passivity based control via the second law. *Computers and Chemical Engineering*, 26:1037–1048.

Yecko, P. A. (2004). Accretion disk instability revisited - transient dynamics of rotating shear flow. *Astronomy and Astrophysics*, 425:385–393.

Yoon, D. Y., Kim, M. C., and Choi, C. K. (2004). The onset of oscillatory convection in a horizontal porous layer saturated with viscoelastic liquid. *Trans. Porous Media*, 55:275–284.

Zangrando, F. (1991). On the hydrodynamics of salt-gradient solar ponds. *Solar Energy*, 46:323–341.

Zhang, K. K. and Schubert, G. (2000). Teleconvection: remotely driven thermal convection in rotating stratified spherical layers. *Science*, 290:1944–1947.

Zhang, K. K. and Schubert, G. (2002). From penetrative convection to teleconvection. *Astrophysical J.*, 572:461–476.

Zhang, X. and Zuazua, E. (2003). Decay of solutions of the system of thermoelasticity of type III. *Comm. Contemporary Math.*, 5:25–84.

Zhao, C. Y., Lu, T. J., and Hodson, H. P. (2004). Thermal radiation in ultralight metal foams with open cells. *Int. J. Heat Mass Transfer*, 47:2927–2939.

Zhou, Y., Rajapakse, R. K. N. D., and Graham, J. (1998). A coupled thermoporoelastic model with thermo - osmosis and thermal filtration. *Int. J. Solids Structures*, 35:4659–4683.

Zorrilla, S. E. and Rubiolo, A. C. (2005a). Mathematical modeling for immersion chilling and freezing of foods. Part I: Model development. *J. Food Engng.*, 66:329–338.

Zorrilla, S. E. and Rubiolo, A. C. (2005b). Mathematical modeling for immersion chilling and freezing of foods. Part II: Model solution. *J. Food Engng.*, 66:339–351.

Index

Applied Mathematical Sciences

(continued from page ii)

Applied Mathematical Sciences

(continued from previous page)